Long-Term Climatic Variations

Data and Modelling

NATO ASI Series

Advanced Science Institutes Series

A series presenting the results of activities sponsored by the NATO Science Committee, which aims at the dissemination of advanced scientific and technological knowledge, with a view to strengthening links between scientific communities.

The Series is published by an international board of publishers in conjunction with the NATO Scientific Affairs Division

A Life Sciences B Physics	Plenum Publishing Corporation London and New York
C Mathematical and Physical Sciences D Behavioural and Social Sciences E Applied Sciences	Kluwer Academic Publishers Dordrecht, Boston and London
F Computer and Systems Sciences G Ecological Sciences H Cell Biology I Global Environmental Change	Springer-Verlag Berlin Heidelberg New York London Paris Tokyo Hong Kong Barcelona Budapest

NATO-PCO DATABASE

The electronic index to the NATO ASI Series provides full bibliographical references (with keywords and/or abstracts) to more than 30000 contributions from international scientists published in all sections of the NATO ASI Series. Access to the NATO-PCO DATABASE compiled by the NATO Publication Coordination Office is possible in two ways:

- via online FILE 128 (NATO-PCO DATABASE) hosted by ESRIN, Via Galileo Galilei, I-00044 Frascati, Italy.
- via CD-ROM "NATO Science & Technology Disk" with user-friendly retrieval software in English, French and German (© WTV GmbH and DATAWARE Technologies Inc. 1992).

The CD-ROM can be ordered through any member of the Board of Publishers or through NATO-PCO, Overijse, Belgium.

Series I: Global Environmental Change, Vol. 22

Long-Term Climatic Variations

Data and Modelling

Edited by

Jean-Claude Duplessy
Marie-Thérèse Spyridakis

Centre des Faibles Radioactivités
Laboratoire Mixte CNRS-CEA
F-91190 Gif-sur-Yvette, France

Springer-Verlag
Berlin Heidelberg New York London Paris Tokyo
Hong Kong Barcelona Budapest
Published in cooperation with NATO Scientific Affairs Division

Proceedings of the NATO Advanced Study Institute on Long-Term Climatic Variations – Data and Modelling, held at Siena, Italy, September 27– October 11, 1992

Additional material to this book can be downloaded from http://extra.springer.com.

ISBN-13:978-3-642-79068-3 e-ISBN-13:978-3-642-79066-9
DOI: 10.1007/978-3-642-79066-9

CIP data applied for

Softcover reprint of the hardcover 1st edititon 1994

Typesetting: Camera ready by authors
SPIN 10091524 31/3130 - 5 4 3 2 1 0 - Printed on acid-free paper

TABLE OF CONTENTS

III - UNDERSTANDING CLIMATIC CHANGES

INTRODUCTION

Climate is the most important component of the Earth's environment and climatic fluctuations have a strong impact on water supplies, soils, vegetation, agriculture, fisheries, energy use, etc. Most generations of mankind have regarded climate as an unreliable, shifting, fluctuating thing, sometimes offering unforeseen opportunities but at other times bringing disaster by famine, flood, or drought. Since the middle of the last century, worldwide economic and technological activities have caused increasingly rapid and potentially stressful changes in our global environment in ways that we are only now beginning to understand. In particular, increasing concentrations of CO_2 and other atmospheric greenhouse gases may significantly alter our climate.

In predicting future global changes, we are thus faced with questions that can be answered only through a better knowledge of the physical processes that drive the evolution of climate and their interaction with biogeochemical cycles and the Earth's ecosystems. It has been widely recognized that in order to make scientific progress on major and important problems, such as those associated with global environmental changes, it is necessary to integrate knowledge from very different disciplines: climatology, oceanography, meteorology, chemistry, ecology, glaciology, geology, astronomy, fluid dynamics and computer sciences.

Climate models have been used to estimate the warming associated with future increases in greenhouse gases. However, the various models differ substantially in their sensitivity and these differences contribute prominently to the uncertainty in estimating the magnitude and rate of future climate change. One possible technique for evaluating model quality and sensitivity is the use of information about past climates. Geological and ice-core data provide enough information to show a clear picture of the entire climatic system in the past and to document its temporal evolution. Together with a modelling effort, the paleo-data permit the analysis of the major processes involved in climatic change.

Establishing collaboration between palaeoclimatologists and physicists studying climate dynamics has been a challenge. This was due to the accuracy of the geological reconstructions being considerably poorer than that of the meteorological data classically used by modellers and also as it relied on much less data. The international effort made first in the CLIMAP (Climate: Long-range

Investigation, Mapping And Prediction) project and then in several international programmes (PAGES/IGBP, EEC Environment programme, ...) showed that a quantitative global reconstruction of the Earth's surface conditions during the recent geological past was a feasible goal. Its results (maps of sea surface temperature during the summer and winter seasons, map of continental albedo changes, composition of the atmosphere and atmospheric greenhouse gas content) have been used by numerous modelling research teams.

However, interdisciplinary interactions are still difficult, because most young scientists are mainly specialised in one scientific field. This makes interaction with specialists from other disciplines difficult. Therefore, the purpose of this Advanced Study Institute on "Long-term climatic variations: Data and modelling" sponsored both by NATO and the Commission of the European Communities was threefold:

• Firstly, to bring together young (PhD students and post-doc) and senior scientists involved in either the reconstruction of past climatic variations or paleoclimate modelling. The remote location of the Certosa di Pontignano near Siena provided a unique opportunity to have long informal discussions and exchanges between scientists working in quite different fields and thus not commonly in communication.

• Secondly, to teach a general course to young students involved in the study of long-term climatic variations, providing a background on both the climate system and its modelling, and on the proxy-data of past climate and their significance.

• Thirdly, to analyse several examples of major climatic changes, which may be used to depict the mechanism by which climate is likely to change rapidly (within a few decades or centuries). The last deglaciation and the massive iceberg discharges which developped during the glaciation have been used as examples of such abrupt climatic changes with large regional or global impacts.

Six lectures were devoted to the presentation of the climatic system and the physical basis for its modelling. The full hierarchy of models, which is now used for modelling climate dynamics, going from 0-D models to sophisticated 3-D models including ocean, atmosphere, ice and possibly a biosphere were presented. The variety of approaches, including also 1-D and 2-D atmospheric models, ocean, ice and asthenosphere models and 2,5-D models of the whole climate system, were well illustrated. In addition, a whole afternoon was devoted to the presentation and discussion of the Paleoclimate Modelling Intercomparison programme (PAMIP), which was launched as a result of the NATO-ARW on "Paleoclimate Modelling" held in Saclay in May 1991.

Twelve lectures summarised the evolution of the global atmosphere, the surficial and deep ocean, the surface of the continents, the biosphere and the ice sheets during the last climatic cycles. They focused on the methods allowing quantitative reconstructions and providing data which permit direct interaction with modellers. The emphasis was placed on recent high resolution reconstructions obtained through advances in continental and ice coring and deep sea sediment drillings.

Eight lectures focused on the understanding of past and future climatic changes, several of them relying on the considerable evolution of atmospheric and oceanic General Circulation Models (GCMs) over the last decade. In addition, special attention was given to the external forcing factors, noticeably astronomical and solar, and on atmospheric CO_2 concentration variations.

At the end of the ASI, it was clear that the broad array of scientists concerned with all the interlinked changes going on in our complicated planet should continue exchanging information from their respective fields, collaborate more closely, jointly plan data gathering experiments, and work together on their physical and biogeochemical interpretation. Such projects would constitute the best elaborated response to the Internatioonal Geosphere-Biosphere Programme, which seeks to describe and understand the interactive physical, chemical and biological processes that regulate the total earth system, the changes that are occurring in this system and the impact of human activities. The active discussions which occurred during the meeting indicated that the problem posed by climatic variations and man-made influence on climate constitute an international interdisciplinary issue in which activities such as NATO's have great and beneficial influence.

The ASI received its major financial support from NATO Special Programme on the science of global environmental change and the Commission of the European Union. The support of the US-NSF, which provided travel funds for several american students, and the University of Siena, which welcomed the ASI to the wonderful site of the Certosa di Puntignano is gratefully acknowledged. Special thanks are due to Carlo Laj and Carlo Alberto Ricci, who greatly contributed to the success of the ASI.

LIST OF AUTHORS and PARTICIPANTS

ABBATE Giulia	ENEA - Area Energia Ambiente e Salute - C.R.E. Casaccia Via Anguillarese 301 00100 ROMA ITALY
BARD Edouard	Laboratoire de Géoscience de l'Environnement Université Marseille III, Case 431 Centre Saint-Jérôme 13397 MARSEILLE Cedex 13 FRANCE
BASSINOT Franck	Laboratoire de Géologie du Quaternaire CNRS Luminy - case 907 13288 MARSEILLE Cedex 09 FRANCE
BAUM Steven	Applied Research Corporation 305 Arguello Drive COLLEGE STATION, Texas 77840 USA
BECQUEY Sabine	Département de Géologie et Océanographie URA CNRS 197, "CRESO", Avenue des Facultés 33405 TALENCE Cédex FRANCE
BERGER André	Institut D'Astronomie& Géophys. G. Lemaitre Université Catholique de Louvain, 2 Chemin du Cyclotron B-1348 LOUVAIN-LA-NEUVE BELGIUM
BERTRAM Caroline	Subdepartment of Quaternary Research, Godwin Laboratory, Cambridge University, Free School Lane CAMBRIDGE CB2 3RS UNITED KINGDOM
BEVERIDGE Neil	Dept. of Earth Sciences University of Cambridge , Downing Street CAMBRIDGE C82 3RS UNITED KINGDOM
BLETTE Veronica	Department of Geology Smith College NORTHAMPTON, Mass. 01063 USA
BOULTON G.S.	Grant Institute of Geology, Edinburgh University West Main Road EDINBURGH EH9 3JW UNITED KINGDOM
BRACONNOT Pascale	LODYC, Tour 14 Case 100, 4, Place Jussieu 75252 PARIS CEDEX O5 FRANCE
BRANCATELLI Alberto	Dipartimento di costruzioni meccaniche e nucleari Via Diotisalvi, 2 56126 PISA ITALY

BROCCOLI Anthony J. — Geophysical Fluid Dynamics Laboratory
Princeton University, P.O. Box 308
PRINCETON, New Jersey 08542 USA

CALDEIRA Ken — Earth System Science Center, 248 Deike Bldg.
The Pennsylvania State University
UNIVERSITY PARK , PA. 16802 USA

CAMPIN Jean Michel — Institut d' Astronomie et de Géophysique G. Lemaître, Université Catholique de Louvain, 2 Chemin du Cyclotron
1348 LOUVAIN-LA-NEUVE BELGIUM

CHALIE Françoise — Laboratoire de Géologie du Quaternaire Faculté des Sciences de Luminy, Case 907
13288 MARSEILLE Cedex 09 FRANCE

CIAIS Philippe — University of Colorado at Boulder
INSTAAR Campus, Box 450
BOULDER, CO. 80309 USA

DALFES Nüzhet — ITU Maden Fakültesi
Genel Jeoloji Anabilimdali
80626 Ayazaga ISTANBUL TURKEY

DALGLEISH Norman — Dept. of Geology & Geophysics
Grant InstituteWest Main Road
EDINBURGH EH9 3JW UNITED KINGDOM

DELECLUSE Pascale — Université Pierre et Marie Curie, LODYC, Tour 15
4, Place Jussieu
75005 PARIS FRANCE

DUCOUDRE Nathalie — DSM / LMCE Bat. 701
Orme des Merisiers, CEN Saclay
91191 GIF-SUR-YVETTE FRANCE

DUPLESSY Jean -Claude — Centre des Faibles Radioactivités, Laboratoire Mixte CNRS-CEA
Avenue de la Terrasse
91198 GIF-SUR-YVETTE Cedex FRANCE

FARRELL John — Department of Geological Sciences
Brown University
PROVIDENCE, RI 02912-1846 USA

FICHEFET Thierry — Institut d'Astronomie et de Géophysique Georges Lemaître, Université Catholique de Louvain, 2 Chemin du Cyclotron
1348 LOUVAIN-LA-NEUVE BELGIUM

GALLUP Christina — Department of Geology and Geophysics , University of Minnesota
310 Pillsbury Dr. S.E.
MINNEAPOLIS, MN 55455 USA

GASSE Françoise	L.H.G.I. Bâtiment 504, Université Paris Sud 91405 - ORSAY Cedex FRANCE
GUIOT Joël	Laboratoire de Botanique Historique et Palynologie, UA 1152 du CNRS, Faculté des Sciences de St Jérôme 13397 MARSEILLE Cedex 13 FRANCE
HALL Nicholas	Department of Meteorology University of Reading, 2 Early Gate, Whiteknights, READING RG6 2AU UNITED KINGDOM
HAXELTINE Alex	Department of Plant Ecology Lund University, Östra Vallgatan 14 S-223 61 LUND SWEDEN
HEWITT Christopher	Room H304, Hadley Centre U.K. Meteorological Office, London Road, Bracknell Berkshire RG12 2SZ UNITED KINGDOM
HOEK Wim	Faculteit der Aardwetenshappen Vrÿe Universiteit , De Boelelaan 1085 1081 HV AMSTERDAM THE NETHERLANDS
HOFFMANN Georg	Max -Planck-Institut für Meteorologie Bundesstrasse 55 D-2000 HAMBURG GERMANY
HOVINE Stéphane	Institut D'Astronomie & Géophys. G. Lemaitre, Université Catholique de Louvain, 2 Chemin du Cyclotron 1348 LOUVAIN-LA-NEUVE BELGIUM
HOWARD Will	Lamont Doherty Geological Observatory Columbia University PALISADES, NY 10964 USA
IMBRIE John	Department of Geological Sciences Brown University PROVIDENCE, RI 02912-1846 USA
IVANOVA Elena	Institute of Oceanology, Shirshov Institute of Oceanology Krasikova street 23 MOSCOW 117218 RUSSIA
JOLLY Dominique	Laboratoire de Géologie du Quaternaire Faculté des Sciences de Luminy , Case 907 13288 MARSEILLE Cedex 7 FRANCE
JOUSSAUME Sylvie	LMCE/LODYC - Bât 709 DSM, Orme des Merisiers, CEN Saclay 91191 GIF-SUR-YVETTE Cedex FRANCE

JOUZEL Jean	DSM/LMCE - Bât 709, Orme des Merisiers, CEN Saclay 91191 GIF-SUR-YVETTE Cedex FRANCE
JUNG Simon J.A.	Geologisch-Paläontologisches Institut der Universität Kiel, Olshausenstrasse 40 2300 - KIEL 1 GERMANY
KNAACK Jensjörg	Geologische-Paläeontologisches Institüt der Universität Kiel, Olshausenstrasse 40 2300 - KIEL GERMANY
KNELLER Margaret	NASA/Goddard Institute for Space Studies 2880 Broadway NEW YORK, NY 10025 USA
KUKLA Georges	Lamont Doherty Geological Observatory Columbia University PALISADES, NY 10964 USA
LAJ Carlo	Laboratoire de Modélisation du Climat et de l'Environnement AND Centre des Faibles Radioactivités, Laboratoire Mixte CNRS/CEA Avenue de la Terrasse91198 - GIF-SUR-YVETTE Cedex, FRANCE
LAUTENSCHLAGER M.	German Climate Computer Center Bundesstrasse 55 D-2000 HAMBURG 13 GERMANY
LENTNER Inigo	Universität Hohenheim, Institut für Botanik (210) Garbenstrasse 30, Postf. 70 05 62 D-7000 STUTTGART 70 GERMANY
MACDONALD Jannette	Dept. of Earth Sciences, University of Cambridge Downing Street CAMBRIDGE CB2 3EQ UNITED KINGDOM
MAGGI Valter	Dipartimento di Scienze della Terra Via Mangiagalli 34 20133 MILANO ITALY
MAIER-REIMER Ernst	Max-Planck Institut für Meteorologie Bundesstrasse 55 D-2000 HAMBURG 1 GERMANY
MARSHALL Shawn	Department of Geophysics & Astronomy The University of British Columbia, 129-2219 Main Mall VANCOUVER, B.C. V6T IZ4 CANADA
MARTIN Philippe	Institute for Remote Sensing Applications C.E.C. Joint Research Center TP 440 2102 ISPRA (VA) ITALY

MARTIN Thierry	Laboratoire de Glaciologie et de Géophysique de l'Environnement Rue Molière, Domaine universitaire, BP 96 38402 St MARTIN d'HERES Cédex FRANCE
MAZZINI Marino	Dipartimento di costruzioni meccaniche e nucleari Via Diotisalvi, 2 56126 PISA ITALY
McMANUS Jerry	Lamont Doherty Geological Observatory Columbia University PALISADES, NY 10964 USA
MILIONIS Alexandros	33, Likiou Street Agia Paraskevi, ATHENS 15341 GREECE
MOMMERSTEG Herman	Hugo De Vries Laboratory, Dep.t of Palynology University of Amsterdam, Kruislaan 318 1098 SM. AMSTERDAM THE NETHERLANDS
OESCHGER Hans	Physikalisches Institut,Universität Bern Sidlerstrasse 5 3012 BERN SWITZERLAND
PELTIER W. R	Department of Physics, University of Toronto 60 St Georges Street TORONTO M5S 1A7, Ontario CANADA
PHILIPPE Marc	Paléobotanique du Mésozoïque, Bât. 401/A Université Claude-Bernard, 43 Bd du 11 novembre 1918 69622 VILLEURBANNE Cédex FRANCE
RAMSTEIN Gilles	DSM / LMCE Bat. 701 Orme des Merisiers, CEN Saclay 91191 GIF-SUR-YVETTE FRANCE
RAVAZZI Cesare	Dipartimento di Scienze della Terra Via Mangiagalli 34 20133 MILANO ITALY
RICH Jonathan	Geological Sciences Department Northwestern University 1847 Sheridan Road EVANSTON, Illinois 60208 USA
ROSELL Antoni	School of Chemistry, University of Bristol Environmental & Analytical Section, Cantocks Close BRISTOL BS8 IT5 UNITED KINGDOM
SADOURNY Robert	Laboratoire de Météorologie Dynamique Ecole Polytechnique, Route Départementale 36 91128 - PALAISEAU Cedex FRANCE

SELTZER Geoffrey	Byrd Polar Research Center 108 Scott Hall, 1090 Carmak Road COLUMBUS, Ohio 43210-1308 USA
SHACKLETON N. J.	University of Cambridge Godwin Laboratory, Free School Lane CAMBRDIGE, CB2 3RS UNITED KINGDOM
SIMONSEN Knud	Nansen Environmental and Remote Sensing Center Edvard Griegsvei 3a N-5037 SOLHEIMSVIK NORWAY
STREET-PERROTT Alayne	School of Geography Mansfield Road OXFORD OX13TB UNITED KINGDOM
T'SIOBBEL Stephen	Geographich Instituut, NRYE Universiteit Brussel Pleinlaan 2 1050 BRUSSELS BELGIUM
TAYLOR Karl E.	Lawrence Livermore National Laboratory P.O. Box 808, L-264 LIVERMORE, Ca 94550 USA
URIARTE Anton	Escuela de Magisterio, Universidad del Pais Vasco Ibaeta s/n. 20002 SAN SEBASTIAN SPAIN
VALDES P.J.	Department of Meteorology University of Reading, 2 Early Gate, Whiteknights, READING RG6 2AU UNITED KINGDOM
VAN'T VEER Ron	Universiteit Van Amsterdam, Hugo De Vries Lab. , Paeobiologie Kruislaan 318 1098 SM. AMSTERDAM THE NETHERLANDS
VAUTRAVERS Maryline	Département Géologie & Océanographie, URA CNRS 197, "CRESO", Avenue des Facultés 33405 TALENCE Cédex FRANCE
VERBITSKY Mikhail	Department of Geology and Geophysics Yale University, P.O. Box 6666 NEW HAVEN, CT 06511-8130 USA
VERSTRAETE Michel M.	Institute for Remote Sensing Applications C.E.C. Joint Research Center TP 440 21020 ISPRA (VA) ITALY
WASHINGTON Richard	Environmental Change Unit University of Oxford, Mansfield Road OXFORD OX13TB UNITED KINGDOM

WILLIAMSON David	Laboratoire de Géologie du Quaternaire Faculté des Sciences de Luminy, Case 907 13288 MARSEILLE Cedex 7 FRANCE
WORSLEY Peter	Postgraduate Res. Institute for Sedimentology The University, P.O. Box 227 Whiteknights, READING RG6 2AB UNITED KINGDOM
YIOU Pascal	DSM / LMCE Bat. 701 Orme des Merisiers, CEN Saclay 91191 GIF-SUR-YVETTE FRANCE
YOUNG Raymond	Hugo De Vries Laboratory, Dept. of Palynology University of Amsterdam, Kruislaan 318 1098 SM. AMSTERDAM THE NETHERLANDS
ZHOU Li-Ping	School of Environmental Sciences University of East Anglia NORWICH NR4 7TJ UNITED KINGDOM

I - THE CLIMATE SYSTEM AND ITS MODELLING

MODELING LONG-TERM CLIMATIC CHANGES

Sylvie Joussaume [1,2]
[1] Laboratoire de Modélisation du Climat et de l'Environnement
DSM, CE Saclay Bat 709, 91191 Gif sur Yvette Cedex, FRANCE
[2] Laboratoire d'Océanographie Dynamique et de Climatologie
Unité mixte CNRS/Université/ORSTOM, Paris

A complete hierarchy of climate models has been developed to help understand past climatic changes, from simple climate models (*Taylor, this volume*) to general circulation models. In the present paper, we overview the use of atmospheric general circulation models and illustrate it with some examples. In part one we give an introduction to the climate system in order to better understand the role of the various components of the climate system and of the various time scales and forcings involved in past climatic changes. More detailed papers can be found in the literature and will be referred to for further reading.

1. THE CLIMATE SYSTEM

As defined in 1975 by the National Research Council, the climate system is composed of five components outlined on Figure 1 : the atmosphere, the hydrosphere, the biosphere, the cryosphere and the lithosphere. Each of them exhibits internal processes characterized by specific spatial and time scales, and interacts in various ways with the other components of the climate system (*Gates*, 1981; *Bergman et al.*, 1981; *Washington and Parkinson,* 1986; *Peixoto and Oort,* 1992; *Trenberth,* 1992). Climate evolution is then governed both by these interactions, or *internal forcings*, and by*external forcings,* defined as those independent of climate, such as solar radiation, land-ocean distribution changes due to continental drift, and aerosol changes due to volcanism.

1.1 Components of the climate system

The atmosphere comprises the gaseous envelope of the Earth and its composition (H_2O, O_3, CO_2, dust...). Forced by the incoming solar radiation, its main role is to redistribute heat from tropical to high latitudes. This is achieved by the mean

NATO ASI Series, Vol. I 22
Long-Term Climatic Variations
Edited by J.-C. Duplessy and M.-T. Spyridakis

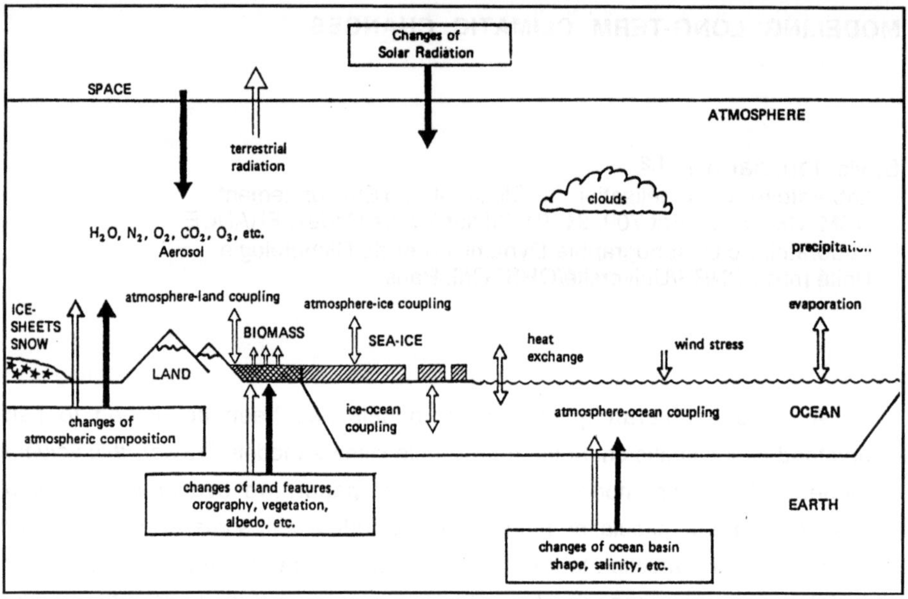

Figure 1: Schematic illustration of the components of the climate system. The full arrows (➡) are examples of external processes, and the open arrows (⇒) are examples of internal processes in climatic changes. Adapted from *GARP* (l975).

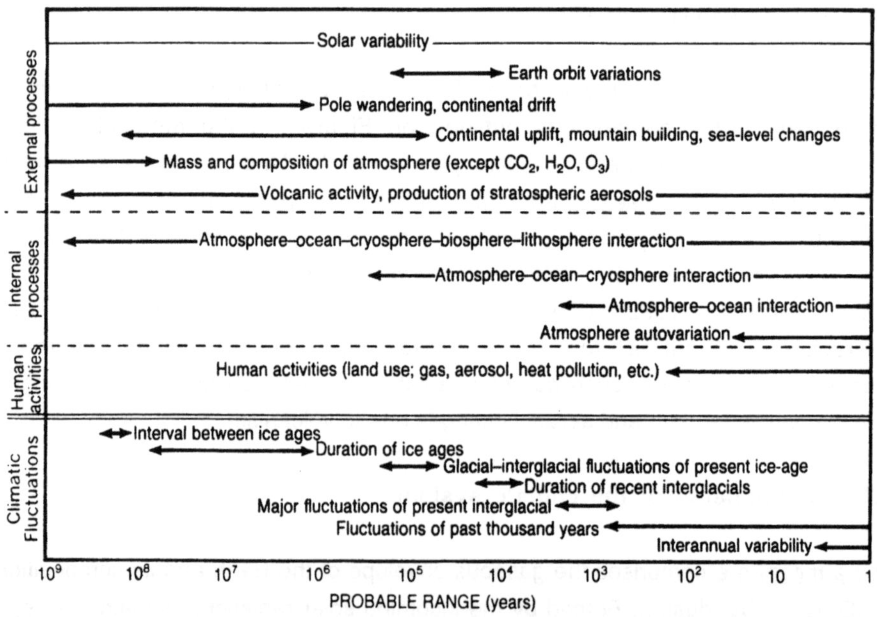

Figure 2 : Time scales of the components of the climate system and of the main external forcings . Adapted from *NRC* (1975) and *Bergman et al.* (1981).

meridional Hadley circulation in the tropics and by the transient and stationnary eddies in mid-latitudes (*Sadourny, this volume*). The atmosphere interacts with the Earth's surface through the turbulent fluxes of heat, momentum and gases (water vapor, CO_2...), through radiation, with surface albedo and thermal emissivity, and through mechanical impact (relief). The atmospheric circulation is thus highly dependent on the state of the other components of the climate system, such as the land-ocean distribution, the ocean surface temperature, relief, snow and ice, and vegetal cover.

The hydrosphere comprises all the liquid water on and beneath the surface, with the most important part being the oceans which cover two thirds of the Earth's surface. Oceans not only constitute an infinite reservoir of water for the atmosphere, but also a strong heat reservoir, thanks to their high absorption of solar energy, and their high specific heat. Oceans play an important role in the redistribution of heat from tropical to high latitude regions. Its contribution is nearly equal to the atmospheric contribution as estimated by *Carissimo et al.* (1985). The oceanic circulation is driven by surface fluxes at the atmosphere-ocean interface and density gradients controlled by salinity and temperature (*Delecluse, this volume*). Oceans interact not only with the atmosphere, but also with sea-ice through water density increased by salt rejection when sea-ice is formed.

The cryosphere comprises snow and ice, either on the sea (sea-ice) or on land (snow, glaciers and ice-sheets). It is, of course, closely related to the atmosphere and hydrosphere through the hydrological cycle. Snow and ice influence the climatic system through their high albedo (from 50% to 90%) and their low thermal conductivity, insulating the atmosphere from the underlying sea or land. These processes tend to decrease the local temperature and therefore to increase the amount of snow-ice, resulting in a net positive feedback.

The lithosphere comprises rocks, soil and sediment on the Earth, and appears to be relatively passive compared to the other components of the climate system. However, mountain building, geographical changes in land-sea distribution, and the evolution of depth and shape of ocean basins resulting from plate tectonics through continental drift and sea-floor spreading, are quite important for the atmosphere-hydrosphere-cryosphere components.

The biosphere includes Earth's plants and animal life. It is quite sensitive to the other components of the climate system. Over land, plant growth depends on solar radiation, temperature, rainfall and carbon dioxide concentration. Inversely, the vegetation cover influences the atmospheric circulation through albedo, surface roughness and water vapor fluxes. In the oceans, primary productivity is limited by the available amount of nutrients brought back to the surface in upwelling circulations. In return, this productivity controls the ocean sink of carbon.

The atmospheric water cycle is a major link between the various components. Not only acting as a source and sink for the various water reservoirs, it plays a crucial role in the Earth's energetics by transporting latent heat energy from the surface to the atmosphere through evaporation over land and oceans. Moreover, water in the atmosphere greatly modifies the radiative transfer, as cloud coverage affects global albedo and as water vapor is a strong greenhouse gas.

1.2 Time scales

The climate system and its external forcings evolve on very broad time scales, as is illustrated in Figure 2. The atmosphere responds within a few months to either changes in the other climatic components (considered as boundary conditions) or to external forcings. In addition, the atmosphere can vary even without modifications in boundary conditions and forcing, displaying autovariations up to several years. For the oceanic circulation, time scales range from months for the surface layer (~50 m) to 100-1000 years for the deeper circulation (down to 4000 m), where thermal and mechanical inertia of water to surface forcing is greater. For the cryosphere, time scales range from seasons for snow and sea-ice, to centuries for glaciers, and 10^2-10^6 years for ice-sheets. The time scale for changes in the vegetation cover range from decades to centuries.

External forcings range on all time scales, from years up to the Earth's age. Changes in solar forcing can result either from changes in the energy radiated by the Sun or from changes in the Earth's orbital parameters which control the seasonality of energy intercepted by the Earth. The solar "constant" (which characterizes the energy radiated by the Sun) varies both on short time-scales, from decades to centuries (*Sadourny, this volume*), and very long time-scales such as the brigthening of the Sun since the Earth's formation. Variations in Earth's orbital parameters range from 10^4 to 10^6 years, with the main periodicities resulting from changes in eccentricity (100 000, 400 000 years), obliquity (41 000 years) and precessionnal effects (19 000 and 23 000 years) (*Berger*, 1988; *Berger, this volume*). As proposed by Milankovitch in the 1920s, these orbital variations are considered to be the main cause of the glacial/interglacial cycles of the Pleistocene (the last 2 million years) (*Imbrie and Imbrie*, 1979; *Imbrie, this volume*). On longer time scales, changes in plate tectonics, such as mountain building and continental drift, become an important forcing and must be accounted for in the mechanisms of climate change (*Frakes*, 1979).

The challenge is to understand how all these internal and external processes have conspired to give the observed past climates, and how they will behave in the future. So complicated a task requires the use of computer simulations.

2- OVERVIEW OF PALEOCLIMATE SIMULATIONS

A complete climate model would explicitly resolve all the physical and dynamical processes of each component of the system at all relevant time scales. Such a global model does not exist, and is far from being achieved, due both to our limited understanding of all the interactions, and to the unrealistic computer requirements.

In the last twenty years, most efforts have been devoted to atmospheric general circulation models (AGCMs), similar to those used for weather forecasting. Detailed descriptions of AGCMs can be found in *Sadourny* (*this volume*), *Washington and Parkinson* (1986), *Schlesinger* (1988), *Henderson-Sellers and McGuffie* (1987).

2.1 Methodology

Having proved their capability to reproduce the main features of modern climate (see for example, *Gates et al.,* 1990), AGCMs can be used under paleoclimatic conditions. Such studies address the mechanisms of climate change, such as changes in insolation and atmospheric composition. Comparisons to paleodata test the reliability of these models under climatic conditions significantly different from the present day. If sufficiently realistic, AGCMs also provide a general description of past atmospheric circulations which complements local information inferred from paleoclimatic data. Finally, the models can be used as tools to help interpret observations by simulating the cycles of tracers, such as water isotopes, whose distribution and evolution are related to paleoclimate change.

Since AGCMs only model the atmospheric component of the climate system, the state of the other components must be specified, i.e. the land-ocean distribution, ice sheets, vegetal cover (mainly through albedo and surface roughness), ocean surface (sea surface temperatures), sea ice extent and the atmospheric composition (mainly CO_2 content), in order to successfully simulate past climatic conditions. Starting from any initial conditions, the numerical simulation then provides the atmospheric circulation in statistical equilibrium with these prescribed *boundary conditions* for a specific *snapshot* period of the past. A full set of boundary conditions is known only for a few past periods, such as the last glacial maximum and mid-Holocene. Even here, some uncertainties remain due to inaccuracies in the estimation of these boundary conditions. In addition, sensitivity of climate to individual forcings can be estimated by modeling the response to changes in boundary conditions such as insolation and land-sea distribution, while holding other climate system components constant, or at most allowing changes in sea surface temperature by including a simple mixed-layer ocean model in order to account for the thermal capacity of the surface ocean. Fully coupled atmosphere-ocean general circulation models will be required in the future but are still under development (*Delecluse, this volume*).

Most studies of past climates using AGCMs concern the last glacial/interglacial cycle, with particular emphasis on the past 20 000 years. A few studies address the pre-quaternary climatic conditions. Some examples are illustrated in the following and more detailed reviews can be found in *Crowley and North* (1991), *Kutzbach* (1985; 1992), *Crowley* (1988), *Barron* (1985).

2.2 The Last Glacial Maximum

Most efforts have been devoted to the Last Glacial Maximum (LGM), 18,000 year ago (radiocarbon date), which corresponds to the last maximum extent and volume of the Laurentide and Fennoscandian ice-sheets. At that time, the incoming solar radiation is not much different than present day, so that changes in the atmospheric circulation are predominantly driven by the different CO_2 atmospheric concentration (200 ppm rather than 350 ppm for today, *Barnola et al.* (1987)), ice-sheets, sea surface temperature, sea-ice and surface albedo (Figure 3), as compiled by *CLIMAP (1976 or 1981).*

Several models have been run under the most recent CLIMAP (1981) boundary conditions [*Kutzbach and Guetter,* 1986; *Rind,* 1987; *Lautenschlager and Herterich,* 1990; *Joussaume,* 1993]. They all give a global annual mean surface temperature of the order of 4°C less than the present climate (Table 1). This cooling, smaller in tropical regions, increases with latitude, reaching maximum values over ice sheets and sea ice in both hemispheres (Figure 4). The cooling is particularly important over the North Atlantic sector, where it can reach 20 to 30°C, and is a result of changes in the atmospheric circulation. The Northern Hemisphere jet is split around the Laurentide ice sheet and brings polar air over the North Atlantic region.

The simulated changes in surface air temperature are in general agreement with paleodata (*Peterson et al.,* 1979; *COHMAP*, 1988) except in tropical regions where models tend to underestimate the cooling. *Rind and Peteet* (1985) argued that this discrepancy could partly be due to the relatively warm SSTs given by *CLIMAP* (1981) for most tropical oceanic areas (Figure 3). Indeed, stronger

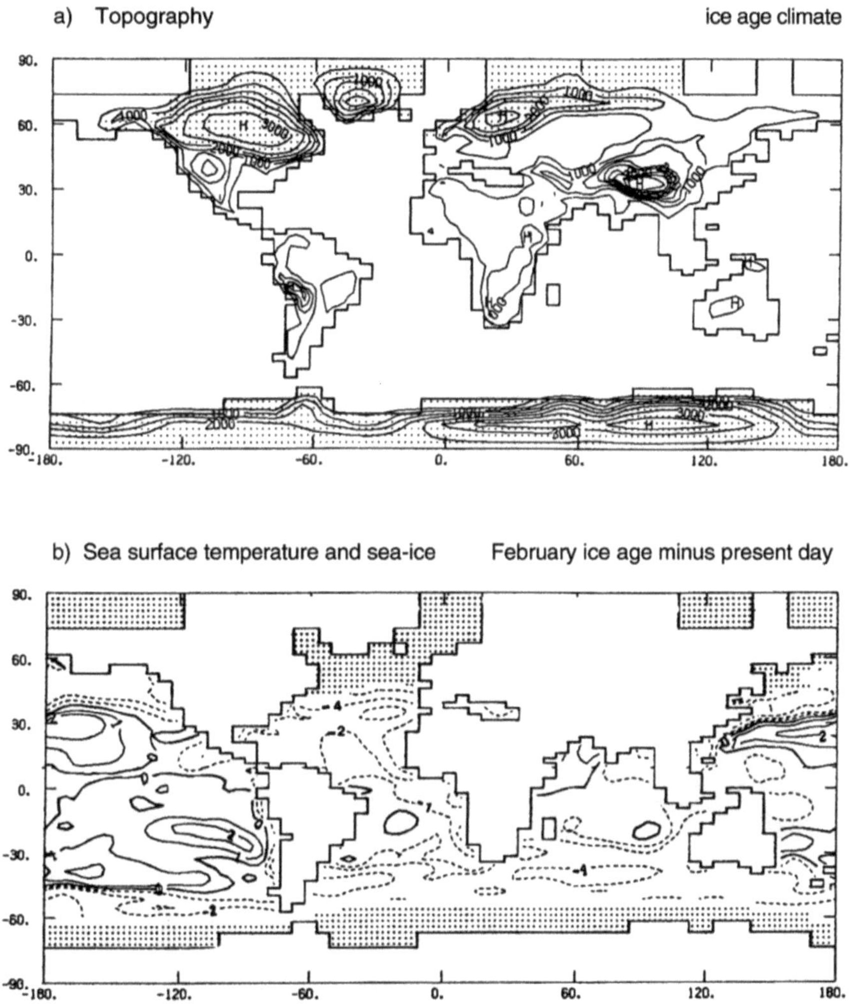

Figure 3: Boundary conditions for the simulation of the last glacial maximum climate as given by *CLIMAP* (1981). a) : ice age topography (at every 500 m), b) : change in February sea-surface temperatures (isolines at every 1°C above -2°C and at every 4°C below -4°C) and sea-ice extent (dotted areas) (from *Joussaume* (1983)).

Annual mean surface air temperature
ice age minus present day

a) Laboratoire de Météorologie Dynamique AGCM

b) National Center for Atmospheric Research Climate Community Model

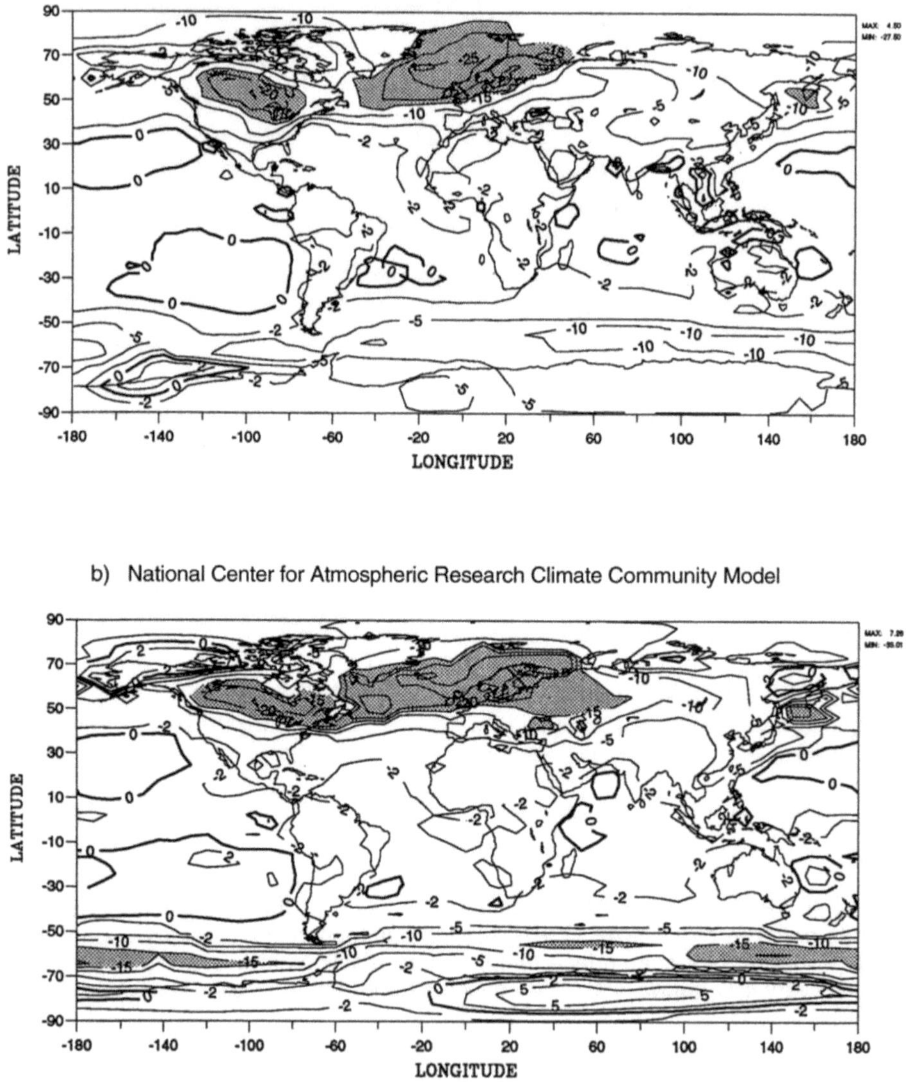

Figure 4 : Simulations of the last glacial maximum climate (18,000 years BP) with atmospheric general circulation models : simulated annual mean changes in surface air temperature relatively to present day. a) LMD AGCM (see *Joussaume*, 1993), b) NCAR CCM (see *Kutzbach and Guetter*, 1986). Isolines at every 5°C and isolines ±2°C, dotted areas for values below -15°C.

Annual mean surface air temperature
ice age minus present day

a) Goddard Institute for Space Studies AGCM

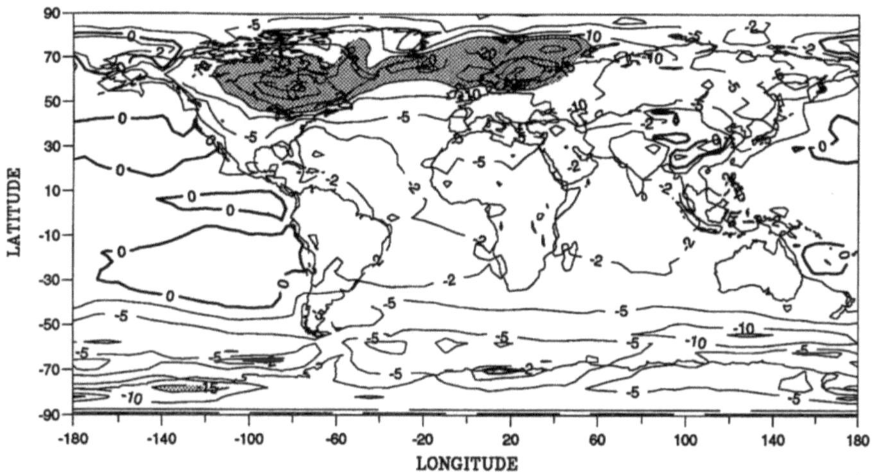

b) Max Planck Institute of Hamburg AGCM

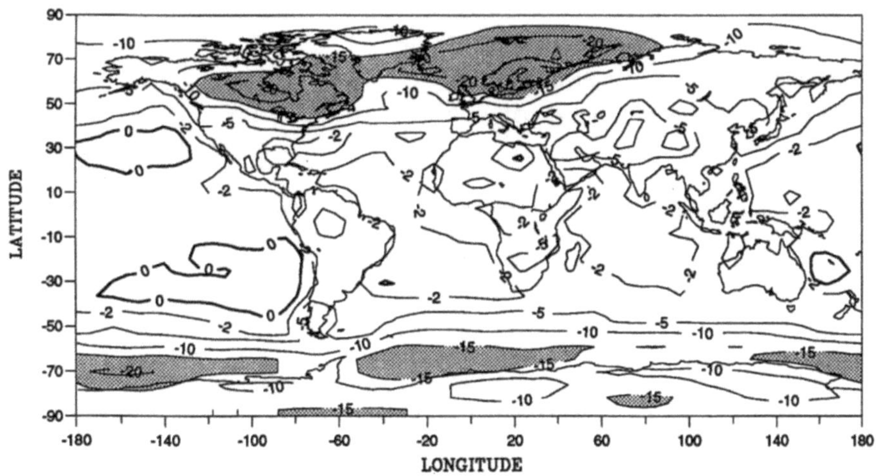

Figure 4 (continued) : c) GISS AGCM (see *Hansen et al.*, 1984), d) MPI AGCM (see *Lautenschlager and Herterich*, 1990). Isolines at every 5°C and isolines ±2°C, dotted areas for values below -15°C.

continental temperature coolings were simulated in numerical experiments based on colder SSTs, as is seen in *Rind and Peteet* (1985) or in the experiments based on the previous *CLIMAP* (1976)'s dataset by *Gates* (1976a and b) and *Manabe and Hahn* (1977). But other reasons may also be responsible for this model-data discrepancy. Model deficiencies are always candidates. In addition, the interpretation of past data may also be questioned. Indeed, in equatorial eastern Africa the use of modern analogues to interpret pollen data has reduced the original temperature depression estimate from 5°-8°C to 4°C ±2° (*Bonnefille et al.,* 1990), improving model-data comparisons.

TABLE 1. Comparison of the simulated changes in global mean surface air temperature for the last glacial maximum obtained by various AGCMs.

AGCMs	Surface air temperature (°C) Ice age minus present	
	NH WINTER	NH SUMMER
Oregon State University (OSU)[1]		-4.9°
Geophysical Fluid Dynamics Laboratory (GFDL)[2]		-5.4°
National Center for Atmospheric Research (NCAR)[3]	-3.7°	-3.9°
Goddard Institute for Space Studies (GISS)[4]	-3.3°	-3.4°
Max Planck Institut für Meteorologie (MPIM)[5]	-4.2°	-5.4°
Laboratoire de Météorologie Dynamique (LMD).[6]	-3.0°	-4.1°

where: [1] *Gates* (1976b), [2] *Manabe and Hahn* (1977), [3] *Kutzbach and Guetter* (1986), [4] *Rind* (1987), [5] *Lautenschlager and Herterich* (1989), [6] *Joussaume* (1993). The OSU and GFDL simulations use the *CLIMAP* (1976) dataset while the others use the *CLIMAP* (1981) dataset.

AGCMs are also appropriate tools to study the atmospheric cycle of key climatic tracers, such as water isotopes (HDO and $H_2{}^{18}0$) and desert dust particles (*Jouzel et al.,* 1993; *Joussaume*, 1993). This kind of study allows a more direct comparison with observed tracer data and helps the interpretation of paleodata in terms of past atmospheric climatic parameters. For example, *Joussaume and Jouzel* (1993) and *Jouzel et al.* (submitted) have shown with two different AGCMs that the mean relationship observed today between the oxygen 18 (or deuterium) content of precipitation and temperature remains practically unchanged under ice age conditions, confirming the use of water isotopes to infer paleotemperatures from ice cores.

Although, AGCM simulations give a 3-dimensional description of the atmospheric circulation, their interpretation is not always straightforward, because climatic features can result from different conflicting processes. It is, therefore, interesting to test the sensitivity of model results to the separate changes in boundary conditions. For example, *Manabe and Hahn* (1977) have found that the increase of land albedo was primarily responsible for the weaker indian monsoon. *Rind* (1987) studied the separate impacts of sea surface temperature, ice-sheet extent (10 m thick), and ice-sheet elevation. *Broccoli and Manabe* (1987) have also performed separate sensitivity experiments to land surface albedo, ice-sheets and CO_2 changes, showing the contribution of both ice-sheets and CO_2 to the global cooling (*Broccoli, this volume*).

2.3 Sensitivity experiments to insolation changes

Many sensitivity experiments to changes in the Earth's orbital parameters have been conducted with AGCMs, some of them including a mixed layer ocean model, in order to study the impact on the atmospheric circulation. A first set of experiments concerns the early Holocene, about 9000 years BP (before present, defined relatively to 1950 AD), when the Northern Hemisphere experienced a stronger seasonal contrast in insolation, up to +7% in June-July-August and -7% in December-January-February. They are complemented by studies at 6000 years BP for which insolation changes are not biased by remnants ice sheets. All the simulations produce an intensified summer monsoon over the Asian and African regions (e.g., *Kutzbach,* 1981; *Kutzbach and Guetter,* 1986; *Kutzbach and Gallimore,* 1988; *Mitchell et al.,* 1988). The continents warm faster than the oceans, increasing the land-ocean contrast. As a result, the Asian thermal low is deeper and moist air penetrates further into the continents (Figure 5). This feature is clearly related to insolation, and in good agreement with data as shown by *Kutzbach and Street-Perrot* (1985) and *Prell and Kutzbach* (1987).

Sensitivity experiments to changes in solar radiation have also been performed at 125 000 and 115 000 years BP, corresponding respectively to +12% and -6% in the July northern hemisphere solar radiation *(Royer et al.,* 1984; *Rind et al.,* 1989). According to the Milankovitch theory, the resulting cooling of continents at 115 000 years BP is thought to have triggered glacial inception by preventing snow melting during summer at northern hemispheric high latitudes. The summer snow cover further intensifies the cooling, leading to the initiation of ice sheets.

Although, results by *Royer et al.* (1984) do not contradict the theory, *Rind et al.* (1989) were not able to prevent complete snow melting. Whether this failure results from model deficiencies, for example in snow hydrology, or from unresolved small scale processes such as mountain glaciers, or from the lack of full ocean coupling is not yet known.

Changes in the other components of the climate system may also influence climate. For example, monsoon changes at 9000 years BP modify the vegetal cover which, in turn, further intensifies monsoon rains over Africa as shown by *Street-Perrot et al.* (1990). Ocean circulation is also an important component. For example, *Broecker and Denton* (1990) argue a role for the thermohaline circulation in abrupt climatic events like the Younger Dryas around 11 000 year BP. However, sensitivity experiments to insolation changes at 11 000 years BP performed by *Rind et al.* (1986), were not able to reproduce the cold observed temperatures in high northern latitudes, even with glacial conditions over the northern Atlantic.

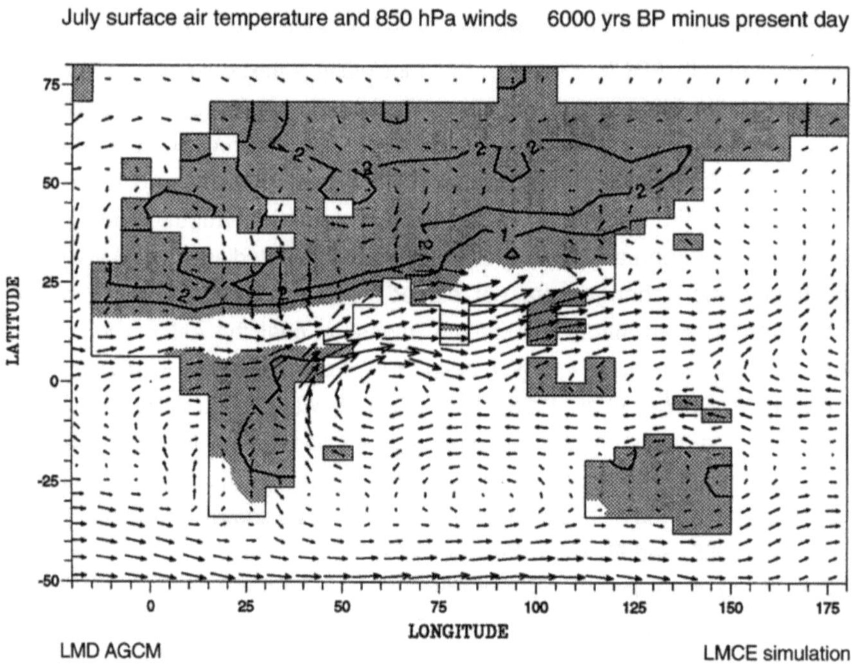

Figure 5 : Sensitivity experiment to insolation changes at 6000 years BP : July changes in surface air temperature and in 850 hPa wind direction. Dotted areas for temperature differences above 0°C and isolines +1°C and +2°C (*de Noblet et al.*, personal communication, 1993).

2.4 Pre-quaternary climates

On time scales of 10^6-10^8 years, Earth's surface changes become important and may well be a major forcing of the climate system. Sensitivity to changes in land-ocean distribution can be tested with AGCMs. *Barron and Washington* (1984) simulated the Cretaceous climate with an AGCM coupled to a simple surface ocean model. The land-ocean distribution was modified as seen from figure 6, resulting in a 20% reduction in total land area. Although the modeled global temperature increases by 5°C compared to present day, this is less than the minimum increase estimated at 6°C from paleodata. In particular high latitude simulated temperatures are too low. Other processes such as enhanced CO_2 due to greater volcanism may have forced Cretaceous warmth. An experiment with a CO_2 quadrupling improves model results, although tropics tend to be overheated. This discrepancy could, in turn, be due to the lack of an explicit inclusion of ocean heat transport. *Barron (1984)* concludes from these modeling studies "that the Cretaceous climate may not be attributable to a single cause (i.e. the area of high-latitude land) but rather may reflect the importance of a whole suite of variables".

Another important question concerns the gradual cooling of the Tertiary, leading to the beginning of the glacial/interglacial cycles 2.5 million years ago. *Ruddiman and Kutzbach* (1989) showed in a series of senstivity experiments to the Rockies and Tibetan Plateau height that the large mountain uplift of five to ten million years ago might have favored cooler conditions. Other processes, such as changes in the ocean circulation (e.g. the closing of the Panama isthmus) and falling CO_2 levels might have also contributed to the general cooling.

CONCLUSION

The examples above illustrate how atmospheric general circulation models can investigate the sensitivity of climate to various forcings. Conversely, past climates can validate the models under climatic conditions significantly different from today. Model results are clearly dependent on the parameterizations used, as seen from CO_2 doubling experiments (*Mitchell et al.,* 1990). Figure 4 shows that similar features also occur for past climates. Although the general cooling trends are similar for all the simulations of the last glacial maximum, significant differences appear on a regional scale, particularly in high latitudes. For example, over east Antarctica, the model-predicted temperature shifts range from -15 to +5°C, whereas estimates from ice cores give values around -6°C (*Jouzel et al.,* 1987). An international project, the *Paleoclimate Modeling Intercomparison Project (PMIP)*,

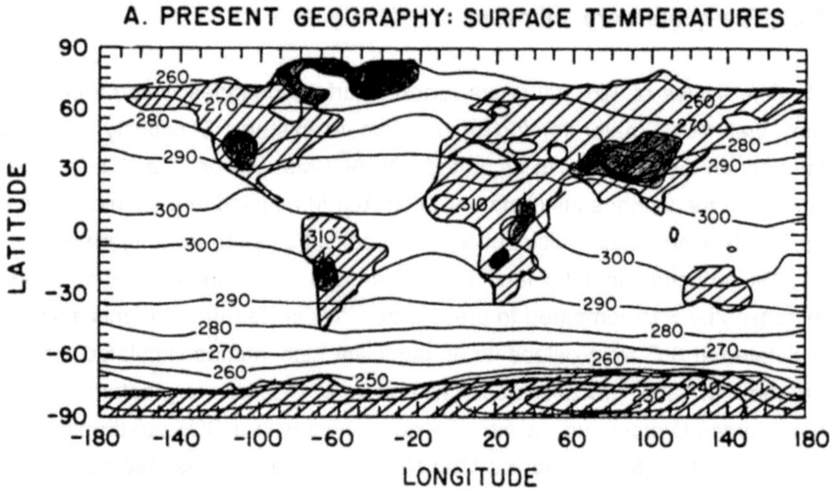

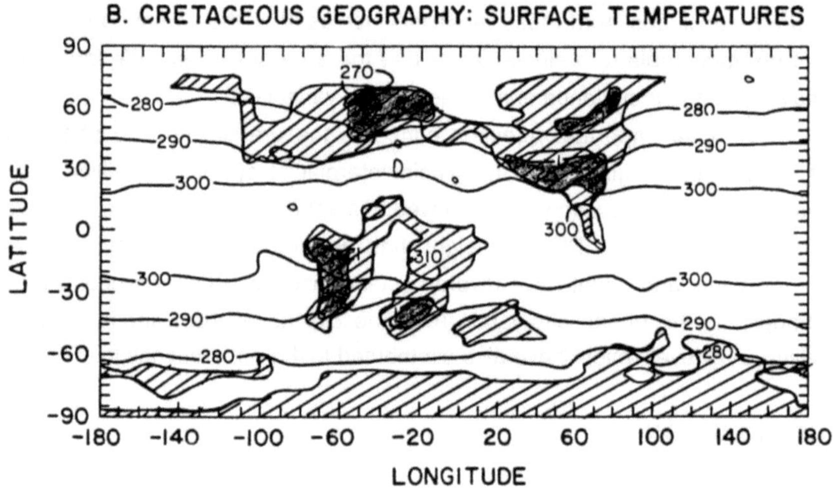

Figure 6: Simulations of the warm Cretaceous (100 million years BP) with the NCAR CCM coupled with a simple ocean surface model. Distributions of surface air temperatures in °K (isolines every 10°) for a) present day geography and b) Cretaceous geography. Topography is drawn in 10^3 meters. Adapted from *Barron* and *Washington* (1984).

coordinated by S. Joussaume and K. E. Taylor and endorsed by PAst Global changES (a core project of the International Geosphere Biosphere Program) and the Working Group on Numerical Experimentation (of the World Climate Research Program) has been initiated to compare model results for ice age and mid Holocene (6000 years BP) conditions. PMIP includes a dozen models. This project will help to increase our understanding of the mechanisms of past climates and to improve our ability to model these mechanisms successfully.

ACKNOWLEDGMENTS

I particularly thank Timothy Hall, Pascale Braconnot and Nathalie de Noblet for their useful comments on the text and Valerie Masson for providing some of the drawings. For figure 4, I also thank Pat Behling, Michael Lautenschlager and Patrick Lonergan for providing model data and Alexandre Armengaud for processing them.

REFERENCES

Barnola, J.M., D. Raynaud, Y.S. Korotkevich and C. Lorius (1987). Vostok ice core : a 160,000-year record of atmospheric C02, *Nature, 329,* :408-414.

Barron, E.J. (1984). Ancient climates : investigation with climate models, *Reports on Progress in Physics:* 1563-1599.

Barron, E.J. and W.M. Washington (1984). The role of geographic variables in explaining paleoclimates: Results from Cretaceous climate model sensitivity studies, *J. Geophys. Res., 89:* 1267-1279.

Barron, E.J. (1985). Climate models : applications for the prepleistocene. In "Paleoclimate analysis and modeling" (A.D. Hecht, ed). Wiley, New-York: 397-421.

Bergman K.H., A.D. Hecht and S.H. Schneider (1981) Climate models, *Physics today* 34:44-51.

Broccoli, A.J. and S. Manabe (1987). The influence of continental ice, atmospheric C02, and land albedo on the climate of the Last Glacial Maximum, *Climate Dynamics, 1*: 87-89.

Broecker, W.S. et G.H. Denton (1989) The role of ocean-atmosphere reorganizations in glacial cycles, *Geochemica and Cosmochimica Acta,* 53: 2465-2501.

Berger, A. (1988) Milankovitch theory and climate, *Rev. Geophysics*, 26: 624-657.

Bonnefille, R., J.C. Roeland, and J. Guiot, Temperature and rainfall estimates for the past 40,000 years in equatorial Africa, *Nature, 346,* 347-349, 1990.

Carissimo B.C., A.H. Oort and T.H. Vonder Haar (1985) Estimating the meridional energy transports in the atmosphere and ocean, *J. Phys. Oceano.* 15: 82-91.

CLIMAP (Climate : Long Range Investigation, Mapping and Prediction) (1976). The surface of the Ice-age Earth, *Science 191*: 1131-1136.

CLIMAP (1981). Seasonal reconstructions of the earth's surface at the Last Glacial Maximum. *Geological Society of America,* Map Chart Series MC-36, Boulder, Colorado.

COHMAP members (1988) Climatic changes of the last 18,000 years: observations and model simulations, *Science*, 241: 1043-1052.

Crowley, T. J. (1988) Paleoclimate Modelling, in *Physically-Based Modelling and Simulation of Climate and Climatic Change- Part II*, M. E. Schlesinger (ed), Kluwer Academic Publishers: 883-940.

Crowley T. J. and G. N. North (1991), *Paleoclimatology*, Oxford University Press, New York.

Frakes, L.A. (1979).*Climates throughout geologic times*, Elsevier.

GARP (1975). "The physical basis of climate and climate modelling", *Garp publications series,* 16, World Meteorological Organization, Geneva

Gates, W.L. (1976a). Modeling the Ice Age climate, *Science* 191: 1138-1144.

Gates, W.L. (1976b). The numerical simulation of Ice-Age climate with a global general circulation model. *J. Atmos. Sci.,* 33: 1844-1873.

Gates, W.L.(1981). The climate system and its portrayal by climate models: A review of basic principles: II). Modeling of climate and climatic change. In *Climatic Variations and Variability: Facts and Theories* (A. Berger et al. eds), Reidel, Dordrecht, Netherlands: 435-460.

Gates, W.L., P.R. Rowntree and Q-C. Zeng (1990), Validation of climate models, in *Climate change : the IPCC Scientifica assessment,* J.T. Houghton et al. (eds), Cambridge University Press: 93-130.

Hansen, J., A. Lacis, D. Rind, G. Russell, P. Stone, I. Fung, R. Ruedy, and J. Lerner (1984) Climate sensitivity : Analysis of feedback mechanisms. In "Climate Processes and Climate Sensitivity" (J.E. Hansen and T. Takahashi, eds.), Maurice Ewing Series N° 5, American Geophysical Union, Washington D.C.: 130-163.

Henderson-Sellers A. and K. McGuffie (1990), *A climate modelling primer*, J. Wiley and sons, New York.

Imbrie, J. and J.Z. Imbrie (1979). *Ices-Ages: Solving the Mystery* Enslow Publishers, Short Hills, New Jersey.

Joussaume S. (1993) Paleoclimatic tracers: an investigation using an atmospheric general circulation model under ice age conditions, part 1: desert dust, *J. Geophys.Res.* 98: 2767-2805.

Joussaume S. and J. Jouzel (1993) Paleoclimatic tracers: an investigation using an atmospheric general circulation model under ice age conditions, part 2: water isotopes, *J. Geophys. Res.* 98: 2807-2830.

Jouzel, J., C. Lorius, J.R. Petit, C. Genthon, N.I. Barkov, V.M. Kotlyakov and V.M. Petrov (1987). Vostok ice core: a continuous isotope temperature record over the last climatic cycle (160,000 years), *Nature,* 329: 403-408.

Jouzel J., S. Joussaume and R. Koster (1993) Use of general circulation models to follow tracers on a global scale, dans "*Global changes in the perspective of the past",* J. A. Eddy et H. Oeschger (eds) : 133-142.

Jouzel, J., R.D. Koster, R.J. Suozzo and G.L. Russell, Stable water isotope behavior during the last glacial maximum: a GCM analysis, *J. Geophys. Res.,* submitted.

Kutzbach, J.E. (1981), Monsoon climate of the early Holocene: climatic experiment using Earth's orbital parameters for 9000 years ago, *Science* 214: 59-61.

Kutzbach, J.E., (1985). Modeling of Paleoclimates, *Advances in Geophysics,* 28A, Academic Press: 159-196.

Kutzbach, J. E. (1992) Modeling Earth system: changes of the past, dans *Modeling the Earth System,* D. Ojima (ed), *UCAR Global Change Institute, vol 3*: 377-404.

Kutzbach, J. E. and R. G. Gallimore (1988) Sensitivity of a coupled atmosphere/ mixed layer ocean model to changes in orbital forcing at 9000 years BP, *J. Geophys. Res., 93:* 803-821.

Kutzbach, J.E. and Guetter P.J. (1986). The influence of changing orbital parameters and surface boundary conditions on climate simulations for the past 18,000 years, *J. Atmos. Sci.,* 43: 1726-1759.

Kutzbach, J.E. and F.A. Stree-Perrot (1985) Milankovitch forcing of fluctuations in the level of tropical lakes from 18 to 0 kyr BP, *Nature,* 317: 130-134.

Lautenschlager, M., and K. Herterich (1989) Climatic response to ice age conditions, part 1: the atmospheric circulation, *Max Planck Institute für Meteorologie Rep. 42,* Hamburg (Germany).

Lautenschlager, M., and K. Herterich (1990) Atmospheric response to ice age conditions --Climatology near the Earth's surface, *J. Geophys. Res.,* 95: 22,547-22,557.

Manabe, S. and D.G. Hahn (1977). Simulation of the tropical climate of an Ice Age, *Journal of Geophysical Research,* 82: 3889-3911.

Mitchell, J. F. B., N. S. Grahame, and K. J. Needham (1988) Climate simulations for 9000 years before present: seasonal variations and effects of the Laurentide ice sheet, *J. Geophys. Res.* 93: 8283-8303.

Mitchell, J.F.B., S. Manabe, T. Tokioka and V. Meleshko (1990), Equilibrium climate change, in *Climate change : the IPCC Scientifica assessment,* J.T. Houghton et al. (eds), Cambridge University Press: 131-172.

NRC Panel on Climatic Variations, U.S. Committee for GARP (1975). "Understanding Climatic change: A Program for Action." National Academy *of Sciences,* Washington, D.C.

Peixoto J. P. and A. H. Oort (1992), *Physics of climate,* American Institute of Physics, New York.

Peterson, G.M., T. Webb III, J.E. Kutzbach, T. Van Der Hammen, T.A. Wijmstra, F.A. and Street. (1979). The continental record of environmental conditions of 18,000yr BP: an initial evolution, *Quaternary Research:* 47-82.

Prell, W.L. and J.E. Kutzbach (1987). Monsoon variability over the past 150,000 years, *J. Geophys.Res.* 92: 8411-8425.

Rind, D. and D. Peteet (1985). Terrestrial conditions at the Last Glacial Maximum and CLIMAP sea-surface temperature estimates : Are they consistent ? *Quaternary Research, 24:* .1-22.

Rind, D., D. Peteet, W. Broecker, A. McIntyre and W. Ruddiman (1986). The impact of cold North Atlantic sea surface temperatures on climate: Implications for the Younger Dryas cooling (11-10K), *Climate Dynamics, 1:* 3-33.

Rind, D., (1987). Components of the ice age circulation. *J. Geophys. Res.,* 92: 4241-4281.

Rind, D., D. Peteet, and G. Kukla (1989) Can Milankovitch orbital variations initiate the growth of ice sheets in a general circulation model, *J. Geophys. Res.* 94: 12,851-12,871.

Royer, J.F., M. Deque and P. Pestiaux (1984). A sensitivity experiment to astronomical forcing with a spectral GCM : simulation of the annual cycle at 125 000 BP and 115 000 BP. In *Milankovitch and Climate,* part 2, (A. Berger et al, eds), Reidel: 733-763.

Ruddiman W.F. and J.E. Kutzbach (1989) Forcing of late Cenozoic northern hemisphere climate by plateau uplift in Southern Asia and the American West, *J. Geophys. Res.,* 94: 18409-18427.

Schlesinger M.E. (1988) *Physically-Based Modelling and Simulation of Climate and Climatic Change, part 1 and 2,* Kluwer Academic Publishers.

Street-Perrot, F.A., J.F.B. Mitchell, D.S. Marchand and J.S. Brunner (1990), Milankovitch and albedo forcing of the tropical monsoons: a comparison of geological evidence and numerical simulations for 9000 years BP, *Transactions of the Royal Society of Edinburgh* 81: 407-427.

Trenberth K. E. (1992), *Climate system modelin*", Cambridge University Press, Cambridge.

Washington W.M. and C.L. Parkinson (1986), *An introduction to three-dimensional climate modeling*, Oxford University Press, New York.

CLIMATE MODELS FOR THE STUDY OF PALEOCLIMATES

Karl E. Taylor
Lawrence Livermore National Laboratory
Livermore, California 94550
U.S.A.

ABSTRACT. Climate models of various complexity have been formulated to study different aspects of climate and climate change. Climate models that have been found to be particularly suitable for paleoclimate studies are described here and examples of applications of each model type are given. Excluded from this survey are atmosphere and ocean general circulation models.

1. INTRODUCTION

For theoretical studies of long term climatic variations scientists can choose among a diverse set of climate models of varying complexity. The models include, to different degrees, the various components and processes that comprise the climate system. Some models simulate the full three-dimensional structure of the atmosphere and ocean, while others can only be used to calculate global-average quantities. Some models simulate the transient changes that take place over decades, centuries or longer time-scales, while others produce equilibrium 'snap-shots' under specified conditions. Models differ also with respect to which of the climate system's components are explicitly included: ocean, atmosphere, biosphere, cryosphere, and in some cases, the upper lithosphere.

Other chapters of this volume describe the formulation and use of atmosphere and ocean general circulation models (GCMs). The purpose of this review is to describe a selection of other types of models that are being used for the purpose of studying paleoclimates and long term climatic variations. The advantages and limitations of each type of model will be indicated and examples of paleoclimate applications will be given. A comprehensive review of all types of climate models is not attempted here. Rather, a representative group of models has been selected, based primarily on how familiar they are to this author. For other reviews of climate models consult, for example, Saltzman (1983), Hecht (1985, see chapters by Barron, Saltzman), Schlesinger (1988, see chapters by Crowley, MacCracken and Ghan, North, Saltzman), Crowley and North (1991), and Weaver and Hughes (1992).

NATO ASI Series, Vol. I 22
Long-Term Climatic Variations
Edited by J.-C. Duplessy and M.-T. Spyridakis

Various kinds of climate models will be described in the following sections. The characteristics of each model will be summarized and the model formulation will be given along with a few sample applications. Some of the terminology that is commonly found in discussions of long term climate variations (e.g., 'fast' vs. 'slow' physics, 'forced' vs. 'free' oscillations, and 'deductive' vs. 'inductive' approaches) will also be briefly defined in relationship to paleoclimate model applications.

2. CLASSIFICATION OF MODELS

Climate models can be grouped according to various attributes (e.g., time-scales treated, climate system components included, spatial dimensions explicitly represented, etc.). In this chapter, the focus is on models of the atmosphere and ocean, although a few examples of coupled models with ice sheets will also be given. See Peltier's chapter in this volume for more discussion of ice-sheet models. Table 1 contains a summary of the resolution and time-scales typically considered in various types of global atmosphere and ocean models.

TABLE 1. Typical Resolution and Time-Scales

Model Type	Resolution: Horizontal (km^2)	Resolution: Vertical (layers)	Time-Scales (years)
Atmospheric Models:			
Atmospheric general circulation models (AGCMs)	10^4	10^1	10^{-4}–10^2
Energy balance models (EBMs)	10^6–10^8	10^0	10^{-1}–10^4
Meridional-plane models (e.g., zonally averaged statistical dynamical models)	10^6	10^0	10^{-1}–10^4
Radiative-convective models	10^8	10^1–10^2	equilibrium
Ocean Models:			
Ocean general circulation models (OGCMs)	10^2	10^1	10^{-2}–10^2
Upwelling-diffusion models	10^6	10^1	10^0–10^4
Meridional-plane thermohaline circulation models	10^6	10^1	10^0–10^4

In terms of spatial and temporal resolution, general circulation models (AGCMs and OGCMs) reside at the top of the hierarchy. These models typically provide useful spatial information on horizontal scales as small as a few hundred kilometers and time-scales as short as an hour or less (although the accuracy of these models becomes progressively poorer at smaller and smaller scales). Because GCMs explicitly represent most of the apparently relevant physical processes (with important exceptions being clouds in the atmosphere and deep water formation in the oceans), they are considered by many to be the most reliable models available for climate studies (although this opinion is by no means universal). The primary limitation in the use of GCMs is the immense computational resources they require. Even without this practical consideration, GCMs may not be ideally suited for a particular application, because other models of possibly lower dimension (e.g., zonally averaged) can sometimes reveal the essential physics that governs climate without the excessive detail inherent in GCM simulations. Moreover, the coupling of AGCMs with OGCMs remains problematic in that without artificial sources and sinks of energy prescribed at the interface between the ocean and atmosphere (so-called 'flux correction'), the models usually tend to drift away from realistic climatic states.

Atmosphere and ocean GCMs are described in more detail in other chapters of this volume, so they will not be discussed further here. We shall consider the other models in the order they are given in Table 1, and we shall also describe examples of applications when two or models (e.g., an atmosphere model and an ocean model) are coupled. A section on an inductive approach to climate modeling is also included.

3. ENERGY BALANCE MODELS (EBMs)

EBMs are based on a single equation for conservation of the energy in a column extending from the top of the atmosphere to the surface (or in many cases the bottom of the ocean). Vertical integration eliminates this dimension, leaving latitude and longitude as the remaining spatial dimensions. In many EBMs, global averaging, or zonal averaging is performed, yielding, respectively, zero-dimensional or one-dimensional models of climate (North *et al.*, 1981; Crowley and North, 1991).

The rate of change of energy E in a column extending from the bottom of the oceans to the top of the is atmosphere is determined by the horizontal energy flux convergence $-\nabla \cdot \underline{F}$ and the fluxes of net shortwave radiation Q entering the atmosphere and longwave radiation R exiting the top of the atmosphere:

$$\frac{\partial E}{\partial t} = -\nabla \cdot \underline{F} + Q - R \tag{1}$$

where in EBMs, typically, the energy content is assumed to be proportional to the surface temperature $\left(E = c_p T\right)$, the horizontal flux is assumed to be proportional to the horizontal temperature gradient $(\underline{F} = -D\nabla T)$, the absorbed solar radiation is computed from the known insolation distribution and the albedo (which may decrease with temperature to account for the increase in surface albedo when snow or ice is present), and the outgoing longwave radiation is assumed to depend linearly on temperature $(R = A + BT)$. With these parameterizations of the important physical processes, (1) can be written as

$$c_p \frac{\partial T}{\partial t} = D\nabla^2 T + Q_0 S(\underline{x},t)[1-\alpha(T)] - A - BT \tag{2}$$

where Q_0 is the solar constant, $S(\underline{x},t)$ describes the spatial and temporal distribution of insolation, and α is the planetary albedo. The planetary albedo may vary spatially and temporally and, as indicated above, is commonly made a function of temperature.

EBMs have been used to explore climate stability and in particular they have been useful in demonstrating the potential importance of ice/snow albedo feedback on the climate system. The hypothesis is that any cooling of climate will cause snow and ice cover to increase, thereby increasing global albedo. This will decrease the shortwave radiation absorbed and tend to reinforce the initial perturbation, constituting a positive feedback. Some of the first studies with energy balance models (Sellers, 1969 and Budyko, 1969) focused on this issue, and, subsequently, even global-average versions of the models have demonstrated interesting aspects of the problem. For example, Cahalan and North (1979) considered the global-average equilibrium version of (2), which requires that there be a balance between shortwave and longwave radiation at the top of the atmosphere (and is the origin of the term, 'energy balance model'):

$$\frac{Q_0}{4}[1-\alpha(T)] = A + BT \ . \tag{3}$$

Assuming the planetary albedo has a step-like decrease (at say −10°C), crudely reflecting the changes that would occur in going from a cold, ice-covered earth to a warmer climate, Cahalan and North (1979) found:

- Multiple equilibrium solutions are possible.
- Not all equilibrium solutions are stable.
- Ice/snow albedo feedback is potentially important.
- A small decrease in solar constant can lead to an ice-covered Earth.

The simplicity of the zero-dimensional (globally averaged) EBMs limits their usefulness. More commonly, one-dimensional (zonally averaged) versions of EBMs are used to study the annual average equilibrium climate as a function of latitude. In one study, for example, Held and Suarez (1974) postulated a critical temperature T_0 below which glacial ice was assumed to appear. They assumed that the albedo could be parameterized by

$$\alpha(\phi) = \begin{cases} 0.4 & \text{for } T(\phi) \geq T_0 \\ 0.7 & \text{for } T(\phi) < T_0 \end{cases} \tag{4}$$

where ϕ is latitude. Held and Suarez then considered how climate might be affected by changes in the Earth's obliquity. Because a single equation is used in this type of energy balance model, mathematical analysis of the model is relatively straight forward, and it is easy to explore the dependence of the solution on model parameters. Among other conclusions, Held and Suarez found that an increase in obliquity causes an ice sheet to recede and also increases climate sensitivity.

Similar EBMs have been used extensively to study other aspects of climate sensitivity. North and Crowley (1985), for example, found that under current conditions, there appears to be a minimum stable size for ice caps. The apparent instability of small ice caps could be part of the explanation for abrupt climate changes (i.e., a strongly nonlinear response to forcing or a change in model parameters).

In another sensitivity study Lindzen and Farrel (1977) argued that the 'diffusion' coefficient D in (2), which accounts for atmospheric and oceanic heat transport, should have a larger value in the tropics than in other zones in order to account more realistically for the Hadley circulation. In contrast to previous studies which found that the Earth might fall relatively easily into an ice-covered state, the more realistic treatment of atmospheric heat transport had an important stabilizing effect. This study reveals one of the primary weaknesses of EBMs: the relatively simple parameterizations, which are fundamental to these models and which permit extensive exploratory studies, may not be wholly reliable representations of the true physics of the atmosphere and ocean.

A major limitation of globally averaged or zonally averaged models is, of course, their lack of spatial detail. In particular, the striking seasonally varying contrasts between continents and oceans cannot be simulated. This limitation has been addressed with energy balance models that divide each zone into a land portion and an ocean portion (see, for example, North and Coakley, 1979). Each of these segments is zonally averaged, so these models are sometimes referred to as $1\frac{1}{2}$-dimensional (with latitude the full dimension and the land-sea contrast the half dimension). Because the land and ocean portions of these models are assigned different heat capacities, the amplitude of the seasonal cycle can be more accurately reproduced. Although these models can account for different fractional areas of land at each latitude, they cannot fully distinguish between different spatial distributions of continents.

Like one-dimensional and $1\frac{1}{2}$-dimensional EBM's, two-dimensional (latitude-longitude) energy balance models retain the assumption that horizontal transport of heat can be formulated in terms of a simple diffusion process. It is not really known how accurate this representation is for changing climatic conditions, but on the largest scales it seems to yield reasonable agreement with observations (North et al., 1983). In a recent paleoclimate application Crowley and Baum (1991) have used a two-dimensional EBM to investigate how the location of a super continent (Gondwanaland, 300-500 Ma BP) affects the area of glaciation. They assumed a 5% decrease in solar luminosity, atmospheric carbon dioxide concentration seven times larger than today, and an orbital configuration that would produce cool summer temperatures. They found that the maximum glaciation occurs when the edge of the continent is located near the pole. In addition they showed that for high CO_2 levels, oceans are needed in the polar region to keep summers cool enough to prevent inland glaciers from melting.

4. MERIDIONAL-PLANE ATMOSPHERIC MODELS

Meridional-plane models (often referred to as 'statistical dynamical' models), yield zonally averaged descriptions of the climate (and in some cases they resolve differences between the surface characteristics of oceans and continents). These models focus on the dynamics of the atmosphere, but sometimes also include a simple ocean model, thermodynamic sea ice model and perhaps a crude glacial ice model. Meridional-plane models are typically used to study the annually averaged or seasonally varying climate. They differ from EBMs in that the equation for conservation of energy that is the basis of EBMs is augmented by equations for the conservation of momentum, mass and sometimes water vapor. Thus the same conservation laws that appear in GCMs are found in meridional-plane models, although it is common to use a quasi-geostrophic approximation to simplify the primitive equations.

In the derivation of the zonally averaged equations used in meridional-plane models (see MacCracken and Ghan, 1988, for a review), the dependent variables are resolved into two components:

$$\psi(\lambda,\phi,z,t)=\overline{\psi}(\phi,z,t)+\psi'(\lambda,\phi,z,t) \tag{5}$$

where the overbar indicates zonal averaging, the prime indicates a deviation from the zonal average, and λ, ϕ, z, and t, represent the longitude, latitude, altitude, and time coordinates, respectively.

The zonally averaged equations are similar to the original equations, except the partial derivatives with respect to longitude are eliminated and additional terms arise, which account for the transport of heat, water vapor and momentum by synoptic-scale eddies. When the thermodynamic energy equation is zonally averaged, for example, a term that accounts for the meridional heat flux is expanded as follows:

$$vT=\overline{v}\overline{T}+\overline{v'T'} \tag{6}$$

The first term on the right hand side is proportional to the heat flux carried by the mean meridional flow, whereas the second term accounts for transport by the eddies. The first term only involves the (zonal-mean) dependent variables, which are directly determined by the solution of the model equations. The second term (known as an 'eddy term'), on the other hand, involves correlations between deviations from the zonal average, which must be represented in terms of the zonal-mean quantities.

The accurate parameterization of the eddy terms is the most challenging task in the development of meridional plane models. Outside of the tropics, eddies are responsible for much of the transport of heat, momentum and water vapor in the troposphere. Eddies are recognizable on weather maps as extra-tropical cyclones with their associated warm fronts and cold fronts, and in satellite photographs of the Earth eddies are often identifiable by a characteristic comma-shaped cloud pattern. The result of the poleward movement of warm air associated with the warm front, and the equatorward transport of cooler air with the cold front is a net transport of heat toward the pole. Eddies of this type are characterized as 'transient' and are fundamentally a result of baroclinic instability. Somewhat realistic, theoretically-based parameterizations have been developed to account for the transport of heat by transient eddies (e.g., Branscome, 1983), but parameterization of momentum transport remains unsatisfactory.

The other component of eddy transport is due to 'stationary' waves (with northward flow at some longitudes and southward flow at others) which are forced by topographical features such as mountain ranges and by uneven heating within a zone (usually associated with land/ocean contrasts in temperature). Although termed 'stationary,' these eddies shift slowly from one season to the next. They are generally less prominent in the Southern Hemisphere (because of the lack of large areas of land there).

Stationary eddies are often neglected in meridional plane models. Although this is clearly not really justifiable, the stationary eddies generally transport less heat than transient eddies, and it is possible that if the stationary eddies did not exist, the transient eddies would compensate somewhat for their absence (Manabe and Terpstra, 1974). In any case a satisfactory parameterization of the effects of stationary eddies is yet to be found.

In contrast to energy balance models, meridional plane models resolve the full vertical structure of the atmosphere, which, as in GCMs, permits more accurate and detailed calculations of radiative transfer. The inclusion of the full set of hydrodynamic equations (again in contrast to EBMs) also provides for an explicit calculation of the mean meridional circulation, including the Hadley cell. Consequently, the meridional plane models are able to reproduce the seasonal migration of the intertropical convergence zone, as well as the precipitation maximum that coincides with it (MacCracken and Ghan, 1988). These models also yield a more realistic solution for the surface temperatures in the tropics, mainly because the explicit simulation of the Hadley cells allows for a more efficient poleward transport of heat in the tropics than is obtained with the usual diffusive heat transport parameterization found in EBMs (Taylor, 1980).

Saltzman and Vernekar (1975) were among the first to use a meridional plane model to study a paleoclimate near the last glacial maximum (18 ka BP). They limited the vertical resolution in their model to two layers, limited their domain to the Northern Hemisphere, and prescribed the ocean temperature below the mixed layer. They compared the climate of 18 ka BP to the climate of today by imposing prescribed changes in ground heat capacity, surface albedo, and soil moisture that reflected the change in glacial ice fraction at each latitude. They found that compared to today's climate, the meridional gradient of surface temperature was enhanced, especially at the ice edge, leading to cooler Arctic temperatures. The larger gradients in temperature led to an increase in baroclinic eddy amplitudes, which enhanced the eddy transport of water vapor and increased precipitation along the leading edge of the glacier. Thus, Saltzman and Vernekar were able to demonstrate a possible mechanism for maintaining or expanding an existing glacier. Only with recent increases in computational speed and computer resources would it be practical to carry out an experiment of this kind with a GCM, which demonstrates one of the real advantages of meridional plane models: they can be used to explore how changes in imposed 'boundary conditions' might affect climate, without consuming excessive amounts of computer time.

5. RADIATIVE-CONVECTIVE MODELS

In radiative-convective models altitude (z) is the only spatial dimension considered. Formally, the governing equation in these models is obtained by averaging the thermodynamic energy equation over longitude and latitude:

$$\frac{\partial \overline{\rho c_p T}}{\partial t} = -\frac{d}{dz}\left(\overline{q_s} + \overline{q_l} + \overline{q_c}\right) \tag{7}$$

where the overbar indicates global averaging, $\rho c_p T$ is the energy per unit volume of the air, and q_s, q_l and q_c are the vertical fluxes of net solar flux, net longwave radiative flux, and convective flux, respectively. Ramanathan and Coakley (1978) provide a review of radiative-convective models.

For annually averaged, steady-state conditions this equation (along with appropriate boundary conditions) implies that

$$\overline{q_s}(z) + \overline{q_l}(z) + \overline{q_c}(z) = 0 \tag{8}$$

Thus, the net vertical flux of energy across any layer of the atmosphere vanishes in steady state, and the two processes that determine Earth's thermal structure (in the global average) are radiation and convection.

The apparent simplicity of the above equation hides the actual complexity of radiative-convective models. The radiative terms, which depend in part on the vertical profile of temperature, must account for the absorption and scattering properties of all the radiatively active constituents (i.e., gases and aerosols) in the atmosphere. This means that the vertical profiles of water vapor, clouds, and other constituents must be independently deduced or prescribed, before solving (8). The convective term accounts not only for thermal convection on scales smaller than ~10 km but also for vertical motion associated with larger scale, organized systems, including motion associated with the general circulation of the atmosphere. The convective heat flux is an important stabilizing process, tending to balance the destabilizing effects of solar radiation which is mostly absorbed at the earth's surface. Typically, convection in these models serves to maintain a 'critical' lapse rate that for present conditions is in rough agreement with observations.

Several limitations of radiative-convective models should be obvious from the brief summary given above. First, by virtue of the spatial averaging, it is impossible to deduce anything about the horizontal variations of climate with these models. Also without explicit dynamics, prognostic determination of the water vapor, cloud and aerosol distributions is not possible. Instead, the profiles for the constituents are typically prescribed, or, in the case of water vapor, deduced given some assumed constraint such as fixed relative humidity. This essentially rules out accurate representation of several potentially important feedbacks in the climate system. Finally, the common assumption in these models that convection should serve to maintain a critical lapse rate that is independent of climatic state may not be valid.

By giving up horizontal resolution in radiative-convective models, one can generally afford to treat radiative transfer in a less approximate way than is practical with more computationally expensive models. The radiation codes in most other models are accurate only when conditions are not too different from the present. Thus they might not be trusted to predict radiative fluxes in an atmosphere with say a ten-fold increase in the concentration of carbon dioxide.

In contrast to other models, radiative-convective models may therefore be particularly appropriate for studying the climates of periods of significantly different atmospheric composition. In one application of a radiative-convective model, for example, Kuhn and Kasting (1983) investigated what has been termed the 'faint young sun paradox' (although it is clearly not a true paradox): the question is why has the earth's surface temperature remained relatively unchanged over the last 4 billion years, even though, according to theory, the solar luminosity has increased by 25% or more during this same period of time? Kuhn and Kasting found, following earlier studies by Hart (1978), that an opposite and much larger trend in atmospheric carbon dioxide concentration could offset the trend in insolation to keep surface temperatures within the range permitted by observations.

Another application of radiative convective models has been in exploratory studies of the implications of a catastrophic impact by an asteroid or large meteorite, which might have caused massive extinctions at the boundary between the Cretaceous and Tertiary geological periods. Using a time-dependent radiative-convective model and accounting for the lower effective heat capacity of continents compared to oceans, Pollack *et al.* (1983) found that enough of the dust thrown up by an impact could remain in the atmosphere long enough and increase the optical depth by enough to strongly cool the continents, but not the oceans. These studies support the idea that large impacts can have major climatic consequences that may affect the evolution of life on Earth.

6. UPWELLING-DIFFUSION OCEAN MODELS

The models described to this point are used primarily to study the equilibrium response of climate to various perturbations. Without a more explicit treatment of the oceans, it is impossible to study the transient response of climate on time-scales of decades to millennia. The simplest models that can account for the uptake (or release) of heat by the oceans in response to climate change on these time-scales are the upwelling-diffusion models.

In the typical upwelling-diffusion model a thin mixed layer sits atop the body of the ocean, where downward diffusion of heat is balanced (in equilibrium) by upward advection of heat by the slow overturning of a thermohaline circulation (although this circulation is prescribed, not solved for explicitly). Like EBM's, upwelling-diffusion models are based on a zonally or globally averaged equation for conservation of energy. The model is forced by prescribed changes in surface temperature, by prescribed changes in circulation, or by coupling it to an atmospheric model.

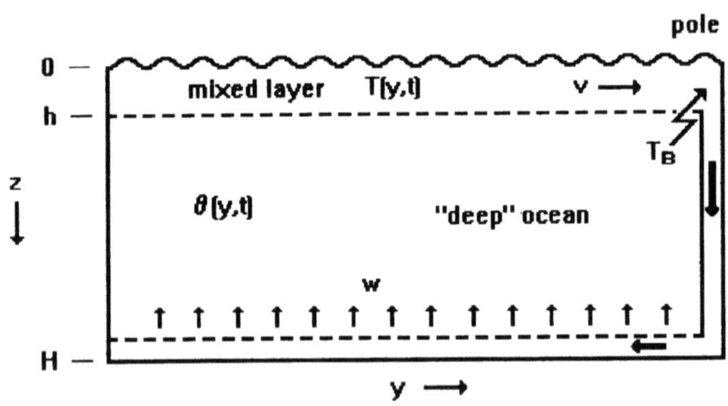

Figure 1. Schematic illustration of an upwelling-diffusion ocean model.

As shown in Figure 1, the bottom water in a zonally averaged upwelling-diffusion model is assumed to be formed at the poleward boundary of the model at a temperature of T_B. The water then upwells uniformly elsewhere and the horizontal flow is neglected in the body of ocean. In the mixed layer the water flows poleward at velocity v to complete the circuit. In upwelling-diffusion models the following equations and boundary conditions govern the deep ocean:

$$\frac{\partial \theta}{\partial t} = w\frac{\partial \theta}{\partial z} + \kappa\frac{\partial^2 \theta}{\partial z^2} \tag{9}$$

$$\theta = T \qquad \text{at } z = h \text{ (continuity condition)} \tag{10}$$

$$\kappa\frac{\partial \theta}{\partial z} + w\theta = wT_B \qquad \text{at } z = H \text{ (no flux condition)} \tag{11}$$

where T and θ are the zonally averaged temperatures of the mixed layer and deep ocean, respectively, w is the upwelling velocity (positive up), and κ the diffusivity. Note that the vertical coordinate, z, is zero at the surface of the ocean and increases with depth.

The mixed layer is governed by an equation of the form :

$$\frac{\partial T}{\partial t} + \frac{\partial vT}{\partial y} = \frac{\kappa}{h}\frac{\partial \theta}{\partial z}\bigg|_{z=h} - \frac{wT}{h} = \boldsymbol{S} \tag{12}$$

where y is the meridional coordinate and $\boldsymbol{S}$ is proportional to the net surface heat flux which is usually imposed to force the model. The horizontal velocity in the mixed layer, v, can be determined from the upwelling velocity through the continuity equation:

$$w = h\frac{\partial v}{\partial y} \tag{13}$$

Upwelling-diffusion models have been used to demonstrate that

- the mixed layer is *not* governed by a single time-constant, initially adjusting rather quickly to a perturbation, but subsequently approaching equilibrium much more slowly as the influence of the lower ocean begins to dominate (Thompson and Schneider, 1979; Hoffert *et al.*, 1980).

- full deep ocean equilibrium requires a few thousand years (see, for example, Hoffert *et al.*, 1980).

- changes in thermohaline circulation can temporarily lead to significant changes in global-average temperature, as well as affect the rate of surface temperature response to perturbed conditions (Harvey and Schneider, 1985; Watts, 1985).

Upwelling-diffusion models are well suited for studying transient climate change on time-scales shorter than a few thousand years. Consequently, they have been popularly used to predict the climate changes that might occur as a result of anthropogenic influences over the next century (Schneider and Thompson, 1981; Wigley and Raper, 1992). They have also been useful in interpreting results from ocean general circulation models (Schlesinger and Jiang, 1990) and in studying techniques for coupling ocean and atmospheric models (Schneider and Harvey, 1986). These models, however, are clearly highly simplified representations of the real ocean; in fact, if the upwelling velocity is held constant, they are linear. Moreover, since the circulation of the oceans is essentially prescribed, they cannot be used to simulate climate change that might be attributed to the complex interactions between surface climate and ocean circulation. For these reasons and because in paleoclimate studies the most important well-known transient forcing of the climate system is the Milankovitch insolation perturbations which have primary time scales longer than 10^4 years, upwelling-diffusion have found limited application.

In one instructive application of an upwelling-diffusion model, Watts (1985) showed that global-average surface temperature changes of a few tenths of a degree can be realized with as little as a 10% change in the rate of bottom water formation. Watts concluded that variations in the rate of bottom water formation could explain some of the temperature variations that have actually occurred, on time scales ranging from a few years to several centuries.

7. MERIDIONAL-PLANE THERMOHALINE CIRCULATION MODELS

Upwelling-diffusion models cannot simulate the dynamic changes in ocean circulation that accompany (and possibly 'cause') climate change because they are based only on the thermodynamic energy equation. Ocean general circulation models, on the other hand, simulate the full three-dimensional behavior of the oceans but require enormous computational resources, which essentially precludes their use in studying climate change occurring over several thousands of years. Meridional-plane models of the thermohaline circulation have been formulated to bridge the gap between the upwelling-diffusion models and ocean general circulation models. These models are based on simplified zonally averaged versions of the full set of hydrodynamic equations, which comprise the conservation laws for heat, momentum, mass and salinity.

In these models an important term in the zonal component of the momentum equation accounts for the east-west pressure gradient across the ocean basin. This term is proportional to the difference in pressure between the eastern and western boundaries of the basin $(P_E - P_W)$ and is inversely proportional to the basin width (D). Since the dependent variables in meridional plane models are zonal average quantities, relationships must be found relating P_E and P_W to the principal unknowns. Marotzke *et al.* (1988) and Wright and Stocker (1991) make certain reasonable assumptions which allow them to derive a relationship between the east-west pressure gradient and the north-south (zonally averaged) pressure gradient:

$$\frac{P_E - P_W}{D(y)} = -\varepsilon \frac{\partial_{\bar{p}}}{\partial_y} \tag{14}$$

where y is the meridional coordinate and the parameter ε is much smaller than 1 for a wide ocean basin without wind stress. Although this relationship cannot be rigorously derived from first principles and although the parameter ε is inversely proportional to a damping time-scale that is not tightly constrained by observations, (14) is necessary to close the set of governing equations. In the absence of truly convincing theoretically-based arguments for the validity of the equations in meridional plane models, Wright and Stocker (1991) have shown nevertheless that except in the equatorial belt and in convectively active areas, the above relationship approximately holds in ocean general circulation model simulations. See the chapter by Hovine and Fichefet in this volume for further details on the formulation and application of meridional-plane models.

In most studies with meridional-plane ocean models, some combination of prescribed heat and salinity (fresh water) fluxes at the surface has been used to force the model. Typically, the surface temperature is forced to relax to some prescribed temperature distribution, while a fresh water flux (which vanishes when globally averaged) is prescribed. In an interesting application of a meridional-plane ocean model, Stocker and Wright (1991) studied the ocean circulation response to hypothetical changes in the fresh water flux carried by the atmosphere from the Atlantic to the Pacific Oceans. First, the model was tuned for present climatic conditions, which required a 0.3 Sv (1 Sv = 10^6 m^3s^{-1}) atmospheric transport of water from the Atlantic to the Pacific Ocean (in reasonable agreement with observations). Under these conditions, according to the model, 20 Sv of deep water would be formed in the North Atlantic, and the simulated temperature and salinity fields compare favorably with observations. When the prescribed atmospheric transport of water from the Atlantic to the Pacific Ocean was reduced to 0.03 Sv, deep water was found to form only in the Southern Ocean. Moreover, the salinity and temperature cross-sections were altered in a way that was not unlike what might have been found during the last glacial maximum. This altered state was stable and characterized by a hysteresis in that to return to the original state, the atmospheric transport of water from the Atlantic to the Pacific Oceans had to be boosted to 0.36 Sv.

Meridional-plane models have also been used to study various types of nonlinear behavior that might be characteristic of the oceans and could possibly explain some of the climate variability implied by the paleoclimate data. Stocker *et al.* (1992) have shown, for example, that their ocean model, when coupled to an energy balance atmospheric model, allows at least four different climatic states. An abrupt transition from one state, characterized by a relatively saline Pacific Ocean, to another state with a relatively saline Atlantic Ocean could be induced by a small change in freshwater discharged into the North Atlantic. The transition from one state to the other occurred over time-scales as short as a hundred years.

8. COUPLED MODELS

Coupled models can be constructed by linking together in various combinations the component models described above (or other models not considered here). A recent example of this type of model is described by Harvey (1992), who has coupled an energy balance model to a meridional plane ocean model and a thermodynamic sea ice model.

An earlier coupled model was constructed by Pollard (1983) in an attempt to simulate the northern hemispheric ice-sheet volume variations that have occurred over the last ~700 ka. Pollard linked a simple ice sheet model to a two-dimensional energy balance model. The ice sheet model included a component accounting for bedrock deformation and linear flow within a thin asthenospheric channel. With these components, the model reproduced realistic time lags between the response of the bedrock depression and changes in ice load (Oerlemans, 1980; Birchfield *et al.*, 1981). Pollard also added an ice-calving mechanism to the model, which has the effect of allowing ice sheets to recede more quickly than they advance.

With this model Pollard found that he could reproduce the pronounced ~100 ka cycle of ice volume, with impressive phase correspondence with variations implied by deep-sea core isotopic records. The bedrock lag and ice calving mechanisms were both important in obtaining this fairly realistic result. Coupling the model to a model accounting for the thermal inertia of the deeper ocean layers caused little change in the simulations.

More recently, Gallée *et al.* (1991, 1992) have coupled a meridional plane atmospheric model to a model of the three main northern ice sheets. This model accounts for bedrock displacement and includes ocean mixed layer and sea ice components. Solution of the model equations relies on an asynchronous coupling method in which given an initial distribution of glacial ice, the atmosphere/mixed layer/sea ice model is integrated forward for 15 or 20 years during which the net snow balance over the ice surfaces is calculated. The net flux of snow is then used to force the ice sheet model, which is integrated forward over a time period of order 1 ka. The new ice-sheet distribution is then fed back to the climate model to begin another cycle. Using this method Gallée *et al.* are able to simulate time intervals of order 100 ka.

Several experiments have been carried out with this model. The focus has been on simulating the last 122 ka and determining what is responsible for the growth and decay of the ice sheets. Gallée *et al.* find that when the ice sheets are prescribed (to be the same as today) and the only forcing is the orbitally determined changes in seasonal insolation, the model simulates changes in the accumulation rate of ice that are significant and very sensitive to the snow ablation process. In particular the changes in ablation rates are highly correlated with June insolation variations.

Next the full coupled model is run and inclusion of the interactive ice sheet produces changes in ice volume that are remarkably similar to the changes inferred from the isotopic record. The success of this model supports the hypothesis that the Milankovitch theory can largely explain the last glacial-interglacial cycle.

9. THE INDUCTIVE APPROACH

The formulation of all the models described so far resulted from a common approach :

1. Begin with the 'fundamental' laws of physics.

2. Simplify the equations, based on approximations and assumptions that appear to be consistent with observations.

3. Parameterize terms that represent processes not accounted for explicitly.

4. Solve the equations.

5. Tune uncertain model parameters to yield a realistic climatic state.

6. Test the model against independent observational data.

7. Make predictions.

This essentially deductive approach works well when the physics of the system is well-understood. For a certain class of problems, however, there is merit in considering an inductive approach which follows a generally different procedure:

1. Begin by collecting and analyzing observations.

2. Identify time-scales, feedbacks, and climate components that might be important.

3. Construct a mathematical model on heuristic grounds (with an eye to reproducing observations and satisfying certain conservation constraints).

4. Solve the equations.

5. Tune uncertain model parameters to yield a realistic climatic state.

6. Test the model against independent observational data.

7. Make predictions.

Note that the last four steps are the same as in the deductive approach. The third step is critical and may require experience and well-developed intuition to arrive at appropriate equations.

One question that has been studied with an inductive approach is whether the ~100 ka cycle in ice volume seen in the observational record of the last ~900 ka can be explained in terms of Milankovitch (astronomical) variations in the Earth's orbital parameters, which vary with periods of roughly 20 ka and 40 ka. Imbrie and Imbrie (1980) attempted to reproduce the ocean sedimentary $\delta^{18}O$ isotope record of climate variations over the last 500 ka. A power spectrum analysis of this record shows power at the Milankovitch forcing frequencies, but stronger variations at 100 ka. The initial model considered by Imbrie and Imbrie was linear and was forced by a function, $f(t)$, which varied over the 500 ka period and was proportional to the insolation changes at a given latitude and season. The model dependent variable y was a measure of total glacial ice volume which could be directly compared to the observed $\delta^{18}O$ record. Imbrie and Imbrie assumed that there was a characteristic adjustment time of the climate system (τ) and that the response would be linear:

$$\frac{dy}{dt} = \frac{f - y}{\tau}. \tag{15}$$

Imbrie and Imbrie found rather poor agreement between the model simulation and observations, even though they attempted to tune their model by choosing different latitudes and seasons for the prescription of f and different values for τ. In particular the model exhibited almost no power at the observed dominant period of 100 ka. Clearly some important characteristic of the climate system had been omitted in this initial linear model.

Based on evidence that large glaciers tend to decay much more rapidly than they grow, Imbrie and Imbrie modified their model by assuming that the response time of the climate system is different during periods of warming and periods of cooling:

$$\tau = \begin{cases} \tau_w & \text{for a warming climate} \\ \tau_c & \text{for a cooling climate.} \end{cases} \tag{16}$$

By tuning τ_w and τ_c (to values of 10 ka and 42 ka, respectively), Imbrie and Imbrie were able to improve the agreement of the model with observations, but there was still not enough power at a period near 100 ka. Thus, although the model demonstrated that the nonlinearity associated with unequal growth and decay times for glaciers could rectify Milankovitch forcing in such a way as to explain some aspects of the observational record, it was not completely successful. Moreover, it was not clear why one should find a response time as long as ~40 ka for glacial ice growth, since characteristic adjustment times for the atmosphere, ocean, and glacial ice are thought to be much shorter than this.

A more complex inductive model has been developed by Saltzman and his co-workers for studying among other things the emergence at about 700 ka BP of ~100 ka period ice age oscillations. Saltzman and Maasch (1990) showed that a slow decrease of atmospheric CO_2 over the last few million years might have reached a critical level that permitted an instability to develop in the climate system and led to near-100 ka period ice-age oscillations. In their model, the 100 ka climatic variations are essentially 'free' oscillations of the nonlinear climate system; that is, they can be present even in the absence of any time-dependent forcing. This is in contrast to the oscillations that must be 'forced' by externally prescribed changes (as in the Imbrie and Imbrie model) and points to the clear possibility that major climate changes might be attributable exclusively to internal interactions within the climate system. Further discussion concerning this essentially inductive modeling approach can be found in Saltzman (1990).

10. CONCLUDING REMARKS

Given the various types of models available for the study of paleoclimates, one might ask, "Which one is best?" The obvious reply is, "It depends on the application." Among the several considerations and constraints that will determine which model is most appropriate are: the time-scales of interest, the spatial-scales of interest, the physical processes that are thought most important, the amount of data available for model validation, and the computer resources available.

The time-scales of interest primarily determine which components of the climate system must be allowed to vary in time, and which components may be held constant. In this regard it is sometimes useful to distinguish between relatively 'fast' and relatively 'slow' physics. The usual convention is to consider variables associated with the atmosphere, ocean (surface) mixed layer, snow and sea-ice as fast response variables which tend to equilibrate within a few decades or less to changes associated with 'slow' physics. The slow response variables are associated with the deep ocean temperature, ocean trace chemical constituents, glacial ice distribution, displacement of the bedrock (by ice loading and unloading), ecosystem evolution, and the concentration of some atmospheric trace gases such as carbon dioxide, all of which may evolve on time-scales of hundreds to thousands of years or more.

For some paleoclimate modeling experiments the 'slow' physics is considered external to the climate system, but important in determining the 'boundary conditions' for the 'fast' physics. In this case the atmosphere and possibly the mixed layer of the oceans are the only components of the climate system that need to be explicitly included. Most studies with GCMs, energy balance models, meridional-plane models of the atmosphere, and radiative-convective models fall into this class. Models of this type can be used to learn about the role of the atmosphere and mixed layer in determining and maintaining a climate consistent with certain specified conditions (e.g., continental and glacial ice distribution and atmospheric carbon dioxide concentration). In this respect, models of this type are normally solved to obtain an equilibrium climate, and thus they are not generally suitable for studying transient climate change where the interactions between 'fast' and 'slow' physics lead to slowly evolving or time-varying states (e.g., ice-age cycles).

The 'fast' physics models can, of course, be used to obtain equilibrium solutions for slowly changing prescribed forcing (e.g., the equilibrium climates consistent with various patterns of insolation, as given by the well-known variations in the Earth's orbital parameters), but in general the 'slow' physics leads to phase lags and feedbacks that cannot be accounted for by such models. These models are also unable to simulate unforced variations of the climate system due to interactions between various slowly responding variables. In order for a model to predict changes that occur on time-scales of hundreds to thousands of years, where at any given time the climate cannot be assumed to be in equilibrium, the model must include prognostic equations for the slow response variables. Of the models discussed in this chapter, inductive models and for some applications upwelling-diffusion ocean models and meridional-plane thermohaline circulation models (possibly coupled to atmospheric and glacial ice models) are appropriate for applications of this kind.

As indicated in this review and as summarized in Table 1, models also differ in their spatial resolution. Compared to GCMs, the models described here are lower in resolution (e.g., globally or zonally averaged), but are computationally more efficient (in that they require much less computer time to simulate a given interval of time). For this reason it is possible to simulate much longer periods of time with these models than is possible with GCMs. Moreover, it is possible to vary parameters or boundary conditions in these models to explore and analyze their behavior more thoroughly than is practical with GCMs.

In reviews of climate models, the term, 'hierarchy,' is often used, with GCMs usually placed at the top because they typically have the highest temporal and spatial resolution and contain the most complete descriptions of the 'fast' physics variables. But for paleoclimate studies of transient climate change, atmospheric GCMs may be completely inappropriate or impractical. In applications of this kind, simplified coupled models of the ocean, atmosphere, and cryosphere would appear to be more complete and advanced than GCMs, and these models might therefore be placed above GCMs in the hierarchy. In other applications where it might be important to treat radiative transfer in as accurate manner as possible, a radiative-convective model might be superior to all others. Thus, perhaps it is somewhat inaccurate to speak of a single model hierarchy. Instead, we might consider the various models available for studying paleoclimates much in the same way an artist considers his palette. The successful artist chooses colors and combines them in ways that contribute to the overall picture. Similarly, the first step in the art of climate modeling is to identify some aspect of the climate system that is of particular interest and then to select from the palette of models the one most appropriate for that application. With the wide choice of models (and combinations of models) available, it is hoped that a clearer picture of how the climate system works will eventually emerge.

Acknowledgments. This work was performed under the auspices of the Department of Energy Environmental Sciences Division by the Lawrence Livermore National Laboratory under contract No. W-7405-ENG-48.

REFERENCES

Birchfield, G. E., J. Weertman, and A. T. Lunde (1981) A paleoclimate model of northern hemispheric ice sheets. *Quat. Res.*, **15**, 126–142.

Branscome, L. E. (1983) A parameterization of transient eddy heat flux on a beta-plane. *J. Atmos. Sci.*, **40**, 2508–2521.

Budyko, M. I., (1969) The effect of solar radiation variations on the climate of the Earth. *Tellus*, **21**, 611–619.

Cahalan, R. F., and G. R. North (1979) A stability theorem for energy balance climate models. *J. Atmos. Sci*, **36**, 1178–1188.

Crowley, T. J., and S. K. Baum (1991) Toward reconciliation of late Ordovician (~440 Ma) glaciation with very high CO_2 levels. *J. Geophys. Res.*, **96**, 22597–22610.

Crowley, T. J., and G. R. North (1991) *Paleoclimatology.* Oxford University Press, New York, 339 pp.

Gallée, H., J.-P. van Ypersele, Th. Fichefet, Ch. Tricot, and A. Berger (1991) Simulation of the last glacial cycle by a coupled, sectorially averaged climate-ice sheet model I. The climate model. *J. Geophys. Res.*, **96**, 13139–13161.

Gallée, H., J.-P. van Ypersele, Th. Fichefet, I. Marsiat, Ch. Tricot, and A. Berger (1992) Simulation of the last glacial cycle by a coupled, sectorially averaged climate-ice sheet model I. Response to insolation and CO_2 variations. *J. Geophys. Res.*, **97**, 15713–15740.

Hart, M. H. (1978) The evolution of the atmosphere of the Earth. *Icarus*, **33**, 23–39.

Harvey, L. D. D.,(1992) A two-dimensional ocean model for long-term climatic simulations: Stability and coupling to atmospheric and sea ice models. *J. Geophys . Res.*, **97**, 9435–9543.

Harvey, L. D. D., and S. H. Schneider (1985) Transient climate response to external forcing on $10^0 - 10^4$ year time scales. Part 1: Experiments with globally averaged, coupled, atmosphere and ocean energy balance models. *J. Geophys. Res.*, **90**, 2191–2205.

Hecht, A. D. (ed.) (1985) *Paleoclimate Analysis and Modeling*. John Wiley & Sons, New York, 445 pp.

Held, I. M., and M. J. Suarez (1974) Simple albedo feedback models of the icecaps. *Tellus*, **26**, 613–628.

Hoffert, M. I., A. J. Callegari, and C.-T. Hsieh (1980) The role of deep sea heat storage in the secular response to climatic forcing. *J. Geophys. Res.*, **85**, 6667–6679.

Imbrie, J. and J. Z. Imbrie (1980) Modelling the climatic response to orbital variations. *Science*, **207**, 943–953.

Kuhn, W. R. and J. F. Kasting (1983) The effects of increased CO_2 concentrations on surface temperature of the early Earth. *Nature*, **301**, 53–55.

Lindzen, R. S., and B. Farrell (1977) Some realistic modifications of simple climate models. *J. Atmos. Sci.*, **34**, 1487–1501

MacCracken, M. C., and S. J. Ghan (1988) Design and use of zonally-averaged climate models. In *Physically-Based Modelling and Simulation of Climate and Climatic Change (Part 2)*, M. E. Schlesinger (Ed.), Kluwer Academic Publishers, Dordrecht, 755–809.

Manabe, S., and T. B. Terpstra (1974) The effects of mountains on the general circulation of the atmosphere as identified by numerical experiments. *J. Atmos. Sci.*, **31**, 3–42.

Marotzke, J., P. Welander and J. Willebrand (1988) Instability and multiple equilibria in a meridional-plane model of the thermohaline circulation. *Tellus*, **40A**, 162–172.

North, G. R., R. F. Cahalan, and J. A. Coakley, Jr. (1981) Energy Balance Climate Models. *Rev. Geophys. and Space Phys.*, **19**, 91–121.

North, G. R., and J. A. Coakley (1979) Differences between seasonal and mean annual energy balance model calculations of climate and climate sensitivity. *J. Atmos. Sci.*, **36**, 1189–1204.

North, G. R., and T. J. Crowley (1985) Application of a seasonal climate model to Cenozoic glaciation. *J. Geol. Soc.*, **142**, 475–482.

North, G. R., J. G. Mengel and D. A. Short (1983) Simple energy balance model resolving the seasons and the continents: Application to the astronomical theory of the ice ages. *J. Gephys. Res.*, **88**, 6576–6586.

Oerlemans, J. (1980) Model experiments of the 100,000-yr. glacial cycle. *Nature*, **287**, 430–432.

Pollack, J. B., O. B. Toon, T. P. Ackerman, C. P. McKay, and R. P. Turco (1983) Environmental effects of an impact-generated dust cloud: Implications for the Cretaceous-Tertiary extinctions. *Science*, **219**, 287–289.

Pollard, D., (1983) A coupled climate-ice sheet model applied to the Quaternary ice ages. *J. Geophys. Res.*, **88**, 7705–7718.

Ramanathan, V., and J. A. Coakley, Jr. (1978) Climate modeling through radiative-convective models. *Rev. Geophys. and Space Phys.*, **16**, 465–489.

Saltzman, B. (1983) Climatic systems analysis. In *Advances in Geophysics*, **25**, B. Saltzman (Ed.) Academic Press New York 173–233.

Saltzman, B. (1990) Three basic problems of paleoclimatic modeling: a personal perspective and review. *Clim. Dyn.*, **5**, 67–78.

Saltzman, B., and K. A. Maasch (1990) A first-order global model of late Cenozoic climatic change. *Trans. Roy. Soc. Edinburgh: Earth Sciences*, **81**, 315–325.

Saltzman, B., and A. D. Vernekar (1975) A solution for the Northern Hemisphere climatic zonation during a glacial maximum. *Quat. Res.*, **5**, 307–320.

Schlesinger, M. E. (Ed.) (1988) *Physically-Based Modelling and Simulation of Climate and Climatic Change*. NATO ASI Series C: Vol. 243, Kluwer Academic Publishers, Dordrecht, 1084 pp.

Schlesinger, M. E., and X. Jiang (1990) Simple model representation of atmosphere-ocean GCMs and estimation of the time scale of CO_2-induced climate change. *J. Climate*, **3**, 1297–1315.

Schneider, S. H., and L. D. D. Harvey (1986) Computational efficiency and accuracy of methods for asynchronously coupling atmosphere-ocean climate models. Part 1: Testing with a mean annual model. *J. Phys. Oceanog.*, **16**, 3–10.

Schneider, S. H., and S. L. Thompson (1981) Atmospheric CO_2 and climate: Importance of the transient response. *J. Geophys. Res.*, **86**, 3135–3147.

Sellers, W. D. (1969) A global climatic model based on the energy balance of the earth-atmosphere system. *J. Appl. Meteor.*, **8**, 392–400.

Stocker, T. F., and D. G. Wright (1991) Rapid transitions of the ocean's deep circulation induced by changes in surface water fluxes. *Nature*, **351**, 729–732.

Stocker, T. F., D. G. Wright, and L. A. Mysak (1992) A zonally averaged, coupled ocean-atmosphere model for paleoclimate studies. *J. Climate*, **5**, 773–797.

Taylor, K. E. (1980) The roles of mean meridional motions and large-scale eddies in zonally averaged circulations. *J. Atmos. Sci.*, **37**, 1–19.

Thompson, S. L., and S. H. Schneider (1979) A seasonal zonal energy balance climate model with an interactive lower layer. *J. Geophys. Res.*, **84**, 2401–2414.

Watts, R. G. (1985) Global climate variation due to fluctuations in the rate of deep water formation. *J. Geophys. Res.*, **90**, 8067–8070.

Weaver, A. J., and T. M. C. Hughes (1992) Stability and variability of the thermohaline circulation and its link to climate. Centre for Climate and Global Change Research Report No. 92-5, McGill University, 56 pp.

Wigley, T. M. L., and S. C. B. Raper (1992) Implications for climate and sea level of revised IPCC emissions scenarios. *Nature*, **357**, 293–300.

Wright, D. G., and T. F. Stocker (1991) A zonally averaged ocean model for the thermohaline circulation. Part I: Model development and flow dynamics. *J. Phys. Oceanogr.*, **21**, 1713–1724.

GLOBAL ATMOSPHERIC MODELLING

Robert Sadourny
Laboratoire de Météorologie Dynamique du CNRS
École Normale Supérieure
75231 Paris Cedex 05
France

1. Introduction

The history of the numerical modelling of atmospheric motion is intimately linked with the history of weather forecasting which started with World War I. The first try in numerical weather prediction was Richardson (1922)'s, who both invented finite difference modelling and... massively parallel processing, as he viewed a network of parallel personal computers (the brains and hands of individuals) passing and receiving individually processed information to their neighbours in a huge mass-production room. With regard to parallel processing, he was in advance by three quarters of a century. With regard to numerical analysis, he was too early by a quarter of a century : his calculation failed because he did not know about the CFL criterion and choose a time step too large for his horizontal resolution. One must add that he envisioned an associated optimum data coverage by dreaming of a Cartesian observing station network in exact coincidence with the computational grid : which promised a brilliant future to the small city of Romorantin, in the centre of France, as a major meteorological observing station.

It is a quarter of a century later, in 1948, that von Neuman, Charney, Smagorinsky, Fjörtoft and Phillips did the actually first successful numerical forecast on the two-dimensional barotropic filtered vorticity equation on the hemisphere using the ILIAC-I computer at Princeton. The prognostic equation was the simplest at hand, but the slowness of the computer only allowed to reach a 24-hour forecast in two days of computing time, so that verification maps were ready far before the "forecast" was actually issued.

The sixties saw the birth of the present-day "primitive equation" global models, starting with the GFDL model of Smagorinsky, Manabe, Miyakoda *et al.* in Washington, D.C., the UCLA model of Mintz and Arakawa, and the NCAR model of Kasahara and Washington. With their coarse resolution, relatively primitive numerical algorithms and grossly simplified physical parametrisations, these general circulation models of the atmosphere began to be used for long-term simulations and the first true climate experiments were run at GFDL (Smagorinsky, 1963). Since that time, we have seen tremendous developments in computer power, numerical

NATO ASI Series, Vol. I 22
Long-Term Climatic Variations
Edited by J.-C. Duplessy and M.-T. Spyridakis

methods and the modelling of physical processes. For a long time, climate modelling and numerical weather prediction have evolved in parallel with relatively little interaction, but the present trend is toward a closer relationship : climate modelling will benefit from intensive testing of parametrisation schemes in operational analysis-forecast mode ; extended-range weather forecasting will benefit from the experience gained on the long-term behaviour of climate models ; and the routine analyses performed in operational weather prediction centres provide a unique reference history of atmospheric evolution against which climate models can be tested.

In the following survey, I shall first recall the basic equations of atmospheric motion ; then I shall describe the various approximations that lead to the widely used so-called "primitive" equations ; later I shall make a few comments on the problem of choosing coordinates and grids, before going into a description of discrete approximation methods ; I shall end by a rather brief survey of physical parametrisations.

2. Basic equations of atmospheric motion

The basic equations of atmospheric motion are first the momentum equation, written in an inertial frame of reference as

$$\rho \frac{D\mathbf{U}_{\mathrm{a}}}{Dt} = -\mathbf{grad}p - \rho\, \mathbf{grad}\Phi_{\mathrm{a}} + \mathbf{div}\,\sigma \quad , \tag{1a}$$

where ρ is density, D/Dt is the Lagrangian derivative, $\mathbf{U}_{\mathrm{a}}$ the absolute velocity, p pressure, Φ_{a} the gravity potential and σ the viscous stress tensor ; or in a frame of reference rotating with the earth,

$$\rho \frac{D\mathbf{U}_{\mathrm{r}}}{Dt} = -\mathbf{grad}p - \rho\, \mathbf{grad}\Phi - 2\Omega \times \rho \mathbf{U}_{\mathrm{r}} + \mathbf{div}\,\sigma \quad , \tag{1b}$$

where $\mathbf{U}_{\mathrm{r}}$ is velocity relative to the earth, $\Phi = \Phi_{\mathrm{a}} - \frac{1}{2}|\Omega \times \mathbf{r}|^2$ is the potential including gravity and the centrifugal force, Ω being the earth rotation vector and $\mathbf{r}$ the radius vector.

As water vapour is mixed with dry air, density is the sum of the densities of both gases : $\rho = \rho_d + \rho_v$. The continuity equation for dry air reads

$$\frac{\partial \rho_d}{\partial t} = -div(\rho_d \mathbf{U}) \quad ; \tag{2}$$

while the continuity equation for water vapour involves sources of water vapour Q_V and Q_S due to local vaporisation and sublimation, and the divergence of the water vapour flux $\mathbf{q}_v$ due to both diffusion and evaporation from the surface :

$$\frac{\partial \rho_v}{\partial t} = -div(\rho_v \mathbf{U} + \mathbf{q}_v) + Q_V + Q_S \quad . \tag{3a}$$

Similar equations hold for the liquid and ice phases of water :

$$\frac{\partial \rho_1}{\partial t} = -div(\rho_1 \mathbf{U} + \mathbf{q}_1) - Q_V + Q_M \quad , \tag{3b}$$

$$\frac{\partial \rho_i}{\partial t} = -div(\rho_i \mathbf{U} + \mathbf{q}_i) - Q_S - Q_M \quad , \tag{3c}$$

where $\mathbf{q}_1$, $\mathbf{q}_i$ are the sedimentation fluxes of liquid water and ice leading to precipitation, and Q_M is the source of liquid water by ice melting.

The energy equation reads in enthalpy form

$$\rho c_p \frac{DT}{Dt} - \frac{Dp}{Dt} = -L_V Q_V - L_S Q_S - div\mathbf{h} \quad , \tag{4a}$$

where $c_p = \rho^{-1}\left(c_p^d \rho_d + c_p^v \rho_v\right)$ is the specific heat of moist air at constant pressure, T is temperature, L_V and L_S are the latent heats of vaporisation and sublimation, and $\mathbf{h}$ the heat flux due to radiation and thermal conduction. The viscous heating term $\delta.\sigma$, where δ is the strain tensor, which should appear on the right-hand side, is always neglected. The equivalent entropy form is

$$\rho c_p \pi \frac{D\theta}{Dt} = -L_V Q_V - L_S Q_S - div\mathbf{h} \quad , \tag{4b}$$

where $\pi = c_p\left(p/p_0\right)^{R/c_p}$ is the Exner function and $\theta = T\left(p/p_0\right)^{-R/c_p}$ the potential temperature (whose logarithm is proportional to the thermodynamic entropy).

Finally, the equation of state of moist air reads

$$p = \rho RT \quad , \tag{5}$$

where $R = \rho^{-1}\left(\rho_d R_d + \rho_v R_v\right)$.

3. Modelling approximations

I shall describe now the approximations which lead to the so-called "primitive equations" most commonly used in numerical weather forecasting and climate modelling.

3.1 *The thin-shell approximation*

The height of the atmosphere (troposphere + stratosphere) is of the order of one hundredth of the earth's radius. A classical approximation is then to neglect the expansion of geometry along **r**. Using spherical coordinates in the inertial reference frame (absolute longitude $\lambda_a = \lambda + \Omega t$, latitude φ, radial distance r, the kinetic energy reads

$$K_a = \tfrac{1}{2}\left(a^2 cos^2\varphi\, \dot{\lambda}_a^2 + a^2\dot{\varphi}^2 + \dot{r}^2\right) \ , \tag{6}$$

where r has been replaced by its mean value a (approximately, the earth's radius), and the dot refers to Lagrangian time derivative. The easiest way to derive the components of (1a) in this coordinate system is to use the Lagrange equations

$$\frac{D}{Dt}\frac{\partial K_a}{\partial \dot{q}_i} - \frac{\partial K_a}{\partial q_i} = -\frac{\partial \Phi_a}{\partial q_i} - \frac{1}{\rho}\left(\frac{\partial p}{\partial q_i} - (\mathbf{div}\,\sigma)_i\right) \ , \tag{7}$$

in terms of the position parameters $\{q_i\} = \{\lambda, \varphi, r\}$. The left-hand side corresponds to the so-called covariant acceleration components, as given by Lagrange's theorem. The right-hand side corresponds to the covariant components of the various forces at work (gravity, pressure and viscosity) ; in particular, $(\mathbf{div}\,\sigma)_i$ are the covariant components of the viscous stress divergence. Expanding (7) and making use of the natural velocity components in spherical coordinates

$$u_a = u + a\Omega\, cos\varphi \ , \quad u = a\, cos\varphi\, \dot{\lambda} \ , \quad v = a\dot{\varphi} \ , \quad w = \dot{r} \ , \tag{8}$$

one gets the three components of the momentum equation in the thin spherical shell approximation :

$$\frac{D}{Dt}\left(a\, cos\varphi\, u_a\right) = -\frac{\partial \Phi}{\partial \lambda} - \frac{1}{\rho}\left(\frac{\partial p}{\partial \lambda} - (\mathbf{div}\,\sigma)_\lambda\right) \ , \tag{9a}$$

$$\frac{Du}{Dt} - \frac{uv}{a} tg\varphi = 2\Omega\, sin\varphi\, v - \frac{1}{a\, cos\varphi}\left[\frac{\partial \Phi}{\partial \lambda} + \frac{1}{\rho}\left(\frac{\partial p}{\partial \lambda} - (\mathbf{div}\,\sigma)_\lambda\right)\right] \ , \tag{9a'}$$

$$\frac{Dv}{Dt}+\frac{u^2}{a}tg\varphi=-2\Omega\, sin\varphi\, u-\frac{1}{a}\left[\frac{\partial\Phi}{\partial\varphi}+\frac{1}{\rho}\left(\frac{\partial p}{\partial\varphi}-(\mathbf{div}\,\sigma)_\varphi\right)\right] , \tag{9b}$$

$$\frac{Dw}{Dt}=-\frac{\partial\Phi}{\partial r}-\frac{1}{\rho}\left(\frac{\partial p}{\partial r}-(\mathbf{div}\,\sigma)_r\right) . \tag{9c}$$

Equation (9a) is the absolute angular momentum equation. Equations (9a',9b,9c) are now expressed in terms of the natural velocity components.

The exact momentum equations would obtain in the same manner, by applying (7) to the exact form of the kinetic energy :

$$K_a=\tfrac{1}{2}\left(r^2cos^2\varphi\,\dot{\lambda}_a^2+r^2\dot{\varphi}^2+\dot{r}^2\right) . \tag{10}$$

They read

$$\frac{D}{Dt}\left(r\, cos\varphi\, u_a\right)=-\frac{\partial\Phi}{\partial\lambda}-\frac{1}{\rho}\left(\frac{\partial p}{\partial\lambda}-(\mathbf{div}\,\sigma)_\lambda\right) , \tag{11a}$$

$$\frac{Du}{Dt}-\frac{uv}{r}tg\varphi+\frac{uw}{r}=2\Omega\left(sin\varphi\, v-cos\varphi\, w\right)-\frac{1}{r\, cos\varphi}\left[\frac{\partial\Phi}{\partial\lambda}+\frac{1}{\rho}\left(\frac{\partial p}{\partial\lambda}-(\mathbf{div}\,\sigma)_\lambda\right)\right] \tag{11a'}$$

$$\frac{Dv}{Dt}+\frac{u^2}{r}tg\varphi+\frac{vw}{r}=-2\Omega\, sin\varphi\, u-\frac{1}{r}\left[\frac{\partial\Phi}{\partial\varphi}+\frac{1}{\rho}\left(\frac{\partial p}{\partial\varphi}-(\mathbf{div}\,\sigma)_\varphi\right)\right] , \tag{11b}$$

$$\frac{Dw}{Dt}-\frac{u^2+v^2}{r}=2\Omega\, cos\varphi\, u-\frac{\partial\Phi}{\partial r}-\frac{1}{\rho}\left(\frac{\partial p}{\partial r}-(\mathbf{div}\,\sigma)_r\right) , \tag{11c}$$

where the left-hand sides are the acceleration components. One sees that the thin-shell approximation simplifies both the Coriolis terms and the sphericity terms : all such terms involving w or appearing in the prognostic equation for w disappear ; this means in particular that, in the thin-shell approximation, the earth's rotation acts only by its vertical component.

Scale analysis arguments also show that the sphericity terms involving exchange of energy between radial and tangential components of velocity are indeed quite small. But the Coriolis terms exchanging energy between the u and w components are not so negligible and may have a systematic effect (for example in the radial velocity equation, due to the prevailing sign of zonal velocity at a given location) ; they are indeed included in some models (like the UKMO unified model : see White & Bromley, 1988). As shown above, however, consistency asks that these terms should be, either all included, or all neglected.

3.2 *The spherical approximation*

In section 2.a, we have used spherical coordinates for convenience, but equipotential surfaces $\Phi = Const.$ are the most natural coordinate surfaces. The spherical approximation consists in taking the gravity acceleration g constant and neglecting the departures of equipotentials from spherical surfaces. The flattening of the earth due to centrifugal forces is of the order of 10 *km*, and the geoid has irregularities reaching 100 *m* at scales of about 1000 *km*. In the spherical approximation we may use the approximate form

$$\Phi = g(r-a) \ . \tag{12}$$

3.3 *The quasi-hydrostatic approximation*

All current general circulation models of the atmosphere used for numerical weather prediction and climate studies are quasi-hydrostatic. Scale analysis of (9c) or (11c) shows that, for large-scale flow, the acceleration term is small compared to the gravity or vertical pressure terms and can thus be neglected in the vertical component of the momentum equation. Using both the thin-shell and the spherical approximations, the radial component of the momentum equation then reduces to the hydrostatic equation

$$\frac{\partial p}{\partial r} = -\rho g \ . \tag{13}$$

The quasi-hydrostatic property is due to the combined effects of aspect ratio, stratification and rotation. Considering a basic stratification $\overline{\rho}(r)$, $\overline{p}(r)$, and perturbations $\rho'(\lambda,\varphi,r,t)$, $\overline{p}(\lambda,\varphi,r,t)$, It can be shown that the relative departure from the perturbation hydrostatic equilibrium scales as

$$\frac{\overline{\rho}\dfrac{Dw}{Dt}}{\rho' g} = O\left(Min(Ro,1)\, \varepsilon^2 Max\left(\frac{\rho'}{\overline{\rho}}, \delta \right) \right) \ , \tag{14}$$

where $Ro = U / fL$ is the Rossby number (ratio of relative vorticity to vorticity of the earth rotation), $\varepsilon = H/L$ is the aspect ratio (ratio of scale height to horizontal scale of motion), and δ is the ratio of divergence to relative vorticity (for large-scale flow, one expects $\delta \ll 1$ because of both rotation and stratification, but δ will in general dominate the last factor on the right-hand side). The quasi-hydrostatic approximation breaks down for $L \approx H \approx 10\ km$; some attempts are being made towards non-hydrostatic general circulation models able to handle mesoscale or even small-scale dynamics ; such models use implicit vertical differencing to avoid instabitlities triggered by vertically propagating sound waves.

3.4. *The "dry mass field" approximation*

Almost all general circulation models do neglect the influence of water vapour on air density, i.e., they neglect ρ_v compared to ρ_d in the continuity equation, which then reads

$$\frac{\partial \rho}{\partial t} = -div(\rho \mathbf{U}) \tag{15}$$

instead of the exact form

$$\frac{\partial \rho}{\partial t} = -div(\rho \mathbf{U} + \mathbf{q}) + Q_V + Q_S \quad , \tag{16}$$

which is the sum of (2) and (3) and integrates in the mass budget evaporation from the surface, diffusion of water vapour, and local vaporisation or sublimation of water droplets or ice crystals ; also in the process, the equation of state reduces to the equation of state of dry air. In this approximation, the water cycle interacts with dynamics only through radiation, latent heat release and the control of surface sensible heat flux by evaporation. A version of the Arpège model at Météo-France is currently the only model to take into account the direct influence of water vapour on the mass field (see Courtier *et al.*, 1991).

4. Coordinates and associated grids

When designing a general circulation model of the atmosphere, the first choice is how to map the fluid domain. Whenever finite differences are used, the choice of mapping determines the choice of the grid, because accuracy is best when the grid is regularly distributed over the chosen mapping.

4.1. *Horizontal coordinates and spherical grids*

Spherical coordinates are the most straightforward choice for mapping the sphere. They have the disadvantage of the convergence of meridians towards the poles, which induces at high latitudes a close packing of the gridpoints in the longitudinal direction. In addition to wasting degrees of freedom which could be beneficial if distributed elsewhere, the resulting very short longitudinal meshsize at high latitudes is detrimental to computational efficiency for explicit Eulerian models because of the Courant-Friedrichs-Lewy stability criterion : $c\Delta t < \Delta x$, where c

is a characteristic velocity or phase velocity of the fluid. Finite-difference models using such grids usually include a longitudinal filter getting rid of the superfluous degrees of freedom. This filter is generally applied to time derivatives rather than to the physical variables themselves ; it is designed to slow down longitudinally propagating gravity waves which would be unstable according to the linear stability criterion.

An optimum mapping of the sphere would be homogeneous and isotropic, with the possibility of zooming if necessary on a given area. Grids with constant increments in latitude and quasi-constant increments in longitudinal distance $(\Delta\lambda \cong \Delta x_0 / a \cos\varphi)$ are no longer in use, because of their non-uniform convergence properties in the vicinity of the poles when the equations are treated in spherical coordinates. The mapping $(\sin\varphi, \lambda)$ used in the LMD model (Sadourny, 1975b), has the advantage that the associated grid is equal-area but suffers from high anisotropy in the vicinity of the poles, with poor meridional resolution there. Another possibility is to map the sphere by projecting a circumscribed regular polyhedron from its center: the resulting grid is quasi-homogeneous and quasi-isotropic. The most homogeneous and isotropic of these is the icosahedral grid, which is a triangulation of the sphere of hexagonal type, with the exception of twelve pentagons centred on the vertices of the icosahedron. Such grids have been tested in the late sixties for the incompressible two-dimensional Euler equations on the sphere (Sadourny, 1968 ; Williamson, 1968), then by the same authors for the shallow water equations ; they are still possible candidates for the next generation of massively parallel computers.

Coordinate transforms can be used to "zoom" the simulation on a particular region of the globe. Such transforms have been used in numerical weather prediction and climate modelling by Staniforth et al. (1978) on the Canadian hemispheric model, by Sharma et al. (1987) on the LMD global model, and by Courtier et al. (1991) on the Arpège Météo-France model. The two former use a product of one-dimensional independent coordinate stretchings, the latter a conformal transform of the sphere introduced by Schmidt (1977).

These various mappings of the sphere are visualised in Figure 1. Due to this wide variety of viewpoints, it is useful to derive the general form of the basic equations of atmospheric motion in arbitrary curvilinear coordinates. As we are interested here in horizontal coordinates, we shall do it for the shallow water model, considered as a prototype of the thin-shell approximation ; also we shall assume that horizontal curvilinear coordinates are orthogonal (which ensures better efficiency). Let us denote them by $\{x^i\}_{i=1,2}$; they are associated with a local orthonormal basis $\mathbf{e}_i = \partial\mathbf{M} / h_i \partial x^i$, where $h_i = |\partial\mathbf{M} / \partial x^i|$ are the linear map factors, the spherical area element being $dA = h_1 h_2 dx^1 dx^2$. The Lagrangian derivatives $\dot{x}^i$ are the contra-

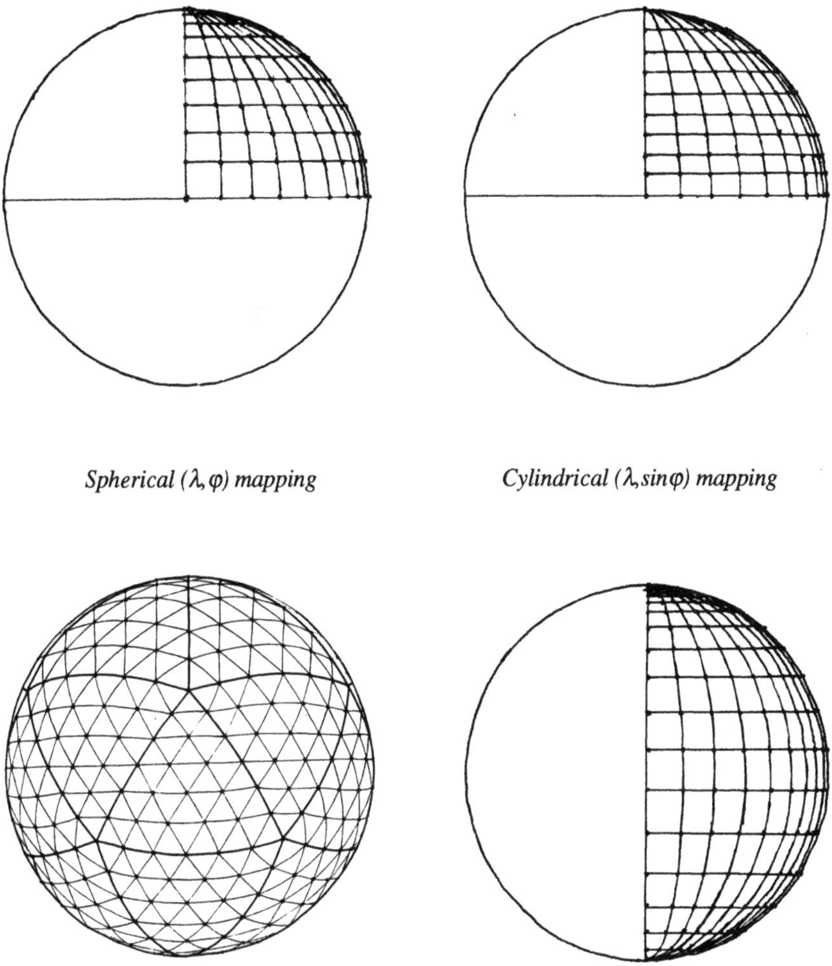

Spherical (λ, φ) mapping

Cylindrical (λ,sinφ) mapping

Icosahedral mapping

Spherical mapping combined with Schmidt's transform centred at poles

Figure 1. *Spherical mappings and grids*

variant components of velocity, from which we derive the natural components $u_i = h_i \dot{x}^i$ and the covariant components $h_i u_i = h_i^2 \dot{x}^i$. The covariant form of the momentum equation can be obtained by applying the Lagrange equations

$$\frac{\mathrm{D}}{\mathrm{Dt}}\frac{\partial K}{\partial \dot{x}^i} - \frac{\partial K}{\partial x^i} = -\frac{\partial \Phi}{\partial x^i} + \varepsilon_{ij} f h_i h_j \dot{x}^j + \frac{1}{\Phi}\left(\mathbf{div}\left(\frac{\Phi}{\rho}\sigma\right)\right)_i \tag{17}$$

to the relative kinetic energy

$$K = \tfrac{1}{2}\left(h_1^2 \dot{x}_1^2 + h_2^2 \dot{x}_2^2\right) \ . \tag{18}$$

(Taking relative kinetic energy, we have added the covariant form of the Coriolis term to the right-hand side of (16) and have used the generalised potential Φ including centrifugal forces ; also, we have adapted the viscous stress term to the shallow water model). The natural components of the acceleration

$$\left(\frac{D\mathbf{U}}{Dt}\right)_i = \frac{1}{h_i}\left(\frac{D}{Dt}\left(\frac{\partial K}{\partial \dot{x}^i}\right) - \frac{\partial K}{\partial x^i}\right) \tag{19}$$

can be written

$$\left(\frac{D\mathbf{U}}{Dt}\right)_i = \frac{Du_i}{Dt} + \frac{1}{h_1 h_1}\left(\frac{\partial h_2}{\partial x^1}u_2 - \frac{\partial h_1}{\partial x^2}u_1\right)\varepsilon_{ij}u_j \ , \tag{20}$$

where the Einstein summation convention is used, $\varepsilon_{12} = -\varepsilon_{21} = 1,\ \ \varepsilon_{11} = -\varepsilon_{22} = 0$, and

$$\frac{D}{Dt} = \frac{\partial}{\partial t} + \frac{u_1}{h_1}\frac{\partial}{\partial x^1} + \frac{u_2}{h_2}\frac{\partial}{\partial x^2} \ . \tag{21}$$

An equivalent and somewhat easier way to derive the components of the momentum equation is to use the intrinsic form

$$\frac{\partial \mathbf{U}}{\partial t} = -(\omega + f)\mathbf{k} \times \mathbf{U} + \mathbf{grad}\left(\Phi + \frac{U^2}{2}\right) + \frac{1}{\Phi}\mathbf{div}\left(\frac{\Phi}{\rho}\sigma\right) \ , \tag{22}$$

where $\omega = rot\mathbf{U}$ refers to the radial component of relative vorticity, f to the Coriolis parameter (or radial component of the earth vorticity), and **k** to the radial unit vector. The two-dimensional velocity divergence $\delta = div\mathbf{U}$ and curl ω read

$$\delta = div\mathbf{U} = \frac{1}{h_1 h_2}\left(\frac{\partial(h_2 u_1)}{\partial x^1} + \frac{\partial(h_1 u_2)}{\partial x^2}\right) , \tag{23}$$

$$\omega = rot\mathbf{U} = \frac{1}{h_1 h_2}\left(\frac{\partial(h_2 u_2)}{\partial x^1} - \frac{\partial(h_1 u_1)}{\partial x^2}\right) , \tag{24}$$

which gives a rather compact form of the shallow water equations :

$$\frac{\partial u_i}{\partial t} = \varepsilon_{ij}(\omega + f)u_j - \frac{1}{h_i}\frac{\partial}{\partial x_i}\left(\Phi + \frac{u_j u_j}{2}\right) + \frac{1}{\Phi}\left(\mathbf{div}\left(\frac{\Phi}{\rho}\sigma\right)\right)_i , \tag{25a}$$

$$\frac{\partial \Phi}{\partial t} = -\frac{1}{h_1 h_2}\left(\frac{\partial(h_2 \Phi u_1)}{\partial x^1} + \frac{\partial(h_1 \Phi u_2)}{\partial x^2}\right) . \tag{25b}$$

The general form of the viscous stress tensor will be considered later in connection with subgridscale lateral diffusion.

4.2. *Vertical coordinate*

Hydrostatic models generally do not use height as vertical coordinate, but rather a function of pressure ; using pressure or a function thereof is more practical from the modelling standpoint, and also for historical reasons related to numerical weather prediction (operational vertical soundings yield temperature and winds as functions of pressure rather than height). The hydrostatic equation (14) allows to rescale the vertical coordinate, using pressure in lieu of geometric height. Using pressure as vertical coordinate has some drawbacks, as the lower boundary becomes time-dependent. This is why Phillips (1957) invented the so-called "sigma" coordinate : $\sigma = p / p_S$, which has the further advantage that it is a terrain-following coordinate, *i.e.* it replaces the complex geometry of the surface $r = r(x^1, x^2)$ by $\sigma = 1$. Like all terrain-following coordinates, however, the sigma-coordinate has the disadvantage that it transports all orographic irregularities into coordinate surfaces well within the flow domain. In fact, the average power spectrum of orographic structures decreases like $k^{-1.5}$, where k is the wavenumber : this means that it is mathematically quite unsafe to estimate space derivatives on sigma-coordinate surfaces over mountains, even using the highest-order approximation like a spectral method. This is why other vertical coordinates have been designed, keeping the sigma structure near the ground, but gradually shifting to un-normalized pressure in the free atmosphere. An example is the so-called "hybrid" coordinate of Simmons and Burridge (1981), a function of p and surface pressure p_S : say, $\zeta(p, p_S)$, verifying

$$\zeta(0, p_S)=0 \; ; \; \frac{\partial \zeta}{\partial p}(0, p_S)=1 \; ; \; \frac{\partial \zeta}{\partial p}(p, p_S)>0 \; ; \; \frac{\partial \zeta}{\partial p}(p_S, p_S)=\frac{1}{p_S} \; ; \; \zeta(p_S, p_S)=1 \quad (26)$$

How does one make use of such artificial coordinate systems ? in fact, quite simply, as a change of vertical coordinate (say, from r to ζ) obeys the following three formal invariance rules :

(i) The Lagrangian derivative is formally invariant :

$$\left(\frac{\partial}{\partial t}\right)_\zeta + \mathbf{U}_H \cdot \mathbf{grad}_\zeta + \dot{\zeta}\frac{\partial}{\partial \zeta} = \left(\frac{\partial}{\partial t}\right)_r + \mathbf{U}_H \cdot \mathbf{grad}_r + \dot{r}\frac{\partial}{\partial r} \quad , \tag{27}$$

where $\mathbf{U}_H$ is horizontal velocity and a derivative indexed with z means that it is taken at constant z ;

(ii) The gravity-pressure term in the equation for $\mathbf{U}_H$ is formally invariant :

$$\mathbf{grad}_\zeta \Phi - \frac{\partial \Phi}{\partial p}\mathbf{grad}_\zeta p = \mathbf{grad}_r \Phi - \frac{\partial \Phi}{\partial p}\mathbf{grad}_r p \quad , \tag{28}$$

which results from the fact that it is a Jacobian form ;

(iii) The continuity equation is formally invariant :

$$\frac{\partial}{\partial t}\frac{\partial p}{\partial \zeta} + div_H\left(\frac{\partial p}{\partial \zeta}\mathbf{U}_H\right) + \frac{\partial}{\partial \zeta}\left(\frac{\partial p}{\partial \zeta}\dot{\zeta}\right) = \frac{\partial}{\partial t}\frac{\partial p}{\partial r} + div_H\left(\frac{\partial p}{\partial r}\mathbf{U}_H\right) + \frac{\partial}{\partial r}\left(\frac{\partial p}{\partial r}\dot{r}\right) \quad , \tag{29}$$

stemming from the fact that, due to the hydrostatic relation, the formal density in any vertical coordinate z is $-\partial p / g\partial z$.

From (28), one can readily visualise the truncation error problem mentioned above for sigma surfaces above steep orographic structures. Both sides of (28) are of the order of the pressure term on the right-hand side (the first term being neglected in the spherical approximation), say, of the order of 10^{-2} *hPa km*$^{-1}$. Near the Andes or on the southern flank of the Himalayas, the pressure term on the left-hand side for a terrain-following coordinate can reach 1 *hPa km*$^{-1}$ or more : it nearly balances the gravity term and this delicate balance will be strongly upset by truncation errors. Another type of vertical coordinate which has the advantages of terrain-following coordinates without the truncation error problem is the "eta" coordinate of Mesinger *et al.* (1988) ; it will not be discussed here.

Still another type of vertical coordinate, quite appealing from theoretical viewpoints, is the entropy (or potential temperature) coordinate.There are a number of advantages pleading for the choice of θ :

- From the entropy equation (4b), the associated vertical velocity vanishes in absence of thermal forcing : in other words, θ has then the nice property of being a Lagrangian coordinate (still in other words, the vertical velocity $\dot{\theta}$ is proportional to the local diabatic heating).

- The gravity-pressure term reduces to the gradient of the Montgomery potential (or dry static energy) $M = \Phi + \theta\pi$:

$$\mathbf{grad}_\theta M = \mathbf{grad}_r \Phi - \frac{\partial \Phi}{\partial p}\mathbf{grad}_r p \;; \tag{30}$$

- The dynamics of Ertel's potential vorticity, in general form

$$\eta = \rho^{-1}\mathbf{rotU.grad}\theta \;, \tag{31}$$

is expressed in a particularly simple (quasi-hydrostatic) form for adiabatic, frictionless motion :

$$\frac{D}{Dt}\frac{rot_\theta \mathbf{U}_H + f}{\partial p/\partial\theta} = 0 \;; \tag{32}$$

The Lagrangian conservation of potential vorticity on isentropes means that, using θ coordinate, we are able to clearly separate quasi-geostrophic, bi-dimensional, isentropic turbulent mixing from other types of cross-isentropic mixing such as convection or three-dimensional small-scale turbulence.

Yet, the θ coordinate is not operationally used, neither for numerical weather prediction nor for climate modelling, because of difficulties in handling near-surface conditions. An additional point is the possible occurrence of statically unstable conditions which would make $\theta(z)$ multivalued. This problem can be avoided by an adequate vertical convection-diffusion scheme (see Section 5.4.), but it remains that the ideal domain for θ is the stratosphere. This is why a mixture of θ and the hybrid sigma-pressure coordinate has been proposed recently by

Zhu *et al.* (1992), and implemented in the UGAMP model at Reading University : it is a function $\zeta(\theta, p, p_S)$ with the following properties

$$\zeta(\theta_T, p_T, p_S) = \theta_T \ ; \ \frac{\partial \zeta}{\partial \theta}(\theta_T, p_T, p_S) = 1 \ ; \ \frac{\partial \zeta}{\partial \theta}(\theta, p, p_S) > 0 \ ;$$

$$\frac{\partial \zeta}{\partial p}(\theta, p, p_S) < 0 \ ; \ \frac{\partial \zeta}{\partial p}(\theta_S, p_S, p_S) = -\frac{1}{p_S} \ ; \ \zeta(\theta_T, p_S, p_S) = 0 \ . \tag{33}$$

The advantage of this coordinate is that, in the upper atmosphere (especially in the stratosphere), it becomes an entropy coordinate ; in the free troposphere, it is close to a pressure coordinate ; and in the lower atmosphere, it becomes a sigma coordinate. The amount of extra complexity is modest : in fact, one can write the identity

$$\mathbf{grad}_\zeta \Phi - \frac{\partial \Phi}{\partial p}\mathbf{grad}_\zeta p = \mathbf{grad}_\zeta \Phi - \theta\, \mathbf{grad}_\zeta \pi \tag{34}$$

If the gravity-pressure term is formulated as the right-hand side, it will automatically shift to $\mathbf{grad}_\theta M$ in the stratosphere, where ζ shifts to θ.

The three types of terrain-following vertical coordinates are visualised in Figure 2.

4.3. *General formulation*

Using arbitrary curvilinear coordinates and vertical coordinate, we make use of (25a) and (28) to write the horizontal momentum equations in final form, using either the logarithm of pressure or the Exner function in the pressure term :

$$\frac{\partial u_i}{\partial t} = \varepsilon_{ij}(\omega + f)u_j - \dot{\zeta}\frac{\partial u_i}{\partial \zeta} - \frac{1}{h_i}\frac{\partial}{\partial x^i}\left(\Phi + \frac{u_j u_j}{2}\right) - \frac{RT}{h_i}\frac{\partial Lnp}{\partial x^i} + \frac{1}{\Phi}\left(\mathbf{div}\left(\frac{\Phi}{\rho}\sigma\right)\right)_i \ , \tag{35a}$$

$$\frac{\partial u_i}{\partial t} = \varepsilon_{ij}(\omega + f)u_j - \dot{\zeta}\frac{\partial u_i}{\partial \zeta} - \frac{1}{h_i}\frac{\partial}{\partial x^i}\left(\Phi + \frac{u_j u_j}{2}\right) - \frac{\theta}{h_i}\frac{\partial \pi}{\partial x^i} + \frac{1}{\Phi}\left(\mathbf{div}\left(\frac{\Phi}{\rho}\sigma\right)\right)_i \ . \tag{35a'}$$

Similarly, the hydrostatic equation can be written either in terms of the logarithm of pressure :

$$\frac{\partial \Phi}{\partial \zeta} = -RT\frac{\partial \mathrm{Ln}p}{\partial \zeta} \tag{35b}$$

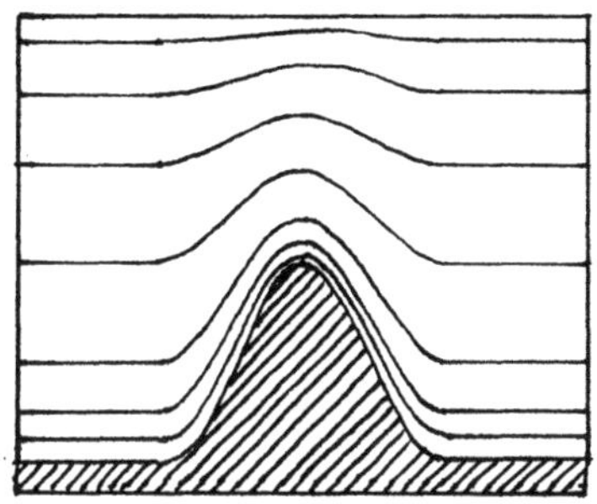

Phillips' sigma coordinate

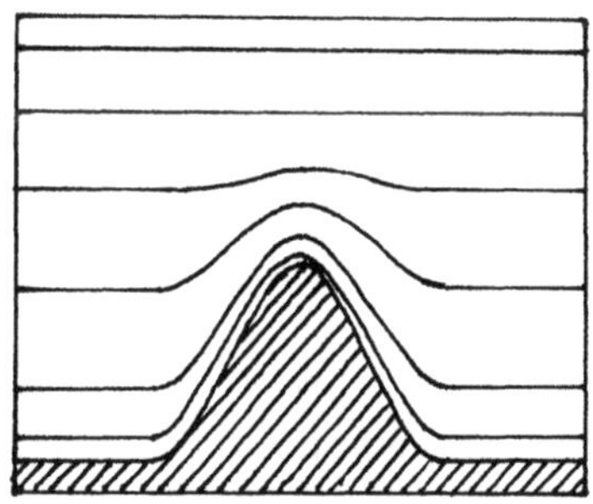

The hybrid pressure coordinate

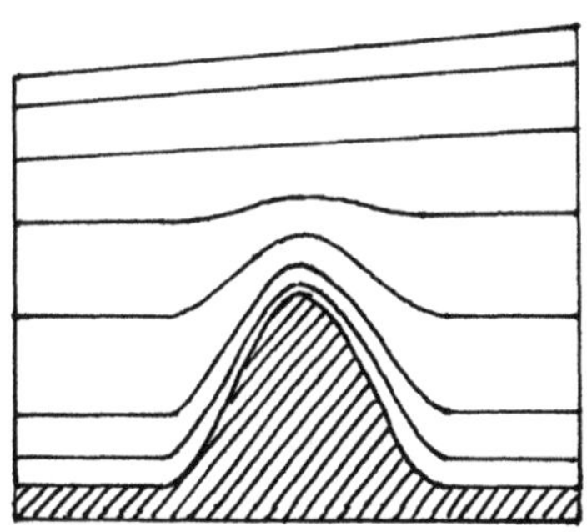

The hybrid pressure-potential temperature coordinate

Figure 2. *Vertical coordinates*

or in terms of the Exner function :

$$\frac{\partial \Phi}{\partial \zeta} = -\theta \frac{\partial \pi}{\partial \zeta} \ , \tag{35b'}$$

which is equivalent to

$$\frac{\partial M}{\partial \zeta} = \pi \frac{\partial \theta}{\partial \zeta} \ . \tag{35b''}$$

It follows from the above arguments that forms (35a',b') will obviously be the proper choice when the vertical coordinate is given by (33).

From (25b) and (29), we get the continuity equation

$$\frac{\partial}{\partial t}\frac{\partial p}{\partial \zeta} = -\frac{1}{h_1 h_2}\left(\frac{\partial}{\partial x^1}\left(h_2 u_1 \frac{\partial p}{\partial \zeta}\right) + \frac{\partial}{\partial x^2}\left(h_1 u_2 \frac{\partial p}{\partial \zeta}\right)\right) - \frac{\partial}{\partial \zeta}\left(\dot{\zeta}\frac{\partial p}{\partial \zeta}\right) . \tag{35c}$$

The vertical integral of (35c) yields the surface pressure tendency equation, as the last term involving the unknown vertical velocity disappears in the integration. This allows to compute the left-hand side of (35c) at any level ζ, and (35c) can then used to compute the vertical distribution of the pseudo-vertical velocity $\dot{\zeta}$.

The form of the energy equation is straightforward from (4a) or (4b) using (21) and we do not need to rewrite it here.

5. Discrete adiabatic formulations

After having chosen a mapping of the sphere and a vertical coordinate, the next step consists in discretising the equations to integrate them on a computer. Current operational weather prediction models and climate models use either finite difference or spherical harmonic formulations in the horizontal, and finite differences or finite elements in the vertical.

5.1. *Finite difference models*

The first global models of atmospheric circulation have been formulated in finite differences, and a few centres still favour this type of approach. Truncation errors are the basic problem of finite differences, leading to strongly detrimental effects like the building up of small scale noise if they are not properly handled. The major step in properly handling finite differences is due to Arakawa (1966) who designed finite difference schemes with inherent conservation properties, first on the barotropic Euler equation, then on the primitive equations.

To control computational noise, the most important property is the global conservation of potential enstrophy, which stems from Lagrangian conservation of Ertel's potential vorticity η in adiabatic, frictionless, thermally nonconducting flow. $D\eta / Dt = 0$ yields

$$\frac{d}{dt}\int_A \eta^2 dm = 0 \quad . \tag{36}$$

where the integration is taken on the whole mass of the atmosphere. This means the conservation of a quadratic norm involving gradients of the velocity field ; enforcing such conservation in a numerical scheme is an efficient way of controlling the generation of small scale noise. The physical meaning is that the potential enstrophy cascade towards small scales is not then artifially enhanced by truncation errors. Another practical by-product is that potential enstrophy-conserving schemes have particularly good stability properties.

A simple example of enstrophy-conserving scheme is the one given in Sadourny (1975a), still used in the HIRLAM and LMD models, although it is not strictly energy-conserving. We give here a variant of that scheme which is both energy- and potential enstrophy-conserving for the shallow water equations, when it is based on a triangulation of the mapping plane. We locate Φ at the nodes of triangles, while velocity components u are defined as the natural components normal to the sides of triangles (in the mapping plane) at their middle points. Indexing of both and orientation of velocity components is done as shown in Figure 3, *i.e.*, for Φ, *i*=0 at the centre and *i*=1,6 counterclockwise at surrounding points ; for u, *i*=1,6 counterclockwise aroud 0, then *i*=7,12 all around. At each u_i location, the map factor colinear with u_i is refered to as h_i, the orthogonal map factor at the same location as h'_i. With such conventions, the energy- and potential enstrophy-conserving scheme is constructed as follows. In the rotation term of (25a), we use potential absolute vorticity

$$\eta_0 = \frac{1}{\Phi_0}\left(f_0 + \frac{2}{3d}\sum_{i=1}^{6} h_i u_i \right) \quad , \tag{37}$$

instead of absolute vorticity, and mass flux

$$U_i = \frac{1}{2}(\Phi_0 + \Phi_i) h'_i u_i \quad , \tag{38}$$

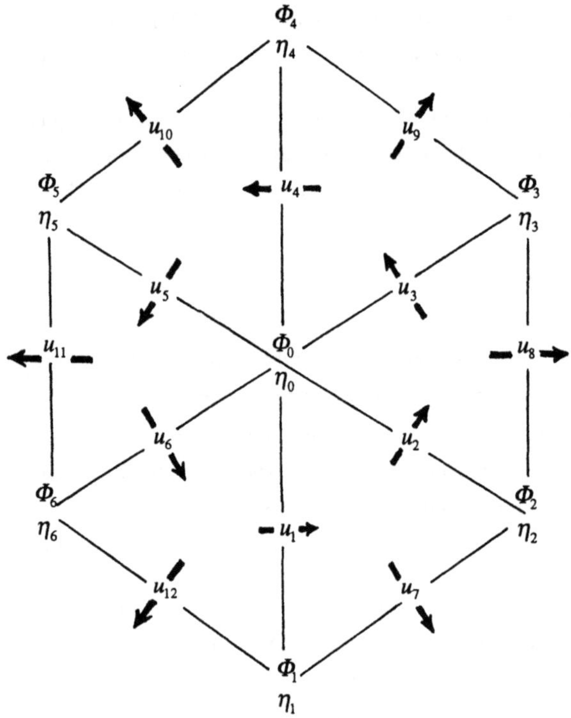

Figure 3. *An example of finite difference structure*

instead of velocity. The rotation term in the u_i tendency then reads

$$R_i = \frac{1}{6\sqrt{3}h_i}\left(\left(\eta_0 + \eta_i + \eta_{(i)+1}\right)\left(-U_{(i)+1} + U_{i+6}\right) + \left(\eta_0 + \eta_i + \eta_{(i-2)+1}\right)\left(U_{(i-2)+1} + U_{(i-2)+7}\right)\right) \tag{39}$$

where the parenthesis in indices means *modulo* 6. The fact that this energy- and potential enstrophy-conserving scheme has no equivalent in square grids is an example of the wide variety of choices in finite differencing. This scheme can be applied, for example, to the icosahedral grid mentioned in section 3.1. In this type of grid, the truncation error is maximum at gridpoints which are immediate neighbours of vertices, getting down again to very small values at the vertices themselves. By preventing artificial generation of small-scale noise, potential enstrophy conservation should control such errors quite efficiently.

A last remark on finite difference schemes : a further dynamical constraint to enforce, in addition to energy- and enstrophy conservation, is the conservation of angular momentum, which turns out to be quite important in climate simulations, as it forces for instance the trade wind structure in the tropics (Held *et al.*, 1980). The global conservation of angular momentum can be forced only on grids which are distributed on parallel circles ; on other types of grids like the icosahedral grid, equivalent local properties can be enforced.

5.2. *Spectral models*

The spectral approach has been favoured by most modelling centres since the mid-eighties, because of their intrinsically lower truncation errors (while finite-difference models are second- or fourth-order, spectral models are formally of infinite order). In the spectral approach, dependent variables are represented by truncated series of surface spherical harmonics

$$Y(x^1,x^2,\zeta,t)=\sum_{n=0}^{N}\sum_{m=-n}^{n} Y_m^n(\zeta,t)\Lambda_m^n(\lambda,\mu) \quad , \tag{40}$$

where λ is longitude, μ the sine of latitude, $(x^1,x^2)\rightarrow(\lambda,\mu)$ is a bijection of the sphere onto itself, and $\Lambda_m^n(\lambda,\mu)$ are spherical harmonics of degree n and order m. The truncation $m=-n,n$; $n=0,N$ is the triangular truncation corresponding to an homogeneous isotropic representation of the Y field on the (λ,μ) sphere. Surface spherical harmonics are defined as the eigenfunctions of the spherical Laplacian

$$\Delta\Lambda_m^n=-\frac{n(n+1)}{a^2}\Lambda_m^n \quad . \tag{41}$$

They are expressed as

$$\Lambda_m^n(\lambda,\mu)=P_m^n(\mu)e^{im\lambda} \quad , \tag{42}$$

where the $P_m^n(\mu)$ are the associated Legendre functions of the first kind

$$P_m^n(\mu)=\sqrt{(2n+1)\frac{(n-|m|)!}{(n+|m|)!}}\frac{1}{2^n n!}\left(1-\mu^2\right)^{|m|/2}\frac{d^{n+|m|}}{d\mu^{n+|m|}}\left(\mu^2-1\right)^n \quad , \tag{43}$$

In these formulas, n is the degree of the spherical harmonic $\Lambda_m^n(\lambda,\mu)$, m being its longitudinal wavenumber (Figure 4). The spherical harmonics form an orthonormal basis in the following sense :

$$\iint_S \Lambda_m^n(\lambda,\mu)\,\Lambda_{-m'}^{n'}(\lambda,\mu)\,d\lambda d\mu = 4\pi\,\delta_{mm'}\,\delta^{nn'} \;, \tag{44}$$

where S is the whole spherical surface and δs are the Kronecker symbols. The complex spectral coefficients $Y_m^n(\zeta,t)$ are calculated as the integrals

$$Y_m^n(\zeta,t) = \frac{1}{4\pi}\iint_S Y\left(x^1,x^2,\zeta,t\right)\Lambda_m^n(\lambda,\mu)\,d\lambda d\mu \;. \tag{45}$$

These integrals are calculated exactly provided a Gaussian grid is used in latitude.

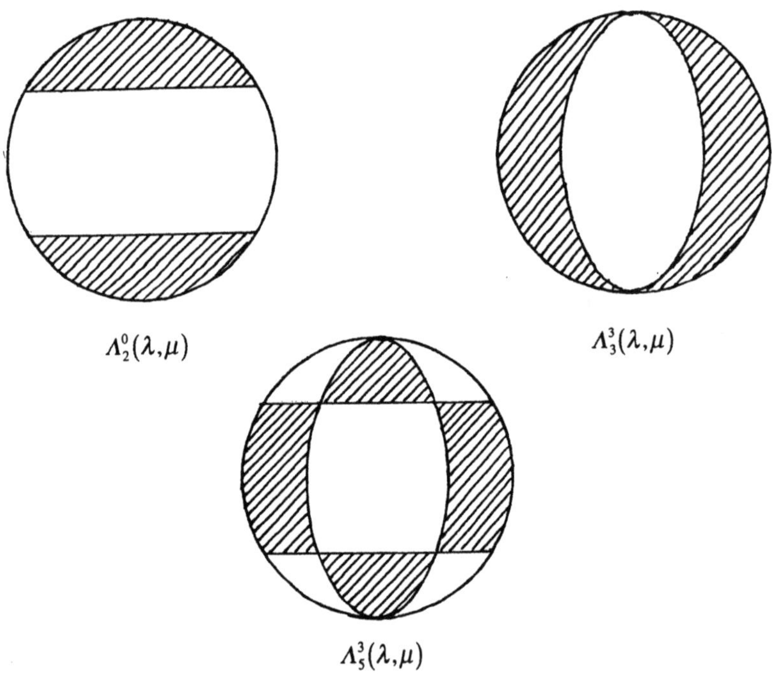

Figure 4. *Spherical harmonics*

The advantages of spectral calculations is their high accuracy in estimating space derivatives, and the homogeneous representation of spherical fields. Purely spectral calculations, *i.e.*, numerical integration of the dynamical equations in spectral space, would however be exceedingly time consuming, because of the convolutions associated with the nonlinear advection terms. Following Orszag (1970), the pseudo-spectral approach has been generalised, where linear operators (the space derivatives) are computed in spectral space and products in physical space ; this approach needs back and forth transforms between spectral and physical space, performed using Fast Fourier Transform in longitude and Gaussian integration in latitude. This approach is however quite sensitive to aliasing (or spectral folding) problems, *i.e.* the fact that the wavenumber domain being doubled in the squaring process, the spectrum of a squared field is misinterpreted by inadequate sampling. The solution to this problem can be described simply on one-dimensional truncated Fourier series. Taking a truncated Fourier representation $[-M',M']$ and a subset $[-M,M]$ of this representation with less resolution $(M<M')$, the square of a function defined on $[-M,M]$ will occupy the spectral domain $[-2M,2M]$. On the other hand, the aliases of $[-M,M]$ on the truncated representation $[-M',M']$ will be the sets of wavenumbers $[2nM'-M, 2nM'+M]$, where n is any positive or negative integer. One readily sees that to avoid aliasing problems, one has to choose $M'>3M/2$; in other words, squaring must be performed on a $M'=3M/2+1$ physical grid for an M^2th-order spectral truncation (this associated grid in physical space is called the collocation grid). The same resolution requirement holds for the Gaussian grid in latitude.

Comparing the respective resolutions of spectral *vs.* finite-difference models is not as obvious as it may look at first. The number of degrees of freedom of a spherical harmonic model with Nth-order triangular truncation is $(N+1)^2$, which corresponds to a uniform equivalent square grid distance of $2\sqrt{\pi}\,a/(N+1)$. Taking the higher resolution of the collocation grid as the actual resolution of the spectral model, even though the physics is calculated on the collocation grid, would be misleading, as all the extra information is lost whenever going back to spectral space. It might be stated, on the other hand, that finite-difference models also suffer from aliasing problems and that, therefore, their formal resolution should not be taken at face value, even though aliasing problems are partly solved by the conservation properties of finite differences. In the end, actual resolution is imposed by the lateral diffusion operator which does act on the Mth-order truncation in spectral models, and on the formal resolution of finite-difference models. We may add that the collocation grid of a spectral model can be reduced by decreasing the number of gridpoints on Gaussian latitude circles as one approaches the poles, keeping a longitudinal mesh-size quasi-uniform in actual distance (Hortal and Simmons, 1991) ; this reduces computing time and has little detrimental effect, as triangular truncation is quasi-homgeneous on the sphere.

5.3. *Vertical discretisation*

Vertical discretisation is generally done in finite differences. The basic constraints to enforce there are

- Conservation of total energy : this means that conversions between potential and kinetic energy associated with baroclinic transient dynamics must be properly treated without spurious energy sources or sinks ;
- Conservation of angular momentum : this relates to the formulation of the discretised hydrostatic equation, and its impact on the pressure gradient term in the horizontal momentum equation. On this point, see for instance the formulation by Simmons *et al.* (1981).

5.4. *Transport problems and semi-Lagrangian schemes*

An important part of atmospheric physics or fluid mechanics in general is the transport problem. Transport of vorticity, for example, is the basic process of turbulence, which imposes all the invariants which govern the flow statistics : the special attention foccused on quadratic invariance properties of quasi-geostrophic turbulence (energy and potential enstrophy) is but an artefact of the spectral (or finite-difference) techniques ; the integral of any function of potential vorticity on the mass of any isentropic surface is indeed an invariant of the Euler quasi-geostrophic equations, and the same occurs for any transported quantity, in the absence of sources or sinks. The transport of heat, moisture, condensed phases of water, radiatively active gases and aerosols are equally important.

The basic properties of a transport equation are :

- positivity : a transported quantity keeps its sign ;
- preservation of shape in a uniform velocity field ;
- conservation of local maxima and minima of mixing ratios : all local maxima or minima remain such in time and conserve their value ; no new local maximum nor minimum are created by transport processes. (A weaker form of this property is the conservation of total variation in dimension one.)

These properties are usually violated by Eulerian models, especially spectral models, due to the Gibbs phenomenon which generates spurious oscillations. Strong research efforts are currently devoted to these problems, both on the numerics side (Prather, 1988) and on the theoretical side, relating to the properties of two-dimensional turbulence (Robert *et al.*, 1991).

A particularly attractive numerical method for the transport equation is the semi-Lagrangian scheme used by Leith and Krishnamurti in the sixties and reintroduced by Robert (1982). It can be considered the simplest scheme of all, as it goes back to the intuitive physical meaning of advection, displacing a quantity from one location to another on the basis of a given velocity field which provides the trajectory. The method requires two steps :

- compute the starting point $\mathbf{M}_{ij}^{-}$ (at $t - \Delta t$) of the trajectory ending at a given gridpoint $\mathbf{M}_{ij}^{+}$, using the velocity field at time t ;
- interpolate the advected field $q(t - \Delta t)$ at $\mathbf{M}_{ij}^{-}$ to obtain the value of $q(t + \Delta t)$ at $\mathbf{M}_{ij}^{+}$.

(Other schemes can be designed, for instance interpolating at the end point rather than at the starting point.) This method, used for the advection terms of the primitive equations, has the advantage that it is inherently stable and thus allows longer time steps ; also, the interpolation scheme can be adjusted to handle some of the properties mentioned above. It remains to be seen, howeber, how both the spectral method and the semi-Lagrangian method will compete with simple finite-differences for high-resolution models running on massively parallel computers.

5.5. *Sub-grid scale diffusion*

Whenever using discrete approximations, one has to keep in mind that an important part of adiabatic dynamics is, so to speak, buried in the sub-grid scales ; and that those adiabatic subgridscale processes must be realistically modelled, because they have a major nonlinear impact on the resolved scales. They are best described in entropy-coordinate models, because macroscopic quasi-geostrophic motion along isentropes is then clearly separated from three-dimensional, cross-isentropic turbulent processes which dominate boundary layer or convective vertical fluxes. Like all sub-grid scale processes, both have to be modelled in a statistical sense.

The simplest boundary layer models are the so-called "eddy-diffusion" models, based on a vertical diffusion operator with a coefficient depending on a local Richardson number. More sophisticated models include the explicit prediction of boundary layer height, or an explicit closure equation for eddy kinetic energy (Mellor and Yamada, 1974).

The effect on resolved scales of the subgridscale horizontal (isentropic) eddies is generally taken care of using diffusion models (Smagorinsky, 1963), although quasi-geostrophic dynamics is somewhat subtler than a simple diffusion process. The basic model is then the Reynolds stress tensor divergence, reformulated in spherical geometry as a viscous stress tensor through the mixing length frame. The deformation tensor reads

$$\mathbf{T} = \begin{bmatrix} \tau & \sigma \\ \sigma & -\tau \end{bmatrix}, \tag{46}$$

where σ and τ refer to shearing and stretching :

$$\sigma = \frac{h_2}{h_1}\frac{\partial}{\partial x^1}\frac{u_2}{h_2} + \frac{h_1}{h_2}\frac{\partial}{\partial x^2}\frac{u_1}{h_1} \quad , \tag{47}$$

$$\tau = \frac{h_2}{h_1}\frac{\partial}{\partial x^1}\frac{u_1}{h_2} - \frac{h_1}{h_2}\frac{\partial}{\partial x^2}\frac{u_2}{h_1} \quad , \tag{48}$$

these expressions are coordinate-dependent but σ and τ can be given an intrinsic meaning taking the formulas above after rotating the local reference frame to get $u_2 = 0$. Sub-grid scale mixing of momentum then reads

$$\left(\frac{1}{\rho}\mathbf{div}(\mu\mathbf{T})\right)_1 = \frac{1}{\rho}\left(\frac{1}{h_1 h_2^2}\frac{\partial(\mu h_2^2 \tau)}{\partial x^1} + \frac{1}{h_2 h_1^2}\frac{\partial(\mu h_1^2 \sigma)}{\partial x^2}\right) \quad , \tag{49}$$

$$\left(\frac{1}{\rho}\mathbf{div}(\mu\mathbf{T})\right)_2 = \frac{1}{\rho}\left(\frac{1}{h_1 h_2^2}\frac{\partial(\mu h_2^2 \sigma)}{\partial x^1} - \frac{1}{h_2 h_1^2}\frac{\partial(\mu h_1^2 \tau)}{\partial x^2}\right) \quad . \tag{50}$$

It is readily verified that this formulation corresponds to a dissipation of kinetic energy

$$\frac{dE}{dt} = -\int_A \left(\sigma^2 + \tau^2\right) dm \quad . \tag{51}$$

On the other hand, it exactly conserves the total angular momentum, a very important constraint on atmospheric modelling. Relating the value of μ to local velocity gradients by using mixing length arguments leads to a nonlinear formulation of sub-grid scale lateral diffusion. This type of consistent formulation turns out however to be expensive in spectral models, where it is usually replaced by a linear formulation based on Laplacian operators, strictly valid for incompressible flow only and non conservative of angular momentum. Another point of iterest is the fact that, in quasi-geostrophic dynamics, total energy is conserved while potential enstrophy is dissipated. This has led many modellers, starting in the late seventies with modellers of two-dimensional turbulence (Basdevant *et al.*, 1981), to use highly scale-selective operators like iterated Laplacians (Δ^3 or even more), or to develop specific energy-conserving, potential enstrophy-dissipating shemes (Sadourny *et al.*, 1985) ; an example of the latter for shallow-water equations is as follows :

$$\frac{\partial \mathbf{U}}{\partial t} = -\left(\frac{\omega_a}{\Phi} - \tau\Delta^{2p+1}\left(\mathbf{U}.\mathbf{grad}\frac{\omega_a}{\Phi}\right)\right)\mathbf{k}\times\mathbf{U} + \mathbf{grad}\left(\Phi + \frac{U^2}{2}\right) \quad , \tag{52}$$

where τ is a dissipation time scale and Δ a nondimensional Laplacian. As it dissipates potential enstrophy while conserving total energy, this type of formulation can actually mimick sub-gridcale baroclinic eddies when applied to the full baroclinic primitive equations. It is of course easy to formulate when using the θ coordinate ; but again, it does not conserve angular momentum.

6. Diabatic processes

I shall limit myself to a rather brief summary of the diabatic processes which drive the general circulation, *i.e.*, radiation and the release of latent heat. Most of these processes are intimately linked to sub-grid scale structures like convective or more generally, all kinds of cloud systems from high-level cirrus to boundary layer-capping stratus.

6.1. *Radiation*

Modelling of the radiative transfer equation is relatively well in hand, at least in the absence of clouds, the main problem there being to reach a satisfactory balance between computational efficiency (line-by-line radiation models are too expensive for being run in genaral circulation models) and precision. Any systematic error in radiation calculations may of course lead to a significant climate drift. A reasonable such balance seems to have been achieved with the most recent general circulation radiation codes (*e.g.*, Morcrette *et al.*, 1986, for long-wave calculations). However, most present models only take into account water vapour, carbon dioxide and stratospheric ozone ; other radiatively active gases like methane, nitrous oxide, tropospheric ozone and chloro-fluorocarbons are being included for climate warming simulations.

6.2. *Moist convection*

Moist convection is another major process in general circulation dynamics. Very simple schemes like the moist adiabatic adjustment of Manabe and Strickler (1964), or the penetrative moist convection scheme of Kuo (1971) and its many variants, are still in use, although mostly perhaps in climate models. The principle of moist adiabatic adjustment is as follows. Whenever a column of atmosphere is both conditionally unstable

$$\frac{\partial}{\partial z}\left(c_p T + L_V q_v + gz\right) < 0 \qquad \text{for} \qquad z_1 < z < z_2 \tag{53}$$

and saturated ($q_v = q_S(T,p)$ for $z_1 < z < z_2$), its vertical profile is brought back to neutral

$$\frac{\partial}{\partial z}\left(c_p T + L_V q_v + gz\right) = 0 \qquad \text{for} \qquad z_1 < z < z_2 \ , \tag{54}$$

while keeping it saturated and conserving its total energy

$$\int_{z_1}^{z_2} \left(c_p T + L_V q_v\right) dz \quad . \tag{55}$$

The excess water vapour is then condensated.

The Kuo penetrative convection scheme acts on a conditionally unstable, unsaturated column. The energy deficit with respect to a neutral profile defined from boundary layer conditions (*BL*) is

$$\Delta E = \left(c_p T + L_V q_v + gz\right)_{BL} \left(z_2 - z_1\right) - \int_{z_1}^{z_2} \left(c_p T + L_V q_v + gz\right) dz \quad , \tag{56}$$

where z_1 refers to the boundary layer top and z_2 is defined as the height where

$$c_p T + L_V q_v + gz = \left(c_p T + L_V q_v + gz\right)_{BL} \quad ; \tag{57}$$

on the other hand, the energy available for penetrative convection is assumed to be the convergence of latent heat in the boundary layer :

$$\Delta E' = -L_V \Delta t \int_{BL} div\left(q_v \mathbf{U}_H\right) dz \quad ; \tag{58}$$

then the proportion of mesh area covered by penetrative convection is assumed to be

$$\alpha = \frac{\Delta E'}{\Delta E} \quad . \tag{59}$$

The energy available for penetrative convection $\Delta E'$ is used to warm and moisten the column from z_1 to z_2, which corresponds to mixing of the penetrative cloud column into its environment.

Following these simple models, more realistic ones have been developed, which will not be described here. Mass flux schemes including entrainment-detrainement effects, based on an efficiency function, have been introduced in the mid-seventies by Arakawa and Schubert (1974) ; and the trend is towards more realistic description by including, for instance, downdraft effects. The importance of modelling shallow convection in the tropics has been demonstrated by Tiedtke who has developed a comprehensive mass flux scheme (1989).

All these schemes, and others like the Betts (1986) and Betts and Miller (1986) scheme, are currently used in a number of general circulation or weather forecasting models. Each of them has its own virtues, the problem there being perhaps that no single parametrisation scheme can accurately describe the macro-structure of moist convection in all possible situations over all the globe.

6.3. *Clouds and cloud radiative forcing*

A domain where major developments are needed is the modelling of the condensed phases of water : *i.e.*, the liquid or ice clouds, and their interaction with radiation. This in fact is a key process in the problem of the sensitivity of climate to external perturbations like the anthropogenic release of greenhouse gases or solar luminosity variations. High clouds like cirrus have a strong greenhouse warming effect, as they tend to be relatively transparent to solar radiation, and at the same time, cold, and thus weakly emitting to space in the infrared ; low clouds, on the other hand, have a strong cooling effect : heavily loaded in water content, they are strong reflectors in the visible, and at the same time, relatively warm and thus strongly emitting in the infrared. The net result under present climate conditions is that clouds have a global net cooling effect of about 15 Wm^{-2}. By changing vertical temperature profiles and water vapour content, a positive external radiative perturbation may upset this balance towards a positive feedback (stronger warming) or a negative one (weaker warming) depending on changes in the vertical structure and radiative properties of clouds. Present-day models are quite disappointing in this respect : their global sensitivity $D\langle T_S\rangle / DR$, where $\langle T_S\rangle$ is the globally average surface temperature in K and R the magnitude of the radiative perturbation in Wm^{-2}, varies by as much as a factor three, from 0.4 to 1.3 (Cess *et al.*, 1991).

Thus we have to put a lot of effort in the realistic modelling of cloud structure and radiative properties. For a long time, clouds have been diagnosed in atmospheric models using such primitive tests as relative humidity tresholds, or a combination of relative humidity with the sign of vertical velocity. Then they the cloud diagnostics have been linked to the condensation processes as actually computed by the model. Now general circulation models tend to go to explicit cloud modelling by incorporating cloud liquid water and ice as separate prognostic variables, after pioneering work by Sundqvist (1978). But the situation concerning the wide dispersion of sensitivity estimates has not improved. What is needed now is more realistic modelling of cloud microphysics, including droplet size, phase change and sedimentation processes.

References

Arakawa, A. (1966) : Computational design for long-term numerical integration of the equations of fluid motion. Part I : Two-dimensional incompressible flow. *J. Comp. Phys.*, **1**, 119-143.
Arakawa, A. , W. H. Schubert (1974) : Interaction of cumulus cloud ensemble with the large-scale environment. *J. Atmos. Sci.*, **31**, 674-701.
Basdevant, C., B. Legras, R. Sadourny, M. Béland (1981) : A study of barotropic model flows : Intermittency, waves and predictability. *J. Atmos. Sci.*, **38**, 2305-2326.
Betts, A. K. (1986) : New convective adjustment scheme. Part I. Observational and theoretical bases. *Quart. J. Roy. Meteor. Soc.*, **112**, 677-691.
Betts, A. K., M. Miller (1986) : New convective adjustment scheme. Part II. Single column test using GATE wave and BOMEX, APEX and Arctic Air Mass data sets. *Quart. J. Roy. Met. Soc.*, **112**, 693-709.
Cess, R.D., G.L. Potter, J.P. Blanchet, G.J. Boer, A.D. Del Genio, M. Déqué, V. Dymnikov, V. Galin, W.L. Gates, S.J. Ghan, J.T. Kiehl, A.A. Lacis, H. Le Treut, Z.X. Li, X.Z. Liang, B.J. McAvaney, V.P. Meleshko, J.F.B. Mitchell, J.J. Morcrette, D.A. Randall, L. Rikus, E. Roeckner, J.F. Royer, U. Schlese, D.A. Sheinin, A. Slingo, A.P. Sokolov, K.E. Taylor, W.M. Washington, R.T. Wetherald, I. Yagai M.H. Zhang, 1990 : Intercomparison and interpretation of climate feedback processes in nineteen atmospheric general circulation models. *J. Geophys. Res.*, 95, 16,601-16,615.
Courtier, Ph., C. Freydier, J.F. Geleyn, F. Rabier ,M. Rochas (1991) : The ARPÈGE Project at Météo France. In : *Numerical Methods in Atmospheric Models*, **II**, 193-231, *European Centre for Medium Range Weather Forecasts Seminar Proceedings.*
Held, I.M., A.Y. Hou (1980) : Nonlinear axially symmetric circulations in a nearly inviscid atmosphere. *J. Atmos. Sci.*, **37**, 515-533.
Hortal, M., A.J. Simmons (1991) : Use of reduced Gaussian grids in spectral models. *Mon. Wea. Rev.*, **119**, 1057-1074.
Krishnamurti, T.N. (1969) : An experiment in numerical prediction in equatorial latitudes. *Quart. J. Roy. Meteor. Soc.*, **95**, 594-620.
Mesinger, F., Z.I. Janjic, S. Mickovic, D. Gavrilov, D.G. Deaven (1988) : The step-mountain coordinate : model description and performance for cases of Alpine lee cyclogenesis and for a case of Appalachian redevelopment. *Mon. Wea. Rev.*, **116**, 1493-1518.
Manabe, S., R.F. Strickler (1964) : Thermal equilibrium of the atmosphere with the convective adjustment. *J. Atmos. Sci.*, **21**, 361-385.
Mellor, G.L., T. Yamada (1974) : A hierarchy of turbulence closure models for planetary boundary layers. *J. Atmos. Sci.*, **31**, 1791-1806.
Morcrette, J.-J., L. Smith and Y. Fouquart (1986) : Pressure and temperature dependence of the absorption in longwave radiation parameterizations. *Beitr. Phys. Atmosph.*, **59**, 455-468.
Orszag, S.A. (1970) : Transform method for calculation of vector-coupled sums : application to the spectral form of the voticity equation. *J. Atmos. Sci.*, **27**, 890-895.
Phillips, N.A. (1957) : A coordinate system having some special advantages for numerical forecasting. *J. Meteorol.*, **14**, 184-185.
Prather, (1988) : Numerical advection by conservation of second-order moments. *J. Geophys. Res.*, **91**, 6671-6681.
Richardson, L.F. (1922) : Weather prediction by numerical process. *Cambridge University Press*, 236 pp.
Robert, J. Sommeria (1991) : Statistical equilibrium states for two-dimensional flows. *J. Fluid Mech.* , **229**, 291-310.
Robert, A.J. (1982) : A semi-Lagrangian and semi-implicit numerical integration scheme for the primitive meteorologicall equations. *J. Meteor. Soc. Japan*, **60**, 319-324.
Sadourny, R., A. Arakawa, Y. Mintz (1968) : Integration of the nondivergent barotropic vorticity equation with an icosahedral-hexagonal grid on the sphere. *Mon. Wea. Rev.*, **96**, 351-356.
Sadourny, R. (1975a) : The dynamics of finite-difference models of the shallow water equations. *J. Atmos. Sci.*, **32**, 680-689.

Sadourny, R. (1975b) : Compressible model flows on the sphere. *J. Atmos. Sci.*, **32**, 2103-2110.

Sadourny, R., C. Basdevant (1985) : Parameterization of sub-grid scale barotropic and baroclinic eddies : Anticipated Potential Vorticity Method. *J. Atmos. Sci.*, **42**, 1353-1363.

Schmidt, F. (1977) : Variable fine mesh in spectral global model. *Beitr. Phys. Atmos.*, **50**, 211-217.

Sharma, O.P., H. Upadhyaya, Th. Braine-Bonnaire, R. Sadourny (1987): Experiments on regional forecasting using a stretched-coordinate general circulation model. *J. Meteorol. Soc. Japan, Special Volume on Short- and Medium-Range Numerical Weather Prediction.*

Simmons, A, D. Burridge (1981) : An energy and angular momentum conserving vertical finite difference scheme in hybrid vertical coordinate. *Mon. Wea. Rev.*, **109**, 758-766.

Smagorinsky, J. (1963) : General circulation experiments with the primitive equations. I. The basic experiment. *Mon. Wea. Rev.*, **91**, 99-164.

Staniforth, A.N., H.L. Mitchell (1978): A variable-resolution finite-element technique for regional forecasting with the primitive equations. *Mon. Wea. Rev.*, **106**, 439-447.

Sundqvist, H. (1978) : Parametrisation for non convective condensation including prediction of cloud water content. *Quart. J. Roy. Meteor. Soc.*, **104**, 677-690.

Tiedtke, M. (1989) : Comprehensive mass flux scheme for cumulus parametrisation in large-scale models. *Mon. Wea. Rev.*, **117**, 1779-1800.

Williamson, D. (1968) : Integration of the barotropic vorticity equation on a spherical geodesic grid. *Tellus*, **20**, 642-653.

White, A.A., R.A. Bromley (1988) : A new set of dynamical equations for use in numerical weather prediction and global climate models. *Meteorological Office, Met O 13 Branch memo.*

Zhu, Z., J. Thuburn, B.J. Hoskins, P.H. Haynes (1992) : A vertical finite difference scheme based on a hybrid s-q-p coordinate. *Mon. Wea. Rev.*, **120**, 851-862.

MODELLING THE OCEAN CIRCULATION

Pascale Delecluse
LODYC - UMR121 (CNRS-UPMC-ORSTOM)
Université Pierre et Marie Curie
Tour 26-4 Boite 100
4 place Jussieu
75252 Paris - Cedex 05
France

Abstract : After a brief presentation of the ocean properties, the set of equations used in global ocean modelling is presented. Different technical choices can affect the solution : the system of coordinates, the grid, the numerical algorithms... The impact of physical choices is then presented. Present capacity of numerical modelling is discussed on two examples : a high resolution model of the tropical Pacific and a global circulation model.

1 - Physical properties of the ocean

1 - 1 Physical structure

The ocean covers 70% of the earth surface and contains 97% of the earth water. It has a considerable "buffer" role in the climate system due to its large heat capacity (2.5 m of ocean water has the same heat capacity than the total atmospheric column above it) and its weight (the density of ocean water is 1.025 x 10^3 kg/m^3 as the air density is 1.2 kg/m^3). The sea water has a temperature range within -1.90 to 32°C for a salinity of open sea varying from 33 to 37 psu (or *‰* : g/kg of sea water) but 75% of the ocean volume is filled with water in a very narrow range of temperature (between 0 and 4°C) and of salinity (between 34.4 and 34.7 psu). In enclosed or semi-enclosed seas, much more variability can be found for temperature and salinity values. The vertical profile of temperature is nearly homogeneous in high latitudes and the constrast between surface temperature and deep temperature increases equatorward. In the equatorial band, a strong and sharp thermocline separates the warm reservoir from the cold deep waters. The thermocline thickens and deepens polewards. These properties are common to the three oceans in winter conditions. The water mass distribution and the ocean circulation are closely linked together and the water mass properties are determined through complex air-sea interactions. It is thus very important to understand and to model how ocean and atmosphere interact and how surface properties are transfered into the deep ocean.

NATO ASI Series, Vol. I 22
Long-Term Climatic Variations
Edited by J.-C. Duplessy and M.-T. Spyridakis

1 - 2 Interaction with the atmosphere

Seasonal cycle of the mixed-layer

The mixed-layer is defined as the thin surface layer of the ocean where turbulent mixing is strong enough to ensure a vertical homogeneity of physical properties. It is the first receptacle of the atmospheric fluxes. When the ocean is gaining heat from the atmosphere, a thin warm layer forms at the surface. Surface wind stress acts as a source of turbulence and permits the deepening of this surface layer. Other sources of oceanic turbulence (as the breaking of internal waves) or the shear of ocean currents contribute to maintain it. The temperature gradient at its bottom is called the seasonal thermocline. As the warm season comes, the seasonal thermocline deepens and the mixed-layer gets warmer. In fall, the heat budget reverses and the mixed-layer temperature decreases as its depth increases. The maximum depth is usually reached during the winter season and coincides with the position of the main thermocline. The high frequency variability of the ocean-atmosphere system is confined within this depth. The mixed-layer depth retreats to the surface level very quickly at the end of the winter season.

Mid-latitude thermocline ventilation

The maximum depth of the seasonal thermocline defines the main thermocline, permanent thermal gradient which separates the warm surface layers from the deep ocean. During the winter season, the main thermocline communicates directly with the atmosphere through a process called the thermocline ventilation or thermal subduction. It occurs in areas where the curl of the wind stress is negative. This curl induces important vertical downward motions, allowing the surface information (momentum and density property) to be transmitted to the main thermocline level. This process is located in sub-polar and mid-latitude regions. It concerns basin circulation and has a typical time scale of tens of years.

Equatorial thermocline

In the equatorial regions, there is no distinction between the seasonal thermocline and the main thermocline. The thermocline there is shallow and the circulation is vigorous. In all seasons, turbulent mixing is strong enough to homogenize the temperature to the level of the main thermocline which is directly affected by the interactions with the atmosphere. It is an area where ocean and atmosphere interactions rapidly develop a basin wide response (like the ENSO - El Niño Southern Oscillation) within a time scale of several months.

Deep convection and sea ice

In high latitudes, the concept of main thermocline disappears : there is no permanent gradient of temperature between the surface layers and the deep waters. During the winter season, very dense surface waters form at the surface : the temperature gets very cold under harsh atmospheric conditions and the formation of sea ice significantly increases the surface salinity by rejection of salt during ice formation. This very dense surface water dives to the deep ocean : it is the process of deep convection. It happens in very limited areas of the world oceans in an intermittent way (a few tens of km^2 during a few weeks). Known areas for formation of the Antarctic bottom waters are the Weddell sea and the Antarctica shelf regions; for the North Atlantic deep water the Labrador sea, the Norvegian Sea and the Greenland Sea. Despite the small area of formation, these waters represent the largest portion of the water masses of the world ocean (up to 75 %). They travel very slowly around the world ocean with typical time scales of a few hundreds of years.

1 - 3 Turbulence of the ocean

The ocean is far from being a smooth viscous fluid. It is affected by a turbulent activity which appears clearly on the variability detected by altimeters (like SEASAT and GEOSAT). This meso-scale turbulence has a typical length scale of 30-50 km for a life time of several weeks. Its contribution to the global circulation is still an open question. In some areas, the amount of heat transported by the eddies is equivalent to the amount of heat transported by the mean circulation. Though the eddies seem to exist everywhere in the oceans, the distribution of the eddy kinetic energy is highly heterogeneous.

2 - The modelling approach

2 - 1 The physical equations

Basic assumptions (refer to Pedlosky - 1987)

The circulation within the oceans is limited to a very thin shell on the earth surface. If H is a typical scale for the ocean depth, L a scale to characterize the horizontal dimension of ocean basin and a is the earth radius, one has :

$H \approx 10^3$ m

$L \approx 10^6$ m

$a \approx 6.4 \times 10^6$ m

When looking at local dynamics, it is possible to neglect sperical effects but this is not valid on a global scale. H/L and H/a are very small quantities used to scale the equations. With these parameters, it is clear that the vertical velocity is very small in front of the horizontal velocity and only the horizontal field of acceleration is kept. Following this scaling, the rotation of the earth acts as if located on the local vertical - tangential effects are neglected. The vertical momentum equation is reduced to the hydrostatic assumption.

The deviation of density from the mean profile is also a small quantity. Density is taken as a constant except in the buoyancy terms. This is the Boussisnesq's assumption.

The sea water is assumed incompressible. This assumption filters sound waves. The vertical velocity is then computed in function of the divergence of the horizontal velocity.

One last important assumption concerns the scale of motion retained by the analysis. The small scales are not treated explicitly. Diffusion within the ocean is performed by turbulent processes and molecular dissipation is neglected. The difficult problem is to find a reasonable parameterization for turbulence.

In a local coordinate system, the acceleration of a fluid particle will be balanced by the Coriolis force, the pressure gradient force and the turbulent Reynolds stresses. A scale analysis shows that in the oceanic interior, acceleration is small in front of the Coriolis term and the ocean is in geostrophic balance, but coastal and equatorial regions do not follow this rule : turbulent processes and advection have there a major contribution.

The Primitive Equations

From the above assumptions, the set of equations retained to describe the ocean circulation is the Primitive Equations - PE (Bryan, 1969). The rigid lid assumption is added in order to filter the external gravity waves which travel at a very high speed compared to the internal waves (200 m/s compared to 1 m/s). The momentum equation expressed the conservation of horizontal momentum. It allows the computation of the two components of the baroclinic speed:

$$\partial_t U_h = - (\mathrm{curl}_z\, U_h + f)\, \mathbf{k} \times U_h - w\, \partial_z U_h - 1/\rho_o\, \mathbf{grad}\, (p + 1/2\, \rho_o\, U_h^2) + \mathbf{F(U)} = \mathbf{G}$$

where **U** is the velocity, U_h its horizontal projection, p the pressure, **k** the local vertical axis, f the Coriolis parameter, ρ_o the reference value for density. The barotropic equation allows to compute the streamfunction ψ of the vertically integrated flow by solving an elliptic equation (H is the ocean depth) :

$$\mathbf{curl}_z \left(\frac{1}{H}\, (\mathbf{k} \times \mathbf{grad}\, \partial_t \psi) \right) = \mathbf{curl}_z \int_{-H}^{0} G\, dz$$

The conservation of heat allows to compute the evolution of temperature T and the conservation of salt, the salinity S :

$$\partial_t T + \text{div}(T \mathbf{U}) = \mathbf{F}(T)$$

$$\partial_t S + \text{div}(S \mathbf{U}) = \mathbf{F}(S)$$

Salinity, temperature, barotropic and baroclinic components of velocity are the prognostic variables of the system. Diagnostic variables are also needed. Pressure, salinity and temperature are combined together in the equation of state (UNESCO 1983) in order to compute the density:

$$\rho = \rho (T, S, z)$$

The hydrostatic equation permits to compute pressure :

$$\partial_z p = - \rho g$$

The assumption of incompressibility allows the computation of the vertical velocity w :

$$\text{div} (\mathbf{U_h}) + \partial_z w = 0$$

Reynolds stresses are parameterized as dissipation and diffusion. The operator is not isotropic in order to take into account the different vertical and horizontal scalings :

$$F (q) = K_H \Delta_H q + \partial_z (K_v \partial_z q)$$

The coordinates system

Most models are written in a geographic grid on the sphere (in function of latitude and longitude), with the vertical as the third coordinate but it is also possible to work with isopycnal models (Oberhüber, 1990) or σ-models (Haidvogel et al., 1990). In isopycnal models, the levels are given in function of density. This is a natural coordinate for oceanic motions which tend to follow isopycnal surfaces with very little diffusion across them. However the intersection of levels with surface or lateral walls is not easy. Another problem comes for the distinct role of salinity and temperature which both have a non-linear contribution to density and evolve differently. In σ-models, the vertical coordinate follows bottom topography. It is stretched above high topography and loose over deep oceans. This is a good way to respect the control by mountain ridges but it gives other problems : the pressure gradient is difficult to define in the neighbourhood of slopes as is the turbulent tensor.

Some models use a distorded grid in order to increase the resolution in specific regions One example is given by the treatment of polar regions in general circulation ocean models. When using geographical coordinates, the zonal grid spacing (Δx) becomes infinitely small as the pole is approached. This raises a numerical instability problem as the first condition for numerical stability is given by the CFL (Courant-Friedrichs-Lewy, 1928) criterion which expresses that : $\Delta x / \Delta t \geq C$ where Δx is the grid spacing in space, Δt is the timestep and C is the maximum phase speed. The respect of the criterion implies strong restriction in timestep in the vicinity of the pole. Either a filtering is used to suppress the numerical noise or the Arctic ocean is not included (Semtner and Chervin, 1988). The filtering solution is not satisfactory in ocean models because it is not acting very well in the presence of coastlines (Bryan, 1987). Another solution is to define a system of coordinates in which the North Pole is not a singular point. This is possible by removing the singular point into a continental area like Canada or Asia. This distorded coordinates system is used in the LODYC (Laboratoire d'Océanographie Dynamique et de Climatologie) general circulation model.

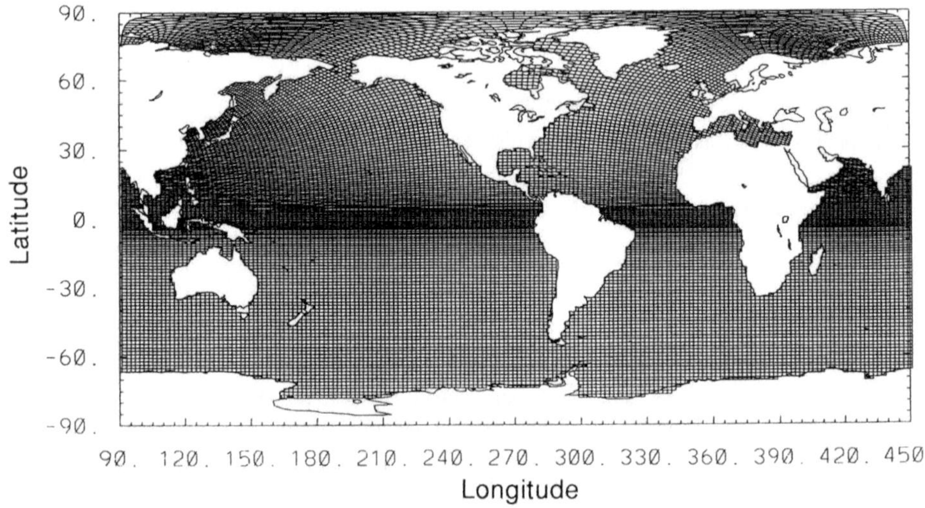

Figure 1 : Numerical grid used in the OPA global model

Many different coordinate systems permit to treat the general circulation of the oceans.

Nevertheless, they all have positive and negative points. The change of coordinates has to respect the fundamental properties of the fluid. When using a variable grid spacing, it is important to respect the conservative properties of the operators from the initial space to the new one. So the correct mathematical way to write the equations is to evidence the vectorial operators (curl, grad, div) and to keep their form invariant under coordinates changes.

Let (y^1, y^2, y^3) be a set of orthogonal curvilinear coordinates on the sphere with y^3 parallel to the local vertical and (x^1, x^2, x^3) where $x^i = x^i(y^1, y^2, y^3)$, the cartesian set of coordinates with the origin located at the sphere centre. The local deformation of the curvilinear coordinate system is simply given by the scale factors e_i :

$$e_i^2 = \left(\frac{\partial x^1}{\partial y^i}\right)^2 + \left(\frac{\partial x^2}{\partial y^i}\right)^2 + \left(\frac{\partial x^3}{\partial y^i}\right)^2$$

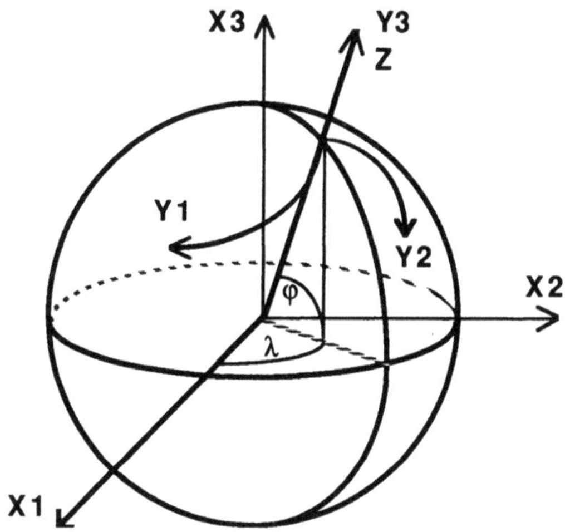

Figure 2 : Transformation of orthogonal curvilinear coordinate system on a sphere

Substituting (y^1, y^2, y^3) by (i, j, k), the scale factors e_i can be expressed as a function of the earth radius a, the latitude $\varphi(i,j)$, longitude $\lambda(i,j)$ and the altitude $z(k)$:

$$\begin{cases} e_1 = (a+z)\left[\left(\dfrac{\partial\lambda}{\partial i}\cos\varphi\right)^2 + \left(\dfrac{\partial\varphi}{\partial i}\right)^2\right]^{1/2} \\ e_2 = (a+z)\left[\left(\dfrac{\partial\lambda}{\partial j}\cos\varphi\right)^2 + \left(\dfrac{\partial\varphi}{\partial j}\right)^2\right]^{1/2} \\ e_3 = \left(\dfrac{\partial z}{\partial k}\right) \end{cases}$$

Since the ocean depth is far smaller than the earth radius, (a + z) can be replaced by a in the above sytem. The scalar and vectorial operators can be written in the tensorial form which is invariant in any orthogonal curvilinear coordinate system transformation :

$$\mathbf{grad}\ f = \begin{pmatrix} \dfrac{1}{e_1}\dfrac{\partial f}{\partial i} \\ \dfrac{1}{e_2}\dfrac{\partial f}{\partial j} \\ \dfrac{1}{e_3}\dfrac{\partial f}{\partial k} \end{pmatrix}$$

$$\mathbf{curl}\ \mathbf{A} = \begin{pmatrix} \dfrac{e_1}{b}\left[\dfrac{\partial(e_3A_3)}{\partial j} - \dfrac{\partial(e_2A_2)}{\partial k}\right] \\ \dfrac{e_2}{b}\left[\dfrac{\partial(e_1A_1)}{\partial k} - \dfrac{\partial(e_3A_3)}{\partial i}\right] \\ \dfrac{e_3}{b}\left[\dfrac{\partial(e_2A_2)}{\partial i} - \dfrac{\partial(e_1A_1)}{\partial j}\right] \end{pmatrix}$$

$$\text{div}\ \mathbf{A} = \frac{1}{b}\left[\frac{\partial}{\partial i}\left(\frac{b}{e_1}A_1\right) + \frac{\partial}{\partial j}\left(\frac{b}{e_2}A_2\right) + \frac{\partial}{\partial k}\left(\frac{b}{e_3}A_3\right)\right]$$

where f is a scalar function, $A = (A_1, A_2, A_3)$ a vector and $b = e_1\, e_2\, e_3$.

This formalism is not often used in models and it has been pointed out recently by Yin and Fung (1991) that variable grid spacings which do not respect this formalism can lose the accuracy of the initial scheme, creating numerical diffusion and artificial upwelling or downwelling motion in case of the vertical advection operator for instance.

2 - 2 Numerical schemes and coding

Discretization in space

The numerical methods to develop the equations are numerous. For atmospheric flows, a widely used method is the spectral technique (the equations are projected on the spherical harmonics). This is not used for ocean general circulation models as the complicated shape of the coastline is not in favor of such a method. In fact a method based on finite elements may be adequate for solving irregular coastlines but most models are based on finite differences and these are the examples we will discuss.

A description of the different properties of numerical schemes can be found in Mesinger and Arakawa (1976). Two different grids have been largely used in ocean modelling. The first one is the B-grid (as in the ocean model developed in GFDL [Geophysical Fluid Dynamics Laboratory], Bryan - 1969) and the second one is the C-grid (as the one used in the model developed in LODYC - see Chartier, 1985). Multiple studies have been done to exemplify the properties of these two grids. The system of primitive equations is complex and it is difficult to find a grid which will keep all the properties of the continuous equations. Certain terms like the kinetic energy and the Coriolis term are naturally computed with a B-grid. On the contrary, the C-grid is more suitable to compute the vertical velocity and the curl of currents.

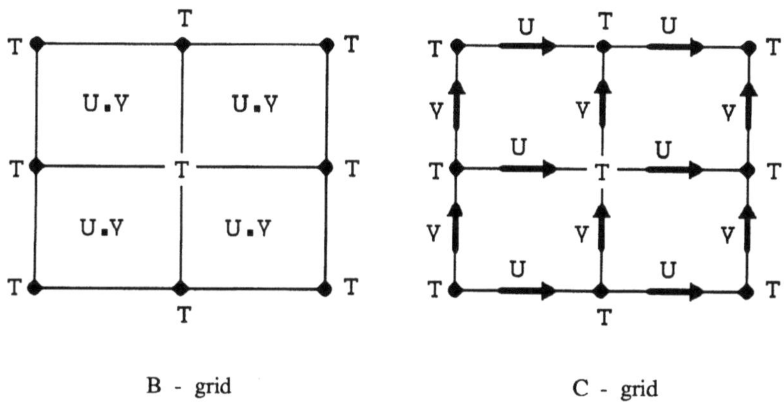

Figure 3 : Arrangement of the variables in the B - grid and in the C - grid.

Extensive studies have been made to describe the properties of these two schemes for the wave propagation (see, for instance, Batteen and Han, 1981 or Wajsowic, 1986). Let us show on a simple example how the discretization in different grid-spacing can affect the wave propagation. Considering the one-dimensional approximation of the shallow water equations (only the x-direction is retained), the continuous system of equations is the following :

$$\partial u / \partial t = - g \, \partial h / \partial x + f v$$

$$\partial v / \partial t = - f u$$

$$\partial h / \partial t = - H \, \partial u / \partial x$$

where u and v designate the zonal and meridional components of velocity, f is the Coriolis parameter, g the acceleration of gravity, H the mean thickness of the layer and h the perturbation of this thickness.

For this system, the relation of dispersion of the inertia gravity waves is the following :

$$\omega^2 = f^2 + gH\,k^2$$

where ω is the frequency and k the wave number.

When the system is discretized on the B-grid, it becomes :

$$\partial u / \partial t = -g\,\delta_x h + f\,v$$
$$\partial v / \partial t = -f\,u$$
$$\partial h / \partial t = -H\,\delta_x u$$

and the relation of dispersion for the inertia gravity waves becomes :

$$\omega^2 = f^2 + 4\ gH\ k^2 \sin^2 (k\,\Delta x / 2)$$

When the system is discretized on the C-grid and f is defined on the u-point, it becomes :

$$\partial u / \partial t = -g\,\delta_x h + f\,\overline{v}^x$$
$$\partial v / \partial t = -\overline{f\,u}^x$$
$$\partial h / \partial t = -H\,\delta_x u$$

and the relation of dispersion for inertia-gravity waves becomes :

$$\omega^2 = f^2 \cos^2 (k\,\Delta x / 2) + 4\ gH\ k^2 \sin^2 (k\,\Delta x / 2)$$

For the C-grid as in the B-grid, the discretized relation is not far from the analytical solution when the wave number is close to zero (that is when the long wavelength is well solved by the equations). As the wave number increases, the discretized relation drifts away from the analytical one : the numerical waves have a much slower wave speed than the analytical one and for a wavelength equals to the grid spacing, the group velocity is zero : the energy is trapped locally and cannot propagate anymore. This kind of behaviour can easily destabilize a numerical scheme. In this case, C-grid and B-grid are not far from one another but their behaviour can be different for other motions like the planetary waves or for 2-dimensional problems.

Discretization in time

Here again it is useful to examine the equations term by term as there is not a single way to differentiate the operators in function of time. The most widely used scheme for integrating the primitive equations is the leapfrog scheme :

$$u(t+\delta t) = u(t-\delta t) + 2\,\delta t\, F(u(t))$$

Three different time steps enter this scheme which is very simple and of second order accuracy. Two following timesteps are necessary to initialize it. It has one major default : it tends to split the solution between the even and the odd timesteps in a physical solution and a numerical one. So it cannot be used alone. In order to prevent this splitting, it is necessary to use a scheme which will mix the odd and even timesteps. Among the different techniques used, one can cite the Matsuno or Euler-backward scheme which is a two-step damping scheme :

$$u(t+\delta t)^* = u(t) + \delta t\, F(u(t))$$

$$u(t+\delta t) = u(t) + \delta t\, F(u(t+\delta t)^*)$$

or one can use the Asselin filter (Asselin, 1972) :

$$u_c(t) = u(t) + \gamma\,(u_c(t-\delta t) - 2\,u(t) + u(t+\delta t))$$

The leapfrog scheme is used for most terms in the primitive equations except for the dissipative terms for which it is unstable. These terms require different methods. The horizontal dissipation is integrated with a forward timestep which is conditionally stable, implying numerical conditions between the timestep, the grid spacing and the coefficient of horizontal mixing.

$$u(t+\delta t) = u(t-\delta t) + 2\,\delta t\, F(u(t-\delta t))$$

Usually the numerical stability conditions implied by the forward scheme give strong constraints for the vertical diffusion term. It may be overcome with an implicit scheme. This last scheme is unconditionally stable but diffusive. As it implies a matrix inversion, it is the last term calculated in the equations :

$$u(t+\delta t) = u(t-\delta t) + 2\,\delta t\, F(u(t+\delta t))$$

2 - 3 Boundary conditions

Lateral and bottom boundaries

Boundary conditions are necessary to close the primitive equations system. Along lateral walls, no flux of heat and salt are permitted (the gradient of these quantities across the boundary is taken to zero). For velocity, different choices are possible. The most widely used conditions are the no-slip condition ($u = v = 0$ at the the boundary) or the slip condition (the velocity normal to the boundary is null but the tangential velocity is conserved). With the first condition, the boundary acts as a source of vorticity. The choice can affect the solution in the vicinity of the boundary where in most cases, the viscous boundary layers are not well solved by the numerical model.

Surface boundary conditions

The surface conditions depend on the overlying atmosphere (see Gill, 1982). For the wind forcing, usually a monthly mean climatology is used as, for instance, the data of Hellerman and Rosenstein (1983) which has been computed from ships data. The sampling of ships presents a very irregular coverage in space and time and the bulk formulae which allow to compute the wind stress from the wind and other parameters of the planetary boundary layer contain large uncertainties (Blanc, 1987). Considered for a long time as a residual field for the atmosphere, the wind stress from atmospheric models could not be validated with in-situ measurements and bears strong systematic biases. It has been considerably improved during the last five years and is often used to force ocean models. This is still a major source of errors in ocean modelling. Scatterometer on satellites will offer an interesting opportunity to increase our knowledge on this quantity.

Heat exchanges between the ocean and the atmosphere are active. The major source of incoming energy in the ocean is the solar flux which is dependent on the cloud coverage. Its penetration depth depends on the quality of the ocean water (which is linked to the biological activity). In return, the ocean radiates to the atmosphere a long wave heat flux which depends on the sea surface temperature through the Stefan's law. In fact, the long wave exchanges between ocean and atmosphere are complex because of multiple reflections between the ocean surface and the clouds.

The sensible heat flux depends on the temperature difference between the two fluids and on the wind speed. This flux is usually one order of magnitude smaller than the solar flux and the latent heat flux but it can locally reach very strong value (over the Gulf Stream for instance).

The most important flux involved in the air-sea exchanges is the latent heat flux which depends on the relative humidity, the saturated vapor pressure in air and sea and the wind speed. This term is always a loss for the ocean, it has the same order of magnitude than the incoming solar flux but its distribution is very different. It includes a strong feedback between ocean and atmosphere and is a powerful regulator : it prevents the ocean to get too warm in tropical areas for instance. However this flux is poorly estimated over the ocean and presents a lot of small scale variability which is difficult to parameterize. If the ocean is forced with a prescribed heat flux which is too warm, a shallow and stable mixed-layer forms which accumulates the excess of heat and is separated by a strong stratification from water below. The sea surface temperature can then reach unacceptable values (> 35°C !). There is no possibility for the ocean to get rid of this heat without interactive heat fluxes. This is the reason why in all OGCM's a restoring term is included in the surface heat budget.

Haney (1971) demonstrated that the heat budget at the ocean surface could be written in the form :

$$Q / (\rho_o C_p) = - \alpha \ (T - T^*)$$

where T is the sea surface temperature and T* a zonal mean equilibrium temperature. ρ_o is the mean ocean density and C_p is the specific heat. α, the restoring coefficient, is the inverse of a time scale (typically 1/30 days). This formulation was recently reviewed by Oberhüber (1988) who proposed to write the heat budget between ocean and atmosphere with the following linearized form :

$$Q = Q_o - \alpha_o (T - T_o)$$

Q_o is the heat budget determined from observations, α_o is the derivative $(- \partial Q_o / \partial T_o)$ and T_o is the observed sea surface temperature. This formulation ensures that, for a simulated sea surface equal to the observed value, the heat exchanged between ocean and atmosphere will be equal to the observed one.

In many general circulation models (see, for instance, Toggweiler, Dixon and Bryan (1989 a - b) - hereafter refered as TDB- or Semtner and Chervin (1988) - hereafter refered as SC), a formulation close to Haney's (1971) is chosen but with some important differences. The heating term is not considered as an external forcing but a restoring forcing. T* is taken as the observed temperature (from the atlas of Levitus - 1982) and the restoring term is supposed to act over a specified mixed-layer. With this approach, the heat flux that the ocean wants to simulate the observed sea surface temperature presents large differences with the climatological heat flux from the observations (see figure 5 in Sarmiento, 1986) as the computed heat flux should equal zero if the simulated temperature is equal to the observed one. Such differences are very large in tropical oceans for instance where much too cold water appears at the surface in low resolution models.

The sea surface salinity depends on the evaporation minus precipitation budget (E) at the ocean surface. The atmosphere is not directly sensitive to salinity - i.e. E is independent of the quantity of salt in the ocean ; there is no local feedback between ocean and atmosphere involving salt except in regions where salinity becomes a controlling factor for the mixed-layer (like in polar regions during ice formation) but it is important to keep in mind that, on large scale, the salt influences the ocean circulation which in turn modifies the atmospheric circulation. E is the least known quantity exchanged between ocean and atmosphere due to the poor description of the precipitations. It is often specified as a restoring term towards the observed sea surface salinity.

2 - 4 Physical parameterizations

Horizontal dissipation and diffusion

The value of the horizontal dissipation coefficient of momentum (K_{HM}) is directly linked to the horizontal grid spacing : its value is mostly chosen for numerical stability reasons. It is much larger than the molecular viscosity within the ocean and is supposed to simulate the large eddy activity which is not explicitly solved. A Laplacian operator (or bi-Laplacian) is a poor choice to translate the turbulent activity as a source of dissipation within the ocean. This is certainly an area where major efforts have to be pursued in order to develop parameterizations more adapted to describe the physics.

A smaller value can be taken for the coefficient of horizontal diffusion on tracers (K_{HH}) as it is not limited by stability conditions. Its order of magnitude can be chosen in accordance with in-situ large scale observations. The choice of a large value for this coefficient (with the same magnitude than in the momentum equation in coarse resolution GCM's) leads to very irrealistic solutions with a wide boundary layer in the western ocean.

Vertical dissipation and diffusion

The coefficients for vertical diffusion and dissipation (K_{VM} and K_{VH}) are several order of magnitude smaller than the horizontal one. For momentum, it is often chosen as a constant with depth; for tracer in global models (like TDB or SC), it is slowly varying with depth, with a larger value in the deep ocean. A small value (0.3 cm^2/s) is chosen in the upper ocean to prevent the slow diffusion of the thermocline. This leads to a very thin surface layer isolated from the deep ocean. As demonstrated in Bryan (1987), the choice of this coefficient is very important for the global balance of the ocean.

In fact vertical mixing within the ocean presents very different regimes that a linear or constant coefficient cannot simulate. Within the mixed-layer, turbulence is very high and, on the contrary, an isopycnal front is very efficient to block turbulence. A constant coefficient is particularly disastrous for the thin equatorial thermocline. A first effort to remedy this problem was done by Pacanowski and Philander (1981) who introduced a parametrization of the vertical turbulent diffusion in function of the value of the local Richardson number. Their formulation which corresponds to a simple diagnostic approach to estimate the local value of the turbulent kinetic energy (TKE) is able to produce large values of K_{VM} and K_{VT} when the stratification is weak or the vertical shear of currents is large but weak values of K_{VM} and K_{VT} in the thermocline region :

$$K_{VM} = \frac{\nu_0}{(1 + \alpha\, Ri)^n} + \nu_{Mb}$$

$$K_{VT} = \frac{K_{VM}}{(1 + \alpha\, Ri)} + \nu_{Tb}$$

where ν_0 is the maximum value authorized for turbulent diffusion coefficient, α is a parameter chosen to 5 and n to 2. ν_{Mb} is the background value for the coefficient of vertical turbulence on momentum (equal to its molecular value 10^{-6} m^2/s) and ν_{Tb} is the background value for the coefficient of vertical turbulence on tracer (equal to its molecular value 10^{-7} m^2/s).

A detailed comparison with observed temperature and currents shows that, though more efficient than a constant value, the Ri parametrization is not able to generate strong enough mixing in the surface layers and the simulated mixed-layer is always too shallow. The vertical shear of currents is strong in the surface layers and the impact of the wind stress forcing in the ocean is concentrated in a very thin surface layer : the Ekman divergence is strongly marked in the surface layer (Reverdin et al., 1991).

In order to improve simulations, more sophisticated formulations can be introduced in numerical models to compute the vertical diffusion in function of the turbulent kinetic energy and a mixing length (Mellor et Yamada, 1982). Let us present, for instance, the 1.5 closure proposed by Blanke and Delecluse (1993). The mixed-layer model is given by the equations developed by Gaspar et al. (1990) (hereafter GGL). It is based on a closure assumption for the turbulent length scales tested by Bougeault and Lacarrère (1989) in atmospheric cases. With the following definition for TKE :

$$\bar{e} = 0.5\, (\overline{u'^2} + \overline{v'^2} + \overline{w'^2})$$

where the various fields are written as a mean value plus a fluctuation part ($x = \bar{x} + x'$), a prognostic equation is developed for TKE

$$\partial_t \bar{e} = -\partial_z(\overline{ew'} + \rho_0^{-1}\,\overline{p'w'}) - \overline{\mathbf{u}_h'w'}\cdot\partial_z \mathbf{U}_h + \overline{b'w'} - \varepsilon$$

where $b = g\,\rho_0^{-1}\,\rho$ is the buoyancy and ε is the dissipation. The vertical turbulent fluxes are parameterized with the usual mixing coefficients K_{VM} and K_{VT} according to :

$$-\overline{\mathbf{u}_h'w'} = K_{VM}\,\partial_z \mathbf{U}_h$$

$$-\overline{b'w'} = -g\,\rho_0^{-1}\,K_{VT}\,\partial_z\rho = K_{VT}\,N^2$$

where N designates the Brunt-Väisälä frequency. The whole vertical flux over TKE, which includes turbulent mixing and pressure effects, requires the introduction of an additional mixing coefficient K_e defined by :

$$-(\overline{ew'} + \rho_0^{-1}\,\overline{p'w'}) = K_e\,\partial_z\bar{e}$$

Though usually taken equal to K_{VM}, a larger value for K_e was chosen ($K_e = 30\,K_{VM}$) for the tropical model configuration. It improves the model stability and allows a convenient TKE homogenization within the surface layers, without affecting the mixed-layer bottom nor the pycnocline. The value of K_e is poorly documented and likely to depend on the spatial scales solved by the OGCM.

Defining a typical dissipation length l_ε and a typical mixing length l_κ, the dissipation is parameterized following Kolmogorov (1942) and the coefficients K_{VM} and K_{VT} are expressed in function of TKE and l_κ :

$$\varepsilon = C_\varepsilon\,\bar{e}^{3/2}\,l_\varepsilon^{-1}$$

$$K_{VM} = C_\kappa\,l_\kappa\,\bar{e}^{1/2}$$

$$K_{VT} = K_{VM} / P_{rt}$$

where the constants $C_\kappa = 0.1$ and $C_\varepsilon = 0.7$ were chosen by GGL to deal with vertical mixing at any depth and where P_{rt} is the Prandtl number. In order to calculate the mixing length, the closure assumption proposed by Bougeault and Lacarrère (1989) for atmospheric models is used : the mixing length is determined by the distance on which a particle converts its TKE into available potential energy. A linearization of their global expression leads to a simple formulation :

$$l_d = l_u = l_\kappa = l_\varepsilon = 2^{1/2}\,\bar{e}^{1/2}\,N^{-1}$$

which is obtained in a stably stratified region with constant values of the Brunt-Väisälä frequency. For tropical simulations, this simplified calculation of the turbulent length scales is a good approximation.

TKE is diagnosed at the surface and bottom of the ocean. The bottom value for TKE is fixed to a numerical minimal threshold ($\bar{e}_{min} = 10^{-11}\ m^2/s^2$) while the surface value includes a dependence on the wind stress :

$$\bar{e}_{surface} = 3.75\ u^{*2} \text{ with } u^{*2} = \tau / \rho_0$$

The time integration of the TKE equation may lead to negative values because the numerical scheme used does not ensure its positivity. A minimum threshold ($\bar{e}_{min} = 10^{-11}$ m^2/s^2) is prescribed to avoid negative values. The computed values of K_{VM} and K_{VT} have to be greater than the molecular viscosity ($10^{-6}\ m^2/s$) and diffusivity ($10^{-7}\ m^2/s$) respectively.

Robust diagnostic modelling

Due to poor parametrization of turbulent mixing and to inadequate grid spacing, global circulation models cannot keep the structure of the thermocline in a long integration. The diffusion of the thermocline is clear in intermediate levels in the ocean (between 500 and 1000 m) where the ocean gets warmer (see figure 9 in Bryan, Manabe and Pacanowski, 1975 or figure 19 in TDB, 1989-b). In order to remedy this flaw, a technique was introduced by Bryan and Sarmiento (1982) which was called the robust-diagnostic method. It proposes to take advantage from the knowledge of the temperature field to prevent the ocean drift. A diagnostic term is added to the temperature and to the salinity equation :

$$\partial T / \partial t + \ldots = \gamma_d\ (To - T)$$

γ_d is the inverse of a time scale and it determines the necessary time to reach equilibrium. It is chosen to 50 years in TDB and 3 years in SC to speed up the equilibrium of the deep ocean (below 710 m). In the global model from LODYC - OPA -, this parameter is not constant but varies in order to respect some considerations :

-a latitudinal variation is imposed in order to have no restoring force in equatorial regions and to reduce it in polar regions.

The first point has been discussed in Fujio and Imasato (1990). Considering a geostrophic balance, if there is an error $\Delta\rho$ in the density field, the error in the velocity field will be ($\Delta\rho / f$). This will give unacceptable large errors close to the equator. It is thus necessary to keep the equatorial regions free from diagnostic forcing.

In polar regions, the problem is different : the knowledge of in situ temperature is very poor and as most measurements were made when the ocean was ice free, the observed temperature presents a warm bias. It is thus necessary to reduce the impact of the observations in these regions.

-a relaxation to zero is imposed in the vicinity of land in order to reduce its impact in areas where the dynamics is swift. Moreover the boundary layers are poorly resolved in the Levitus' file where the radius of influence is large.

-a variation with depth is made as in TDB. This variation, proposed by Sarmiento and Bryan (1982), has the purpose to reduce the restoring factor in the deep ocean where the typical time scale is longer.

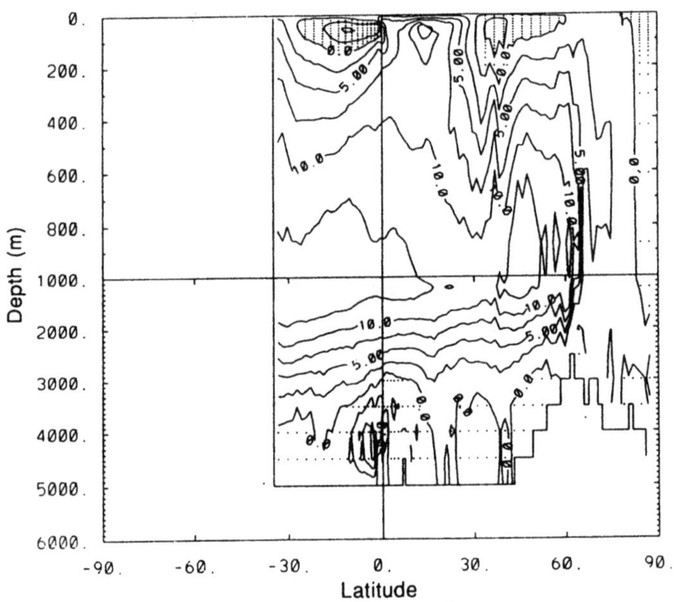

Figure 4 : Streamfunction of the zonally integrated currents in the North Atlantic in OPA. Dotted areas correspond to negative values. The contour interval is 2.5 Sv (1 Sv = 10^6 m^3/s)

The robust diagnostic technique is very efficient to prevent the temperature drift in intermediate waters but it produces artificial sources of heat within the ocean. In that respect, it cannot be kept to study the climatic evolution of the ocean. It has also undesirable effects on the ocean dynamics : it tends to prevent the deep water formation by convection by stabilizing too much the water columns (see the figure 4 in TDB-1989a which contrasts the overturning circulation for the global ocean with the prognostic model which has a strong temperature drift but an active convection and the robust diagnostic model which has reduced convective processes). Though it is true that convective processes are strongly limited by the robust diagnostic technique, the quantity of North Atlantic deep water transported to the South can be restored by the diagnostic term. If one compares the overturning circulation in the North Atlantic in the OPA model (figure 4) and the SC model, the transport of North-Atlantic deep water to the South is very active (more than in TDB) though both these models have a robust diagnostic equation for the temperature and moreover the SC model does not have the Arctic ocean.

The robust diagnostic technique may be helpful to stabilize the deep ocean to present conditions. It should not be kept in areas strongly interacting with the atmosphere or where drastic changes are suspected in a climate variability study.

Acceleration of convergence

The heat capacity of the ocean is one thousand times larger than the heat capacity of the atmosphere so that the ocean adjusts to a heating anomaly one thousand times more slowly than the atmosphere : one is tempted to speed up the oceanic time integration by a factor of one thousand. However the baroclinic speed which controls the internal adjustment within the ocean is 30 times slower in the ocean than in the atmosphere : the speed up of the ocean may only reach a factor of 30.

In order to take advantage of this discrepancy, Bryan (1984) designs a numerical technique which allows to accelerate the spin up to convergence by working with a distorded physics. Two parameters were introduced. The first one α permits to work with a distorded time scale between the momentum equation and the tracer equation. The distorded time t' is equal to : $t' = t / \alpha$

The second parameter γ corresponds to a distorded stratification and it allows the timestep in the tracer equation to vary on a same vertical : $N'^2 = N^2 / \gamma$

The consequences of this distorded physics was analyzed for typical ocean motion :

-the free waves have a slower phase speed

-their meridional structure is slightly modified

-the growth rate of baroclinically unstable waves is reduced but there is a wider range of instability.

The technique is efficient for searching an equilibrium state in coarse resolution models. However its application is not suitable for many oceanic problems :

-it cannot be used for transient or time evolving problems (in particular, it is very questionable to keep this technique when studying a mean seasonal cycle)

-it cannot be used in high resolution models where baroclinically unstable processes are important.

-the use of γ variable with depth does not keep the conservation of T and S in a water column.

3 - Some examples of ocean modellings

The modelling dilemma : space or time ?

The first problem faced by oceanographers is linked to the current limited capacity of supercomputers. To solve the ocean circulation with the typical length scale of ocean dynamics requires a grid spacing of at least 10 km. To bring the ocean thermohaline circulation to equilibrium requires time scales of a few hundreds of years : it is impossible to respect both these criteria on present day computers. 100 000 grid-points integrated during 1000 years represent several years of CRAY 2. This is the main reason why the regional modelling, favouring process studies with high resolution, has been so popular in oceanography. The multiple aspects of the dynamical regional modelling which have considerably helped our understanding of the ocean will not be discussed here but one example will be given for the circulation in equatorial basin. Then the global ocean circulation will be discussed.

3 - 1 Simulation of the equatorial Pacific circulation

As pointed above, regional modelling has been intensely developed in recent years. Equatorial regions present a very specific density profile. The surface layers contain very warm water and are separated from the cold waters below by a sharp thermocline. The incoming solar radiation is very large in these regions which globally return a large flux in sensible and latent heat to the atmosphere. These fluxes are important in the trade winds regions and in the regions of gustiness (convective areas). The net heat budget is close to zero in warm areas with an equilibrium temperature near 29°C (Newell, 1986) but the equatorial oceans act as a very efficient heat pump in the upwelling areas where the trade winds maintain a divergent ocean circulation. In these regions, cold water goes up to the surface and the latent heat flux comes to zero. These areas gain a tremendous amount of heat rapidly exported poleward by the divergent circulation.

The equatorial oceans can adjust very rapidly to a modification in the atmospheric circulation for the equator is a wave guide where the phase speed is very large : 1m/s for a Rossby wave, 3 m/s for a Kelvin wave. Both the nullification of the Coriolis force at the equator and the sharp equatorial stratification contribute to define a large internal radius of deformation (300 km) and a quick phase speed. 200 days only are necessary to bring the Pacific ocean close to equilibrium. This has to be compared to the adjustment time in mid-latitudes basins which is of the order of tens of years ! The internal radius -300 km- is nicely solved with a grid spacing of 30 km at the equator and the turbulent activity is explicitly simulated. Moreover it takes a reasonable computing time to bring the ocean to equilibrium.

Another favourable factor is the recent and numerous observational programs in tropical regions so that the data sampling is more suitable for validation exercise than in mid-latitudes. Theoretical advances have provided the needed framework to understand the ocean processes. Moreover the equatorial Pacific undergoes a large scale climatic anomaly each 2 to 10 years which is the most spectacular evidence of the natural and large scale air-sea coupling.

For these reasons, tropical modelling has experimented a bunch of efforts in recent years. A intercomparison study on the mean seasonal cycle in the Pacific ocean (Stockdale et al., 1993) has shown that some biases appeared very systematically : warming of the summer hemisphere, cold upwelling tongue, warming of eastern coastal areas ... These errors are neither specific of the Pacific ocean nor of the selected atmospheric forcings : they also appear (though with different amplitude) in tropical Atlantic modelling (Reverdin et al, 1991) and they more or less remain the same when running with different atmospheric forcings (Dandin, 1991). It is therefore natural to look at some of these biases as deficiencies in the physics of the ocean model.

In order to investigate these points, the LODYC Pacific model was run with a penetrative solar radiation and two different parameterizations of vertical mixing; the first one in function of the local Richardson number (RI) and the second one in function of the Turbulent Kinetic Energy (TKE). The SST difference, in annual mean, between the two runs (TKE minus Ri), shows that the ocean is colder over large interior areas by 0.5 to 1.0 °C, but the cold tongue is warmer, indicating a less active upwelling. The decrease in SST can reach 1 to 2°C along the eastern coasts. This is mainly due to a better efficiency in upper layer mixing which prevents the formation of warm surface ponds isolated from the deeper layers by a strong stratification : a true mixed-layer, with nearly homogeneous temperature profile, exists in the first layers of the simulated ocean. The TKE formulation also prevents strong vertical diffusion at the basis of the thermocline : the thermocline structure is very pinched along the equator.The Ekman drift is no more concentrated in a thin surface layer but presents a much deeper extension. Consequently, the surface currents are reduced : this is particularly important for the equatorial divergence which is nearly divided by a factor of 2 with this new scheme. Figure 5 contrasts the streamfunction of the currents integrated in longitude with the physics dependent on the Richardson number and with the TKE physics. Changes are very important within a few degrees of the equator : the upwelling circulation is drastically reduced and there is a deepening of the overturning cells with TKE. Despite a less intense vertical circulation, surface water is still cold as it comes from deeper levels ; the equatorial upwelling is displaced eastward.

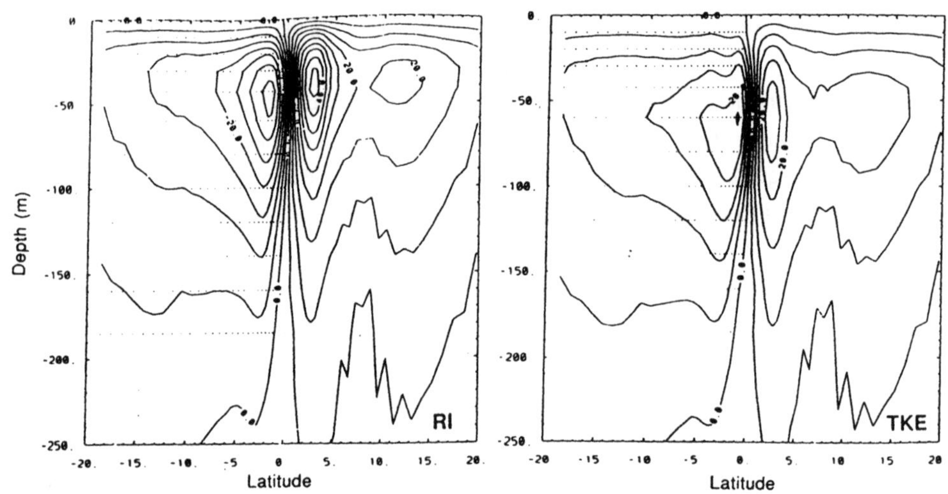

Figure 5 : Streamfunction of the zonally integrated currents in the tropical Pacific simulated with the vertical diffusion coefficient depending on the RI and on TKE. The contour interval is 5 Sv. Dotted areas correspond to negative values.

With the new parametrization, currents are strongly modified. Westward surface currents have an increased vertical penetration in the central and western equatorial Pacific due to the enhancement of vertical mixing in regions where mixed-layer is deep. On the contrary, they are reduced in the eastern Pacific, where the low vertical mixing below the thermocline does not damp the strength of the equatorial undercurrent, permitting an erosion by the vertical advection of eastward momentum.

The most spectacular signature appears on the Equatorial Undercurrent (EUC) This current is very sensitive to small changes in the equatorial balance. In the western and central parts of the basin, a deeper mixed-layer leads to a deepening of the zonal pressure gradient which is a powerful term to accelerate the EUC. The increase of vertical turbulence is more or less compensated by the zonal pressure gradient. As the EUC moves eastward, the thermocline is upwelled to the surface and the vertical mixing is decreasing at the EUC core level : the acceleration of the EUC persists further and maximum speed is reached in the eastern tropical ocean.

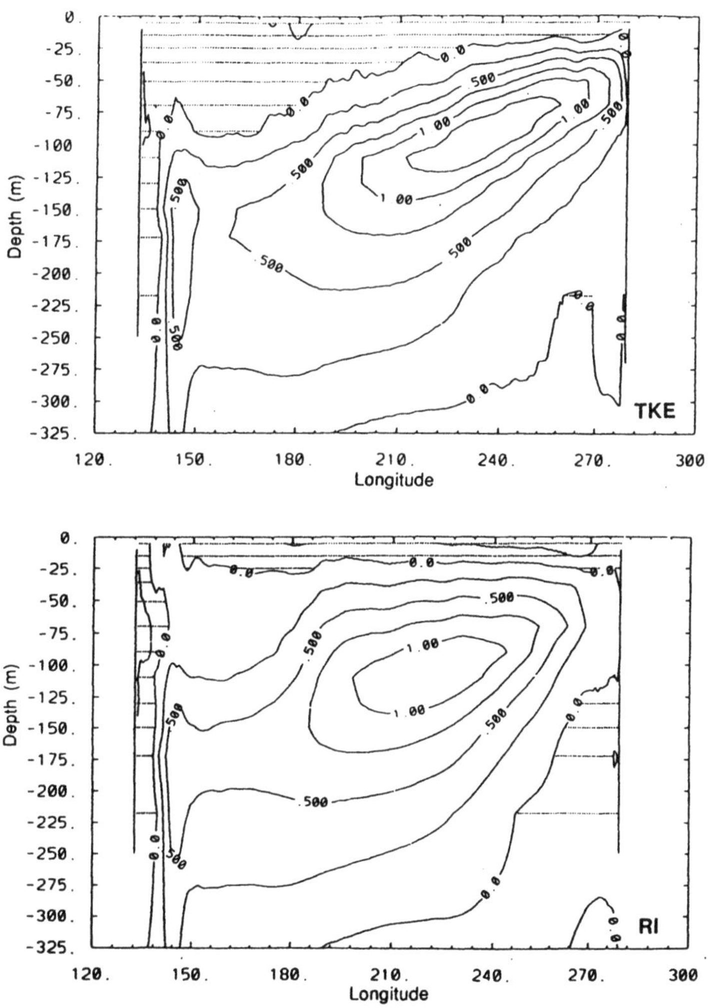

Figure 6 : Equatorial section of the annual mean of zonal current simulated with the vertical diffusion coefficient depending on the local Richardson number and on TKE. The contour interval is 0.25 m/s and dotted areas correspond to westward current.

By modifying the distribution of the vertical turbulent fluxes of heat and momentum in the ocean, significant changes are induced in the circulation. Surface currents are slightly reduced with a deeper extension of the Ekman circulation, sub-surface currents are intensified, vertical circulation is reduced but reaches deeper levels. Consequently, a new balance for the surface heat reservoir appears and this highlights the sensitivity of the ocean response to internal parametrization. The response of a coupled model, where the air-sea feedbacks increase small differences, may evolve very differently with different physical parametrizations.

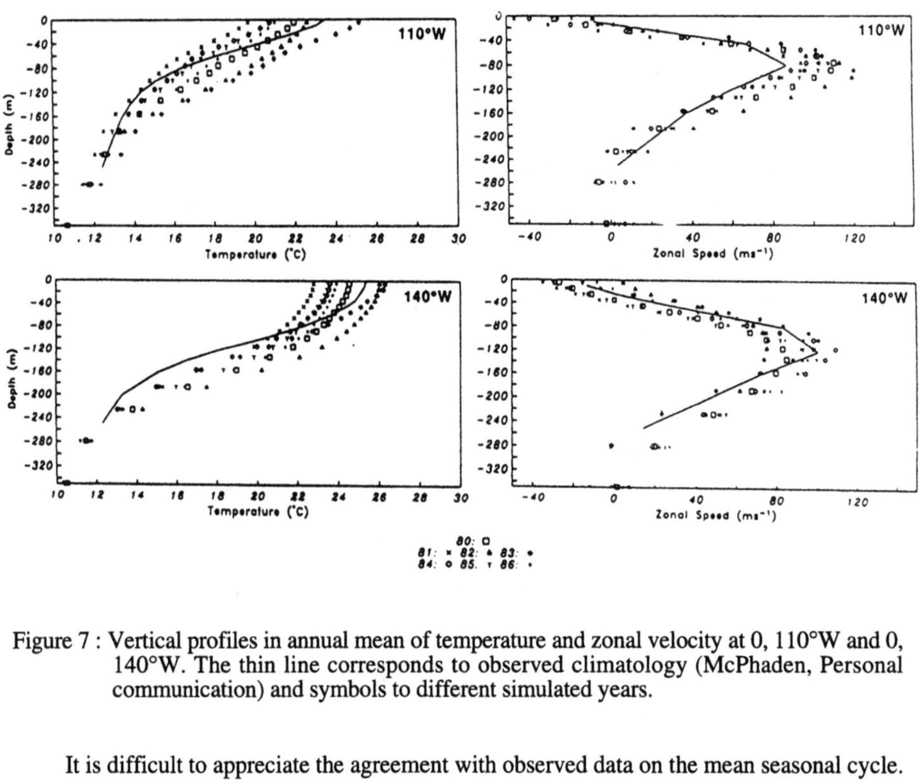

Figure 7 : Vertical profiles in annual mean of temperature and zonal velocity at 0, 110°W and 0, 140°W. The thin line corresponds to observed climatology (McPhaden, Personal communication) and symbols to different simulated years.

It is difficult to appreciate the agreement with observed data on the mean seasonal cycle. A reduction of surface divergence seems in agreement with surface drifters but the sampling is barely adequate to study the seasonal variability. In order to appreciate the performances of the Pacific model with TKE, some comparisons were made between simulated profiles at two different locations -140°W and 110°W- along the equator and moorings data (McPhaden and McCarty, 1992)- see figure 7.

The simulation exhibits a cold upwelling season due to strong zonal wind stress component during the boreal summer. This cold bias affects the seasonal and the interannual cycle and is clearly linked to a deficiency in the atmospheric forcing (here, the outputs from the EMERAUDE climate model from Meteofrance forced by observed SST from 1979 to 1987). The ocean model has a large interannual variability in the surface and subsurface fields. The temperature profile shows more than 5°C of variabilty in the upper thermocline and simulated profiles surround the observed profile. 1983 is the warmest porfile, in agreement with observations. The velocity profiles are also in good agreement with the observed ones. The shear is very strong between the SEC and the EUC. Magnitude and depth of the EUC core are close to the observed ones but the EUC seems too strong in the far east of the equatorial Pacific. The surface current can quickly reverses from west to east and this reversal is particularly frequent during 1982 and 1983. This run is presently used to describe the El-Niño episode of 1982-83.

3 - 2 A global ocean circulation model

In order to qualify what can be expected from present models of global circulation, some characteristics of the model OPA built in LODYC will be discussed and explained. The resolution adopted for this global ocean model can be contrasted by two different kinds of models also developed to study the large scale ocean circulation : the low resolution global ocean model developed in GFDL for large scale climate studies (Toggweiler, Dixon and Bryan (1989 a - b)) and the high resolution ocean model developed by Semtner and Chervin (1988) for studying the present simulation with a resolution of half a degree. The last model is not global but extends to 65°N. Neither of these models has a high enough resolution to describe the meso-scale activity within the ocean. SC model solves part of the spectrum but half a degree is too large to explicitly solve the dynamics in mid and high latitudes.

The numerics of these models are different : SC and TDB are written on a B-grid and OPA on a C-grid. This last model uses the tensorial formalism and works on a distorded grid where the coordinate pole is displaced in Asia out of the computational domain (figure 1). In order to avoid the pole problem, SC stop the model at 65°N and TDB use a Fourier filtering.

For OPA, the ocean mesh follows the coastline as closely as possible : very shallow seas are included in the model. This choice may be a source of worries for it is difficult to solve the barotropic streamfunction on steep topography, but it is worth to keep shallow waters where rapid water transformation can happen. Moreover this choice is in better agreement with the land-ocean distribution adopted in atmospheric GCM's. The land is separated from the ocean by a mask technique. As these 3 models have different horizontal resolutions, large differences appear in the definition of topography. This is a crucial problem for ocean circulation as basins communicate through very thin passages. The width of the Drake Strait can be doubled for numerical reasons and this bears consequences on the solution : the control of the water volume depends on the topographic section and water masses properties can be largely affected by the wide extent of a strait. Difficult areas are the straits in Indonesia between the Pacific ocean and the Indian ocean, the straits between the Nordics seas and the North Atlantic ocean, the Bering strait... The first control exerted over the flow is given by the contours of toppography : it is important to have a good resolution of ocean ridges and straits.

The rigid lid approximation which is used in these three models implies the computation of the barotropic potential on each islands of the domain. This potential may be fixed as a constraint or freely evolving calculated with an elliptic equation (Semtner, 1974). The barotropic streamfunction is firstly computed for a domain where all the land points have the same potential. Then, the potential of each island is evaluated. When an island is linked to coast, the flux between the island and the coast is assumed to be zero. Main connections between the oceans have to evolve freely. In OPA, there are many other islands : New-Zealand, Madagascar, West Indies, New Guinea, Philippines, Kalimantan, Iceland, Spitzbergen... and the Bering strait is open. In TDB, Madagascar and New-Zealand are not separated islands. TDB consider only two islands and SC three (Antarctica, New zealand, Australia-New guinea.

	TDB	OPA	SC
Δ x (in degree)	3.75	.5 to 2.	0.5
Δ y (in degree)	4.5	.5 to 2.	0.5
Levels number	12	30	20
K_{HM} (in m^2/s)	250 000	2 000-15 000	1 000
K_{HH} (in m^2/s)	1 000-500	2 000	1 000
K_{VM} (in cm^2/s)	20	TKE	$100 / (1+5 \, Ri)^2$
K_{VH} (in cm^2/s)	250 000	TKE	$K_{VM} / (1+5 \, Ri)$
$1/\gamma_d$ (in years)	50	1 to 0.15	3 (above 710 m)
Acceleration $\alpha\gamma$	6 to 180	8 to 11	7

Table 1: Characteristics and parameters used in three different models of the ocean global circulation.

A summary of the different characteristics of these models is given in table 1. A large number of vertical levels appears in OPA to treat the TKE equation. These levels are not equidistant but the distance between levels increases with depth. This choice is made to respect the temperature structure in the oceans, with a thin upper layer overlying a stratified region and a thick bottom layer nearly homogeneous.

Analyses of the ocean global circulation model are an important and difficult task. The main difficulty provides from the fact that we are far from being able to describe the global circulation as our knowledge of the oceanic fields is poor. The results can be discussed by comparison with in-situ observations (temperature, currents, salinity) or by confronting simulated fields with theories on ocean circulation. The best observed variable in the ocean is the temperature field. However, as was noticed before, a prognostic integration of the temperature leads to diffuse temperature profiles and sluggish circulation. When using the robust diagnostic technique, temperature is no more available to validate the simulation and the ocean circulation has to be evaluated through the currents distribution. Direct estimations of currents are scarce and local. Integrated transport (horizontal or vertical) can be calculated and compared to estimations. It is also possible to make an indirect description of the ocean circulation from analyses of passive tracers (like the CFC or the tritium/helium).

Barotropic streamfunction

The barotropic streamfunction should verify the model proposed by Sverdrup (1947), in presence of an eastern coastline. The figure 8 shows the isolines of the barotropic streamfunction simulated from the model currents.

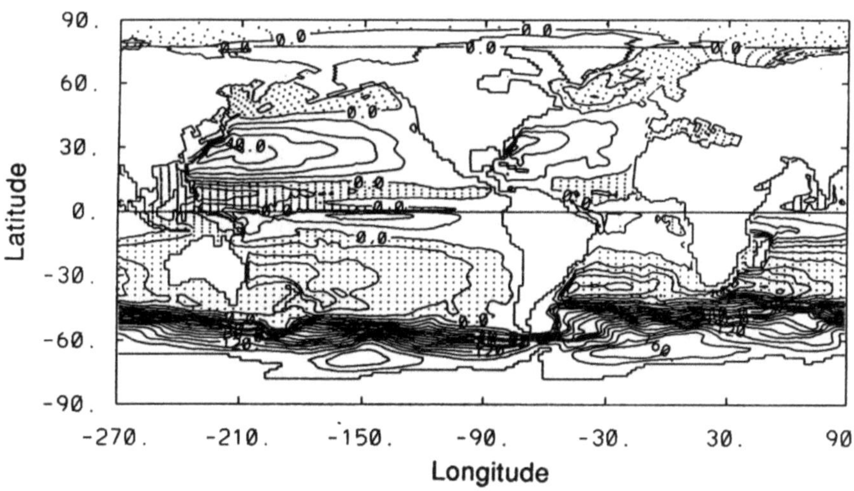

Figure 8 : Streamfunction of the vertically integrated currents simulated by OPA. Dotted areas correspond to negative values. The contour interval is 10.0 Sv (1 Sv = 10^6 m^3/s).

Over most of the oceans, the agreement with theory is quite good except in the Indian ocean, south of 10°S where the analytical Sverdrup's model cannot take into account the contribution of the throughflow between the Indian and the Pacific oceans. Of course, there is no value from this theory in the Southern Oceans. Close to the western boundaries, the transport values calculated from the Sverdrup's balance correspond to a linear return flow, which is clearly too small compared to observations ; the values simulated by the model have the same bias. They correspond to a linear estimation of the circulation along the western boundary and weak values of coastal currents. This is a common drawback of all global ocean models. In the oceanic interior, the flow is basically in geostrophic balance as expected. The streamfunction reaches 140 Sv in the Drake passage, 11 Sv in the throughflow and 0.8 Sv in the Bering strait - in clear agreement with observations.

Vertical overturning

In a global model, heat, salt and water transport must reach a global balance with the external forcings. A common way to evaluate the models is to look at the streamfunction of the zonally averaged circulation which represents the vertical overturning in the ocean though only a qualitative picture can been drawn from the data. The figure 9 represents the overturning streamfunction calculated in the OPA model and should be compared to the figure 4 in TDB-1989a and to figure 2 in SC.

The global surface circulation is marked by an equatorial upwelling which brings up 70 Sv to the surface. The main contribution comes from the Pacific ocean (60 Sv). These figures are clearly smaller than previous estimations from global oceans. In fact, the drastic reduction of the equatorial upwelling is due to the use of the TKE scheme to compute vertical diffusion in OPA.The magnitude of this upwelling is in better agreement with in-situ estimation (Wyrtki - 1981- advanced between 37 and 51 Sv for the equatorial Pacific). Upwelled water comes from depths until 600 - 1000 m, especially in the southern ocean.

30°N and 30°S mark the position of the convergence where water dives to intermediate levels. Between 1000 and 2000 m the circulation is mainly towards the South and indicates the presence of the North Atlantic deep water. Its contribution is mostly important in the Atlantic ocean where the southward transport between 1000 and 2000 m can reach 15 Sv. Note that, as discussed before, part of this water is not formed in OPA by convective processes. It is globally form in the northern Atlantic within the water masses in order to adjust the observations. Despite this unphysical mechanism, its path through the world ocean is well marked.

Deeper in the ocean, the bottom water created along the Antarctica shelf is slowly moving to the North. It represents nearly 15 Sv of water moving to the North. Once again the source of water formation is only partly represented but its effect on the global circulation seems in agreement with estimation.

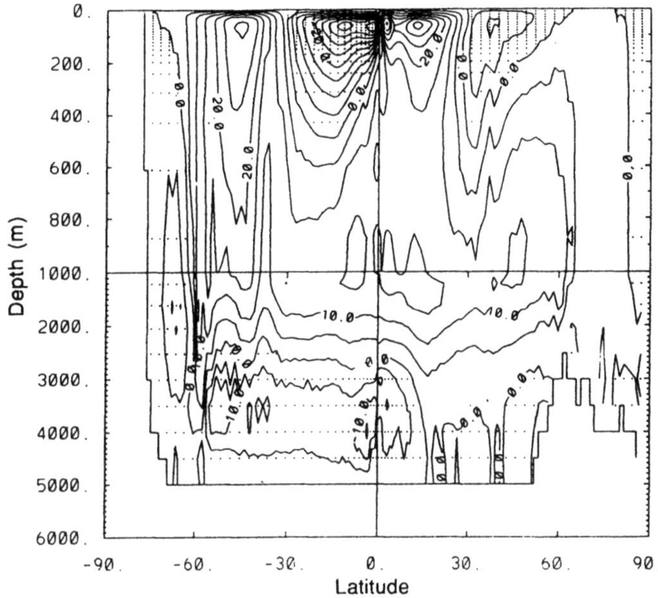

Figure 9 : Streamfunction of the zonally integrated overturning in the global ocean in OPA. Dotted areas correspond to negative values. The contour interval is 5.0 Sv (1 Sv = 10^6 m^3/s)

SC model develop a similar simulation. Its equatorial upwelling is much stronger but the North Atlantic deep water and the bottom water have the same order of magnitude than in OPA. Both of them are stronger than the prognostic model of TDB.

Northward heat transport

The ocean circulation transports as much heat as the atmospheric circulation. It has been evaluated by different methods : as a residual from the heat budget of the overlying atmosphere, from hydrographic sections in the ocean and from ocean models. All these evaluations give a heat transport which is poleward in the Pacific (between 1 and 3 Petawatts), southward in the Indian Ocean and northward in the whole Atlantic ocean but large discrepancies exist on its magnitude. In the North Atlantic where the in situ data have certainly the densest coverage, the uncertainty on the transport estimation is as large as the transport itself. Nevertheless, its sign is consistent.

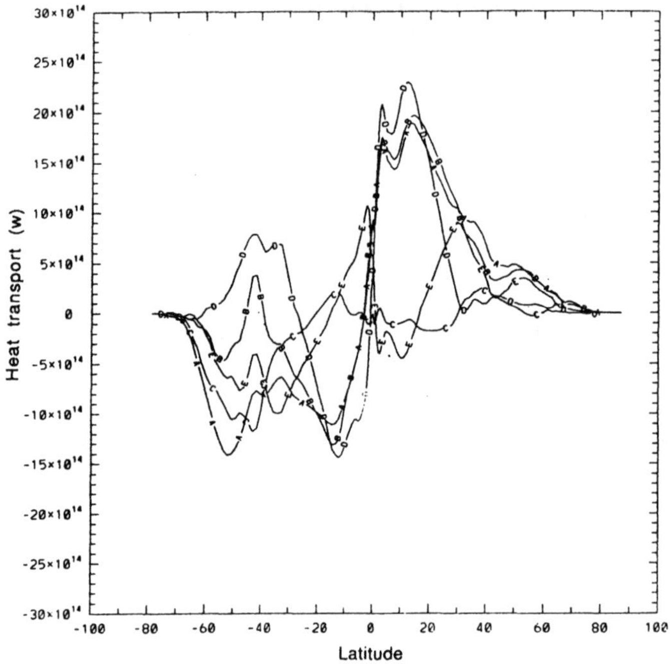

Figure 10 : Northward oceanic heat transport for the global ocean in OPA. Curve A corresponds to the total transport, curve B to the advective contribution, curve C to the diffusion contribution, curve D to the overturning contribution and curve E to the horizontal gyre contribution.

OPA (figure 10) and SC have very similar global heat transport. The equator marks the boundary between northward and southward transports. The maximum of transport is reached in subtropical latitudes and attain 1.5 PW (1 Petawatt = 10^{15} Watts) in the northern hemisphere and nearly 1.0 PW in the southern hemisphere. There is an increase in the southward transport South of 50°S. The major contribution to this transport comes from the overturning term (the vertical circulation). This term corresponds to the equatorial divergence of heat in surface layers and drops drastically the transport in the convergence areas at 30°S and 30°N. South of 30°S, its contribution is opposite to the mean transport. On the contrary, the gyre circulation (the horizontal circulation) tends to oppose the mean transport in tropical region and to positively contribute to it in southern latitudes. The contribution of eddies to the heat transport in the oceans is much debated. In situ measurements (Bryden - 1979 and 1982) indicate that the eddies can locally transport as much heat as the mean circulation. However numerical studies made by Bryan (1986) seem to indicate that the time averaged contribution of the eddies to the heat transport is compensated by modifications of the mean global circulation so that an internal compensation exist in global models : coarse resolution models and high resolution models seem to have nearly the same heat transport (though not for the same physical reasons).

The resolution may not be the most important factor to determine the mean heat transport. In fact, recent studies from Bryan (1987) have shown that the northward heat transport is very sensitive to the parametrization of the vertical mixing. Its value is nearly proportional to the value of the vertical mixing coefficient used in the ocean model when the heat flux is able to adjust. The adjustment of vertical diffusion within the ocean will be a critical task for coupled climate model.

4 - Conclusion

20 years of experiments with numerical models of the ocean circulation based on the primitive equations have considerably helped our understanding of their behaviour and have built increasing confidence in their outputs. Primitive equation models are very successful in regional modelling at high resolution. They are particularly adapted and extensively used to study the equatorial circulation. They are also used in multiple process studies and in theoretical approaches.

Nevertheless, the global approach of the ocean circulation remains a very difficult task. Though the knowledge and the understanding of the simulated ocean circulation is growing, the results are still far from the observed ocean and there is a crucial need to improve physical processes in numerical codes. Among many problems, one can refer to vertical turbulent diffusion, convection and static stability, topography and continental shelf... and one major problem which is the horizontal turbulent diffusion. Present computers cannot support to bring to equilibrium a global ocean model with the tenth degree resolution required to solve the meso-scale activity. It is thus necessary to search for adequate parameterizations but the present understanding of oceanic turbulence makes it difficult. Besides physical improvements, a careful check has to be done on the numerical algorithms used in different codes as the physics depends on the numerical tools used to write it.

Modelling the global ocean still offers a challenge to numerical mathematics and supercomputers. How can one speed up an ocean code ? These codes require long integration and large memories. Moreover the computations have to be accurate. New architectures like the massively parallel computers may offer an alternative way to accelerate the integrations.

Yet through long term experiments and high resolution modelling, present computers have brought to the oceanographers huge amounts of data which have considerably modified the understanding of the ocean. Regional and scattered observations are no more considered for themselves but also as pieces of a global framework. Global observing and modelling of the ocean is the challenge nowadays.

Acknowledgments. The author like to thank Gurvan Madec, Maurice Imbard, Claire Lévy, Bruno Blanke, Philippe Dandin and Olivier Marti who have helped to run the different model cases presented here. Support for computations were provided by the Conseil Scientifique of the CCVR (Centre de Calcul Vectoriel pour la Recherche).

References

Asselin R (1972) Frequency filter for time integration, Mont. Weath. Rev., vol. 100, N° 6, 487-490.

Blanke B and P Delecluse (1993) Low frequency variability of the tropical Atlantic ocean simulated by a general circulation model with mixed layer physics, J. Phys. Oceanogr., 23, 1363-1388.

Batteen M L and Y-J Han (1981) On the computational noise of finite-difference schemes used in ocean models, Tellus, 33, 387-396.

Blanc T V (1987) Variation of bulk-derived surface flux, stability, and roughness results due to the use of different transfer coefficient schemes, J. Geophys. Res., 92, 3867-3876.

Bougeault P and T Lacarrère (1989) On the stability of the third-order turbulence closure for the modelling of the stratocumulus-topped boundary layer, J. Atmos. Res., 88, 4579-4592.

Bryan F (1987) Parameter sensitivity of primitive equation ocean general circulation models, J. Phys. Oceanogr., 17, 970-985.

Bryan K (1969) A numerical method for the study of the circulation of the world ocean, J. Comput. Phys., 4(3), 347-376.

Bryan K (1984) Accelerating the convergence to equilibrium of ocean-climate models, J. Phys. Oceanogr., 14, 666-673.

Bryan K (1986) Poleward buoyancy transport in the ocean and mesoscale eddies, J. Phys. Oceanogr., 16, 927-933.

Bryan K, S Manabe and R C Pacanowski (1975) A global ocean-atmosphere climate model. Part II. The oceanic circulation, J. Phys. Oceanogr., 5, 30-46.

Bryden HL (1979) Poleward heat flux and conversion of available potential energy in the Drake Passage, J. Mar. Res., 37, 1-22.

Bryden HL (1982) Sources of eddy energy in the Gulf Stream recirculation region, J. Mar. Res., 40(4), 1047-1068.

Chartier M (1985) Un modèle numérique tridimensionnel aux équations primitives de la circulation générale de l'océan, Thèse de l'université Pierre et Mairie Curie, CEA Report R-5372, 111 pages.

Courant, Friedrichs and Lewy (1928) Über die partiellen differenzengleichungen der mathematischen physics, Math. Annalen, 100, 32-74.

Dandin P (1991) Modélisation de l'océan Pacifique tropical. Rapport interne, LODYC.

Fujio S and N Imasato (1990) Diagnostic calculation for circulation and water mass movement in the deep Pacific, J. Geophys. Res., 96, 759-774.

Gaspar P, Y Gregoris and JM Lefevre (1990) A simple eddy-kinetic-energy model for simulations of the ocean vertical mixing : tests at station Papa and Long-Term Upper Ocean Study Site site, J. Geophys. Res., 95, 16179-16193.
Gill A E (1982) Atmosphere and ocean dynamics. Academic Press, 662 pp.
Haidvogel D B, J Wilkin and R Young (1991) A semi-spectral primitive equation ocean circulation model using vertical sigma and orthogonal curvilinear horizontal coordinates, J. Comput. Phys., 94, 151-185.
Haney R L (1971) Surface thermal boundary condition for ocean circulation models, J. Phys. Oceanogr., 1, 241-248.
Hellerman S and M Rosenstein (1983) Normal monthly wind stress over the world ocean with error estimates, J. Phys. Oceanogr., 13, 1 093-1 104.
Kolmogorov A N (1942) The equation of turbulent motion in an incompressible fluid, Izv. Akad. Nauk. SSSR, Ser. Fiz., 6, 56-58.
Levitus S (1982) Climatological atlas of the world ocean, NOAA Professional paper, 13, Washington D.C.
McPhaden M J and M McCarty (1992) Mean seasonal cycle and interannual variations at 0, 110°W and 0, 140°W during 1980-1991. Tech. Rept, 95, ERL-PMEL, NOAA.
Mellor G L and T Yamada (1982) Development of a turbulent closure model for geophysical fluid problems, Rev. Geophys and Space Phys., 20, 851-875.
Mesinger F and A Arakawa (1976) Numerical methods used in atmospheric models, GARP Publication, N° 17.
Newell R E (1986) An approach towards equilibrium temperature in the tropical eastern Pacific, J. Phys. Oceanogr., 16, 1338-1342.
Oberhüber J M (1988) An atlas based on the 'COADS' data set: The budget of heat, buoyancy and turbulent kinetic energy at the surface of the global ocean, Max-Planck-Institut für Meteorologie, Hamburg, Rept 15.
Oberhüber J M (1990) Simulation of the Atlantic circulation with a coupled sea-ice-mixed layer-isopycnal General Circulation Model. Max-Planck-Institute für Meteorologie, Rept 59.
Pacanowski R and S G H Philander (1981) Parametrization of vertical mixing in numerical models of tropical ocean, J. Phys. Oceanogr., 11, 1443-1451.
Pedlosky J (1987) Geophysical Fluid Dynamics, Springer Verlag, 710 pp.
Reverdin G, P Delecluse, C Lévy, P Andrich, A Morlière and J M. Verstraete (1991) The near surface Atlantic in 1982-1984: results from a numerical simulation and a data analysis, Prog. Oceanogr., 27, 273-340.
Sarmiento J L (1986) On the north and tropical Atlantic heat balance, J. Geophys. Res.,91, 11677-11698.
Sarmiento J L and K Bryan (1982) An ocean transport model for the North Atlantic, J. Geophys. Res., 87, 394-408.
Semtner A J Jr (1974) An oceanic general circulation model with bottom topography, Tech. Rep. 9, 99 pp., Dep. of Meteorol., Univ. of Calif., Los Angeles.
Semtner A J Jr and R Chervin (1988) A simulation of the global ocean circulation with resolved eddies, J. Geophys. Res., 93, 15 502-15 552.
Stockdale T, D Anderson, M Davey, P Delecluse, A Kattenberg, Y Kitamura, M Latif and T Yamagata (1993) Intercomparison of tropical Pacific ocean GCM's. WMO Rep.
Sverdrup H U (1947) Wind-driven currents in a baroclinic ocean : with application to the equatorial currents in the eastern Pacific, Proceedings of the National Academy of Science, 33, 318-329.
Toggweiler J R, K Dixon and K Bryan (1989) Simulation of radiocarbon in a coarse-resolution world ocean model, 1, Steady state prebomb distributions, J. Geophys. Res., 94, 8217-8242.
Toggweiler J R, K Dixon and K Bryan (1989) Simulation of radiocarbon in a coarse-resolution world ocean model, 2, Distributions of bomb-produced carbon 14, J. Geophys. Res., 94, 8243-8264.

Unesco (1983) Algorithms for computation of fundamental property of sea water, UNESCO Tech. Paper in Marine Science 44, 53 pp.
Wajsowicz R C (1986) Free planetary waves in finite-difference numerical models, J. Phys. Oceanogr., 116, 1-17.
Wyrtki K (1981) An estimate of equatorial upwelling in the Pacific, J. Phys. Oceanogr., 11, 1205-1214.
Yin F L and I Y Fung (1991) Net Diffusivity in General Circulation Models With Nonuniform Grids, J. Geophys. Res., 96, C6, 10 773-10 776.

PRECESSION, ECCENTRICITY, OBLIQUITY, INSOLATION AND PALEOCLIMATES

A. Berger and M.F. Loutre
Université Catholique de Louvain
Institut d'Astronomie et de Géophysique G. Lemaître
2 Chemin du Cyclotron
B-1348 Louvain-la-Neuve, Belgium

Abstract. Different parameters related to the solar radiation received at the top of the atmosphere are derived and their long-term variations analysed in the frequency domain. The annual mean energy received by the whole Earth is varying in time according to the variations of the mean distance from the Earth to the Sun, i.e. as a function of $(1-e^2)^{-\frac{1}{2}}$. The so-called 100 ka period of the eccentricity (and the other periods as well) are originating from a combination of the fundamental periods of the climatic precession parameters. The insolation at a given latitude and for a fixed longitude of the Earth on its orbit is a function of obliquity through the factor related to the zenith angle of the Sun and of precession through the distance factor. Moreover a deeper analysis of the spectrum of the distance factor shows that it contains also, with much less power, half precession periods, eccentricity periods and combination tones between eccentricity and precession. Over the Quaternary, the latitudes of the polar and the tropical circles, varying with obliquity, are situated respectively between 65.5° and 68° and between 22° and 24.5°. Their present-day motion towards north is estimated to be 14.4 m per year. Finally, it is shown that in most insolation parameters, the precessional signal dominates the obliquity one, except in high polar latitudes mainly of the winter hemisphere, although the power of the obliquity signal increases from low to high latitudes.

1 Insolation frequencies in paleoclimatic records

Since the late 1960's, judicious use of radioactive dating and paleomagnetic techniques gradually clarified the Pleistocene time scale (Broecker et al., 1968; Bard et al., 1990). Better instrumental methods came on the scene for determining oxygen isotopes for ice-age foraminifera relics (Emiliani, 1966), ecological methods of core interpretation were perfected (Imbrie and Kipp, 1971), global climates in the past were reconstructed (CLIMAP, 1976, 1981) and atmospheric general circulation models and climate models became available (Kutzbach, 1985; Crowley, 1988; COHMAP members, 1988; Berger et al., 1990; Gallée et al., 1992). With these improvements in dating and interpreting geological data in terms of paleoclimates, with the increased use of computers and the development of astronomical and climate models, a more critical and deeper investigation of the astronomical theory is now possible.

NATO ASI Series, Vol. I 22
Long-Term Climatic Variations
Edited by J.-C. Duplessy and M.-T. Spyridakis

According to the astronomical theory of paleoclimates (Berger, 1988), the perturbations in the Earth's orbital parameters change sufficiently the latitudinal distribution and the seasonal pattern of insolation received at the top of the atmosphere to trigger climate variations at the 10 to 100 ka time scales. The Milankovitch theory (1941) is a particular version of this more general astronomical theory of paleoclimates, Milankovitch using the caloric seasons to explain the long-term climatic variations during the Quaternary.

From spectral analysis of climate sensitive indicators extracted from selected deep-sea records of the southern Indian Ocean, Hays, Imbrie and Shackleton demonstrated for the first time in 1976 that the astronomical frequencies (corresponding to the 100, 41, 23 and 19 ka periods) are significantly superimposed upon a general red noise spectrum. It is the geological observation of the bipartition of the precessional peak (23 and 19 ka were found instead of the usual 21 ka), confirmed in the astronomical computations made independently by Berger (1977a), which was one of the first most delicate and impressive of all tests for the Milankovitch theory. This landmark paper also showed that the 100-ka period contribution to the total variance far exceeds that expected from a simple linear relationship between insolation and ice volume and that there is a fairly consistent phase relationship between insolation, sea surface temperature and ice volume, each preceded the next by 2 to 4 ka.

Since 1976, spectral analysis of climatic records of the past 800 ka or so, has provided substantial evidence that, at least near the frequencies of variation in obliquity and precession, a considerable fraction of the climatic variance is driven in some way by insolation changes accompanying the perturbations of the Earth's orbit (Imbrie and Imbrie, 1980; Imbrie et al., 1984; Imbrie et al., 1989; Berger, 1989a).

However, the shape of the spectrum depends also upon the location of the core and the nature of the climatic parameter analysed (Hays et al., 1976). For example, analysis of geologic time series from the subpolar North Atlantic (Ruddiman and McIntyre, 1984) indicates that SST fluctuations, a monitor of the oceanic polar front, varied with a 41-ka period, a response consistent with a dominant influence by the high latitude tilt signal of the Milankovitch caloric insolation. Farther south, along the latitude of the southern terminus of the Laurentide Ice Sheet (about 45°N), SST varied on a 23-ka period. Milankovitch cycles can also be found in eolian sediments and mid-and high-latitude pollen records. Over the North Pacific, a long time series of eolian transport from interior eastern Asia records significant fluctuations at the 100-ka and 41-ka periods. Janececk and Rea (1984) suggest that this 41-ka period result from tilt variations that cause changes in equator-to-pole temperature gradients (and hence wind). A pollen series from Grande Pile, France (Woillard, 1978), yields spectra dominated by 23-ka precession power, but also with indications of significant higher frequencies (Molfino et al., 1984). A series of analyses indicate that North Atlantic Deep Water also fluctuated at 41-ka periods with NADW lagging obliquity by about 8,000 years (Boyle and Keigwin, 1985/1986).

In low latitudes, climate fluctuations have more 23-ka power on both land and sea. In the Arabian Sea, an upwelling index related to the Asian monsoon has 23 ka fluctuations that lag precession by about 5 000 years (Prell, 1984). An index of African aridity has been developed based on abundances of african freshwater diatoms in equatorial Atlantic sediments which fluctuation have clear 23-ka power (Pokras and Mix, 1987). McIntyre et al. (1989) found very strong 23-ka power in a foraminifera record from the eastern equatorial Atlantic. There is also 41-ka power in this series, a response that may reflect a tilt peak at the equator according to Short et al. (1990).

There is also a growing body of information available as to fluctuations of various components of the carbon cycle. For example, the ice core CO_2 record has significant power in Milankovitch bands especially at 21 ka, although the temperature spectrum indicates periods around 40 and 20 ka, with the 40 ka dominating the 20 ka (Barnola et al., 1987; Jouzel et al., 1987). Shackleton and Pisias (1985) demonstrated that benthic $\delta^{13}C$, a measure of whole-ocean carbon storage, increased in the glacials with the fluctuations coherent and in phase with ice volume changes. Their measures of ocean productivity have significant 23-ka power, although equatorial Atlantic productivity indices measured by Curry and Crowley (1987) have 100-ka and 41-ka signals but no 23-ka power.

In the Mediterranean Pliocene, rhythmic lithological variations in the Trubi and Narbone Formations of Sicily and Calabria show cycles that could be related to precession and eccentricity (Hilgen, 1987). In particular, the precession cycle corresponds well with the mean duration of the deposition of basic rhythmites, which comprise small-scale, bipartite or quadripartite marl rhythms in the Trubi and individual sapropelitic layers in the Narborne Formation. The 100 and 400 ka cycles of eccentricity would match the average duration of both layer-scale marl rhythms in the Trubi, which approximately comprise 5 and 20 basic rhythmites respectively, as well as small-scale and large-scale clusters of sapropelitic layers in the Narbone Formation. However, consistent discrepancies in the Rossello composite section (Pliocene of Sicily, Hilgen and Langereis, 1989) could argue for an age re-assignment of the Gilbert and Gauss chrons based on phase relations between the sedimentary and supposedly corresponding orbital cycles.

A similar conclusion is shown by Shackleton et al. (1990) who proposed that the currently adopted radiometric dates for the Matuyama-Bruhnes boundary, the Jaramillo and Olduvai Subschrons and the Gauss-Matuyama boundary underestimate their true astronomical age by between 5 and 7 %. The basis for such a revised time scale is the stronger contribution from the precession signal in the record from ODP Site 677.

There is also evidence that the orbital variations were linked to climate at periods shorter than 100 ka during the past few hundred million years (Fischer, 1986; Berger, 1989b). This appeared at times when major ice masses were probably absent. Walsh power spectra of the Blue Lias Formation (basal Jurassic) show two cycles with duration less than 93 ka which may record changes in orbital precession and obliquity (Weedon, 1985/86). Carbonate production in pelagic mid-Cretaceous sediments, quantified by calcium carbonate and optical densitometry time series, reflects the orbital eccentricity and precessional cycles (Herbert and Fisher, 1986). Fourier analysis of long sections of the Late Triassic Lockatong and Passaic formation of the Newark Basin show periods in thickness corresponding roughly to the astronomical periodicities (Olsen, 1986). All these interesting results has encouraged research on the stability of the solar system in order to determine to which extent the changing Earth-Moon distance, for example, influenced the length of the main astronomical periods (Berger et al., 1989).

Besides the usual 100, 41, 23 and 19 ka-periods which characterize most of the late Pleistocene data, Ruddiman et al. (1986) succeeded in finding in the geological records one of the secondary astronomical periods that was already predicted by Berger (1977a), the 54 ka one. A similar period of 58 ka was found in a 400 ka record of the paleomagnetic field from Summer Lake in South-central Oregon (Negrini et al., 1988). Finally, combination tones have also been found in deep-sea cores with a high sedimentation rate (Pestiaux et al., 1988) and in the Vostok ice core over the last 140,000 years (Yiou et al., 1991).

A fairly coherent phase relationship was also reasonably well defined between insolation and ice volume in Kominz and Pisias (1979) where obliquity consistently lead the ^{18}O

record by about 10,000 years, whereas precession seemed to be in phase with the 23 ka geological signal. However, the recent results obtained by SPECMAP show that these leads and lags are more complicated. CLIMAP (1976) and more recently the SPECMAP (Imbrie et al., 1984) teams have indeed shown that the phase lags in the climate response to orbital forcing depend upon the nature of the climatic parameters themselves and upon their geographical location. For example in their data, the sea-surface temperature of the southern oceans seems to lead the response of the northern hemisphere ice sheets by roughly 3,000 years. The results of the SPECMAP Atlantic study reveal the lead of southern hemisphere responses ahead of the northern hemisphere and highest sensitivity of the high-latitude North Atlantic in all astronomical bands (Imbrie et al., 1989). Spectral analysis of a number of Atlantic time series illustrates quite well this spatial dependence of the frequency response of SST records, most of them showing distinct frequencies in the primary Milankovitch bands (Imbrie et al., 1989). In the obliquity band, there is relatively little response in low latitudes, except in a few records near the equator. In the precession band, there is a suggestion of a phase shift accross the equator; South Atlantic temperature changes lead oxygen isotopes (ice volume) and North Atlantic responses are either in phase or lag ice volume. The southern hemisphere phase leads also occurs in the 100-ka band. All (but one) SST phases at 100-ka are within 12,000 years of ice volume and eccentricity maxima.

This complexity is not too surprising as these spectra depend upon the way the climate system reacts to the insolation forcing and upon which type of insolation it is sensitive too. Contrary to the well-received Milankovitch idea that the high polar latitudes must record the obliquity signal (as shown in the Vostok core, for example, Jouzel et al., 1987) whereas low latitudes record only the precessional one, the latitudinal dependence of the insolation parameters is more complex. Clearly the mid-month high-latitude summer insolation displays a stronger signal in the precession band than in the obliquity one (Berger and Pestiaux, 1984).

2 The 100-ka Cycle

As already shown by Hays et al. (1976), the variance components centered near a 100 ka cycle, which dominates most upper Pleistocene climatic records, seem to be in phase with the eccentricity cycle (high eccentricity at low ice volume). Unfortunately, the interpretation of the data is not always as clear. The 100 ka cycle, so dominant a feature of the late Pleistocene record, does not exhibit a constant amplitude over the past 2-3 million years. This periodicity seems to disappear before 10^6 years ago, at a time the ice sheets were much less developed over the Earth, reinforcing the idea that the growth of the major ice sheets may have played a role in the modulation of the 100 ka cycle. Spectral analysis of DSDP Hole 552A reveals indeed a dominant quasi-periodicity associated with obliquity-induced temperature variations in surface water and weaker peaks at the eccentricity and precession periods (Backman et al., 1986). More precisely, Shackleton et al. (1988) found a 100 ka-peak for the period 0 to 1 Ma BP, a 100 ka and a 41 ka-peaks for 0.5 to 1.2 Ma BP and mainly a 23 ka-one for 2.5 to 3.5 Ma BP, this section displaying however significantly less variance.

In any case, the amount of insolation perturbation at 100 ka is not enough to cause a direct climate change of ice-age magnitude. The exceptional strength of this cycle calls, therefore, for a stochastic (Hasselman, 1976; Kominz and Pisias, 1979) amplification of the insolation forcing, or for a non-linear amplification through the deep ocean circulation, the carbon dioxide, the ice sheet related feedbacks (Birchfield and Weertman, 1978; Pollard, 1984), the isostatic rebound of the lithosphere (Oerlemans, 1980) and/or the ocean-ice interactions. The 100 ka climatic cycle can indeed be explained both (i) from the eccentricity signal directly, provided an amplification mechanism can be found (as in the double potential theory of Nicolis 1980, 1982 and Benzi et al., 1982); and/or (ii) by a beat between the two main precessional components as shown by Wigley (1976) and Imbrie and Imbrie (1980) from non-linear climate theory. It is therefore very important to clearly explain the spectrum of the insolation and of the astro-climatic parameters which are used in climate model.

Recent ice sheet models show also that the 100 ka cycle can be simulated with (Ghil and Le Treut, 1981; Saltzman et al., 1984) or without (Lindzen, 1986) internal free oscillations related to resonances when astronomically forced. It is significantly reinforced when isostatic rebound (Hyde and Peltier, 1985) and iceberg calving are taken into account (Pollard, 1982).

Finally, it must be remembered that Milankovitch requested a high eccentricity for an ice age to occur - which is just the reverse of the correlation claimed by Hays et al. (1976). Considering only the caloric insolations developed to the first degree of eccentricity, Milankovitch viewed indeed the effect of the eccentricity through the precessional parameter alone.

Facing all these problems related to the stability of the frequencies in space and time and to the origin of the 100 ka cycle, related to eccentricity or to precession, it seems worth to give some fundamentals about the formulas used to compute the insolations which force directly the climate system. This is expected to help identifying more easily the origin of the frequencies found in geological data.

3 Astronomical origin of the eccentricity cycles

The eccentricity, e, of the Earth's orbit around the Sun is a measure of its shape given by : $e^2 = \frac{a^2-b^2}{a^2}$, where a and b are the semi-major and semi-minor axes of the elliptical orbit. It is presently 0.016 and it has changed between 0 and 0.07 over the Quaternary with a highly variable periodicity which mean is roughly 100 ka (Berger, 1976). Its computation is obtained from solving the Lagrange equations of the motion of the planets around the Sun (Berger, 1977b). The trigonometrical series derived by Berger (1978), from the Bretagnon (1974)'s analytical solution of this system, is given by:

$$e = 0.0287 + \Sigma E_k \cos(\lambda_k t + \phi_k) \quad (1)$$

where the amplitudes, E_k, the mean rates, λ_k, and phases, ϕ_k are given in Table 3 of Berger (1978). The periods of the six terms corresponding to the largest amplitudes are recalled in Table 1. It is also interesting to give the corresponding periods of a more recent solution (Berger and Loutre, 1991) obtained from Laskar (1988)'s integration of the planetary system. The comparison between the two sets of values show the excellent accuracy with which these periods are presently determined.

Table 1: Period of the most important terms in the serie expansions of eccentricity, e, $e \sin\pi$, climatic precession, $e \sin\tilde{\omega}$ (Berger, 1978) and eccentricity again from Berger and Loutre (1991).

	Berger, 1978			Berger-Loutre, 1991
	eccentricity	$e \sin\pi$	$e \sin\tilde{\omega}$	eccentricity
1	412 885	308 043	23 716	404 178
2	94 945	176 420	22 428	94 782
3	123 297	72 576	18 976	123 818
4	99 590	75 259	19 155	98 715
5	131 248			130 615
6	2 035 441			2 379 077

This series (1) is originating from the expansion of

$$e = \sqrt{(e \sin\pi)^2 + (e \cos\pi)^2} \qquad (2)$$

where the two terms $e \sin\pi$ and $e \cos\pi$ involve the longitude, π, of the perihelion measured from a fixed vernal equinox of reference (taken as the one of 1950 A.D.; for the explanation of the astronomical parameters see Figures 1 and 2). For the 1978 solution, the expansion of $e \sin\pi$ comes from the solution by Bretagnon (1974) and the amplitudes, mean rates and phases were calculated and given in Table 4 of Berger (1978). The four most important ones are recalled in Table 1. It can easily be shown that each period, PE, of the eccentricity is obtained as a combination of 2 periods, PP, of $e \sin\pi$, through:

$$\frac{1}{PE_k} = \frac{1}{PP_i} - \frac{1}{PP_j} \qquad (3)$$

So the first six periods PE_k ($1 \leq k \leq 6$) of the eccentricity in (1) are respectively given by the combination of the periods PP numbers 2 and 1, 3 and 1, 3 and 2, 4 and 1, 4 and 2, 3 and 4.

But e can also be obtained through a relation similar to (2) involving the climatic precession, $e \sin\tilde{\omega}$, where $\tilde{\omega}$ is the longitude of the perihelion measured from the vernal equinox at a given date, t:

$$e = \sqrt{(e \sin\tilde{\omega})^2 + (e \cos\tilde{\omega})^2} \qquad (4)$$

This climatic precession is a measure of the Earth-Sun distance at the summer solstice (see later formula 25). Its mean periodicity is 21 ka but its analytical expansion (or

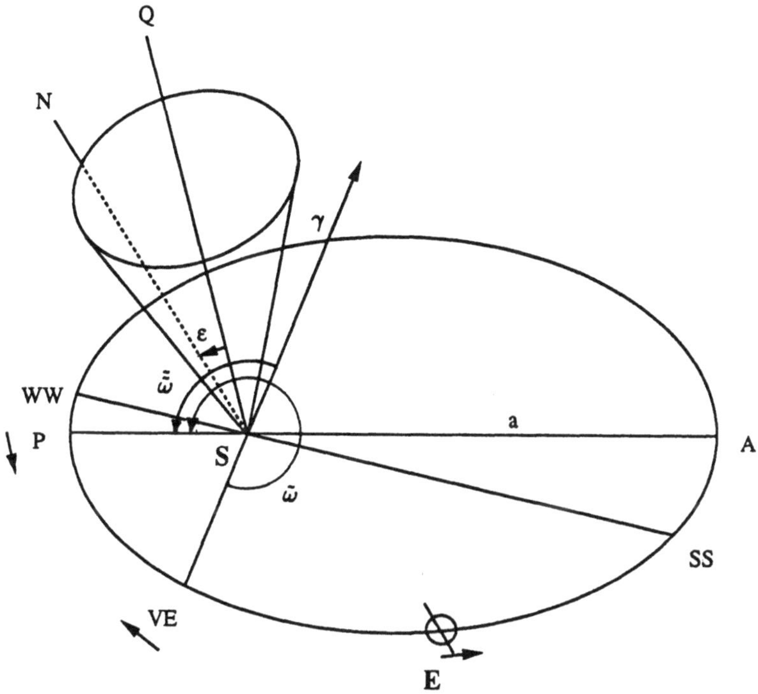

Figure 1: Elements of the Earth's orbit.

The orbit of the Earth, E, around the Sun, S, is represented by the ellipse PEA, P being the perihelion and A the aphelion. Its eccentricity e is given by $\frac{(a^2-b^2)^{1/2}}{a}$, a being the semimajor axis and b the semiminor axis. WW and SS are the winter and summer solstices, respectively; γ is the vernal point. WW, SS, and γ are approximately located where they are today. SQ is perpendicular to the ecliptic, and the obliquity ε is the inclination of the equator upon the ecliptic; i.e., ε is equal to the angle between the Earth's axis of rotation SN and SQ. Parameter $\tilde{\omega}$ is the longitude of the perihelion relative to the moving vernal equinox, and is equal to $\pi + \psi$. The annual general precession in longitude, ψ, describes the absolute motion of γ along the Earth's orbit relative to the fixed stars. The longitude of the perihelion, π, is measured from the reference vernal equinox of A.D. 1950 and describes the absolute motion of the perihelion relative to the fixed stars. $\tilde{\tilde{\omega}} = \tilde{\omega} - 180°$, is the longitude of the perihelion measured from γ; its numerical values are calculated in Berger (1978).

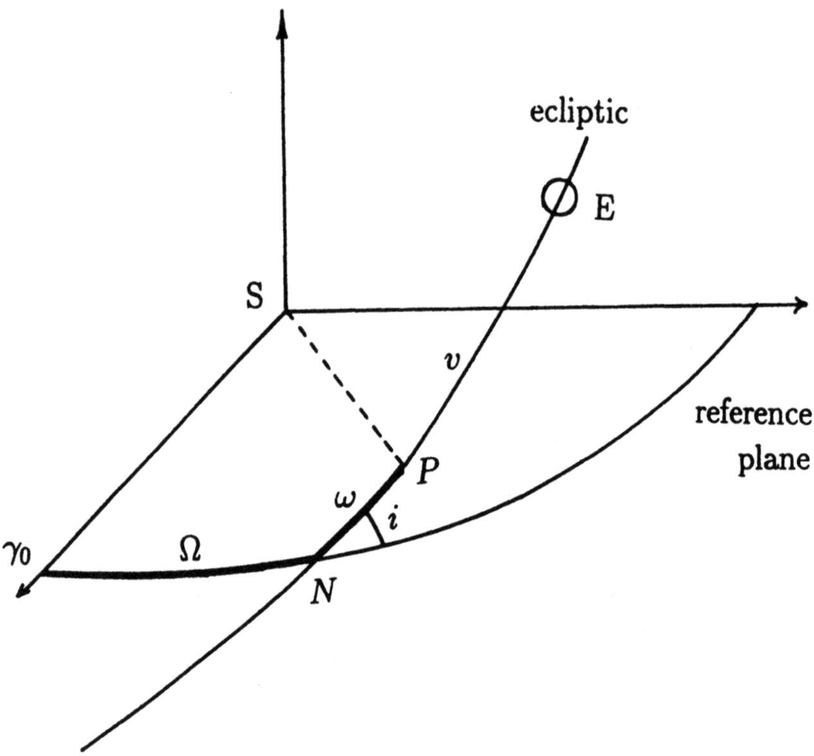

Figure 2: Position of the Earth (E) around the Sun (S).

In astronomy, it is usual to define an orbit and the position of the body describing that orbit by six quantities called the elements. Three of them define the orientation of the orbit with respect to a set of axes, two of them define the size and the shape of the orbit (a and e respectively), and the sixth (whith the time) defines the position of the body within the orbit at that time. In the case of a planet moving in an elliptic orbit about the Sun, it is convenient to take a set of rectangular axes in and perpendicular to the plane of reference, with the origin at the centre of the Sun. The x-axis may be taken towards the ascending node N, the y-axis being in the plane of reference and 90° from x, while the z-axis is taken to be perpendicular to this reference plane so that the three axes form a right-handed coordinate system. γ_0 is the reference point from where the angle are measured. As the reference plane is usually chosen to be the ecliptic at a particular fixed date of reference (named epoch of reference in celestial mechanics; Woolard and Clemence, 1966), γ_0 is, in such a case, the vernal point at that fixed date (the vernal point is also referred to as the First Point of Aries indicating the position of the Sun when it crosses the celestial equator from the austral to the boreal hemisphere). P is the perihelion; Ω the longitude of the ascending node; ω, the argument of the perihelion; $\pi = \Omega + \omega$ the longitude of the perihelion; i the inclination; v the true anomaly; $\lambda = \pi + v$ the longitude of the Earth in its orbit.

spectrum in the frequency domain) given by:

$$e \, \sin\tilde{\omega} \;=\; \Sigma \, P_k \, \sin(\alpha_k t \;+\; \zeta_k) \qquad (5)$$

shows clearly two main periods averaging close to 23 ka and 19 ka (see Table 1 of this paper or Table 2 of Berger, 1978).

The calculation of $\tilde{\omega}$ is more complex than of π as it requests to consider also the motion of the Earth's axis of rotation. This involves the astronomical precession (the so-called general precession in longitude, ψ) and the obliquity, ϵ. The astronomical precession is describing the absolute motion of the vernal equinox in time and its period to circle around the Earth's orbit is roughly 26,000 years. The combination of the absolute motion (counterclockwise) of the perihelion given by π and of the absolute motion (clockwise) of the vernal equinox given by ψ provides the relative motion of the perihelion against the vernal equinox, $\tilde{\omega} \;=\; \pi \;+\; \psi$. On the other hand, the obliquity is the tilt of the equator on the Earth's orbit. Its present value is $23^o27'$. It has varied over the Quaternary from $22^o30'$ to 25° approximately, with a mean period of 41 ka.

Although the origin of the period of each term in (5) is difficult to trace back (Berger and Loutre, 1990), it is also possible to use a relation similar to (3) where the PP's are now those related to $e \, \sin\tilde{\omega}$ to find the origin of the PE_k. These PE_k ($1 \leq$ k ≤ 6) related to the α_k of (5) are obtained respectively from the combination of the following periods of $e \, \sin\tilde{\omega}$: 2 and 1, 3 and 1, 3 and 2, 4 and 1, 4 and 2, and 3 and 4. This technique was used by Berger (1973), noted partly from numerical investigation by Stothers (1987) and demonstrated fully in a comprehensive paper by Berger and Loutre (1990).

So very clearly, the eccentricity periods in celestial mechanics can be expressed as combinations of the precessional periods themselves.

4 Energy received by the whole Earth over a full year

If S_a is the solar energy received by unit of time on a surface of unit area perpendicular to the Sun rays and situated at the distance a from the Sun (a is equal to the semi-major axis of the Earth's orbit), the total energy W received by the Earth at the distance r

from the Sun is equal to

$$W_E = S_a \left(\frac{a}{r}\right)^2 \pi R^2 \tag{6}$$

R being the Earth's radius.

Over a complete year of length T, the energy, W_E^T, received by the whole Earth, will be

$$W_E^T = \int_0^T W_E dt = S_a \, \pi R^2 \int_0^T \left(\frac{a}{r}\right)^2 dt \tag{7}$$

As we see from (7), we must compute the mean value of $(\frac{1}{r})^2$ over the whole year:

$$< \frac{1}{r^2} > = \frac{1}{T} \int_0^T \frac{1}{r^2} dt$$

This can easily be done by using the second law of Kepler

$$\frac{r^2}{2} \frac{dv}{dt} = \frac{\pi a^2}{T} \sqrt{1-e^2} \tag{8}$$

where v is the true anomaly of the Earth on its elliptical orbit (Figure 2).

Replacing (8) in (7) leads to:

$$W_E^T = S_a \, \pi R^2 \int_0^{2\pi} \frac{T}{2\pi} \frac{1}{\sqrt{1-e^2}} dv$$

Assuming that e is constant over one year (we recall that e varies between 0 and 0.07 in 100,000 years at the maximum), we obtain:

$$W_E^T = \frac{S_a}{\sqrt{1-e^2}} \pi R^2 T \tag{9}$$

As a is secularly invariable (at least to an excellent accuracy as demonstrated in celestial mechanics), T is also invariable as seen from the third law of Kepler:

$$\frac{4\pi^2}{T^2} a^3 = const.$$

Therefore, W_E^T is only a function of the energy output from the Sun through S_a and of the eccentricity of the Earth's orbit through $(1-e^2)^{-\frac{1}{2}}$.

The integration of (6) shows that the mean value of $(\frac{a}{r})^2$ over one year is equal to $(1-e^2)^{-\frac{1}{2}}$. If we denote r_m the constant value of r which satisfies the following relation:

$$< \frac{a^2}{r^2} > \equiv \frac{1}{T} \int_0^T \left(\frac{a}{r}\right)^2 dt = \frac{1}{\sqrt{1-e^2}} \tag{10}$$

we obtain

$$r_m^2 = a^2 \sqrt{1-e^2} \tag{11}$$

r_m can also be considered as the radius of a circle which has the same area as the elliptical orbit of the Earth around the Sun. r_m is thus the mean value of r, energetically speaking, because the total energy received by the Earth over a whole year (9) is the same as that received by a fictitious Earth revolving around the Sun at a constant distant r_m (see Appendix 1 for the arithmetic mean of r). If we denote S_0 the energy received by unit of time on a unit area perpendicular to the Sun rays and situated at this mean distance r_m from the Sun, S_0 is the so-called solar constant and we have:

$$4\pi a^2 \, S_a = 4\pi r_m^2 \, S_0$$

which means:

$$S_a = S_0 \, (\frac{r_m}{a})^2$$

Through (11) we obtain $S_a = S_0 \sqrt{1-e^2}$

or

$$S_0 = \frac{S_a}{\sqrt{1-e^2}} \tag{12}$$

S_0 is, actually, not a true constant and it would be more appropriate to call it solar parameter, for example. It depends indeed (see (12)) upon the solar output, which is determined from observations and studied from theoretical investigations of the solar activity and luminosity, and upon the mean distance from the Earth to the Sun.

The energy received by the whole Earth over one year becomes

$$W_E^T = S_0 \, \pi R^2 T \tag{13}$$

(13) is evidently equivalent to (9) except that (13) is an implicit function (not explicit as in (9)) of e through (12).

If we denote by $\overline{W}_E$ the energy received per unit of the Earth's surface and per unit of time calculated for a full year, (9) and (13) give:

$$\overline{W}_E = \frac{S_0}{4} = \frac{S_a}{4\sqrt{1-e^2}} \tag{14}$$

which is presently estimated to be roughly 340 Wm^{-2} (Lean, 1991).

Although the absolute effect of the factor $(1-e^2)^{-1/2}$ is relatively small (a few per mil at the most), this term increases the annual global insolation at high eccentricity and decreases it at low eccentricity. As compared to present-day value ($e = 0.016$), W_E increases by 0.27 % (3.7 Wm^{-2} over 1360 Wm^{-2}) for $e = 0.075$ and decreases by 0.01 % (0.17 Wm^{-2} over 1360 Wm^{-2}) for $e = 0$. This behaviour is coherent with the Hays et al. (1976) and others' result showing, over most of the upper Quaternary, low ice volume at high eccentricity. It is obvious that if we simplify (9) or (14) by omitting the term in e^2, this kind of behaviour cannot be seen: in such a case, r_m is approximated by a, as it was unfortunately too frequently assumed.

Let us finally note that the expansion of $(1-e^2)^{-\frac{1}{2}}$ to the second degree in eccentricity leads to:

$$\overline{W_E} = \frac{S_a}{4} \left(1 + \frac{e^2}{2}\right) \tag{15}$$

and therefore the variation around a reference value denoted by the subscript 0 can be approximated by:

$$\Delta\overline{W}_E = \overline{W}_E - \overline{W}_{E,o} = \frac{S_a}{8} \Delta(e^2) = \frac{S_a}{4} \bar{e} \Delta e \tag{16}$$

where $\bar{e} = \frac{e+e_o}{2}$ and $\Delta e = e - e_o$

5 Insolation of the Earth at a given instant of time and latitude

At a given instant during the course of the year, corresponding to a declination δ of the Sun and to a distance r, the insolation, W, received on a horizontal surface located at a latitude ϕ is given by (Figure 3 and Berger et al. (1993b) for more details):

$$W = S_a \left(\frac{a}{r}\right)^2 \cos z \tag{17}$$

where z is the zenith distance of the Sun given by:

$$\cos z = \sin\phi \sin\delta + \cos\phi \cos\delta \cos H \tag{18}$$

- δ is related to the true longitude of the Sun, λ, by:

$$\sin\delta = \sin\lambda \sin\epsilon \tag{19}$$

Over one year, δ varies between two extreme values, $-\epsilon$ and $+\epsilon$, whereas λ varies from 0 to 360°. But because of (19) and the long-term variations of ϵ, the insolation

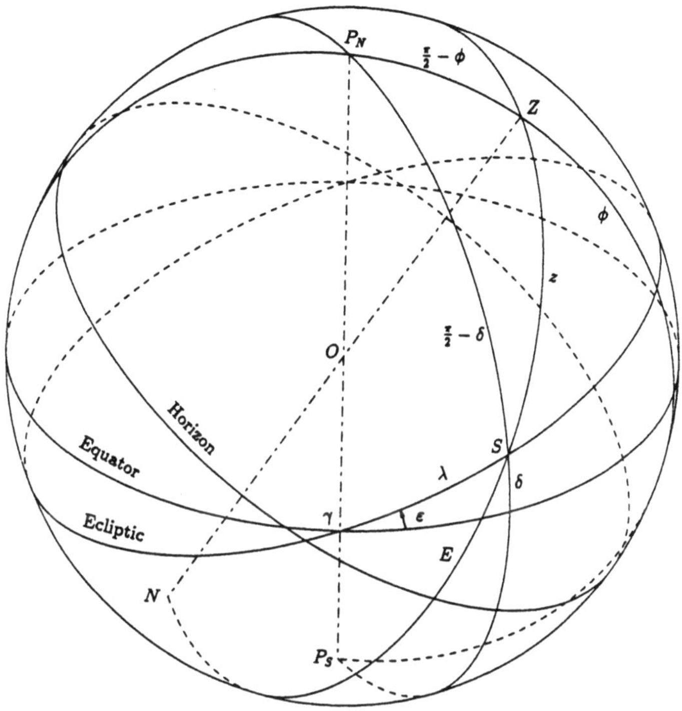

Figure 3: Position of a point (S) on the celestial sphere (Woolard and Clemence, 1966). The astronomical horizon is the great circle in which the celestial sphere is intersected by the plane that passes through the observer (O) and is perpendicular to the direction of the local gravity. The zenith (Z) is the point vertically upward or overhead. The nadir (N) is the point diametrically opposite the zenith. The vertical circles are the great circles through the zenith and nadir, and therefore necessarily perpendicular to the horizon. The two diametrically opposite points on the celestial sphere, which mark the ends of the axis of the apparent diurnal rotation of the sphere are called celestial poles (P_N, P_S). The great circle of the celestial sphere midway between the poles is the equator. Great circles through the poles are called secondary circles. The zenith distance (z) of any point on the celestial sphere is the angular distance from the zenith measured along the vertical circle through the given point. It varies from 0 to 180°. It is complementary to the altitude E. The position of any point on the celestial sphere can also be measured by the angle between the meridian (great circle through the poles, the zenith and the nadir) and the secondary through the point. This angle is called hour angle (H). The declination (δ) is the angular distance of the point from the equator measured on the secondary. The latitude of the observer, ϕ, is the angular distance from the equator to the zenith measured on the meridian. The longitude of the point (S) on its orbit is the angular distance from the vernal point on the ecliptic. The obliquity, ϵ, is defined as in Figure 1.

at a given δ fixed through time does not correspond always to the same position of the Earth about the Sun given by λ.

- H is the hour angle of the Sun which defines the time of the day. In time units, it is related to the legal time, TL, through

$$H = TL + FH - WT - 12 - LG + ET \tag{20}$$

where FH is the number of the time-zone (0 for Greenwich and positive to the west), WT allows to define the summer time and the winter time (e.g., in Europe $WT = 1$ for summer and 0 for winter), LG is the geographical longitude expressed in time units and measured positively to the west and ET the equation of time which is independent of the latitude and is only a function of the day (presently, its minimum value is -14 minutes on 11 February and its maximum is reached on 3 November with +16 minutes).

- r, the Earth-Sun distance, is given by the ellipse equation:

$$r = \frac{a(1-e^2)}{1+e\ \cos v} \tag{21}$$

- v being the true anomaly related to the true longitude, λ, of the Earth by

$$\lambda = v + \tilde{\omega}\ . \tag{22}$$

As in this formula, $\tilde{\omega}$ is measured from the vernal equinox, 180° has to be added to the numerical value calculated by Berger (1978). This is related to the fact that, in practice, the numerical calculations are done using as a reference the direction in which the Sun is seen from the Earth at the beginning of the spring, the so-called vernal point γ defining therefore the angle $\tilde{\tilde{\omega}}$ on Figure 1 with clearly $\tilde{\omega} = \tilde{\tilde{\omega}} + 180°$.

The following relationships obtained from (21):

$$r_{\text{perihelion } (v=0°)} = a(1-e) \tag{23}$$

$$r_{\text{aphelion } (v=180°)} = a(1+e) \tag{24}$$

show that the normalized distance (r/a) from the Earth to the Sun varies in the course of the year by $2e$, which means that the corresponding energy received by the Earth varies as $4e$, reaching a maximum of 30 % roughly (for $e = 0.075$) over the Quaternary.

Moreover

$$r_{\text{summer solstice } (\lambda=90^o)} = \frac{a(1-e^2)}{1+e\ \sin\tilde{\omega}} \sim a(1\ -\ e\ \sin\tilde{\omega}) \tag{25}$$

$$r_{\text{winter solstice } (\lambda=270^o)} = \frac{a(1-e^2)}{1-e\ \sin\tilde{\omega}} \sim a(1\ +\ e\ \sin\tilde{\omega}) \tag{26}$$

$$r_{\text{spring equinox } (\lambda=0^o)} = \frac{a(1-e^2)}{1+e\ \cos\tilde{\omega}} \sim a(1\ -\ e\ \cos\tilde{\omega}) \tag{27}$$

$$r_{\text{fall equinox } (\lambda=180^o)} = \frac{a(1-e^2)}{1-e\ \cos\tilde{\omega}} \sim a(1\ +\ e\ \cos\tilde{\omega}) \tag{28}$$

For $e = 0.016$ and $\tilde{\omega} = 282^o$ which are present day conditions, we have therefore $\frac{r}{a}$ equals to 0.984 at the perihelion, 1.016 at the aphelion, 1.016 at summer solstice, 0.984 at winter solstice, 0.996 at the vernal (spring) equinox and 1.003 at the autumnal (fall) equinox. The seasonal contrast of $\frac{r}{a}$ measured by the difference between the winter and summer solstices is therefore equal to $2\ e\ \sin\tilde{\omega}$, which represents today a value slightly over 3 %, but can amount up to 15 %. The maximum and minimum contrasts are reached at high excentricity and for the summer solstice occuring respectively at perihelion and at aphelion.

With (18) and (21), (17) becomes:

$$W = S_a \frac{(1+e\ \cos v)^2}{(1-e^2)^2} (\sin\phi\ \sin\delta\ +\ \cos\phi\ \cos\delta\ \cos H) \tag{29}$$

Over a given year, ϵ, e and $\tilde{\omega}$ are assumed to be constant; over a given day, λ and δ are assumed to be constant and H varies from 0 at solar noon to 360° (or from 0 to 24h).

The long-term variation for a given latitude, day and hour thus clearly depends upon e, $\tilde{\omega}$ which appear in (29) through (22) and ϵ which is implicitely contained in δ according to (19). Moreover, the behaviour of each factor in (29) is governed by a different orbital parameter: ϵ drives the long term variation of $\cos z$, $\tilde{\omega}$ the long term variation of $(1 + e\ \cos(\lambda - \tilde{\omega}))^2$ and e the long term variation of $(1-e^2)^2$. This last factor is particularly interesting and it must be recognized that the daily insolation is depending upon $(1-e^2)^{-2}$ and not upon $(1-e^2)^{-1/2}$ which drives the total energy received by the Earth over one year (see (14)).

However, W depends also upon e through $(1 + e \cos v)^2$. In the formula (29), the distance factor $\rho^{-2} = (\frac{a}{r})^2$ can indeed be written:

$$\frac{1 + 2e\cos v + e^2\cos^2 v}{(1-e^2)^2} = (1 + 2e\cos v + \frac{e^2}{2}\cos 2v + \frac{e^2}{2})(1-e^2)^{-2}$$

and $(1-e^2)^{-2}$ can be developed to an excellent accuracy:

$$(1-e^2)^{-2} \sim 1 + 2e^2 + \text{terms of order 4 and over}$$

which leads finally to:

$$\left(\frac{1 + e\cos v}{1-e^2}\right)^2 \sim 1 + 2e\cos v + \frac{e^2}{2}\cos 2v + \frac{5e^2}{2} + 4e^3\cos v \tag{30}$$

neglecting all terms of power 4 and over in eccentricity (this corresponds to an accuracy of 10^{-5} at the minimum).

Being given the definition of $v = \lambda - \tilde{\omega}$ and for a given value of λ, the spectrum of ρ^{-2} must therefore be dominated by the climatic precession ($e\sin\tilde{\omega}$ or $e\cos\tilde{\omega}$) displaying mainly 23 and 19 ka periods (Berger, 1977a). Moreover, this spectrum must also show, with a much less power, half precession periods (11.5 and 9.5 kr) because we have in (30) a term in $e\cos 2(\lambda - \tilde{\omega})$ weighted by $\frac{e}{2}$. Small spectral power at the eccentricity periods (400 and 100 ka) are also expected through the term $\frac{5e^2}{2}$. Finally, combination tones are awaited from the third term on. These analytical results are confirmed by spectral analyses which shows clearly, for spring equinox ($v = -\tilde{\omega}$) and summer solstice ($v = 90^o - \tilde{\omega}$), strong precessional peaks, weak peaks at the eccentricity and half precession periods and some other combination tones (Figures 4 and 5).

Finally, this formula (29) has the advantage to show that, for a given λ, the precession and obliquity signals originate from two distinct factors with a clear physical meaning: the precession signal originates from the distance factor (ρ^{-2}) and the obliquity signal comes from the inclination factor ($\cos z$). Therefore, for any fixed distance from the Earth to the Sun, there is only an obliquity signal in the insolation through geological times and for a fixed zenith distance (or altitude) there can only be a precessional signal. Consequently, for a given hour of a particular day, defined by a given value of λ, W is mainly a function of both precession and obliquity, the spectral amplitude at these frequencies depending upon the latitude and upon the time of the year ; it is also a function of eccentricity but with much less power.

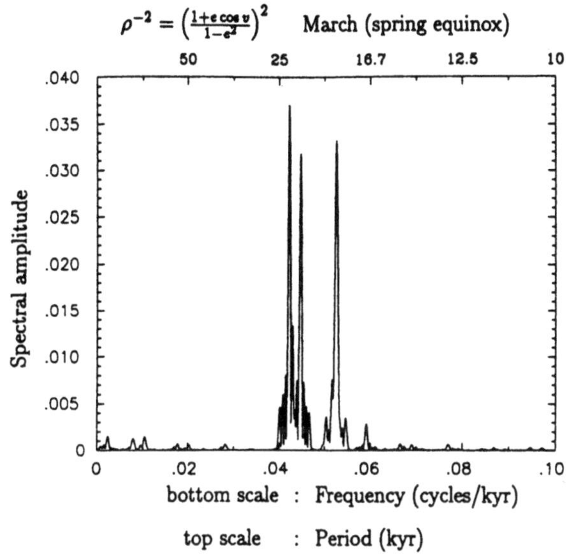

Figure 4: Spectral amplitude in the Thomson multi-taper harmonic analysis, at spring equinox (March), of $\rho^{-2} = \left(\frac{1+e\cos v}{1-e^2}\right)^2$.

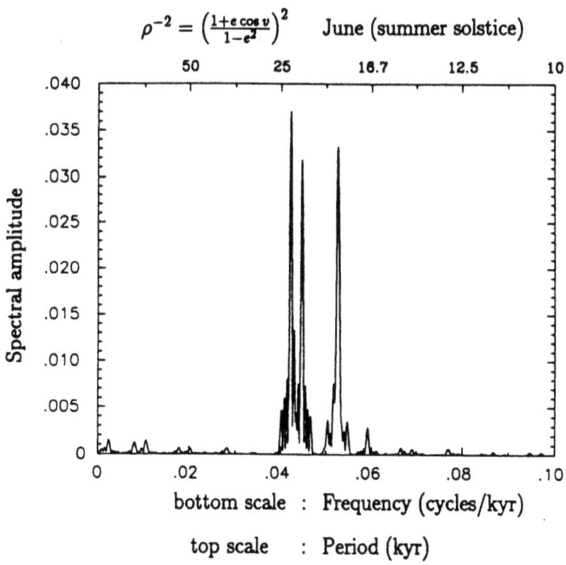

Figure 5: Spectral amplitude in the Thomson multi-taper harmonic analysis,at summer solstice (June), of $\rho^{-2} = \left(\frac{1+e\cos v}{1-e^2}\right)^2$.

However, if we would like to compare, for different geological times, the irradiances received on an horizontal surface for the same inclination of the sunrays, we must consider the insolation at fixed z (or altitude h) which clearly does not correspond to a fixed H; H being related to z through (18), we have indeed:

$$\cos H = \frac{\cos z - \sin\phi\sin\delta}{\cos\phi\cos\delta} \tag{31}$$

For any λ, except $\lambda = 0$ and 180°, the variation through time of the hour angle corresponding to a fixed z shows only a strong obliquity quasi-periodicity.

For the sake of comparing insolations in the same physical and astronomical contexts, it is the insolations at a fixed value z which must be intercompared from one geological time to another. It means, that the spectrum of (17) is strictly the one of ρ^{-2}. Again, as (30) shows for a fixed λ, there is no power of obliquity in such calculation.

In such conditions, the instantaneous insolation given by (17), i.e. $W = \frac{S_a}{\rho^2}\cos z$, can be computed for each value of z situated between 90° and z_{min} for each latitude for which there is a sunrise and a sunset ($z = 90^o$). z_{min}, which corresponds to $H = 0$, is given by $\cos z_{min} = \cos(\phi - \delta) = \sin\phi\sin\lambda\sin\epsilon + \cos\phi\sqrt{1-\sin^2\lambda\sin^2\epsilon}$ as it can be deduced directly from (18) and (19). For the long day during which the Sun does not set, the maximum value of z is reached at midnight for $H = 180^o$ (or 12h), which means:

$$\cos z_{max} = -\cos(\phi + \delta) = \sin\phi\,\sin\lambda\,\sin\epsilon - \cos\phi\,\sqrt{1-\sin^2\lambda\sin^2\epsilon}$$

However if the diurnal cycle of insolation has to be calculated, as it is the case in climate model, irradiance for different values of H will be used. As we just pointed out, for a fixed ϕ and λ, one given value of H does not obviously correspond to the same z all the time, the relation depending upon the obliquity. According to (18) and (19) for the fixed values of H, the corresponding $\cos z$ can indeed be calculated from:

$$\cos z = \sin\phi\sin\lambda\sin\epsilon + \cos\phi\sqrt{1-\sin^2\lambda\sin^2\epsilon}\cos H \tag{32}$$

6 Perihelion and the astronomical seasons

Among all possible values of v over one year ($0 \leq v \leq 360^\circ$), the following ones illustrate the total effect of the eccentricity alone:

$$\text{at the perihelion } (v = 0^\circ): \quad W^P = S_a \frac{1}{(1-e)^2} \cos z \qquad (33)$$

$$\text{at the aphelion } (v = 180^\circ): \quad W^A = S_a \frac{1}{(1+e)^2} \cos z \qquad (34)$$

$$\text{at orthogonal distance } (v = 90^\circ \text{ or } 270^\circ): \quad W = S_a \frac{1}{(1-e^2)^2} \cos z \qquad (35)$$

Taking 0.075 as a maximum value of e, the factors containing e vary with a maximum amplitude of respectively 17, 13 and 1 %. Although it is interesting to see that these insolations are only functions of obliquity and eccentricity and not of precession, one must recognize that the perihelion drifts continuously through all the seasons according to the long term variation of its longitude.

By fixing arbitrarily the spring equinox at 21 March, the Earth is indeed at the perihelion, for example at 14 September, 1 February, 22 April, 13 July, 3 September and 3 January for respectively 122 ka, 20 ka BP, 15 ka BP, 10 ka BP, 7 ka BP and now. This is also reflected by the long term variation of the length of the astronomical seasons which is only a function of the climatic precession (Berger and Pestiaux, 1984; see also Annex 2). If we consider the astronomical spring, summer, fall and winter, their length varies over the last 10^6 years between 82.5 and 100 mean solar days. Some examples are given for the same dates as above in Table 2.

These examples illustrate how complex the relationship between the insolations and the orbital parameters is. It also underlines the importance that the eccentricity can have individually; terms containing e alone cannot be ignored, especially since they reinforce the impact on climate in the explanation of the strength of 100 ka cycle. On the other hand, the insolations are involved, in intricated ways, in a number of non-linear processes in the climate system. These can distord the input signal both in amplitude and frequency. The physical mechanisms through which the climate system would response to such a given forcing are thus very difficult to conceive and cannot be deduced by only comparing the spectra of the input (insolation) and of the output (the geological records) signals. Physical models of the climate system will provide a complementary and necessary information to analyse the complex way through which the system is responding to the astronomical forcing (Gallée et al., 1992; Berger et al., 1993a).

Table 2: Length of the astronomical spring, summer, fall and winter in mean solar days for 6 selected times of the past

date ka	Spring $0 \leq \lambda < 90^o$	Summer $90^o \leq \lambda < 180^o$	Fall $180^o \leq \lambda < 270^o$	Winter $270^o \leq \lambda < 360^o$
122	95.7	86.2	87.0	96.4
20	91.5	94.5	91.1	88.2
15	88.2	92.0	94.5	90.6
10	90.2	88.3	92.4	94.3
7	92.8	88.6	89.8	94.0
0	92.8	93.6	89.8	89.0

7 Daily cycle of insolation

Sunset or sunrise correspond to z=90°, which means from (18):

$$\sin\phi\ \sin\delta\ +\ \cos\phi\ \cos\delta\ \cos H\ =\ 0$$

This gives the absolute value of the hour angle, H_o, at sunrise ($H = -H_o$) and sunset($H = H_o$):

$$\cos H_o\ =\ -\ \tan\ \phi\ \tan\ \delta \tag{36}$$

H_o does exist only for: $-1\ \leq\ \tan\ \phi\ \tan\ \delta\ \leq\ +1$ which means that the latitudes for which there is a daily sunset and sunrise are given by

$$-(90^\circ - \mid\delta\mid)\ \leq\ \phi\ \leq\ (90^\circ - \mid\delta\mid) \tag{37}$$

Therefore, the latitudes where there is no sunset (H_o = 12h) are defined by:

$$\mid\phi\mid > 90^\circ - \mid\delta\mid \text{with} \begin{cases} \phi > 0 & \text{if}\ \ \delta > 0 \\ \phi < 0 & \text{if}\ \ \delta < 0 \end{cases} \tag{38}$$

and the latitudes where there is no sunrise ($H_o = 0$)

$$\mid\phi\mid > 90^\circ - \mid\delta\mid \text{with} \begin{cases} \phi > 0 & \text{if}\ \ \delta < 0 \\ \phi < 0 & \text{if}\ \ \delta > 0 \end{cases} \tag{39}$$

For $\delta = 0$, the length of the day is everywhere 12 hours, except at the poles which are singular points where H_o is set to 0. For $\phi = 0$ the length of the day is always equal to 12 hours according to (36). The solar zenith distance, z, (or altitude above horizon, $h = 90^o - z$) varies during the course of a day between extreme values which correspond to H equal to 0 (at solar noon) and 180° (at midnight), if we account for the values of z and h corresponding also to the Sun below the horizon (in such a case $0 \leq z \leq 180^o$ and $-90^o \leq h \leq 90^o$).

For any value of ϕ and δ, we have

$$\text{for } H = 0: \quad z_{min} = |\,\phi - \delta\,| \qquad h_{max} = 90^o - |\,\phi - \delta\,| \tag{40}$$

$$\text{for } H = 180^o: \quad z_{max} = 180^o - |\,\phi + \delta\,| \quad h_{min} = -90^o + |\,\phi + \delta\,| \tag{41}$$

where z_{max} and h_{min} are therefore giving numerical values also for the Sun below the horizon. (We must point out that for the days where there is a sunset and sunrise, the values of z and h traditionally given are those for which $-H_0 \leq H \leq H_0$; in such cases, $z_{max} = 90^o$ and $h_{min} = 0$).

Allowing a complete cycle for $0 \leq H \leq 360^o$, the following relationships might be useful:

$$\begin{aligned}
h(\phi, \delta, H) &= h(-\phi, -\delta, H) \\
h_{\max}(\delta, \phi) &= h_{\max}(-\delta, -\phi) \\
h_{\min}(\delta, \phi) &= h_{\min}(-\delta, -\phi) \\
h_{\max}(\delta, \phi) &= -h_{\min}(\delta, -\phi) \\
h_{\max}(\delta, \phi) &= -h_{\min}(-\delta, \phi)
\end{aligned}$$

Figure 6 illustrates how h is varying all over the day ($-180^o \leq H \leq 180^o$) for any positive value of δ. For $\delta < 0$, a similar figure can be drawn easily from:

$$h(\phi, \delta, H) = h(-\phi, -\delta, H)$$

We want to stress that the derivative of this function $h(\phi, \delta, H)$ with respect to H at a given time in the year (δ) does not exist in two cases: when $\phi = \delta$ at $H = 0$ and when $\phi = -\delta$ at $H = 180^o$.

Figure 7 gives the variation of $h_{\max}$ and $h_{\min}$ at a particular date given for a positive value of δ (here $\delta = 23^o27$, the present-day value of ϵ). For a day with $\delta < 0$, we have

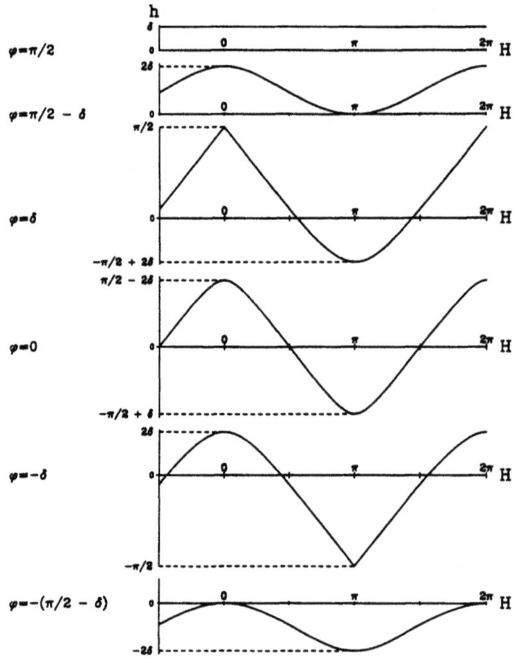

Figure 6: Variation of the altitude (h) all over the day ($-180^o \leq H \leq 180^o$) for any positive value of δ. For δ negative the same figure can be drawn easily from: $h(\phi, \delta, H) = h(-\phi, -\delta, H)$.

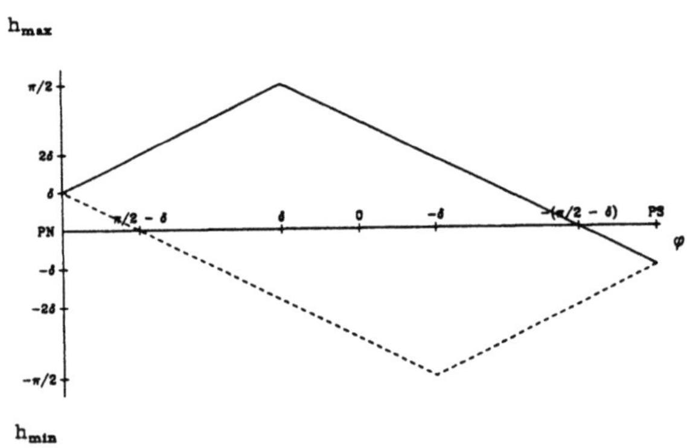

Figure 7: Variation of h_{max} and h_{min} at a particular date given for a positive value of δ. For $\delta < 0$, we have the same drawing with $h_{max}(-\delta, -\phi) = -h_{min}(\delta, -\phi)$ and $h_{min}(-\delta, -\phi) = -h_{max}(\delta, -\phi)$.

the same drawing but with

$$h_{\max}(-\delta,\phi) = -h_{\min}(\delta,\phi)$$
$$h_{\min}(-\delta,\phi) = -h_{\max}(\delta,\phi)$$

However, in the calculation of the energy received from the Sun, it is obvious that the maximum value allowed for z is 90° and the minimum value allowed for h is 0, both occuring for $H =\mid H_0 \mid$, when the Sun sets or rises. The advantage of (41) is nevertheless to allow an easy calculation of the minimum value of the altitude of the Sun above the horizon during the course of the long day in polar latitudes.

8 Insolation at solar noon

At solar noon (i.e. for $H = 0$), W is given by:

$$W_{Noon} = S_a \frac{(1+e \cos v)^2}{(1-e^2)^2} \cos(\phi - \delta) \tag{42}$$

The obliquity signal is only present in the $\cos(\phi - \delta)$ factor if we choose a day of the year defined by a given value of λ:

$$\cos(\phi - \delta) = \sin\phi \, \sin\epsilon \, \sin\lambda + \cos\phi \sqrt{1 - \sin^2\epsilon \, \sin^2\lambda}$$

If the chosen day is defined by a fixed value of δ, there is no contribution of $\cos(\phi-\delta)$ to the spectrum of W_{Noon} and according to (30), $\frac{(1+e\cos v)^2}{(1-e^2)^2}$ will provide a spectral power at the frequencies of eccentricity and precession and at linear combinations between frequencies of eccentricity, precession and obliquity (due to λ which varies at the obliquity frequencies for a given δ according to (19)).

Moreover, as ϵ is only varying within a small range, W_{Noon} is dominated by precession everywhere, except maybe at high latitudes for small eccentricity values where obliquity contributes also significantly to the variation of W_{Noon} (Figure 8).

1. This is trivial for $\phi = \delta$, i.e. for those latitudes and days where the Sun reaches the zenith (z=0) at noon (for such a day δ, the maximum energy is reached over the Earth for $\phi = \delta$). For z=0, (42) becomes:

$$W_{Noon} = S_a \frac{(1+e \cos v)^2}{(1-e^2)^2} \tag{43}$$

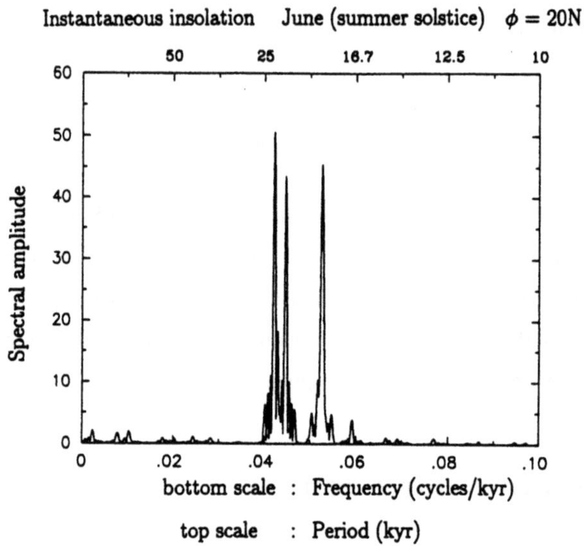

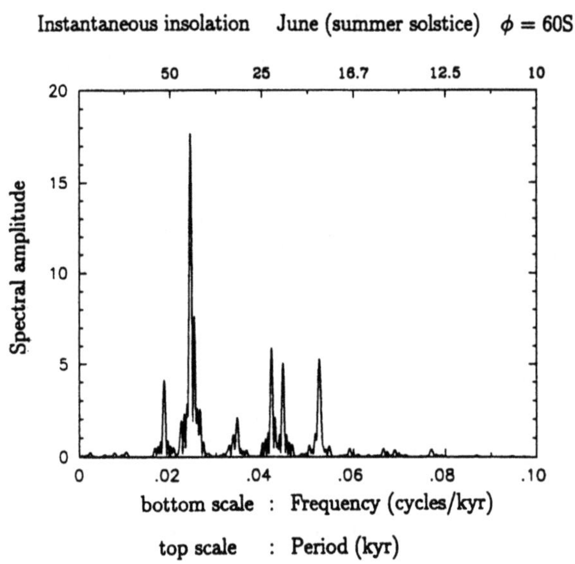

Figure 8: Spectral amplitude in the Thomson multi-taper harmonic analysis for summer solstice (June) of the instantaneous insolation at noon at 20N (top panel) and at 60S (bottom panel).

and there is obviously no obliquity signal in W_{Noon} for all those latitudes situated between the tropics and defined by $\mid \phi \mid \leq \epsilon_{min}$, provided the time of the year is selected according to $\delta = \phi$. The behaviour of (43) is dominated by precession, slightly modulated by the eccentricity. However, as δ is a function of the true longitude of the Sun through the obliquity (see 19), the position of the Earth relatively to the Sun (for a given value of δ) is not always the same through geological times, except for the equinoxes where $\lambda = 0$ and 180° correspond always to $\delta = 0$. For the solstices $\lambda = 90^o$ and 270°, $\delta = \epsilon$ and $-\epsilon$ respectively and varies at the astronomical time scale. In other words, the latitudes for which the Sun is at the zenith at noon at the solstices (the latitudes of the tropics defined by $\mid \phi \mid = \epsilon$) changes therefore in time with ϵ.

2. For the other latitudes, Table 3 gives the relative contribution of $\cos(\phi - \delta)$ to W_{Noon}, both at the equinoxes and northern hemisphere summer solstice, for two extreme values of ϵ.

   For $\delta = 0$, the change related to $\cos(\phi - \delta)$ is evidently strictly equal to zero for a given fixed ϕ; it is not for the polar and tropical circles which do not correspond to a fixed ϕ but varies in time with ϵ. ($\cos(\phi - \delta)$ is respectively equal to $\sin \epsilon$ and $\cos \epsilon$). The present-day latitudes of the polar circles are $66^o33'$ N and S. Over the Quaternary, as we have roughly $22^o \leq \epsilon \leq 24^o\ 30'$, the southernmost position of the arctic circle is $65^o\ 30'$ and its northernmost position is 68°. At the birth of Christ, 2000 years ago, the latitude of the arctic circle was $66^o\ 18'$; within 2000 years it will be $66^o\ 49'$, which means that its motion towards north is presently around 14.4 m per year.

   At the summer solstice for which $\delta = \epsilon$ is varying in time, the absolute change of W_{Noon} due to obliquity factor is 7 % at the maximum (0.64 to 0.71); this maximum change is reached for the tropical circle of the winter hemisphere where $\cos(\phi - \delta) = \cos 2\epsilon$. The relative change can however reach more than 10 %. This is the case for the northern pole for which $\cos(\phi - \delta) = \sin \epsilon$ and for the high southern polar latitudes, north of the antarctic polar circle. The impact of the variation of $\cos(\phi - \epsilon)$ is indeed maximum for the maximal value of $\sin(\phi - \epsilon)$, it means for $\phi = -90^o + \epsilon$, the latitude of the antarctic circle. Consequently a small variation of the altitude of the Sun above the horizon, due to the variation of the obliquity, has much more influence in the frequency domain on $\cos(\phi - \epsilon)$ and therefore on insolation, when

Table 3: Contribution of $\cos(\phi - \delta)$ to the insolation at solar noon for some latitudes, at the spring equinoxe ($\delta = 0$) and summer solstice ($\delta = \epsilon$), for two extreme values of ϵ. Δ gives the relative change in % of this contribution when going from ϵ_{min} to ϵ_{max}.

	ϕ	$\delta = 0$		Δ	$\delta = \epsilon$		Δ
		$\epsilon = 22^o30'$	$\epsilon = 25^o$	(%)	$22^o30'$	25°	(%)
pole	90°	0	0	0	0.38	0.42	10
polar circle	$90^o - \epsilon$	0.38	0.42	10	0.71	0.77	8
middle	45°	0.71	0.71	0	0.92	0.94	2
tropical circle	ϵ	0.92	0.91	-1	1	1	0
equator	0	1	1	0	0.92	0.91	-1
tropical circle	$-\epsilon$	0.92	0.91	-1	0.71	0.64	-9
middle	-45	0.71	0.71	0	0.38	0.34	- 10
polar circle	$-90 + \epsilon$	0.38	0.42	10	0	0	0
pole	-90°	0	0	0	0	0	0

the Sun is low above the horizon than when it is high. Finally, it can be shown that the latitudes above the tropical circle of the summer hemisphere receive less energy when ϵ is smaller, which increases the latitudinal contrast, a result also used by Milankovitch (1941) through his caloric insolations to force an ice age. All latitudes south of the tropical circle of the winter hemisphere receive more energy when ϵ decreases.

During the time obliquity is going through a full cycle, precession goes to about 2 cycles. Over each cycle, the precessional factor, $(1 + e \cos v)^2$, changes between two extrema: $(1-e)^2$ and $(1+e)^2$, which are reached respectively for $\lambda - \tilde{\omega} = 180^o$ (Earth at the aphelion) and $\lambda - \tilde{\omega} = 0^o$ (Earth at the perihelion). For example, we have:

$$\text{for } \lambda = 0\ (\delta = 0) \quad \tilde{\omega} = 180^o \quad \text{(autumnal equinox at perihelion)}$$
$$\tilde{\omega} = 0^o \quad \text{(vernal equinox at perihelion)}$$

$$\text{for } \lambda = 90^o\ (\delta = \epsilon) \quad \tilde{\omega} = 270^o \quad \text{(winter solstice at perihelion)}$$
$$\tilde{\omega} = 90^o \quad \text{(summer solstice at perihelion)}$$

This change between the two extremes of the precessional factor is equal to $4e$. It reaches for e=0, 0.02, 0.04 and 0.075 respectively 0, 8, 16 and 30 %. It must be stressed here that these extrema within the precessional cycle reinforce considerably the spectral power at the eccentricity periods, in particular for those originating from a beat between one of the 23 ka-periods and one of the 19 ka-periods As shown in section 2 of this paper, those

"23-19" combinations lead to beat periods located between 95 and 130 ka contributing to the so-called 100 ka-eccentricity cycle (so-called 100 ka period) because most of the time the lack of resolution does not allow to discriminate between 95 and 130 ka. For the other beats leading to the eccentricity periods numbers 1 and 6 in Table 1, the problem is more complicated because one would have to be able to discriminate between the precessional periods numbers 1 and 2, where the difference amounts only 1300 yr or between numbers 3 and 4 where the difference is less than 200 years. Moreover, the time series would need also to be long enough to detect the 400 ka and 2,300 ka periods).

But, these extrema of the precessional factor are weighted by the eccentricity factor $(1-e^2)^{-2}$, which means that they become respectively $(1+e)^{-2}$ and $(1-e)^{-2}$. They contribute to a change from 1 to 1 (0 %), 0.96 to 1.04 (8 %), 0.92 to 1.09 (17 %) and 0.86 to 1.17 (31 %), for the four eccentricity values just considered. As the long term variation of e is roughly 5 times slower than that of the climatic precession, this eccentricity factor contributes very little to the modulation of the precession factor over one precessional cycle. But over one full eccentricity cycle, $(1-e^2)^{-2}$ can change W_{Noon} by a maximum of 1 %.

All these comparisons between the relative contribution of ϵ, $e \sin\tilde{\omega}$ and e hold for all latitudes and days for which $W \neq 0$. For any eccentricity value above roughly the present-day value, the precessional term dominates everywhere, but mainly in middle and low latitudes. It is only in high polar latitudes, where W_{Noon} is smaller, that the obliquity plays relatively a more significant role.

These results are difficult to generalise for all hours of the day. The reason lies in $\cos z$ (see 18), $\cos H$ weighting $\cos\delta$ but not $\sin\delta$. Being given the range within which δ is allowed to vary $(-\epsilon \leq \delta \leq \epsilon)$, $\cos\delta$ is at least more than twice as large as $\sin\delta$, which according to the value of H will give ϵ more or less power. This will also depend upon the latitudes through the $\sin\phi$ and $\cos\phi$ factors. Finally, the value of $\cos z$ will depend upon the signs of $\sin\phi\ \sin\delta$ and of $\cos\phi\ \cos\delta\ \cos H$, which needs to consider whether or not ϕ and δ have the same sign and $\mid H \mid$ is greater than 90° (6 hours). This can be seen numerically through spectral analysis of the instantaneous insolation values at noon summer solstice for all latitudes (e.g. Figure 9 or other examples in Berger et al., 1993b).

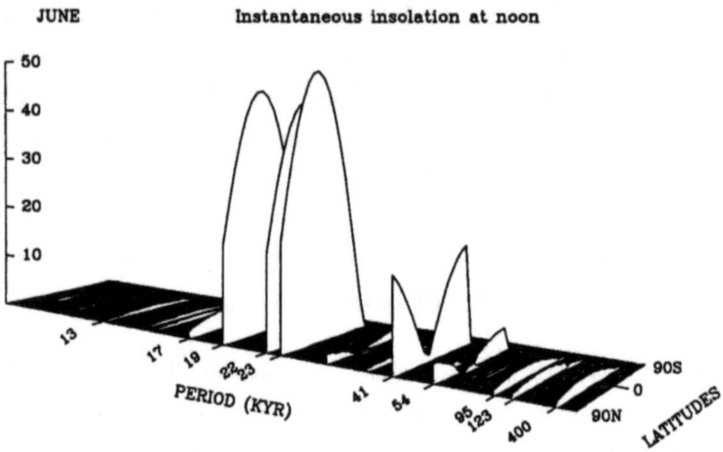

Figure 9: Spectral amplitude in the Thomson multi-taper harmonic analysis of the instantaneous insolation at noon for summer solstice (June) and for each latitude between the north pole and the south pole.

9 Equator, Poles and Equinoxes

Simplifications are occuring in three particular cases related to this instantaneous insolation at a given time t: insolation at the equator W_{EQ}, insolation at the pole W_{PO}, and insolation at the equinoxes (vernal and autumnal) W_{VA}.

At the equinoxes, $\delta = 0$ implies that $H_0 = 90^o$ (or 6 hours) for all latitudes as it can be deduced from (36). For the instantaneous insolation, we have:

$$W_{VA} = S_a \left(\frac{a}{r} \right)^2 \cos\phi \, \cos H \tag{44}$$

which shows that whatever the time of the day we consider, W_{VA} does not depend at all of obliquity. It is strictly a function of precession and eccentricity as it can be deduced from a geometrical point of view: at the equinoxes, the axis of rotation of the Earth lies in a plane normal to the solar beam. The relative importance of the precessional and eccentricity components is the same as already discussed in the previous sections, for any latitude and time of the day.

At the equator, $\phi = 0$ implies that $H_o = 90^o$ (or 6 hours) for all days of the year as it can be deduced from (36). For the instantaneous insolation, we have:

$$W_{EQ} = S_a \left(\frac{a}{r} \right)^2 \cos\delta \cos H \tag{45}$$

Except at the equinoxes, W_{EQ} is therefore a function of ϵ through $\cos\delta$:

$$\cos\delta = \sqrt{1 - \sin^2 \epsilon \sin^2 \lambda}$$

whatever the time of the day we consider. Its maximal influence is felt for $\lambda = 90^o$ or 270°, it means at the solstices, but remains very weak, as it can be seen from Table 3. At such solstices, the equator receives more energy when ϵ is small, as already concluded from Table 3:

$$W_{EQ} = S_a \left(\frac{a}{r} \right)^2 \cos\epsilon \cos H \tag{46}$$

At the pole, $\phi = 90^o$ and (18), leads to

$$\cos z = \sin\delta$$

which shows that the altitude of the Sun (above the horizon) does not change very much over 24 hours during the long polar day. In such a case, for any instantaneous time during that long day, we have:

$$W_{PO} = S_a \left(\frac{a}{r} \right)^2 \sin\delta = S_a \left(\frac{a}{r} \right)^2 \sin\epsilon \sin\lambda \tag{47}$$

At the poles, there is either :

no sunrise:	$H_o = 0$	if $\phi.\delta \leq 0$
or no sunset:	$H_o = 180^o$ (or 12h)	if $\phi.\delta > 0$

This means that at the northern pole, $W_{PO} \neq 0$ for $0 < \lambda < 180^o$ with a maximum weight of ϵ for $\lambda = 90^o$. At that time, the pole receives less energy when ϵ is small (as shown again in Table 3):

$$W_{PO} = S_a \left(\frac{a}{r} \right)^2 \sin\epsilon \tag{48}$$

(46) and (48) lead clearly to an intensification of the latitudinal gradient of insolation at small ϵ.

10 Total daily insolation

The total daily insolation is simply obtained by integrating (17) over 24 hours of true solar time, t_s. But we must represent the diurnal march of irradiation as a function of the mean solar time, t, which, by definition, is regular, the true solar time being not because of the elliptical shape of the orbit and the second law of Kepler. The relation between the two is given by the equation of time, ET, provided in the Astronomical Ephemeris for each day (see 20)

$$t_s = t + ET$$

Therefore

$$\frac{dt_s}{dt} = 1 + \frac{dET}{dt}$$

and neglecting the small variations of ET compared to 1, we have $dt \sim dt_s$. As the solar hour-angle, H, in time units is called t_s, we have

$$H = \frac{2\pi}{\tau} t_s$$

where τ is the interval of 24 hours. Therefore,

$$dH = \frac{2\pi}{\tau} dt_s \simeq \frac{2\pi}{\tau} dt \qquad (49)$$

and assuming r is constant over the day, the integration of (17) becomes:

$$\overline{W}_d = \frac{1}{\tau} \int_{24h} W \; dt = \frac{S_a}{2\pi} \left(\frac{a}{r}\right)^2 \int_{-H_o}^{H_o} \cos z \; dH$$

which gives:

- for the latitudes where there is a daily sunrise and sunset:

$$|\, \phi \,| < 90^o - |\, \delta \,|$$

$$\overline{W}_d = \frac{S_a}{\pi} \left(\frac{a}{r}\right)^2 (H_o \sin\phi \sin\delta + \cos\phi \cos\delta \sin H_o) \qquad (50)$$

$$\text{length of the day} = \frac{24h}{\pi} \cos^{-1}(-\tan\phi \tan\delta)$$

- for the latitudes $|\, \phi \,| \geq 90^o - |\, \delta \,|$

  - the long polar night is defined by: $\phi.\delta \leq 0$

$$\text{and we have :} \quad \overline{W}_d = 0 \qquad (51)$$

$$\text{length of the day} = 0$$

- the long polar day is defined by: $\phi.\delta > 0$

$$\text{and we have :} \quad \overline{W}_d = S_a \left(\frac{a}{r}\right)^2 \sin\phi \, \sin\delta \tag{52}$$

$$\text{length of the day} = 24h.$$

This has been discussed in length in Berger (1978). The spectral behaviour of this daily insolation was presented in Berger and Pestiaux (1984) and Berger et al. (1993b). It can be summarized as follows

1. for the equinoxes:

$$\begin{matrix}\text{vernal} \\ \text{autumnal}\end{matrix} \text{ equinox} : \overline{W}_d = \frac{S_a}{\pi} \frac{(1 \pm e \cos\tilde{\omega})^2}{(1-e^2)^2} \cos\phi \tag{53}$$

Formula (53) holds for all latitudes.

2. for the solstices:

$$\begin{matrix}\text{summer} \\ \text{winter}\end{matrix} \text{ solstice} : \overline{W}_d = \frac{S_a}{\pi} \frac{(1 \pm e \sin\tilde{\omega})^2}{(1-e^2)^2} (\pm H_o \sin\phi \sin\epsilon + \cos\phi \cos\epsilon \sin H_o) \tag{54}$$

with $\cos H_o = \mp \tan\phi \tan\epsilon$ for the daily sunrise and sunset. But formula (54) holds also for the two following particular cases:

- the long polar night: $|\phi| \geq 90^o - \epsilon, \;\; \phi.\delta \leq 0 \; : \; H_o = 0$

- the long polar day: $|\phi| \geq 90^o - \epsilon \;\; \phi.\delta > 0 \; : \; H_o = 180^o$

Again we must stress that the numerical value of $\tilde{\omega}$ is "the numerical value obtained from the astronomical serie which development is given in Berger (1978)" plus 180°.

(53) clearly shows that $\overline{W}_d$ at the equinoxes is only a function of precession, as already concluded for the instantaneous insolation W_{VA}. At the solstices, (54) shows that although the ϵ effect is more important at high than at low latitudes, the precession effect still dominates except at the high latitudes of the winter hemisphere (close to the polar circle).

Moreover, just as in the case of instantaneous insolation, for a given time λ in the year, the equation (50) shows that the precession and obliquity signals in the daily insolation originate from two distinct factors: the precession signal arises through

$$(1 + e \cos(\lambda - \tilde{\omega}))^2 \tag{55}$$

in the distance factor and the obliquity signal comes from the factor

$$H_0 \sin\phi \sin\delta + \cos\phi \cos\delta \sin H_0 \tag{56}$$

This characteristic allows to follow more easily the impact of precession and obliquity variations in both hemispheres: the obliquity plays the same role in both hemispheres during the same local season. Indeed equation (56) takes the same value for a latitude ϕ and a time λ in the year (corresponding to δ) than for a latitude $(-\phi)$ and a time $(\lambda + 180^o)$ (corresponding to $-\delta$). On the contrary precession has an opposite effect on both hemispheres. Indeed for a latitude ϕ and a season corresponding to $\lambda(\delta)$ the variation is given by (55). For a latitude $(-\phi)$ and a season corresponding to $(\lambda + 180^o)$ $(-\delta)$, the variation reverses as $e \cos(\lambda - \tilde{\omega})$ in (55) changes its sign and (56) holds.

11 Conclusions

The analytical expressions providing the insolation at given latitude, day of the year and time of the day have confirmed the spectral analysis of the numerical values which can easily be computed through the usual formulas. Some particular dates and times of the day have provided examples which illustrate clearly that in most occasions, the precessional signal dominates the obliquity one, except in high polar latitudes mainly of the winter hemisphere.

Other parameters than the instantaneous and daily insolations can be analysed. These are, for example, irradiation received during some particular times of the day (Cerveny, 1991 and Berger et al., 1993b), or integrated over a specific part of the year and over a given zonal band (Berger, 1975; Godart, 1986; Loutre, 1993). Their spectral behaviour give some interesting information which in addition to the results presented in this paper allow to analyse how the climate system might respond to the insolation forcing. However, as there are many non-linear feedbacks intervening, it is necessary to go well beyond the comparison of the spectra of the insolations (input to the climate system) and of the geological records (proxy for the output of the system). Climate models, and in particular those which are able to reproduce the transient response of the full climate system to the insolation forcing (e.g. Berger et al., 1990; Gallée et al., 1992) are urgently needed to better understand the mechanisms through which the astronomical frequencies (amplitudes and phases) are transformed.

12 References

Backman J., Pestiaux P., Zimmerman H. and O. Hermelin, 1986. Palaeoclimatic and palaeoceanographic development in the Pliocene North Atlantic: Discoaster accumulation and coarse fraction data. In: C.P. Summerhayes and N.J. Shackleton (eds), *North Atlantic Palaeoceanography*, Geological Society, Special Publ. n° 21, pp. 231-242.

Bard E., Hamelin B., Faribanks R.G. and A. Zindler, 1990. Calibration of the ^{14}C timescale over the past 30,000 years using mass spectrometric U-Th ages from Barbados corals. *Nature*, 345, pp. 405-410.

Barnola J.M., Raynaud D., Korotkevitch Y.S. and C. Lorius, 1987. Vostok ice core: a 160,000 year record of atmospheric CO_2. *Nature*, 329(6138), pp. 408-414.

Benzi R., Parisi G., Sutera A. and A. Vulpiani, 1982. Stochastic resonance in climatic change. *Tellus*, 34, pp. 10-16.

Berger A., 1973. Théorie astronomique des paléoclimats. *Thèse de doctorat*, Faculté des Sciences, Université Catholique de Louvain. Unpublished manuscript.

Berger A., 1975. Astronomical theory of paleoclimates: a cascade of accuracy. *WMO-IAMAP*, Norwich, WMO n° 421, pp. 65-72, Genève.

Berger A., 1976. Obliquity and precession for the last 5,000,000 years. *Astronomy and Astrophysics*, 51, pp. 127-135.

Berger A., 1977a. Support for the astronomical theory of climatic change. *Nature*, 268, pp. 44-45.

Berger A., 1977b. Long term variations of the Earth's orbital elements. *Celestial Mechanics*, 5, pp. 3-74.

Berger A., 1978. Long term variations of daily insolation and Quaternary climatic changes. *Journal of Atmospheric Sciences*, 35(12), pp. 2362-2367.

Berger A., 1988. Milankovitch theory and climate. *Review of Geophysics*, 26(4), pp. 624-657.

Berger A., 1989a. Pleistocene climatic variability at astronomical frequencies. *Quaternary International*, 2, pp. 1-14.

Berger A., 1989b. The spectral characteristics of pre-Quaternary climatic records, an example of the relationship between the astronomical theory and geo-sciences. In: A. Berger, S. Schneider and J.Cl. Duplessy (eds), *Climate and Geo-Sciences, a Challenge for Science and Society in the 21st Century*, pp. 47-76, Kluwer, Dordrecht, Holland.

Berger A. and P. Pestiaux, 1984. Accuracy and stability of the Quaternary terrestrial insolation. In: A. Berger, J. Imbrie, J. Hays, G. Kukla and B. Saltzman (eds), *Milankovitch and Climate*, pp. 83-112, D. Reidel Publ. Company, Dordrecht, Holland.

Berger A., Loutre M.F. and V. Dehant, 1989. Influence of the changing lunar orbit on the astronomical frequencies of pre-Quaternary insolation patterns. *Paleoceanography*, 4(5), pp. 555-564.

Berger A. and M.F. Loutre, 1990. Origine des fréquences des éléments astronomiques intervenant dans le calcul de l'insolation. *Bulletin Sciences*, 1-3/90, pp. 45-106, Académie Royale des Sciences, des Lettres et des Beaux-Arts de Belgique.

Berger A., Gallée H., Fichefet Th., Marsiat I. and C. Tricot, 1990. Testing the astronomical theory with a coupled climate-ice sheet model. In: L.D. Labeyrie and C. Jeandel (eds), Geochemical Variability in the Oceans, Ice and Sediments. *Palaeogeography, Palaeoecology*, 89(1/2), *Global and Planetary Change Section*, 3(1/2), pp. 125-141.

Berger A. and M.F. Loutre, 1991. Insolation values for the climate of the last 10 million years. *Quaternary Science Reviews*, 10 n°4, pp. 297-317.

Berger A., Gallée H. and C. Tricot, 1993a. Glaciation and deglaciation mechanisms in a coupled 2-D climate - ice sheet model. *Journal of Glaciology*, (in press).

Berger A., Loutre M.F., and C. Tricot, 1993b. Insolation and Earth's orbital periods. *J. Geophys. Res.*, (in press).

Birchfield G.E. and J. Weertman, 1978. A note on the spectral response of a model continental ice sheet. *Journal of Geophysical Research*, 83(C8), pp. 4123-4125.

Boyle E.A. and L.D. Keigwin, 1985/1986. Comparison of Atlantic and Pacific paleochemical records for the last 215,000 years: Changes in deep ocean circulation and chemical inventories. *Earth Plan. Sci. Lett.*, 76, pp. 135-150.

Bretagnon P., 1974. Termes à longues périodes dans le système solaire. *Astronomy and Astrophysics*, 30(1), pp. 141-154.

Broecker W.S., Thurber D.L., Goddard J., Ku T., Matthews R.K. and K.J. Mesolella, 1968. Milankovitch hypothesis supported by precise dating of coral reefs and deep sea sediments. *Science*, 159, pp. 297-300.

Brouwer D., and G.M. Clemence, 1961. *Methods of Celestial Mechanics.* Academic Press, New York, 598p.

Cerveny R.S., 1991. Orbital signals in the diurnal cycle of radiation. *J. Geophys. Res.*, 96(D9), pp. 17,209-17,215.

CLIMAP Project Members, 1976. The surface of the Ice-Age Earth. *Science*, 191, pp. 1131-1137.

CLIMAP Project Members, 1981. Seasonal reconstruction of the Earth's surface at the Last Glacial maximum, McIntyre A. and Cline R. (eds), Geological Society of America. Map and Chart Series MC-36, Boulder, pp. 1-18.

COHMAP Members, 1988. Climatic changes of the last 18,000 years: Observations and model simulations. *Science*, 241, pp. 1043-1052.

Crowley T.J., 1988. Paleoclimate modelling. In: M. Schlesinger (ed.), *Physically-Based Modelling and Simulation of Climate and Climatic Change*, pp. 883-949, Kluwer Academic Publishers, Dordrecht, Holland.

Curry W.B. and T.J. Crowley, 1987. The $\delta^{13}C$ of equatorial Atlantic surface waters: implications for ice-age pCO_2 levels. *Paleoceanography*, 2, pp. 489-517.

Emiliani C., 1966. Isotopic paleotemperatures. *Science*, 154(3751), pp. 851-857.

Fischer A.G., 1986. Climatic rhythms recorded in strata. *Ann. Rev. Earth Planet. Sci.*, 14, pp. 351-376.

Gallée H., van Ypersele J.P., Fichefet Th., Marsiat I., Tricot C. and A. Berger, 1992. Simulation of the last glacial cycle by a coupled, sectorially averaged climate - ice-sheet model. II. Response to insolation and CO_2 variation. *Journal of Geophysical Research*, 97(D14), pp. 15,713- 15,740.

Ghil M., and H. Le Treut, 1981. A climate model with Cryodynamics and Geodynamics. *Journal of Geophysical Research*, 86, pp. 5262-5270.

Godart O., 1986. Geometrical insolation of a planet. *Annales de la Société Scientifique de Bruxelles*, T. 100 II, pp. 53-85.

Hasselman K., 1976. Stochastic climate models, part I. *Tellus*, 28, p. 473.

Hays J.D., Imbrie J. and N.J. Shackleton, 1976. Variations in the Earth's orbit: pacemaker of the Ice Ages. *Science*, 194, pp. 1121-1132.

Herbert T.D. and A.G. Fischer, 1986. Milankovitch climatic origin of mid-Cretaceous black shale rhythms, Central Italy. *Nature*, 321(6072), pp. 739-743.

Hilgen F.J., 1987. Sedimentary rhythms and high resolution chrono-stratigraphic correlations in the Mediterranean Pliocene. *Newsletters Stratigraphy*, 17(2), pp. 109-127.

Hilgen F.J. and C.G. Langereis, 1989. Sedimentary cycles in the Mediterranean Pliocene: discrepancies with the quasi-periods of the Earth's orbital cycles ? In: D.G. Smith, A. Berger, P.L. de Boer (eds), *Milankovitch Cyclicity in the Pre-Pleistocene Stratigraphic Record, Terra Abstracts*, 1(1), p. 241.

Hyde W.T. and W.R. Peltier, 1985. Sensitivity experiments with a model of the ice age cycle. The response of harmonic forcing. *Journal of Atmospheric Sciences*, 42(20), pp. 2170-2188.

Imbrie J., and J.Z. Imbrie, 1980. Modelling the climatic response to orbital variations. *Science*, 207, pp. 943-953.

Imbrie J., and N.G. Kipp, 1971. New micropaleontological method for quantitative paleoclimatology: application to a Late Pleistocene Caribbaen Core. In: K.K. Turekian (ed.), *Late Cenozoic Glacial Ages*, pp. 71-81, Yale University Press, New Haven.

Imbrie J., Hays J., Martinson D.G., McIntyre A., Mix A.C., Morley J.J., Pisias N.G., Prell W.L. and N.J. Shackleton, 1984. The orbital theory of Pleistocene climate: support from a revised chronology of the marine ^{18}O record. In: A. Berger, J. Imbrie, J. Hays, G. Kukla and B. Saltzman (eds), *Milankovitch and Climate*, pp. 269-305, D. Reidel Publ. Company, Dordrecht, Holland.

Imbrie J., McIntyre A. and A. Mix, 1989. Oceanic response to orbital forcing in the Late Quaternary: observational and experimental strategies. In: Berger A., Schneider S., and Duplessy J.Cl. (eds), *Climate and Geo-Sciences*, pp. 121-164, Kluwer, Dordrecht, Holland.

Janecek T.R. and D.K. Rea, 1984. Pleistocene fluctuations in Northern Hemisphere tradewinds and westerlies. In: A. Berger, J. Imbrie, J. Hays, G. Kukla and B. Saltzman (eds), *Milankovitch and Climate*, pp. 331-347, D. Reidel, Dordrecht, Holland.

Jouzel J., Lorius Cl., Petit J.R., Genthon C., Barkov N.I. and V.M. Kotlyakov and V.M. Petrov, 1987. Vostok ice core: a continuous isotope temperature record over the last climatic cycle. *Nature*, 329(6138), pp. 403-408.

Kominz M.A. and N.G. Pisias, 1979. Pleistocene climate: deterministic or stochastic? *Science*, 204, pp. 171-173.

Kutzbach J.E., 1985. Modeling of paleoclimates. *Adv. Geophys.*, 28A, pp. 159-196.

Laskar J., 1988. Secular evolution of the solar system over 10 millions years. *Astronomy and Astrophysics*, 198, pp. 341-362.

Lean J., 1991. Variations in the Sun's radiative output. *Review of Geophysics*, 29(4), pp. 505-535.

Lindzen R.S., 1986. A simple model for 100K-year oscillations in glaciation. *Journal of Atmospheric Sciences*, 43(10), pp. 986-996.

Loutre M.F., 1993. Paramètres orbitaux et cycles diurne et saisonnier des insolations. *Thèse de doctorat*, Faculté des Sciences, Université Catholique de Louvain, Louvain-la-Neuve. Unpublished manuscript.

McIntyre A., Ruddiman W.F., Karlin K. and A.C. Mix, 1989. Surface water response of the equatorial Atlantic Ocean to orbital forcing. *Paleoceanography*, 4, pp. 19-55.

Milankovitch M., 1941. *Kanon der Erdbestrahlung.* Royal Serbian Academy, Spec. publ. 132, section of Mathematical and Natural Sciences, vol. 33 (published in English by Israel program for Scientific Translation, for the U.S. Department of Commerce and the National Science Foundation, Washington D.C., 1969).

Molfino B., Heusser L.H. and G.M. Woillard, 1984. Frequency components of a Grande Pile pollen record: Evidence of precessional orbital forcing. In: A. Berger, J. Imbrie, J. Hays, G. Kukla and B. Saltzman (eds), *Milankovitch and Climate*, pp. 391-404, D. Reidel, Dordrecht, Holland.

Negrini R.M., Verosub K.L. and J.O. Davis, 1988. The middle to late Pleistocene geomagnetic field recorded in fine-grained sediments from Summer Lake, Oregon, and Double Hot Springs, Nevada, USA. *Earth and Planetary Science Letters*, 87, pp. 173-192.

Nicolis C., 1980. Response of the Earth-atmosphere system to a fluctuating solar input. In: *Sun and Climate*, pp. 385-396, CNES-CRS-DGRST, Toulouse, October 1980.

Nicolis C., 1982. Stochastic aspects of climatic transitions-response to a periodic forcing. *Tellus*, 34, pp. 1-9.

Oerlemans J., 1980. Model experiments on the 100,000-yr glacial cycle. *Nature*, 287, pp. 430-432.

Olsen P.E., 1986. A 40-million-year lake record of Early Mesozoic orbital climatic forcing. *Science*, 234, pp. 842-848.

Pestiaux P., van der Mersch I., Berger A. and J.Cl. Duplessy, 1988. Paleoclimatic variability at frequencies ranging from 1 cycle per 10,000 years to 1 cycle per 1,000 years: evidence for non-linear behavior of the climate system. *Climatic Change*, 12(1), pp. 9-37.

Pokras E.M. and A.C. Mix, 1987. Earth's precession cycle and Quaternary climatic changes in tropical Africa. *Nature*, 326, pp. 486-487.

Pollard D., 1982. A simple ice sheet model yields realistic 100 kyr glacial cycles. Nature, 296, pp. 334-338.

Pollard D., 1984. Some ice-age aspects of a calving ice-sheet model. In: Berger A., Imbrie J., Hays J., Kukla G. and Saltzman B. (eds), *Milankovitch and Climate*, pp. 541-564, Reidel Dordrecht, Holland.

Prell W.L., 1984. Monsoonal climate of the Arabian Sea during the late Quaternary: A response to changing solar radiation. In: Berger A., Imbrie J., Hays J., Kukla G. and Saltzman B. (eds), *Milankovitch and Climate*, pp. 349-366, Reidel, Dordrecht, Holland.

Ruddiman W.F. and A. McIntyre, 1984. Ice-age thermal response and climatic role of the surface Atlantic Ocean, 40°N to 63°N. *Geol. Soc. Am. Bull.*, 95, pp. 381-396.

Ruddiman W.E., Shackleton N.J. and A. McIntyre, 1986. North Atlantic sea-surface temperatures for the last 1.1 million years. In: Summerhayes C.P. and Shackleton N.J. (eds), *North Atlantic Palaeoceanography Geological Society Special Publication*, 21, pp. 155-173.

Saltzman B., Hansen A.R. and K.A. Maasch, 1984. The Late Quaternary glaciations as the response of a three-component feedback system to earth-orbital forcing. *Journal of Atmospheric Sciences*, 41(23), pp. 3380-3389.

Shackleton N.J. and N.G. Pisias, 1985. Atmospheric carbon dioxide, orbital forcing, and climate. In: E.T. Sundquist and W.S. Broecker (eds), *The Carbon Cycle and Atmospheric* CO_2*: natural variations Archean to Present*, pp. 303-317, Geophys. Mono. 32. Am. Geophys. Union, Washington D.C.

Shackleton N.J., Imbrie J. and N. Pisias, 1988. The evolution of oceanic oxygen-isotope variability in the North Atlantic over the past three million years. *Phil. Trans. R. Soc. London*, B318, pp. 679-688.

Shackleton N.J., Berger A. and W.R. Peltier, 1990. An alternative astronomical calibration of the lower Pleistocene timescale based on ODP site 677. *Transactions of the Royal Society of Edinburgh: Earth Sciences*, 81, pp. 251-261.

Short D.A., Mengel J.G., Crowley T.J., Hyde W.T. and G.R. North, 1991. Filtering of Milankovitch cycles by Earth's geography. *Quaternary Research*, 35, pp. 157-173.

Stothers R.B., 1987. Do slow orbital periodicities appear in the record of Earth's magnetic reversals ? *Geophysical Research Letters*, 14(11), pp. 1087-1090.

Weedon G.P., 1985/1986. Hemipelagic shelf sedimentation and climatic cycles: the basal Jurassic (Blue Lias) of South Britain. *Earth and Planetary Science Letters*, 76, pp. 321-335.

Wigley T.M.L., 1976. Spectral analysis: astronomical theory of climatic change. *Nature*, 264, pp. 629-631.

Woillard G.M., 1978. Grande Pile peat bog: A continuous pollen record for the last 140,000 years. *Quaternary Research*, 9, pp. 1-21.

Woolard E.W. and G.M. Clemence, 1966. *Spherical Astronomy*. Academic Press, New York, London.

Yiou P., Genthon C., Ghil M., Jouzel J., Le Treut H., Barnola J.M., Lorius Cl. and Y.N. Korotkevitch, 1991. High-frequency paleovariability in climate and CO_2 levels from Vostok ice-core records. *Journal of Geophysical Research*, 96 n° B12, pp. 20,365-20,378.

ANNEX 1 - Arithmetic mean of r

Let us calculate the mean value of the Earth-Sun distance over one year:

$$< r > = \frac{1}{T}\int_0^T r\,dt$$

If we introduce the eccentric anomaly, E, we have

$$r = a(1 - e\cos E)$$

E is an angle which varies with v and is used in the parametric equation of the ellipse (Brouwer and Clemence, 1961):

$$dv = \frac{a}{r}\sqrt{1-e^2}\,dE$$

and

$$\frac{dE}{dt} = \frac{2\pi}{T}\frac{a}{r}$$

which leads to:

$$< r > = \frac{a}{2\pi}\int_0^{2\pi}(1 - e\cos E)^2 = a\left(1 + \frac{e^2}{2}\right)$$

to compare with $r_m = a(1-e^2)^{\frac{1}{4}} \sim a(1 - \frac{e^2}{4})$. As an example for $e = 0.02$, $< r > = 1.0002\ a$ and $r_m = 0.9999\ a$. Energetically speaking the weighted averaged value of r is smaller than a, whereas the arithmetic mean is larger than a.

ANNEX 2

Length of the seasons.

It might be interesting to look for the position of the solstices and autumn equinox relative to a spring equinox arbitrarily fixed at March 21, for example. This might be done from the relationship between the real Sun and the mean Sun and the definition of the equinoxes and solstices. If λ is the longitude of the true Sun along the ecliptic, this longitude being measured counterclockwise from the vernal equinox, we have:

$$\sin\delta = \sin\lambda \sin\varepsilon$$

and the beginning of the astronomical seasons defined relative to the northern hemisphere is given by:

$$\begin{array}{lll} \lambda = 0 & \text{for} & \text{spring} \\ \lambda = 90^o & \text{for} & \text{summer} \\ \lambda = 180^o & \text{for} & \text{autumn} \\ \lambda = 270^o & \text{for} & \text{winter} \end{array}$$

If λ_m is the longitude of the mean Sun, we have (Brouwer and Clemence, 1961; Berger, 1978):

$$\lambda_m = \lambda - 2\left[\beta(1+\cos\varphi)\sin v - \beta^2(\frac{1}{2}+\cos\varphi)\sin 2v + \beta^3(\frac{1}{3}+\cos\varphi)\sin 3v - ...\right] \quad (1)$$

and

$$\begin{aligned} \lambda = \lambda_m & + \left(2e - \frac{1}{4}e^3 + \frac{5}{96}e^5 + \frac{107}{4608}e^7\right)\sin M \\ & + \left(\frac{5}{4}e^2 + \frac{11}{24}e^4 + \frac{17}{192}e^6\right)\sin 2M \\ & + \left(\frac{13}{12}e^3 + \frac{43}{64}e^5 + \frac{95}{512}e^7\right)\sin 3M \\ & + \; ... \qquad (2) \end{aligned}$$

with v and M being the true and the mean anomalies of the Sun, and

$$\begin{aligned} \cos\varphi &= \sqrt{1-e^2} \\ \beta &= \frac{1}{e}\left(1-\sqrt{1-e^2}\right) \end{aligned}$$

The mean solar longitude λ_m, is computed from (1) for the spring equinox ($\lambda = 0$) arbitrarily fixed at March 21; λ_m can then be calculated for each calendar day defined from an increment of the mean longitude given by $\Delta\lambda_m = \frac{360^o}{365}$, the length of the year being assumed to be 365 mean solar days. From these successive fixed values of λ_m, λ can be computed from (2) to allow the computation of the insolation for that particular calendar day. It must be noted that there is a drift artificially created by introducing the calendar day instead of the theoretical solar day defined by the longitude, λ, of the true Sun. This can be seen when comparing Figure A.2.1 showing the daily irradiation computed at the autumn equinox (September) and Figure A.2.2 showing the daily irradiation computed at September 21. This drift is due to the fact that, at that fixed calendar date, the Earth is not occupying the same position relative to the Sun for different times of the past. 114 kyr ago, September 21, was corresponding to $\lambda = 174^o$; 103 kyrs ago it was corresponding to $\lambda = 189^o$; presently, it is 178°.Considering a given λ instead of a given λ_m insures that identical physical situations are intercompared through geological times as far as the position of the Earth relative to the Sun is concerned. Unfortunately, it is either true that a given interval $\Delta\lambda$ will not correspond to the same absolute time length at the different point of the Earth's orbit (due to second law of Kepler), a problem solved by considering λ_m instead.

On the other hand, for each particular value of λ defining the equinoxes and the solstices, λ_m can be computed from (1). Therefore, from a increment, $\Delta\lambda_m = \frac{360^o}{365}$, it is possible to compute the calendar day which corresponds to the autumn equinox (September) and to the solstices for any time of the past and of the future, or equivalently the length of the astronomical seasons (see Table 2 and Figure A.2.3).

These lengths are totally related to precession as it is confirmed by their hereafter analytical expression. Indeed the second law of Kepler tells us that the area swept over by the line joining the Sun and a planet is proportional to the time elapsed. If T_1 and T_2 are the time elapsed while the radius vector is sweeping over the area A_1 and A_2 respectively, we have

$$T_1 = \frac{A_1}{A_2} T_2$$

Let us now consider the particular case of the astronomical seasons (defined as hereover):

$$\begin{aligned} A_{\text{year}} &= \frac{1}{2} \int_{v=-\tilde{\omega}}^{v=2\pi-\tilde{\omega}} a^2 \rho^2 dv \\ A_{\text{spring}} &= \frac{1}{2} \int_{v=-\tilde{\omega}}^{v=\pi/2-\tilde{\omega}} a^2 \rho^2 dv \end{aligned}$$

$$
\begin{aligned}
A_{\text{summer}} &= \frac{1}{2}\int_{v=\pi/2-\tilde{\omega}}^{v=\pi-\tilde{\omega}} a^2\rho^2 dv \\
A_{\text{autumn}} &= \frac{1}{2}\int_{v=\pi-\tilde{\omega}}^{v=3\pi/2-\tilde{\omega}} a^2\rho^2 dv \\
A_{\text{winter}} &= \frac{1}{2}\int_{v=3\pi/2-\tilde{\omega}}^{v=2\pi-\tilde{\omega}} a^2\rho^2 dv
\end{aligned}
$$

Neglecting the terms in e^2 and higher in ρ^2, the length of the seasons is finally given by:

$$
\begin{aligned}
T_{spring} &\cong \frac{T}{4}\left(1 - \frac{4e}{\pi}(\sin\tilde{\omega} + \cos\tilde{\omega}\right) \\
T_{summer} &\cong \frac{T}{4}\left(1 - \frac{4e}{\pi}(\sin\tilde{\omega} - \cos\tilde{\omega})\right) \\
T_{autumn} &\cong \frac{T}{4}\left(1 + \frac{4e}{\pi}(\sin\tilde{\omega} + \cos\tilde{\omega})\right) \\
T_{winter} &\cong \frac{T}{4}\left(1 + \frac{4e}{\pi}(\sin\tilde{\omega} - \cos\tilde{\omega})\right)
\end{aligned}
$$

where T is the length of the year. This clearly shows that the length of the astronomical seasons are only a function of the climatic precession.

For the "half-year" astronomical seasons of the northern hemisphere, we can therefore write:

$$
\begin{aligned}
T_{summer} &= \frac{T}{2}\left(1 - \frac{4e\ \sin\tilde{\omega}}{\pi}\right) \\
T_{winter} &= \frac{T}{2}\left(1 + \frac{4e\ \sin\tilde{\omega}}{\pi}\right)
\end{aligned}
$$

if $0 \leq \tilde{\omega} \leq 180^o$, the half-year northern hemisphere, summer is shorter than winter and longer for $180^o < \tilde{\omega} < 360^o$.

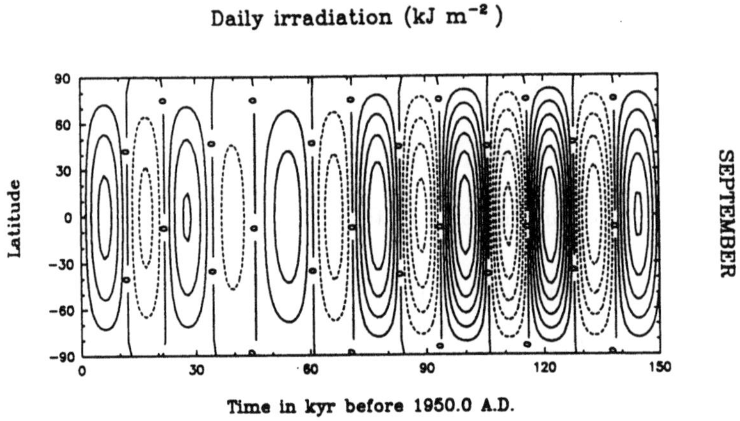

Figure A.2.1: Long term variation of the deviation from present day values for Autumn equinox (September) of the daily irradiation ($kJ\,m^{-2}$), contour interval 500 $kJ\,m^{-2}$. Solid lines are positive and dashed lines are negative deviations.

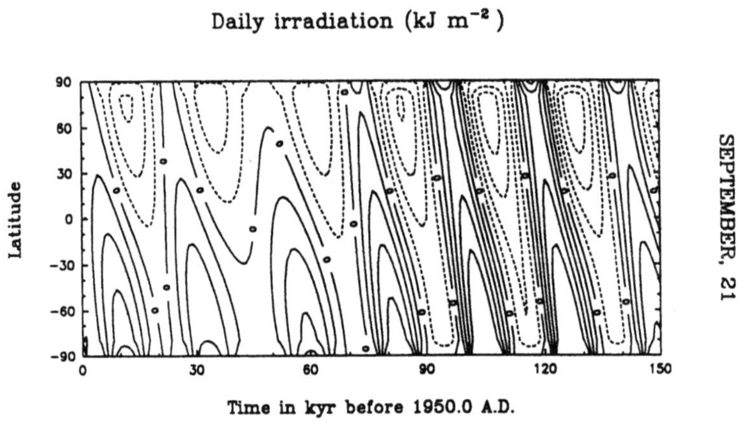

Figure A.2.2: Long term variation of the deviation from present day values for calendar day September 21 of the daily irradiation ($kJ\,m^{-2}$), contour interval 1000 $kJ\,m^{-2}$. Solid lines are positive and dashed lines are negative deviations.

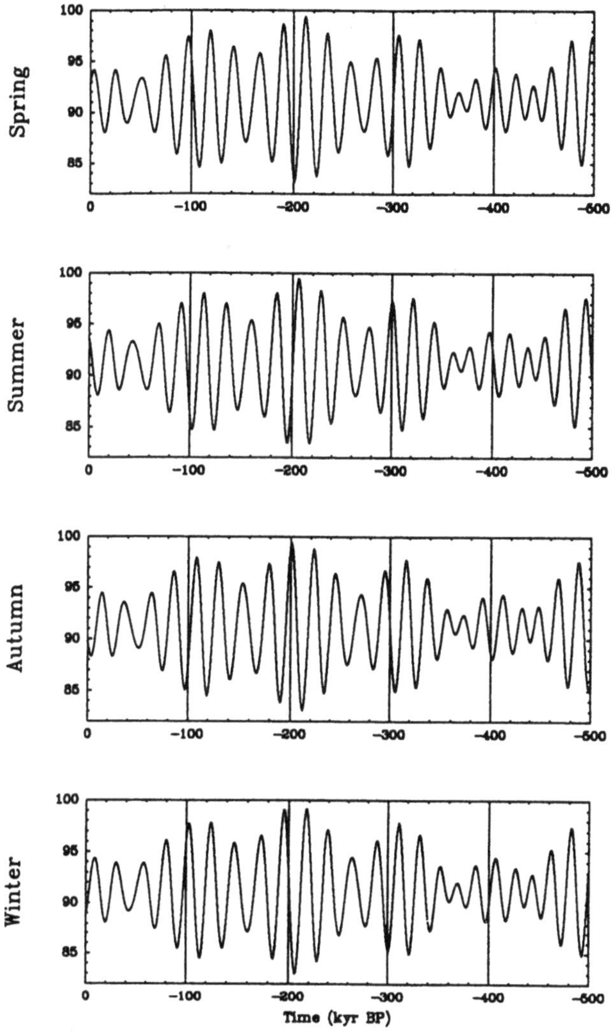

Figure A.2.3: Long-term variations of the length of the seasons over the last 500 kyr; from top to bottom: spring, summer, autumn, winter.

BIOSPHERE MODELING FOR CLIMATE STUDIES

Michel M. Verstraete and Philippe Martin
Institute for Remote Sensing Applications
CEC Joint Research Centre, TP 440
I-21020 Ispra (VA)
Italy

Abstract. The role of the biosphere in the global climate system is briefly outlined, and their interaction is investigated through two specific modeling exercises. First, since vegetation is sensitive and slowly adjusting to climate fluctuations, its composition and diversity reflect the interplay of the laws of succession under the constraints of climatic and environmental stress. This interaction can be studied with the help of dynamic gap forest models. A particular instance of this class of models is used to investigate whether the forest affects ecological succession through a modification of its own microclimate. It is found that such feedbacks are not strong enough in the wet climate of Minnesota, and this implies that the pollen record truly represents the overall climatic conditions of the region. This paper also proposes a new approach to the exploitation of trees as a climate archive. Dendroclimatology has traditionally resorted to empirical correlations between tree-ring characteristics and climate in the recent past, and extrapolated such relations beyond the period of instrumental records. An explicit modeling approach provides a powerful and flexible tool, but its application requires the technique of model inversion. A simple tree-ring model is constructed, and it is shown that this method retrieves the climatic information of a particular year from an analysis of the variability of rings between different trees subject to the same climate, rather than from the ring-to-ring variations in the same tree.

1 Introduction

The atmosphere is the component of the climate system with the fastest response time. The atmosphere is also the locus of repeated severe events with catastrophic consequences for human activities, such as tropical cyclones, tornadoes, violent storms, droughts, floods, etc. It is not surprising, therefore, that the bulk of the effort to understand the state and evolution of our environment focused first on the atmosphere, and especially on weather forecasting. At the same time, it is well known that the local weather conditions at any particular location, or the overall features of the general circulation and of the climate as a whole are intimately dependent on the exchanges of momentum, energy, and a variety of chemical compounds at the surface (*e.g.*, Ojima, 1990).

The atmosphere is in direct contact with all other components of the climate system, namely the oceans, the terrestrial environments, and the cryosphere. A growing interest in surface-atmosphere interactions has been evident for the past few decades, and the

NATO ASI Series, Vol. I 22
Long-Term Climatic Variations
Edited by J.-C. Duplessy and M.-T. Spyridakis

role of the biosphere in particular has been investigated in some detail. This results from a deeper understanding of the nature and function of various biological processes, such as photosynthesis, respiration, and evapotranspiration, but also of the side effects of the presence of vegetation, such as a modification of the albedo and roughness of the surface (*e.g.*, Verstraete and Dickinson, 1986; Avissar and Verstraete, 1990).

Beyond these scientific questions, substantial issues need to be addressed with respect to the maintenance of a living environment, not only for human life but also for countless animal and plant species. Indeed, the biosphere not only influences the atmosphere, but is also affected by it. Climatic changes expected to result from the increase in greenhouse gases concentrations will affect the distribution and state of health of the biomes, and these effects are largely unknown (*e.g.*, Bolin, 1980; Houghton *et al.*, 1990; Houghton *et al.*, 1992).

By the same token, the biosphere has been significantly affected by past climatic changes, and these effects are still noticeable today. For example, the stress caused by drought or temperature extremes is still perceptible in the annual rings of trees. These impacts are not restricted to plants, of course, and can be seen in lake varves, or ice accumulation in glaciers (*e.g.*, Bradley, 1991). Nevertheless, some features of the plant kingdom can be interpreted in terms of past climates and add to the interest of climatologists for the biosphere.

At least two difficulties need to be addressed when investigating the relations between the atmosphere and the biosphere. First is the extreme diversity of biological processes and the intrinsic complexity of the living world. Indeed, living organisms obey not only the fundamental laws of physics but also are the locus of biological and ecological processes. The diversity of shape and function, the variety of species, the laws of genetics, competition and evolution attest of a world of staggering complexity, which is far from being understood and therefore difficult to model.

The second difficulty in modeling these interactions has to do with the rather large differences in the characteristic space and time scales at which the physical processes in the atmosphere and climate system, and the biological processes in the organisms, individuals and ecosystems take place: For a given time scale, biological processes tend to occur at much smaller spatial scales than do the physical processes in the atmosphere. One consequence of this mismatch is that only very crude attempts have been made so far to represent the biosphere in atmospheric general circulation models.

In this paper, we would like to expose the reader to two particular forms of biological modeling, not so much to show how to represent the biosphere in climate models, but, rather, to point to the possible exploitation of the biosphere as a climatic archive, provided the proper tools can be designed for the retrieval of the hidden information. Two main conditions are required in order to meaningfully interpret observable environmental records in terms of past climatic fluctuations: (1) the system under study must have been sensitive to these climate variations, so that significant changes in one or more observable variables are expected to result from changing weather patterns, and (2) the preservation of the signal over long time periods must be ensured, either because the system itself has a long life time, or because other processes are taking place to protect the information. Such records exist both in the living world (*e.g.*, trees, corals) and in the physical environment (*e.g.*, lake varves, glaciers). Paleoclimatologists rely almost exclusively on the interpretation of these natural archives to reconstruct past environmental conditions.

In the first section below, we use a gap-phase forest dynamics model to see whether a forest can sufficiently alter its own microclimate over periods of hundreds of years to modify its ecological succession, by altering the availability of light, water and nutrients. In the following section, we present a new perspective on the retrieval of quantitative information from tree-rings, based on the inversion of explicit tree-ring models.

2 Forests as natural climatic archives

The boundary layer surrounding an object determines how easily it can exchange mass, energy and momentum with its environment. The corresponding aerodynamic resistances of a single tree are several orders of magnitude lower than those of a forest. An isolated individual is consequently much more exposed to weather stresses than the same tree in a wood or a forest. Vegetation therefore has the capacity to alter its own microclimate in a way which can affect its physiology, and the evapotranspiration of a forest is not equal to the sum of the rate of water loss that would be experienced by each individual tree outside of the forest environment. Conversely, only the trees located near the edge of a forest are fully affected by varying climatic conditions: they constitute a more sensitive climatic archive than the more protected trees inside the forest.

At a larger scale, can vegetation modify its own microclimate sufficiently to influense ecological succession? And if so, what kind of climate does pollen analysis yield? Climatically sensitive gap-phase forest dynamics models are designed to investigate these questions. Such a model is presented below, and results from a few simulations will be discussed.

2.1 Description of a climatically sensitive gap-phase forest dynamics model

The efficient description of the establishment, growth and mortality of trees that make up the dynamics of the unmanaged forest requires advanced models which combine both deterministic and stochastic processes. For instance, gap-phase dynamics models of the type pioneered by Botkin *et al.* (JABOWA, 1972) track the life cycle of each individual tree in a series of forest stands and use Monte Carlo simulations to describe secondary succession processes. Similarly, tree establishment and growth are functions of stand structural constraints. The latter determine the amount of light available for photosynthesis and of limitations in resources, such as moisture, nutrients and temperature as well as random forcings. Mortality is a stochastic function of the species-specific expected longevity of each tree and its growth history.

Over the years, gap-phase dynamics models have been developed by a number of authors (*e.g.*, West *et al.*, 1981 and Shugart, 1984). They now encompass a wide variety of geographical locations and forest types. In this category of models, the Energy, water, and momentum eXchange, and Ecological dynamics model (EXE; Martin, 1990, 1992) constitutes a first attempt to couple a physically and physiologically based water budget and an explicit treatment of ecological dynamics. EXE can be used in conjunction with actual climatic data or driven by a General Circulation Model.

EXE simulates the transfer of water and energy (1) through the stomatal apertures into the aerodynamic boundary layer surrounding the leaves, (2) within the canopy, across the aerodynamic boundary layer at the soil surface, and (3) between the canopy and the air above it. Radiative transfer in the canopy is computed using a two-stream approximation (*e.g.*, Dickinson, 1983, p. 317). The temperature as well as the vapor pressure, at the soil surface, of the canopy, and of the air within the canopy, are calculated simultaneously with the amount of water stored in the canopy, the evaporation from wet leaf surfaces, the evaporation from dry leaf surfaces, and the evaporation from the ground. Infiltration is computed iteratively as a function of throughfall and the specific humidity of the first soil layer. The diffusion of heat and water are calculated by solving the Richards equation (*e.g.*, Hillel, 1982, p. 221).

The forest microclimate is computed once at daytime and once at nighttime for each of the 365 days of the year. Environmental conditions, as perceived by the trees, are summarized at the end of each simulation year. This information is then transmitted to the forest ecology part of the model.

EXE uses as its biological core a modified version of the LINKAGES model developed by Pastor and Post (1986). The modifications include in particular diagnostic routines tracking the birth, growth and death of trees and the relative importance of the environmental factors affecting them. EXE also incorporates additional biological parameterizations representing, for instance, root growth and distribution in the soil profile.

Carbon (C) and nitrogen (N) cycling are treated explicitly. Different time steps are used for different parts of the model: growth and decomposition are computed monthly, whereas the birth and death of trees is updated yearly. The Monte Carlo simulation is performed on 10 plots of 0.08 ha each and up to 72 eastern North American tree species can be considered. Each year (including year one) and for each plot, the model selects seedlings from this pool (Pastor and Post 1985, p. 30–34). In its present formulation, the model only includes trees. Inputs to this part of EXE are the soil hydraulic properties, its initial carbon and nitrogen contents and 26 species-specific parameters, such as maximum age and decomposition rates for each litter type and species. The four biological prognostic variables for each plot are the number of individuals of each species, the increment in diameter at breast height for each individual tree, the foliar mass of each individual tree and the C to N ratio of each ecosystem component.

Each model simulation produces a large number of variables, which can be analyzed statistically. These include the characteristics of the ecosystem (total above ground biomass, leaf litter production, leaf litter nitrogen content, above ground net primary production, available nitrogen, soil CO_2 emissions, soil organic matter, and species above ground biomass), agroforestry indicators (growing season degree days and growing season drought days), the physical condition of the forest (minimum leaf area index, maximum leaf area index, fine root biomass, rooting depth, and maximum tree height), the forest micrometeorological status (the aerodynamic and physiological resistances, the temperature and the vapor pressure of the air above the canopy, of the air in the canopy, at the surface of the canopy, and at the surface of the soil, precipitation, canopy interception, throughfall, snow accumulation, runoff, the radiative forcing on the ground, the radiative forcing on the canopy, the soil latent heat flux, the dry canopy latent heat flux, the wet canopy latent heat flux, the total canopy latent heat flux, the specific water content of each of the ten soil layers, and the soil temperature profile), the demography of the forest (birth and mortality rates, as well as statistics about the factors causing them), and growth limitation by light, temperature, moisture, or nitrogen.

The present climate is simulated using daily micrometeorological data for incoming solar radiation, ambient air temperature, ambient air humidity, and precipitation from the *Typical Meteorological Year* dataset (National Climatic Center, 1981).

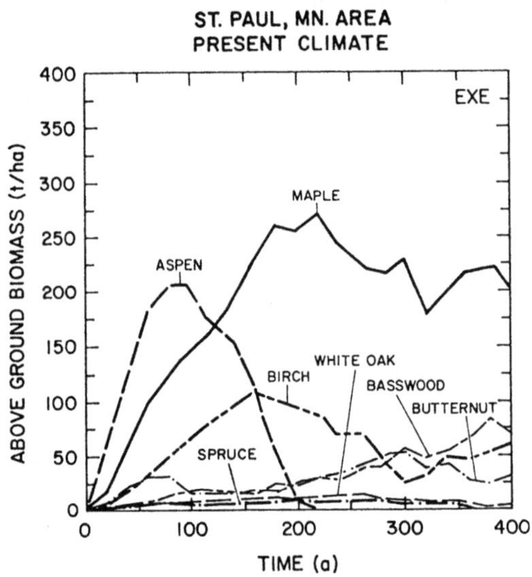

Figure 1: Simulation of forest growth for the St. Paul, MN, area, under present climatic conditions.

EXE performs very satisfactorily in the two reference regions, the areas around St Paul and Duluth, MN (Martin, 1992). The soil is assumed to be a uniform clay loam (U.S. Department of Agriculture, 1983). Küchler's map of potential vegetation (U.S. Geological Survey, 1970, pp. 90–91) was used to assess the validity of the forests being simulated. In the St Paul area, three types of vegetation may be found, namely oak savanna (*Quercus-Andropogon*), maple-basswood (*Acer-Tilia*), and northern flood-plain forests (*Populus-Salix-Ulmus*) along the Minnesota river. As shown in Figure 1, EXE produces the maple-basswood forest typical of this region (*e.g.*, *cf.*, Vankat, 1979, pp. 136, 138–139, and 142; and Braun, 1950, p. 327). In addition, Braun (1950, p. 335) states that "the early abundance of aspen and birch followed by maple-basswood association as maple, basswood, and oak seedlings develop" is well documented in large disturbed areas.

In the Duluth area, there are four typical vegetation assemblages. Towards the North, Great Lakes pine (*Pinus banksiana, P. resinosa, P. strobus*) forests may be found (*e.g.*, Braun, 1950, p. 343). In less northern locations, Great Lakes spruce-fir (*Picea-Abies*) forests are encountered (*e.g.*, Braun, 1950, p. 383). Towards the South, maple-basswood forest may grow in local pockets. Finally, in mesic (*i.e.*, moist) areas conifer bogs develop (*Picea, Larix, Thuja*).

EXE produces a spruce (*Picea glauca, P. mariana*) dominated forest with some yellow birch (*Betula alleghaniensis*) in the understory (Figure 2). During the first 150 years of the simulation, aspen (*Populus grandidentata, P. tremuloides*) is very abundant.

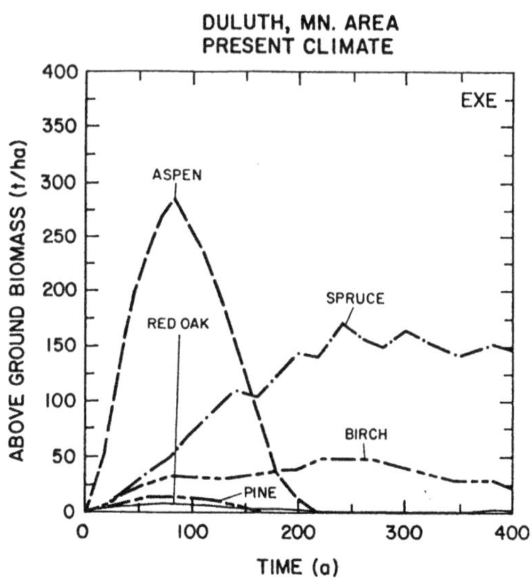

Figure 2: Simulation of forest growth for the Duluth, MN, area, under present climatic conditions.

2.2 Studies of forest microclimatic feedbacks

As mentioned above, the biological and physical parameters which determine the transfer of energy, water, and momentum change as the forest develops. When these changes are significant, they can alter the development of the forest, which could then further modify the transfer parameters themselves.

Two types of phenomena are most likely to occur. First, as the forest matures, it increases its leaf area index and develops a deeper and fuller root system. In a moist climate, this may enable it to evaporate more water than a grassland. A general decrease in soil moisture availability can then result. Second, differences in the canopy temperature due to differences in solar radiation interception and in evapotranspiration may change the dates at which the growing season begins and ends.

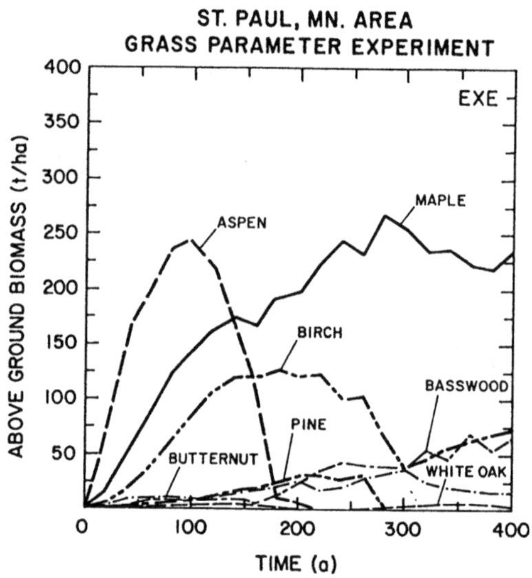

Figure 3: Simulation of forest growth using the grassland physical parameters for the St. Paul, MN, area, under present climatic conditions.

To examine the possibility of a biological feedback on vegetation, two simulations are performed, for the climate conditions of St Paul, MN, and Duluth, MN. The forest is first simulated using the physical parameters characteristic of the onset of secondary succession. This first case is referred to as the "grassland parameter experiment". The simulation is then repeated with the physical parameters of a mature forest, and this second case is referred to as the "mature forest parameter experiment". The interesting differences between these two scenarios concern the leaf area index, the rooting depth and the fine root biomass, and the height of the vegetation. Possible differences in the stomatal behavior are ignored. If the microclimate of the model forest has little or no effect on its development, the results of the grassland and mature forest parameter experiments should be indistinguishable.

For St Paul, MN, with grassland parameters, the yearly averaged difference in degree day sum is −71. The dynamics of the forest is significantly altered during the first 300 years (Figure 3). However, the forest ultimately converges toward an ecological state similar to that of the control (Figure 1). The growth limiting factors and the number of individuals per species per plot are the same. The yearly averaged difference in degree-day sum between the control simulation and the simulation with mature forest parameters is +41 for St Paul, MN. Because of the higher temperature during the first 120 years, the above-ground biomass of the aspen trees colonizing the plots is much lower (Figure 4). However, the forest converges toward the same ecological state as the control, as in the previous case.

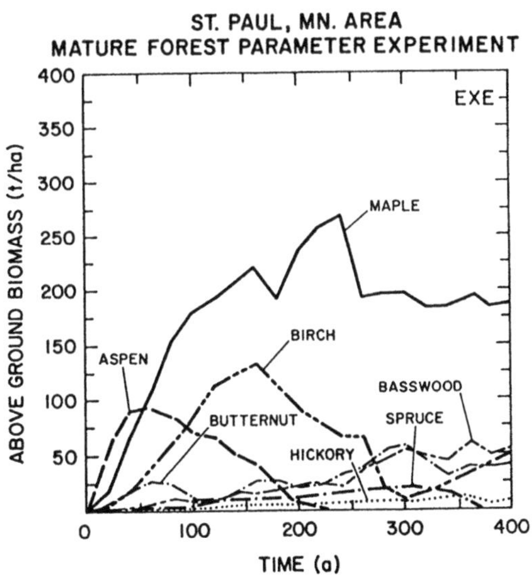

Figure 4: Simulation of forest growth using the mature forest physical parameters for the St. Paul, MN, area, under present climatic conditions.

For Duluth, MN, with grassland parameters (Figure 5), the yearly averaged difference in degree day sum is −5, whereas with mature forest parameters (Figure 6), there is virtually no difference. As a result, in both cases, the behavior of the forest is not significantly different from that of the control (Figure 2).

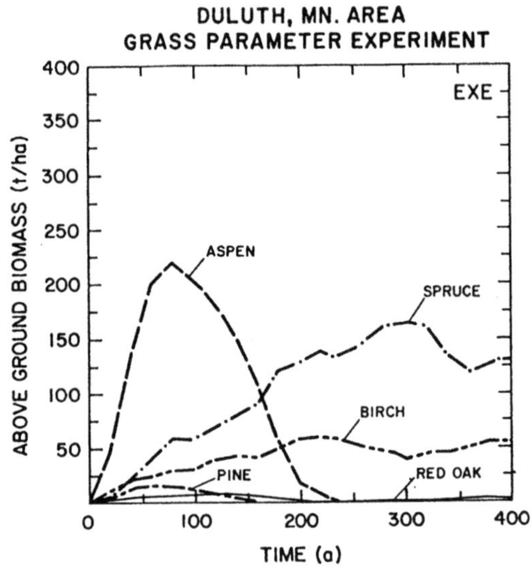

Figure 5: Simulation of forest growth using the grassland physical parameters for the Duluth, MN, area, under present climatic conditions.

2.3 Discussion

Minnesota has a relatively wet climate. Hence, it does not provide the best location to examine vegetation feedbacks on its microclimate. Also, because this is a one-way forcing scheme, changes in the forest microclimate do not feedback on the atmosphere: They cannot affect the mesoscale climate. Variations in canopy temperature caused by biological feedbacks occasionally produced noticeable differences in the early development of the forest, but the simulated forests converged toward an ecological state similar to that of the control. Studies in drier areas would therefore be necessary to fully assess the importance of biologically mediated micrometeorological feedbacks.

The capacity of the forest to affect its evolution by modifying its microclimate has implications for the retrieval of paleoclimatic information from pollen analysis: If the

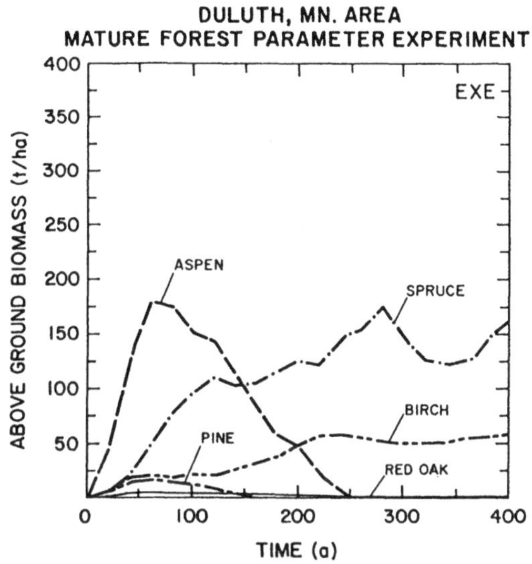

Figure 6: Simulation of forest growth using the mature forest physical parameters for the Duluth, MN, area, under present climatic conditions.

local microclimatic modifications are limited, the pollen distribution may be representative of the general climate of the area. If, however, the forest does modify appreciably its own environment, it is likely that the pollen distribution reflects these local conditions, and the interpretation of the record may be more difficult.

3 A new look at tree-ring analysis

The existence of an internal structure in trees has been observed and exploited since antiquity by wood craftsmen, but it is only in the early years of this century that A. E. Douglass suggested this wood structure could be used to reconstruct past climates. Subsequently, significant work has been accomplished to try to retrieve qualitative or quantitative information about past climates from the width (or, more recently, from other physical or chemical characteristics) of these rings (*e.g.*, Fritts, 1976; Schweingruber, 1988).

3.1 The classical approach

Most temperate trees normally grow a single ring each year through cell division in the xylem. In the first part of the growing season (spring and early summer), relatively large thin-walled (hence lighter) cells are produced. Later growth results in smaller cells with thicker walls, which appear darker (late wood). This differentiation is the cause of the visible aspect of the rings. There are significant exceptions to this annual pattern: trees in tropical regions either do not seem to produce visible rings at all, or to produce them at irregular intervals. Also, mid-latitude trees sometimes develop additional (or 'false') rings for reasons that are not always entirely clear. Because of these difficulties, the discussion below will be restricted to typical temperate trees.

So far, the presence of rings in tree sections has been exploited in the following manner: First, the rings are identified and counted. Assuming a single ring is produced each year, the year of formation of each ring can be established as soon as the date of one ring is known. This condition is fulfilled if a core is taken from a living tree. When dealing with older materials, ring chronologies of trees in the same area can be compared and correlated, so that multiple events are identified in both series, and the undetermined series can be dated. Next, the width of each ring is measured. Nowadays, other ring characteristics can be measured also, including ring density, cell count, chemical or isotopic composition, etc. In any case, this procedure results in the creation of one or more time series of annual measurements for each tree core or wood sample. The third step consists in interpreting the year-to-year variations in these measurements in terms of interannual changes in climate. This can be done empirically by comparing the series to actual measurements of the relevant meteorological variables over the recent past, and extrapolating the relation further back to retrieve estimates of the same variables for the years covered by the tree-ring series. An abundant literature is available on these methods: in addition to the seminal works mentioned above, further information and additional references can be found, for instance, in Hughes (1991) and Lamb (1977).

It is worthwhile reflecting on the conceptual process implied by this approach: ring width (or any other such variable) surely depends both on the type of tree species, the age and health status of the tree, and on various environmental parameters. Since all observations in the time series refer to the very same tree, the data can be 'normalized' to account for these effects, or the interpretation effort can focus on relative variations from year to year. The main source of interannual variability is assumed to be weather elements, whose impact on the tree is represented by a 'response function'. The latter describes how the tree is affected by changes in weather conditions, and may or may not be known explicitly. It is, however, assumed to apply equally well at all times.

3.2 The modeling approach

To the extent these response functions are empirical correlations between observed recent ring series and measured weather events, they may provide a way to characterize past climates but do not contribute to a more fundamental understanding of the relevant geophysical or ecological processes responsible for the observations. From a scientific research point of view, it would be more satisfying to develop a mechanistic model of tree-ring to describe how a tree actually grows a ring, as a function of the available resources (*e.g.*, nutrients) and environmental constraints. To our knowledge, few authors have ventured in this direction (*e.g.*, Fritts *et al.*, 1991).

The implication of this approach, however, has not been fully investigated yet, and is the focus of the rest of this paper. A mechanistic model of ring growth allows the retrieval of past climatic information from a different perspective, namely by inversion of this model against the observed data. It will be seen that this approach attempts to determine which weather patterns must have occurred in a given year to result in the production of the rings observed in different trees, rather than to find the sequence of weather events that best correlates with the observed ring chronology in a single tree.

If ring width was a simple function of a single climatic parameter (say, temperature), we could directly invert that function, analytically or numerically, to find the temperature corresponding to each observed ring. However, ring width is a function of multiple variables, operating on a variety of space and time scales, such as air temperature, soil water content, health and age of the individual, or even some stable parameters such as slope and aspect of the terrain, tree species, etc. Even if we had a comprehensive model that predicted ring width as a function of all the relevant variables, we could not invert it mathematically against a single measurement of ring width because we have only one equation and many unknowns.

This inverse problem occurs in a number of fields of science, and the standard approach consists in numerically searching for the solution to a slightly different but closely connected problem. We first attempt to compensate for the lack of equations by taking multiple measurements. A number of mathematical conditions must be realized for this to work, but if all the variables that affect the growth of rings except for the climatic parameters are known or constant, we assume that the variability in the observations unaccounted for by these variables must be due to the climate.

We then ask the question differently: instead of requiring the exact value of the climatic parameters (say, air temperature and soil moisture) that resulted in the growth of an observed ring, we require a 'best estimate' of these parameters, where 'best' is understood as the 'most likely' value, given our current understanding of the dependency of ring growth on climatic parameters. Cast in this way, we reduce the problem to one of optimization, where we search for the minimum of a figure of merit function (typically a least mean square estimate of the difference between the observed and the predicted ring widths). The values of the climatic parameters which correspond to this minimum are those that best account for the observed variance in ring width, and are assumed to be the values that most likely occurred at the time.

3.3 A toy tree-ring model

This approach and the issues it raises will become clearer with an example. Since the focus of this discussion is mostly methodological, we construct a very simple model, which will be called our toy tree-ring model. Let's assume that the width of a ring depends on the tree species, and on the growing season average air temperature and water availability in the soil. Specifically, the size of the ring will be taken to increase non-linearly with temperature, up to a species-dependent threshold, and to decrease beyond this optimal temperature. Since many biological processes exhibiting optimum temperature conditions (such as photosynthesis and primary productivity) appear to be well described by a second-degree equation, we represent the temperature dependency of the ring width as follows:

$$\lambda_T = \left[1 - \left(1 - \frac{T}{T_{opt}}\right)^2\right] \tag{1}$$

where T_{opt} (in C) is the optimal average operating temperature for the particular species of tree being observed, and T (in C) is the average air temperature during the growing season. This formal dependency is shown in Figure 7.

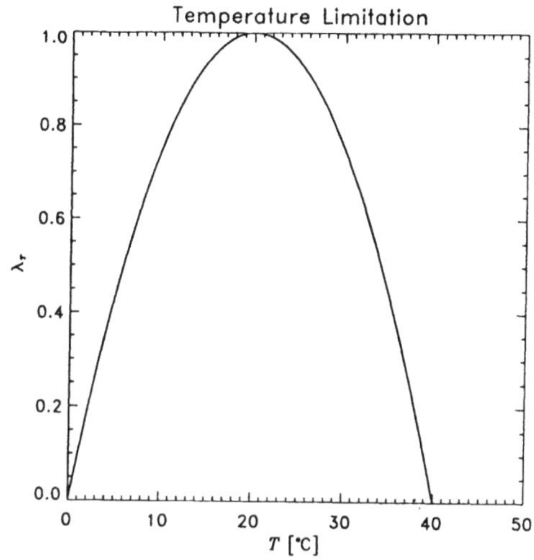

Figure 7: Temperature-dependent growth limiting factor in the toy tree-ring model.

Simultaneously, ring width is assumed to be proportional to an index of soil water availability, defined as the relative excess of water above the wilting point, with no further dependency beyond the field capacity. The wilting point is the volumetric soil water amount for which plants start to wilt, while the field capacity is the volumetric soil water amount which remains in the soil some time (say a few days to a week) after it is saturated and left to drain naturally. Better defined concepts should be used for detail work on soil-plant water relations (see *e.g.*, Hillel, 1982), but these will suffice for our purpose. We will further assume that the ring does not grow at all if the average soil water content during the growing season does not exceed the wilting point, and that any soil water in excess of the field capacity does not further affect the width of the ring, either positively or negatively. Mathematically, these conditions are expressed as follows:

$$\begin{aligned} \lambda_W &= 0 \qquad \text{if } \theta < \theta_{wilt} \\ \lambda_W &= \frac{\theta - \theta_{wilt}}{\theta_{field} - \theta_{wilt}} \qquad \text{if } \theta_{wilt} < \theta < \theta_{field} \\ \lambda_W &= 1 \qquad \text{if } \theta_{field} < \theta \end{aligned} \tag{2}$$

where θ_{wilt} and θ_{field} (both in $m^3\,m^{-3}$) are the wilting point and the field capacity, two observable soil properties, and θ is the average volumetric soil water content (in $m^3\,m^{-3}$) over the growing season. This function is displayed in Figure 8.

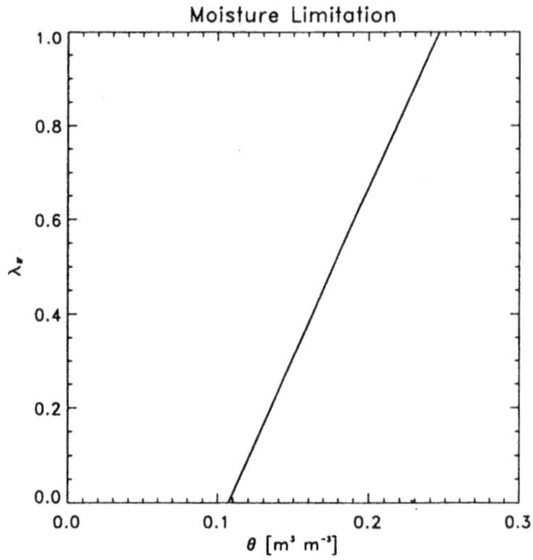

Figure 8: Soil moisture-dependent growth limiting factor in the toy tree-ring model.

The complete toy tree-ring width model is then given by

$$\rho = \rho_0 \, \lambda_T \, \lambda_W \tag{3}$$

where ρ (in μm) is the model-predicted tree-ring width, and ρ_0 (in μm) is the yearly tree-ring width under optimal conditions ($T = T_{opt}$ and $\theta = \theta_{field}$).

This model therefore describes the annual width (ρ) of a tree-ring in terms of two species-dependent variables (ρ_0 and T_{opt}), two soil characteristics (θ_{wilt} and θ_{field}), and two climatic parameters (T and θ). Of the six variables used in the right hand side of the model equation, four are known *a priori*, in the sense that they can be objectively determined when the tree core is obtained, and we are left with a single equation in two unknowns.

Of course, the growth of a ring in a given year does not depend only on the climatic conditions during that year, but also on the past history of the tree and the sequence of stresses it has sustained. These effects could be included in the model by making it time-dependent, but the added complexity would obscure the main point of this discussion. We will therefore proceed with this static model.

3.4 Generation of a synthetic data set

Since we will need multiple ring widths 'observations' to invert the model, we first generate a synthetic data set by using the model in *direct mode*. Clearly, it does not help to keep measuring the same ring over and over again: no new information is gained, except perhaps a better statistical estimate of the actual width. It does not help either to measure successive rings of the same tree, because the principle of the inversion procedure is to retrieve *constant* model parameters. We must therefore naturally measure the widths of rings in different trees in the neighborhood of each other, *i.e.*, in trees which have experienced the *same* climatic stress.

Table 1 shows how the 'observed' data set is being constructed for a single year. The first two columns display the properties of the trees, namely the expected ring width ρ_0 under optimal meteorological conditions, and the optimal temperature T_{opt} for the given species. The next two columns describe the soil in terms of its hydraulic properties: the water content at the wilting point θ_{wilt} and at field capacity θ_{field}. The values given here are meant to be realistic but do not result from actual measurements.

Table 1: Assumed tree, soil, and climatic data; observed ring widths

ρ_0 [mm]	T_{opt} [C]	θ_{wilt} [$m^3\,m^{-3}$]	θ_{field} [$m^3\,m^{-3}$]	T [C]	θ [$m^3\,m^{-3}$]	ρ_{obs} [mm]
1.80	20.00	.170	.332	22.00	.200	0.330
1.80	20.50	.170	.332	22.00	.200	0.332
2.00	21.00	.107	.246	22.00	.200	1.336
2.00	21.50	.107	.246	22.00	.200	1.338
2.20	22.00	.107	.246	22.00	.200	1.472
2.20	22.50	.107	.246	22.00	.200	1.472
2.40	23.00	.146	.320	22.00	.200	0.744
2.40	23.50	.146	.320	22.00	.200	0.742
2.60	24.00	.146	.320	22.00	.200	0.802
2.60	24.50	.146	.320	22.00	.200	0.798

The fifth and sixth columns represent the actual average weather conditions (temperature T and soil moisture θ) assumed for the growing season of the considered year. These are the values we will be interested in recuperating from an analysis of the tree-ring record by inverting the model. The last column of Table 1 exhibits the actual ring width of each tree for that particular year.

In a realistic situation, we know which tree is cored, so we have access *a priori* to the information contained in the first four columns of this Table (nature of the tree and properties of the soil). Furthermore, the last column, the actual ring width, is the measurable variable. We would like to retrieve information about the climate of the selected year from these observations.

3.5 Inversion of the toy tree-ring model

This is achieved as follows: First, we guess an arbitrary (but reasonable) value for the climate parameters of the year under study, for example, that the average temperature and soil moisture during the growing season were 25 C and 0.24 $m^3\,m^{-3}$, respectively. Given these guesses, we can use the model to predict a 'theoretical value' of ring width, *i.e.*, the ring width that would have been produced by the trees if the weather conditions had been as guessed. This operation is repeated for each tree in the sample, and the sum of the squares of the differences between the observations (ρ_{obs}) given in Table 1 and the model-predicted values (ρ_{mod}) is computed:

$$\delta^2 = \sum_{i=1}^{M} [\rho_{obs} - \rho_{mod}]^2 \tag{4}$$

where $M = 10$ is the number of observations (tree-rings) used. Computed ring widths will most probably not correspond to the 'observed' ones, but we aim to identify which combination of average air temperature and soil water content would best describe the observed variability in ring width by selecting slightly different values of the climatic parameters, and repeating this operation, until we identify the combination which corresponds to the smallest value of δ^2. Standard numerical methods are available to locate the minimum of a multivariate function (*e.g.*, Press *et al.*, 1992). When that minimum is found, the current values of T and θ are taken to be the most probable or likely climate parameters for that year, because they best account for the variance in the data set, given the form of the model.

Table 2 summarizes the initial guess and the values actually retrieved by the numerical algorithm. It can be seen that the latter is very close indeed to the 'true values' that are also indicated. The Table also shows the final value of δ^2 and the root mean square error at the minimum of this function. It is important to realize that these true values were not used in the retrieval.

Table 2: Initial guess, retrieved and actual climatic parameters

	Initial guess	Retrieved	Actual
Temperature [C]	15.00	21.96034	22.0
Soil moisture [$m^3\,m^{-3}$]	0.22	0.20003	0.20
δ^2	$1.39\ 10^{-5}$		
RMS	$4.17\ 10^{-3}$		

3.6 Discussion and conclusion

Numerous mathematical and biological issues would need to be discussed further in connection with this model inversion approach, such as the optimal design of the tree-ring model, the requirements that need to be satisfied for the inversion to be possible, the numerical methods of optimization, etc. Some of these issues are discussed by Verstraete and Pinty (1992), and Pinty and Verstraete (1992), in the context of satellite remote sensing. We would like to close this paper, however, with a short discussion of the methodological differences between this method and the classical approach to dendroclimatology.

The principle of a mechanistic model of tree-ring is that the growth of the ring can be estimated explicitly in terms of a set of variables describing the conditions of observations (type of tree species, characteristics of soil, etc.) and a set of climate parameters. These variables and parameters are linked to the measurable ring quantity (in our case ring width) through the model equation. The inversion approach, for its part, attempts to analyze the variability in the observed variable, resulting from a diversity of conditions of observations, to retrieve the most probable unknown model parameters. For this to work, it is essential that these parameters be constant over the period of observation, otherwise the inversion may produce undesirable results. For instance, if the observations of rings from multiple years are combined for the inversion, then only the average climate parameters for the entire period can be retrieved, given the tree and soil properties. To put it differently, the inversion of a model against a data set should only be attempted to retrieve the values of constant parameters.

Since the weather changes from year to year, the retrieval of past climatic information by inverting a tree-ring model such as described above requires to work one year at a time on a number of trees, rather than one tree at a time for a number of years. In fact, if the model were inverted against a ring chronology from a single tree, given a weather history, it should logically be able to retrieve the tree species or whatever species-specific characteristics the model uses. Since we usually know or can determine this information, this would not be of great interest to the paleoclimatologists.

The implication of this approach is that tree-ring data should be collected differently, for example by acquiring cores from as many tree species as possible. And if the soil hydraulic status turns out to be an important driving variable in ring growth, soil properties should also be collected together with the cores.

Another major point needs to be mentioned here, even though we will not be able to discuss it in detail. It is very difficult to validate an empirical model because any correlation found between two data sets may be applicable only to the particular (local and temporal) conditions in which these data were collected: There is no *a priori* guarantee that such a relation would hold at other places, for other times, or with other tree species. By contrast, since the parameters of a physically-based model are precisely defined measurable quantities (in our case, the average temperature and soil moisture during the growing season), and since the inversion procedure retrieves the best estimate of these parameters from an analysis of the ring data, it is possible to validate the model by comparing the results of the inversion to the observed historical values. In other words, in this explicit modeling approach, the climatic data accumulated during the last 50 years or so, is used to validate the tree-ring model, not to statistically fit an empirical relation through the observations.

Tree-ring modeling merits to be investigated further, especially by tree physiologists and dendroclimatologists, because it will provide both a more fundamental understanding of the relation between tree growth and climate, but also improve our capability to analyze objectively the existing record in terms of past climates. This approach may require to look at the paleoclimatic problem from a different perspective, and to acquire different data than is currently done, but it provides an objective and very powerful way to validate the tree-ring model.

Acknowledgments

This new approach of tree-ring analysis was originally proposed by Michel Verstraete at the 5th Annual Climate and Forest Workshop on 'Water Resources and Global Change', organized jointly by the National Center for Atmospheric Research and the US Forest Service in Boulder, 8–10 October 1991. The comments of the participants in this workshop were instrumental in encouraging us to investigate this issue further. The comments of Dr. Malcolm Hughes on the final draft of this paper and the help of Stéphane Flasse in setting up the programs to invert the simple tree-ring model are greatly appreciated. The financial support of the Exploratory Research programme of the Joint Research Centre is acknowledged.

References

Avissar, R. and M. M. Verstraete (1990) 'The representation of continental surface processes in mesoscale atmospheric models', *Reviews of Geophysics*, **28**, 35–52.

Bolin, B. (1980) *Climatic Changes and Their Effects on the Biosphere*, WMO No. 542, World Meteorological Organization, Geneva, 49 pp.

Bradley, R., Editor (1991) *Global Changes of the Past*, Proceedings of the OIES Global Change Institute, 24 July–4 August 1989, Snowmass, Office for Interdisciplinary Earth Studies, Boulder, Colorado, 514 pp.

Fritts, H. (1976) *Tree Rings and Climate*, Academic Press, London.

Fritts, H. C., E. A. Vaganov, I. V. Sviderskaya, and A. V. Shashkin (1991) 'Climatic variation and tree-ring structure in conifers: empirical and mechanistic models of tree-ring width, number of cells, cell size, cell-wall thickness and wood density', *Climate Research*, **1**, 97–116.

Houghton, J., G. Jenkins, and J. Ephraums, Editors (1990) *Climate Change: The IPCC Scientific Assessment*, Cambridge University Press, Cambridge, 365 pp.

Houghton, J. T., B. A. Callander, and S. K. Varney, Editors (1992) *Climate Change 1992: The Supplement Report to the IPCC Scientific Assessment*, Cambridge University Press, Cambridge, 200 pp.

Hughes, M. (1991) 'The tree-ring record', in *Global Changes of the Past*, Edited by Bradley, R., Proceedings of the OIES Global Change Institute, 24 July–4 August 1989, Snowmass, Office for Interdisciplinary Earth Studies, Boulder, Colorado, 117–137.

Lamb, H. (1977) *Climate: Present, Past and Future, Vol. 2: Climatic History and the Future*, Methuen, London, 835 pp.

Ojima, Dennis, Editor (1992) *Modeling the Earth System*, Proceedings of the OIES Global Change Institute, 16–27 July 1990, Snowmass, Office for Interdisciplinary Earth Studies, Boulder, Colorado, 488 pp.

Pinty, B. and M. M. Verstraete (1992) 'On the design and validation of bidirectional reflectance and albedo models', *Remote Sensing of Environment*, **41**, 155–167.

Press, W. H., S. A. Teukolsky, W. T. Vetterling, and B. P. Flannery (1992) *Numerical Recipes*, Second Edition, Cambridge University Press, 963 pp.

Schweingruber, F. (1988) *Tree Rings: Basics and Applications of Dendrochronology*, D. Reidel, Dordrecht, 276 pp.

Verstraete, M. M. and R. E. Dickinson (1986) 'Modeling surface processes in atmospheric general circulation models', *Annales Geophysicae*, **4**, 357–364.

Verstraete, M. M. and B. Pinty (1992) 'Extracting surface properties from satellite data in the visible and near-infrared wavelengths', Proceedings of the TERRA-1 Conference on Understanding the Terrestrial Environment, Taylor and Francis, London, 203–209.

I I - GEOLOGICAL EVIDENCES

MID-LATITUDE ICE SHEETS THROUGH THE LAST GLACIAL CYCLE : GLACIOLOGICAL AND GEOLOGICAL RECONSTRUCTIONS

Geoffrey S. Boulton and Tony Payne
Department of Geology and Geophysics
University of Edinburgh
Kings Buildings
Edinburgh EH9 3JW

The principal role of ice sheets in the climate system is to modulate the low frequency component of long-term change. They are coupled to other parts of the system by interactions across their upper, basal and marine boundaries. Numerical models of ice sheet behaviour require a description of the physics of time dependent thermo-mechanically coupled flow, specification of the boundary conditions and an appropriate time dependent climatic forcing function. The basis of such a model is described.

The behaviour of ice sheets in time generates products which are preserved in the geological record. This relationship between forcing function, glacial process and geological product can be used as a basis for forward and quasi-inverse approaches to modelling in which glaciological theory and geological data are integrated. In the forward model, geological data is used to test the assumptions of model and forcing. In the quasi-inverse model, the data is used to constrain the model and to infer the forcing. These approaches are used to reconstruct the European ice sheet through the last glacial cycle and to reconstruct the European and Laurentide ice sheets' final decay. The model predicts the time-dependent discharge of meltwater from storage to the ocean, the pattern of deglacial sea level change and the rates of net ablation through time. Whereas the landward margins of ice sheets may primarily respond to atmospheric forcing, the activity of the marine margins may be strongly influenced by internal, dynamic instabilities. We examine the collapse of the marine margins of the last British ice sheet and the instability of the marine margins of ice sheets in general, suggesting a mechanism which could play a role in generating "Heinrich layers".

1.0 THE ROLE OF ICE SHEETS IN THE CLIMATE SYSTEM

The sun is the ultimate source of energy which drives the climate system. However, the spatial and temporal variation of energy fluxes through the system, and the magnitude of the earth's response

NATO ASI Series, Vol. I 22
Long-Term Climatic Variations
Edited by J.-C. Duplessy and M.-T. Spyridakis

to solar insolation changes is determined by the geography of the earth's surface and the interactions between the oceans, atmosphere and ice sheets and the biogeochemical processes which determine atmospheric chemistry and the magnitude of the "greenhouse" effect. These components of the Earth's surface system, which determine its response to insolation changes, are strongly coupled and are sometimes referred to as the *climate system.* Understanding the nature of the coupling between the components of this system lie at the heart of contemporary attempts to understand climate change.

The spatial variation of climate depends on the efficiency with which incident solar energy, which is greatest in the equatorial regions and least in polar regions, is redistributed by the the earths fluid envelope of oceans and atmosphere. Because of this, the earths topography, largely determined by the tectonic response to the pattern of internal heat dissipation, controls the large scale pattern of circulation of the atmosphere and oceans, which in turn determines the rate and pattern of poleward heat transport by atmosphere and oceans.

Figure 1 shows the annual mass fluxes between the three major components of the system and the water masses in storage. The transfer of mass from the oceans to build late Quaternary ice sheets required an increase in the net mass flux of no more than 5% over 10,000 years and required no more than 0.04% of the earth's absorbed solar radiation in a similar time to destroy them.

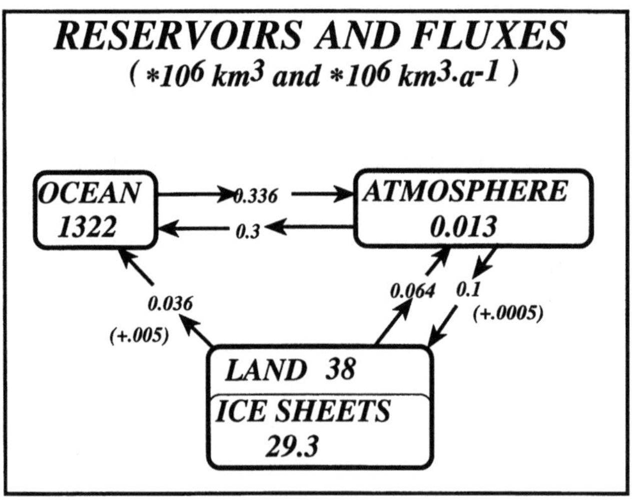

Figure 1 *Water storage in the three principal components of the climate system and the annual fluxes between them.*

The ice sheets play a fundamental role in the operation of the climate system through the coupling which exists at their three boundaries.

2.0 ICE SHEET COUPLIMG WITH OTHER PARTS OF THE CLIMATE SYSTEM

2.1 At the atmospheric boundary

a) **Albedo feedback** whereby ice sheet growth in response to cooling increases the earths reflectivity, thereby leading to further atmospheric cooling.

b) **Topographic coupling** with the atmosphere whereby ice sheet growth leads to change in the atmospheric flow around the ice sheet, changing the mosaic of accumulation on it and thereby influencing its further pattern of growth.

2.2 At the ocean boundary

a) **Meltwater coupling** whereby ice sheet decay, produced either by climate warming or dynamic instability in the ice sheet, generates a meltwater flux which cools the ocean surface sufficiently to produce a major atmosperic impact which can slow down or reverse ice sheet decay.

b) **Sea level coupling** whereby increasing sea level at the marine margin of an ice sheet flowing into the ocean will tend to increase the flow rate and drawdown of the ice sheet surface, leading to ice sheet decay. Conversely, sea level lowering can produce ice sheet advance without any change in mass balance.

2.3 At the basal boundary

a) **Basal instability** may be produced where the friction at the bed of the ice sheet is reduced relatively rapidly so that the ice sheet surface slope, which determines the basal shear stress, is reduced and the ice sheet expands. Its surface may be lowered to such an extent that enhanced surface melting causes the ice sheet to decay. Reduction in basal friction may occur because of :
- melting at the hitherto frozen bed of the ice sheet;
- expansion of the ice sheet over deformable sediment;
- an increase in accumulation rate which increases glacier flow which, in turn, increases the melt rate sufficiently to reduce basal friction.

b) **Lithosphere deflection** occurs when mass is transferred from the oceans, which have an almost global distribution, to ice sheets, which are more localised. Recent glacial periods have been associated with a reduction of the order of 100m in the depth of the global ocean, and a concomitant creation of ice sheets of the order of

1000-3000m in thickness in a few areas of the continent surface. This redistribution of mass is associated with flow in the asthenosphere away from the centres of growing ice sheets, and flexure of the lithosphere. The interaction between ice sheets and lithosphere is highly coupled. Sinking of the lithosphere beneath the ice sheet will tend to lower the ice sheet surface, thus affecting its surface mass balance.

3.0 ICE SHEET MODELLING

Numerical models which attempt to simulate the behaviour of ice sheets in time and space within the context of the climate system require three ingredients:

- a mathematical description of the temperature-dependent **physics of flow** and possible sliding of the ice sheet over its bed.
- specification of the **boundary conditions** (surface, bed and possible marine boundary) through which the ice sheet is coupled to other components of the climate system (see 2.0).
- time dependant **forcing functions** at the boundaries (generally the upper boundary), which drive ice sheet variation in time and space.

3.1 Temperature-dependent flow

The temporal evolution of the three dimensional form of an ice sheet can be described using the continuity equation for ice thickness (Mahaffy, 1976):

$$\frac{\partial h}{\partial t} = -\frac{\partial}{\partial x}(\bar{u})(H) + b \qquad (1)$$

where (H) is ice thickness, (t) is time and (x) is the horizontal coordinate. The rate of ice thickness change at any point is a function of local mass balance (b) and the local convergence/divergence of horizontal flow, which is found from the horizontal variation of ice discharge. The time evolution is found by integrating (1) through time. This requires the mass balance (b), which is prescribed, and vertically-integrated horizontal ice velocity (u) to be known at each point.

The shear stress (τ)which drives flow is given by:

$$\tau_{xz} = -\rho_i g\,(H+h-z)\frac{\partial}{\partial x}(H+h) \qquad (2)$$

(Budd and Jenssen,1975), where (ρ_i) is the density of ice. The strain rate is derived from Glen's(1955) flow law for ice, from which the velocity (u) in the ice mass can be deduced:

$$u = -2(\rho_i g)^n \left[\frac{\partial(H+h)}{\partial x}\right]^{n-1} \left(\frac{\partial(H+h)}{\partial x}\right) \int_h^z A(T)(H+h-z)^n \, dz + u_h \qquad (3)$$

where (n) is a constant in Glen's Law, (A) is a temperature-dependent proportionality factor, (T) is ice temperature and (u_h) is the vertical velocity at the ice/bed interface (effectively the melting rate). Sliding occurs at the basal boundary when the temperature there is at the melting point. We use a simple sliding law of the form:

$$u_s = a\tau_b{}^n \left(\frac{1}{N}\right)^m \qquad (4)$$

where u_s is the basal sliding velocity, τ_b is the basal shear stress, N is the effective normal stress at the ice/bed interface and a and m are parameters.

Flow and temperature are highly coupled because of the temperature dependence of flow, because advection plays a major role in determining the temperature distribution in the ice sheet and because flow generates frictional and strain heating. The temperature model is based on the conservation of energy:

$$\frac{\partial T}{\partial t} = \frac{k_i}{\rho_i c_i}\frac{\partial^2 T}{\partial z^2} - u\frac{\partial T}{\partial x} - w\frac{\partial T}{\partial z} + \frac{g}{c_i}(H+h-z)\frac{\partial u}{\partial z}\frac{\partial}{\partial x}(H+h) \qquad (5)$$

where (c_i) is the specific heat capacity of ice and (k_i) its thermal conductivity.

When ice temperature is at the pressure melting point, melting will occur at a rate (S):

$$S = \frac{k_i}{L\rho_i}\left(\frac{\partial T}{\partial z_c} - \frac{\partial T}{\partial z_b}\right) \qquad (6)$$

where subscripts (c) and (b) refer to the basal temperature before and after melting (Budd and others, 1971) and (L) is the latent heat of ice. The temperature gradient is determined by the geothermal heat flux, frictional heating at the bed, strain heating in the ice and the conductivity of the medium. Ritz (1987) has shown that the temperature evolution of the top 2km of the lithosphere interacts with that of the ice sheet.It is therefore included in the model.

3.2 Boundary conditions

a) Basal boundary

The existence of sliding at the basal boundary depends upon the temperature of the basal ice. When this is at the melting point, sliding occurs, and is described in the model by equation (4).

Time dependent lithosphere deflection beneath the ice load depends upon the diffusivity of the asthenosphere, which determines the rate of deflection, and the stiffness of the crust, which determines the shape of deflection. If we assume zero lithosphere rigidity, diffusion in the asthenosphere will determine the impact of ice sheet load on the elevation of the crust (h(x,t)):

$$\frac{\partial h}{\partial t} = D_a \frac{\partial^2}{\partial x^2}\left(h - h_0 + \frac{\rho_i}{\rho_m} H\right) \qquad (7)$$

where D_a is the diffusivity of the asthenosphere and $h_0(x)$ is the relaxed bedrock elevation.

b) Upper boundary

For the upper boundary we require a mass balance condition and a mean annual temperature for every point on the ice sheet surface. We use a parameterised mass balance distribution, but could equally well use output from a GCM or energy balance model. The equation used to derive a mass balance is obtained from the form of empirical mass balance distributions with elevation on modern glacier surfaces (Boulton and others, 1984):

$$\begin{array}{ll} b = \alpha(H{=}h-\lambda) - \beta(H+h-\lambda)^2 & \text{if } H+h-\lambda \angle 2000\text{m} \\ b_{max} & \text{if } H+h-\lambda \geq 2000\text{m} \end{array} \qquad (8)$$

where b_{max} is the maximum allowed positive mass balance, (λ) is the ELA and (α) and (β) are constants.

Ice sheet surface temperature (T_S) is assumed to depend solely on sea level air temperature (SLAT) and on the lapse rate (ψ) :

$$T_s = SLAT - \psi(H+h) \qquad (9)$$

c) Marine boundary

The behaviour of the marine boundary is determined by three processes:

- possible floating of the glacier terminal tongue
- increased tendency to sliding as buoyant forces reduce the effective stress at the ice/bed interface
- increased ablation through the iceberg calving mechanism

The grounding line is the point at which flotation begins. It is defined by:

$$h_b = H \frac{\rho_i}{\rho_w} \qquad (10)$$

where (h_b) is the depth of water at the grounding line, (H) the ice thickness at the grounding line and (ρ_w) the density of water.
Sliding will be influenced by buoyancy through the effective stress in equation (4).
The iceberg calving rate is assumed to be a function of buoyancy (B) at the grounding line, where:

$$B = \rho_i\, g\left(h_B - \frac{\rho_i}{\rho_w} H\right) \qquad (11)$$

where h_b is the elevation of the bed in relation to sea level and where (C), the calving rate, is ,

$$\bar{C} = dB^f \qquad (12)$$

(d) and (f) are parameters.

3.3 Forcing functions

The model is forced by varying the surface temperature (SLAT and/or Ψ) and the equilibrium line altitude (ELA) on the upper boundary. We realise that these characteristics are a reflection of more complex atmosphere/ice sheet interactions, and such interactions can be included if the ice sheet model is coupled with an atmosphere/ice sheet energy balance model. In the examples used in this article, we have taken proxy records of palaeoclimate as means of generating time dependent change in SLAT and ELA functions.
Conditions on the basal boundary are prescribed.
If the model were used to follow the concurrent evolution of all ice sheets, relative sea level at the marine margin would be part of the model output. However, in the examples below, we follow the evolution of single ice sheets, and thus a component of sea level change due to the other ice sheets is prescribed. This is added to the relative sea level component due to the modelled ice sheet.

4.0 RESPONSE OF AN ICE SHEET TO CHANGE IN ITS BOUNDARY CONDITIONS

Changes in the overall geometry of an ice sheet and in the location of ice divides come about because of changes in the surface, basal or marine boundary conditions. Figure 2 shows the nature of the response to changes in these three boundary conditions:

2a - **Atmospheric changes** redistribute the mass balance mosaic on the ice sheet surface, causing a shift in the ice divide towards the area of highest net accumulation. The precise location of the ice sheet margin is highly sensitive to changes in net mass loss in the ablation area.

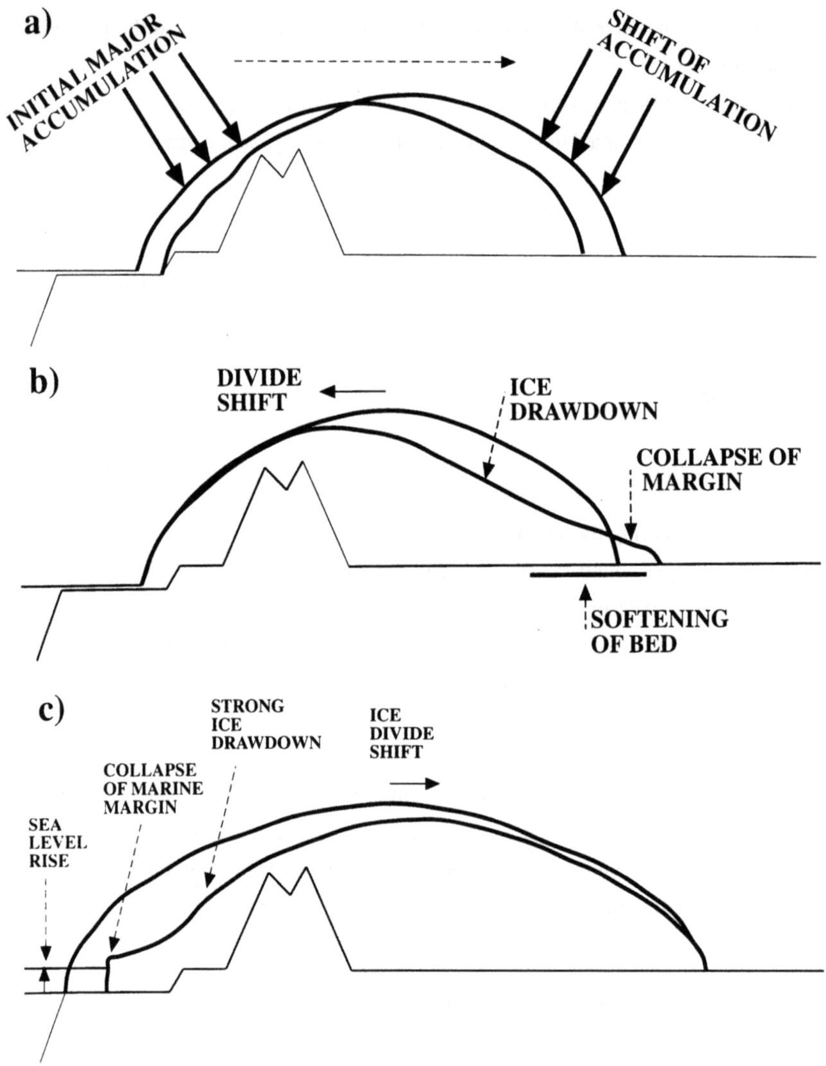

Figure 2a-c *Large-scale response of an ice sheet to the following changes in boundary conditions: a) Change in the distribution of mass balance on the surface. b) Reduction in friction at the basal boundary. c) Increase in relative sea level, and therefore of buoyancy, at the marine boundary.*

2b - **Basal boundary** changes in, for instance, thermal regime at the bed from a frozen to a melting condition, or the build up of deformable sediment at the ice /bed interface, may reduce the basal friction and thus cause flattening of the ice sheet profile. In addition, sinking of the lithosphere beneath the ice sheet may cause lowering of its surface. In both cases there will be an effect on the surface mass balance, which will effect in turn the ice sheet's dynamics.

2c - **Marine boundary** changes may have a major influence on ice sheet geometry and dynamics. For instance, an increase in sea level will increase the calving rate of icebergs through increased buoyancy at the marine margin. This will tend to draw down the marine margin, and which, if large enough, may cause the ice divide to retreat. The effect may be sufficiently large to influence the non-marine margin.

Most geological investigations of past ice sheet behaviour concentrate upon the evidence of fluctuations in their land margins. Interpretations tend to be similarly restricted, generally invoking changes in the mass balance of the terminal zone as a cause of ice sheet margin change. Shifts in centres of mass or changes in the marine or basal boundary condition, which could equally account for changes in the margin, tend to be ignored. Geologists should seek to find evidence which will discriminate between these causes.

5.0 INTEGRATION OF GEOLOGICAL THEORY AND GEOLOGICAL DATA IN ICE SHEET SIMULATION

5.1 General

Ice sheet models can be run as parts of very much larger models of the whole climate system (e.g. Berger and others,1990), in which case they tend to be relatively simple in comparison with the atmospheric component and produce relatively poor simulations of the behaviour of individual ice sheets.

In this article, we concentrate on glaciological reconstructions of ice sheets which attempt to simulate the direct geological evidence of past ice sheet behaviour as closely as possible. In such an approach, it is computationally impracticable at present to attempt to model the whole coupled system. We therefore consider the direct forcing of the ice sheet by uncoupled changes in temperature and mass balance at the upper surface and sea level at the marine margin.

The behaviour of an ice sheet in time and space leaves evidence the geological record (figure 3). Provided that an ice sheet model is associated with sub-models which can simulate aspects of this glacial geology, a model can provide a theoretical connection between variation in the climatic boundary conditions and the ice sheet's geological output. We can use this connection in two ways:

The forward model : starting from a given climate forcing function and predicting glacial geology.

- **The inverse model**: using glacial geology to constrain the ice sheet model and determining the form of climate forcing function which is compatible with glacial geology

In the forward model, we use glacial geology to test the model and the climate forcing function. In the inverse model, we use the model to infer the climate forcing function which is compatible with glacial geology.

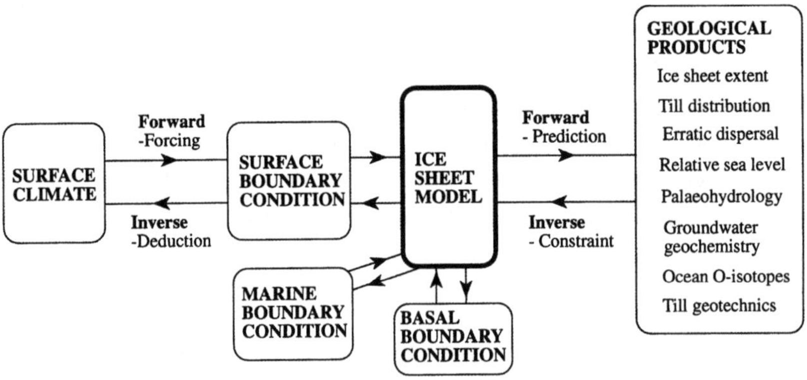

Figure 3 *The logic of glaciolological/geological modelling and the products of glacial action which leave evidence in the geological record. The geological evidence can be used to test hypotheses about ice sheet dynamics or the climate drive, or used to constrain models so as to deduce the climate drive.*

Unlike many other proxy climate indicators which, like pollen or marine microfaunas, tend to reflect localised conditions, ice sheets display integrated behaviour on a continent-wide scale, and thus tend to reflect large scale patterns of climate variation, although they can, like other parts of the climate system, show behaviour which is a reflection of internal dynamics alone. Climate is often regarded solely as an atmospheric phenomenon. Whereas it is possible to take this view for short timescales, on which little change in the nature and geometry of ocean ciculation or ice sheet dynamics occurs, on longer timescales, oceanic and ice sheet changes tend to condition the possible range of atmospheric variability. Thus, a geological view of climate requires analysis of the whole climate system. In such a view internal dynamic changes are also climatic changes. The real contrast lies between internally- and externally-forced behaviour, both of which are attributes of the climate system.

The forward and inverse approaches have been applied to the ice sheets which developed over North America and Europe during the last glacial cycle, illustrated largely by the example of the

European ice sheet. Further details of the method are contained in Boulton and Payne (1992 and in press).

5.2 The forward model applied to the last glacial cycle in Europe

In this approach, we utilise a temporal function which describes the evolution of climate through the last glacial cycle to drive the model. Output from the model which leaves evidence in the geological record includes:

- fluctuation of the ice margin in time and space
- the time dependent pattern of crustal flexure
- the distribution of till generated by the ice sheet

We use the sea surface temperatures in the north-east Atlantic derived by Ruddiman and others (1986) as a guide to the general trend of climate through the cycle. We assume a maximum

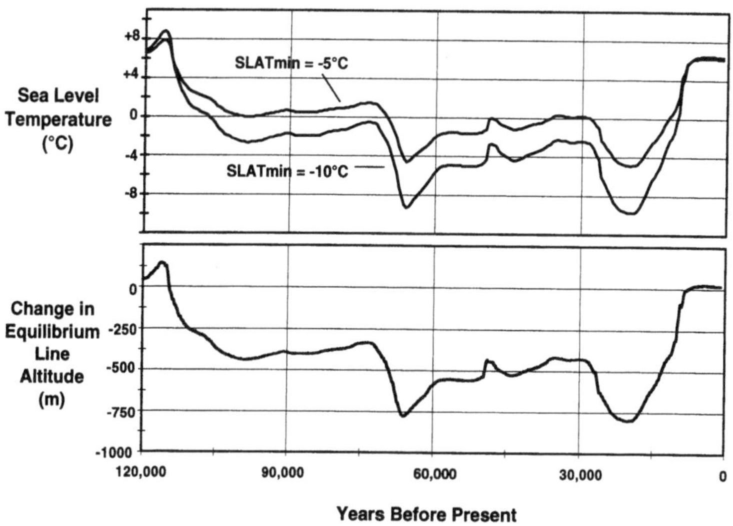

Figure 4 *Changes of sea level air temperature (SLAT) and equilibrium line altitude (ELA) over the ice sheet through the last glacial cycle derived from contemporary sea surface temperatures in the N.E.Atlantic from Ruddiman and others (1986). It shows two versions of SLAT, one based on a glacial maximum temperature of -10°C and another of -5°C.*

temperature depression for north-west Europe compared with the present day from estimates produced by Zagwijn (1985), and an equilibrium line altitude (ELA) depression which is enough to cause the ice sheet to advance to its known maximum. The form of the ELA surface is that of the snowline for the present day (Büdel,1949). A simple linear relationship is assumed between sea level temperature (SLAT), ELA and the SST curve of Ruddiman and others (1986). The procedure is described in Boulton and Payne (1992) and the results are shown in figure 4.

As the ELA descends, it first intersects the mountains of Norway, where ice sheet growth begins in the northern and south-western sectors. The model then solves for the temperature dependent flow of the ice sheet (Boulton and Payne, 1992) and analyses interactions at the surface, basal and marine boundaries in the way summarised in sections 2 and 3.

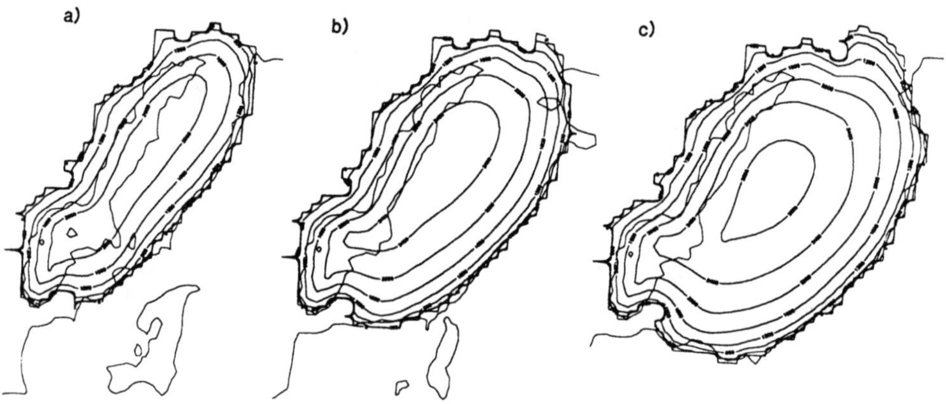

Figure 5 *Modelled pattern of growth of the European ice sheet at: a) 7,500 b) 10,000 and c) 15,000 years after initiation. Contours are plotted for 400m intervals. The modelled pattern simulates well the patterns reflected in the geological record. Note the major embayment in the SW sector of the ice sheet produced by strong calving at the head of the deep Norwegian Channel.*

The modelled form of the ice sheet during its expansion is shown in figure 5.

Two features are particularly important:

As the ice sheet expands to the west from the initial nucleation areas, strong calving and drawdown prevent the ice sheet from expanding beyond the continental shelf edge. However, the ice sheet is free to expand in a easterly and south-easterly direction. As a consequence (c.f. figure 2), the ice divide migrates away from the mountains in a south-easterly direction, and flow directions change. In the north, they rotate in an anticlockwise direction, whilst in the south they rotate in a clockwise direction during glacier expansion. Evidence of such shifts of flow direction has been found (Hirvas and others, 1986; Böse, 1990).

An area of deep water which follows the coast of south-western and southern Norway, the Norwegian Channel, has a major effect on the ice sheet. It is so efficient in drawing down the margin that the ice sheet is unable to expand to the south-west. As a consequence, a great embayment occurs in this sector of the ice sheet which is very clearly reflected in the geological evidence of patterns of advance and retreat (figure 15).

In order to illustrate further aspects of ice sheet behaviour, a transect through the ice sheet has been chosen, extending from the western continental shelf of Norway to northern Germany, which approximates to an ice sheet flowline for most of the last glacial cycle. The character of the ice sheet at the last glacial maximum along this flowline is illustrated in figures 6 and 7.

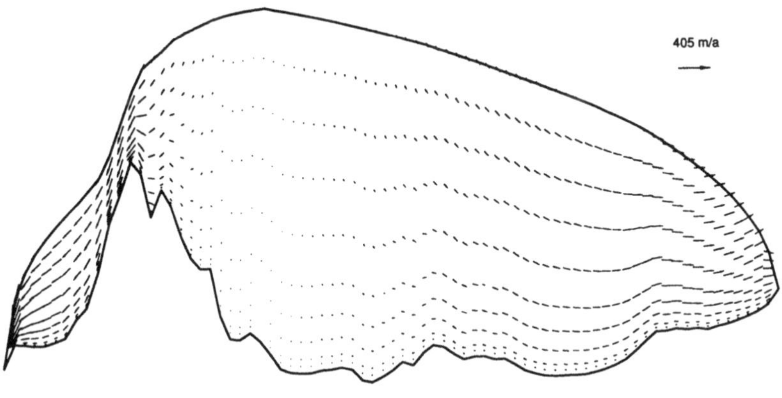

Figure 6 *Modelled cross-section across the European ice sheet from the western Norwegian continental shelf to northern Germany at the last glacial maximum showing flow vectors. (Note that the form of the bed of the ice sheet includes the computed lithosphere flexure in response to ice sheet load.)*

Figure 6 shows flow vectors within the ice sheet. Beneath the ice divide and in the central zone of accumulation, velocities are low and vectors directed down into the ice sheet. In the terminal zone of ablation, they are directed upwards towards the surface. Velocities increase outwards, as the net accumulation flux which the ice sheet must transport increases with distance from the divide On the western flank, constriction of flow across the mountains produces a zone of high velocities, which probably accounts for the strong erosion which occurs in parts of the mountain area, even though the total flux in this zone is small.

Figure 7 shows the modelled internal temperature distribution. It shows low temperatures in the core and relatively high temperatures at the base and margin. If there were no flow, there would be a simple temperature distribution from warm at the base to cold at surface.

Horizontal advection of cold ice from the core however, cools the middle to lower layers of the ice sheet to produce a strong kink in the vertical temperature gradient. Much of the outer part of the bed of the ice sheet is at the pressure melting point, although several tens of kilometres at the margin lie below the melting point. Cooling of the bed towards the centre is enhanced by penetration of the mountain mass into higher, colder levels of the ice sheet. If the mountains were not present, the melting zone would be more extensive.

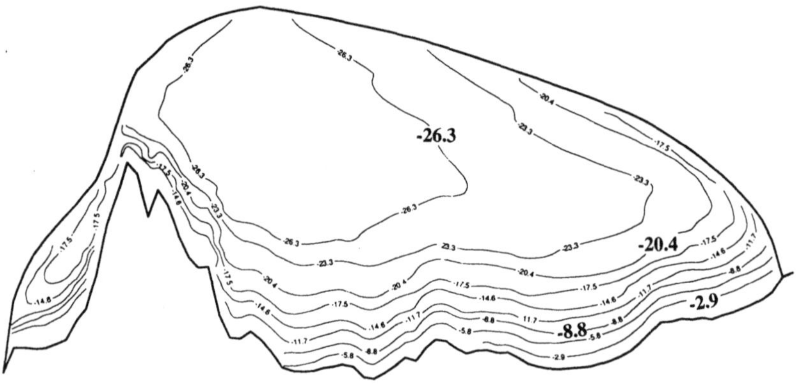

Figure 7 *Modelled cross-section at the last glacial maximum showing the internal temperature distribution.*

Figure 8 shows the variation of simulated ice sheet surface profile along the transect from western Norway to northern Germany through the whole glacial cycle, driven by the ELA variation shown in figure 4. How does this agree with geological evidence of ice sheet behaviour through the cycle? Figure 9 shows a tentative reconstruction of the fluctuations of the ice sheet along the transect based on stratigraphic and geomorphological evidence such as that compiled by Boulton and others (1985), Lundqvist (1986) and Mangerud (1991). The general form of the modelled pattern of fluctuation is similar to that derived from geological evidence. The principal disparity is that the model ice sheet does not respond to climatic forcing as rapidly as the real ice sheet appears to have done. This reveals itself in two features:

- the modelled peaks lag behind the real peaks
- the modelled ice sheet shows little or no retreat during periods of interstadial warming

Moreover, the magnitude of crustal flexure as revealed in Lateglacial and Postglacial relative sea level records in northern Europe is less than predicted by the model (Boulton and Payne, 1992).

We thus conclude that although the model reproduces well the spatial patterns associated with glaciation in Europe and produces a temporal pattern of ice sheet variability which has a general similarity to that reconstructed from geological evidence, a better match should be sought, and one which also reproduces better the pattern of lithosphere flexure.

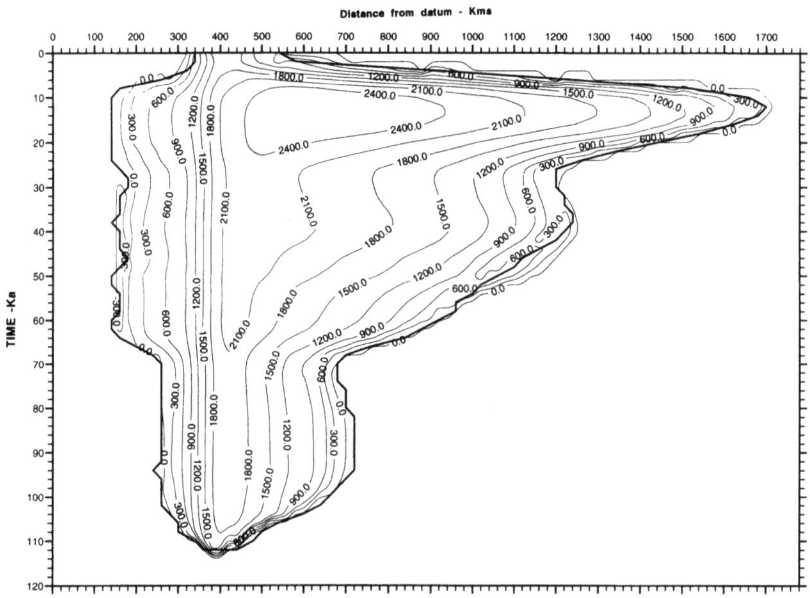

Figure 8 *Modelled ice surface elevation and ice marginal position along the transect in space and time through the last glacial cycle. The western edge of the Norwegian continental shelf is to the left, northern Germany to the right.*

5.3 A quasi-inverse approach to reconstruction of the European ice sheet through the last glacial cycle.

A mismatch between simulation and reality may arise because of errors or omissions in the physical or mathematical construction of the model, errors in parameter values or in use of an inappropriate forcing function. The construction of the model is similar to that of models which satisfactorily reconstruct the form and dynamics of modern ice sheets (e.g. Huybrechts,1986). However, it is possible that a time-dependent property, such as time-dependent rheology, may cause the nature of the response of an ice sheet to a given forcing to diverge progressively from the behaviour of an ice sheet with a time-invariant rheology. Nevertheless, we assume for the moment that the physical and mathematical basis of the model is adequate, and that errors arise from inappropriate parameter values or incorrect forcing functions.

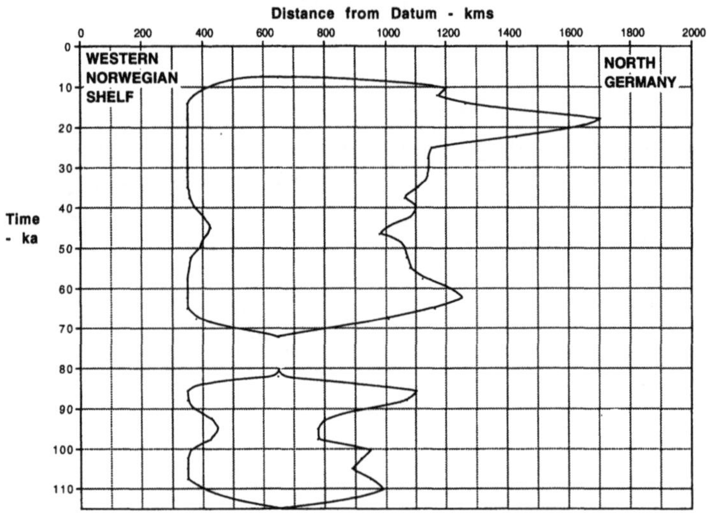

Figure 9 *A tentative reconstruction of the position of the margin of the European ice sheet in time and space through the last glacial cycle from geological evidence.*

We therefore undertake a large number of forward runs of the model in which we specifically attempt to match both the history of ice sheet fluctuations shown in figure 9 and the pattern of crustal flexure reflected in relative sea level curves (Boulton and Payne,1992). This is not a true inverse approach, which we regard as currently intractable, but in starting from a match with the geological evidence, we start from the product (figure 3) and attempt to establish the range of forcing which is compatible with it.

If we assume that the physics and the numerical routines are correct, sensitivity tests suggest that the ice sheet model is most sensitive to the assumed atmospheric lapse rate (Hindmarsh and others, 1989), to the basal boundary condition and to the ELA forcing function. If the lapse rate is very low however, a large ice sheet is unable to sustain itself because of inadequate mass accumulation. Our results therefore suggest that the lapse rate must have been within the range of modern lapse rates (0.6 - 1.7 $^0C.m^{-1}$). The only ways in which we are able to generate the smaller ice thickness required by the relative sea level data and the stronger fluctuations suggested by the geological evidence in figure 10, is to use a "softer" basal boundary condition (by adjusting parameter a in equation 4) and a stronger ELA fluctuation during the glacial cycle.

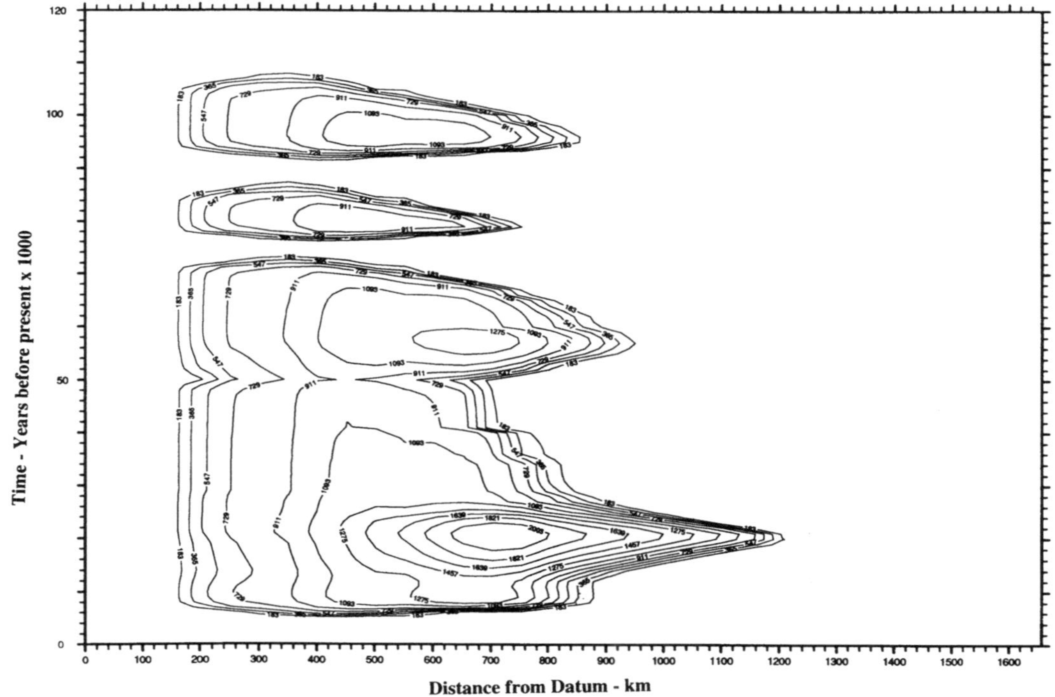

Figure 10 *Best fit model of the variation of surface elevation of the European ice sheet through the last glacial cycle using a quasi-inverse approach. Note that the vertical time axis has been plotted in the opposite sense compared with figures 8 and 9.*

Figure 10 illustrates the the model which presents the best fit to the pattern of fluctuation shown in figure 9, and one which is compatible with the relative sea level data. It shows a much thinner ice sheet whose margin fluctuates much more strongly than that of the forward model. The problem now arises that in using an ELA function which produces model behaviour which fits the tempo of rapid final deglaciation, the model also shows much stronger fluctuations in the early part of the glacial cycle, decaying completely along the line of transect during the equivalents of isotope stages 5a and 5c.

Figure 11 shows the pattern of ELA fluctuation required to fit the model to the pattern in figure 10. Increase in the ELA during the warming stages within the glacial cycle are much greater than in the forward model and greater than might be expected from conventional reconstructions of contemporary palaeotemperature based on palynology (e.g. Zagwijn, 1985). This may simply reflect the relatively poor basis for botanically-based palaeotemperature reconstructions during glacial periods when there are few plant taxa. The evidence of coleoptera however, which are more responsive to climate change than floral taxa, suggests that interstadial warming may have been very much greater than inferred from the floral record (Coope,1977), and thus more in keeping with the inferences drawn here.

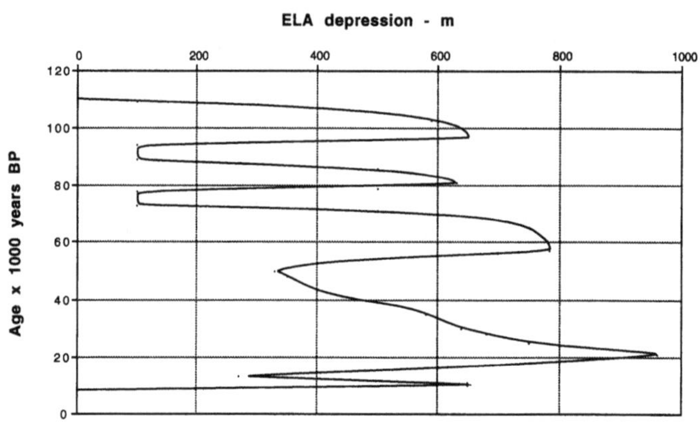

Figure 11 *Pattern of equilibrium line altitude (ELA) fluctuation required to drive the model in figure 10.*

6.0 PALAEOHYDROLOGY THROUGH THE GLACIAL CYCLE AND THE MELTWATER DISCHARGE TO THE OCEANS

6.1 European ice sheet meltwater through the last glacial cycle

Large scale ice sheet fluctuations impact strongly on the oceans. Ice sheets grow by accumulation of ice ultimately derived from ocean water, thus reducing the global water mass and lowering mean sea level. This lowering was about 120m during the last glacial cycle (Fairbanks,1989). As ice sheets melt, water is released from storage and sea level rises. As they grow, the reverse occurs. The gain or loss of water from storage, and therefore the changing global ocean water mass, can be derived directly from the models illustrated above. Figure 12 shows the release of meltwater from storage by the European ice sheet through the whole glacial cycle for the model in figure 8. This does not however give the total meltwater discharge. It omits the mass which accumulates during the winter accumulation season but which is immediately lost during the following season of summer melting, and which contributes neither to sea level change nor to isotopic change in the ocean.

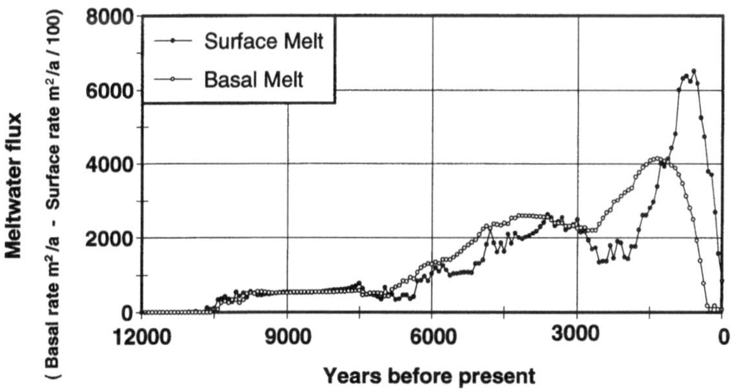

Figure 12 *The flux of meltwater from storage in the European ice sheet through the last glacial cycle in the model shown in figure 8. The diagram shows both water released by melting of the surface and from the ice sheet sole.*

6.2 Isotopic variation in the European ice sheet through the last glacial cycle and its oceanic impact

The isotopic fractionation produced by changes in ice sheet storage strongly effect the isotopic composition of the oceans, which is used as a fundamental index of global change. The net loss of water from the oceans occurs through a distillation process which reduces ocean salinity. The change in O-isotope composition of the oceans as recorded in the skeletons of fossil foraminifers is a reflection of this salinity change. It occurs because of the fractionation process which concentrates more of the light isotope in the evaporated air mass compared with the parent ocean water. Further precipitation from the air mass enhances this trend.

Isotopic composition of precipitation on ice sheet surfaces has been empirically related to the mean annual surface air temperature (T_a) by Dansgaard (1964):

$$\delta_{18SMOW} = \eta T_a - \gamma \qquad (10)$$

where (δ_{18SMOW}) is defined with reference to Standard Mean Ocean Water (SMOW), and (η) and (γ)are constants.

It is possible to use this empirical relationship in evaluating quantitatively the impact of ice sheet growth and decay on oceanic isotopic composition. Our model is partly driven by a sea level temperature distribution through the glacial cycle. If we assume a constant lapse rate through the cycle, we can calculate the mean annual temperature at any point on the ice sheet surface and at any time, from which we can determine the isotopic composition of precipitation. We then model the transport of isotopes through the ice sheet using:

$$\frac{\partial \delta_{18}}{\partial t} = -u\frac{\partial \delta_{18}}{\partial x} - w\frac{\partial \delta_{18}}{\partial z} \qquad (11)$$

where (δ_{18}) is the isotopic content of glacier ice. We then assume transport to be by advection alone as diffusion within compacted glacier ice is very slight (Johnsen, 1977). From this we compute the distribution of isotopic composition through the ice sheet, its mean isotopic composition and the isotopic composition of meltwater. (Details of numerical treatment are given in Boulton and Payne,1992) The depletion of global isotopic composition ($\Delta\delta_{18\,OCEAN}$) is given by:

$$\Delta\delta_{18\,OCEAN} = \frac{\delta_{18\,MEAN} \bullet V_{ICE}}{V_{OCEAN} - V_{ICE}} \qquad (12)$$

where V_{ice} is the volume of the simulated ice sheet and V_{ocean} is the volume of ocean water.

Figure 13 shows the isotopic composition across the European ice sheet transect at several time slices. It is important to note a significant hysteresis effect, in which similar ice volumes during growth and decay have quite different isotopic distributions and mean isotopic compositions. Much of the ice in a growing ice sheet accumulated at a time when the ice sheet was smaller, its surface elevation lower and the precipitation isotopically relatively heavy. Much of the ice in a decaying ice sheet accumulated when the ice sheet was larger, its surface elevation greater and the precipitation was isotopically relatively light.This hysteresis effect is shown in figure 14 for both ice sheet and ocean composition. The ocean effect is relatively small because of the great difference in size between the ice sheet and ocean reservoirs.

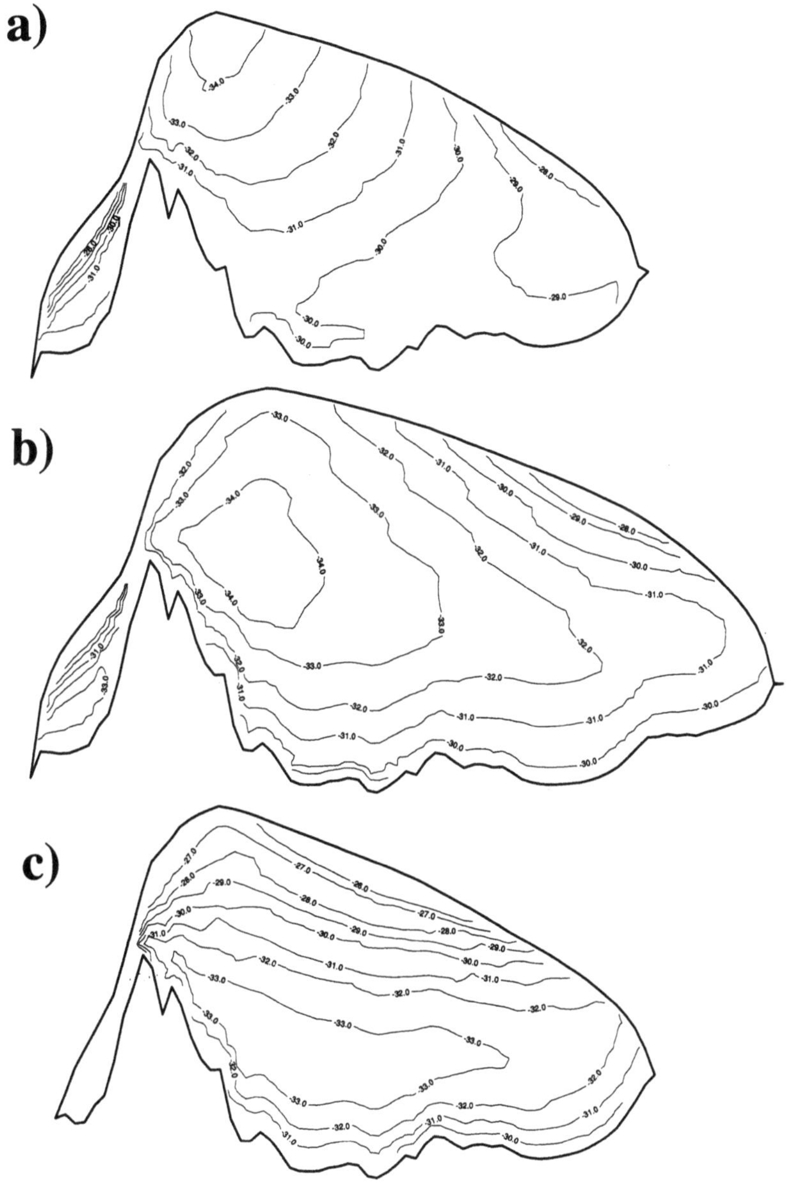

Figure 13 *Modelled transects across the European ice sheet during: a) build-up; b) the maximum extent; c) during decay; showing the distribution of isotopic composition in the ice sheet. Note that the growing ice sheet is isotopically heavier than the ice sheet of equivalent volume during decay.*

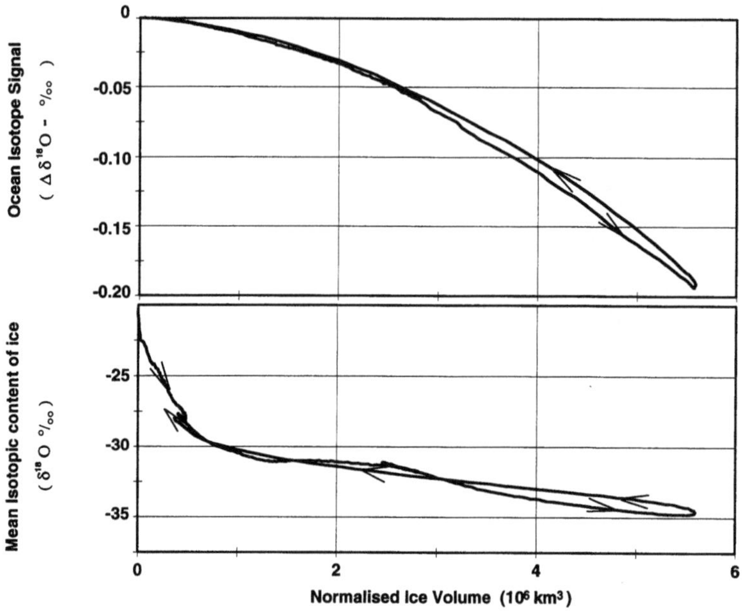

Figure 14 *The isotopic composition/volume hysteresis of mean isotopic composition for the European ice sheet a), and its oceanic effect b), during the last glacial cycle. The time during the cycle is shown by dots on the curves.*

7.0 FINAL DECAY OF THE NORTHERN HEMISPHERE MID-LATITUDE ICE SHEETS

A quasi-inverse approach to reconstruction of final decay has been applied to both the European and Laurentide ice sheets.The model has been fitted as precisely as possible to best estimates of their areal decay.

Figure 15 shows a reconstructions of the pattern and tempo of decay of the European ice sheet based largely on Boulton and others (1985) and Lundquist (1986), but with a timescale derived from the U/Th calibrations of the ^{14}C scale by Bard and others (1993). A Laurentide model has also been created to match the tempo of retreat suggested by Boulton and others (1985). The pattern of decay has been matched, for Europe, with the model parameterised as shown in figure 10, and with a similar model for the Laurentide ice sheet. The pattern of areal decay can thus be changed into a pattern of volumetric decay which, in turn, can be used to derive a pattern of relative sea level change. Figure 16 not only shows this best estimate of volume change but also the range of ice sheet volumes which the parameter study, unconstrained by geological data, suggests could be possible.

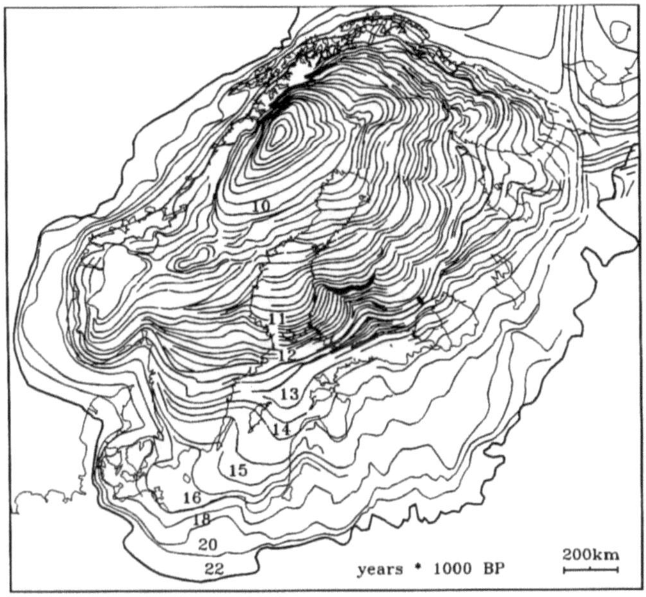

Figure 15 *Pattern of decay of the European ice sheet at the end of the last glacial cycle, with estimated dates of deglaciation in years before present corrected from a ^{14}C chronology using the data from Bard and others (1993). From Boulton and Punkari (in prep).*

The modelled reconstructions clearly show sea level trends similar to the relative sea level curves generated for Barbados by Fairbanks (1989). If the European and Laurentide volumes are added together, they produce an equivalent sea level curve approximately parallel to the Barbados curve. The optimal model gives a sea level equivalent about 30m less than the Barbados curve at 18ka and about 15m less at the time of final Laurentide deglaciation. This parallelism suggests the important conclusion that the Laurentide ice sheet could be taken as a volumetric proxy for the behaviour of global ice sheets during deglaciation. The discrepancy between the optimal ice sheet equivalent sea levels and the Barbados curve decreases rapidly from 30m to 20m between 18ka and 15ka. Current thinking would suggest that this reflects the early decay of marine-based ice sheets on Eurasian continental shelves (e.g.Tushingham and Peltier,1991).

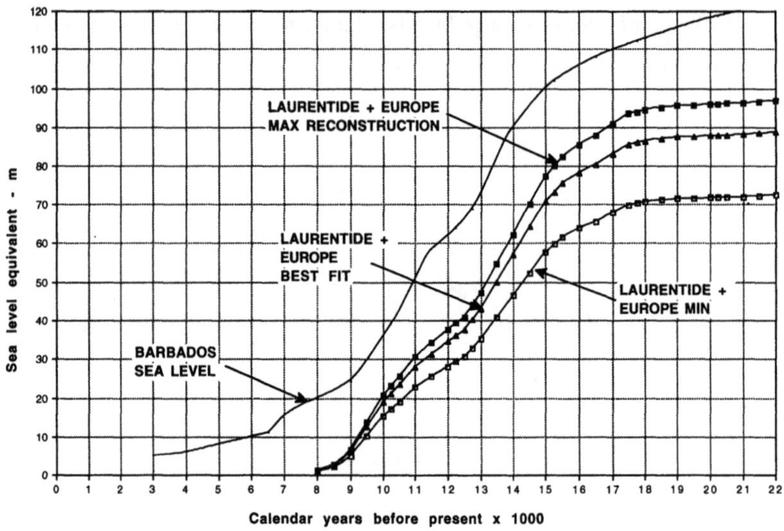

Figure 16 *Equivalent sea level changes generated by the quasi-inverse models of the European and Laurentide ice sheets, showing maximum and minimum models and a "best fit" model. The inferred relative sea level curves are compared with the Barbados sea level curve of Fairbanks (1989).*

The discharge to the oceans from storage in the European and Laurentide ice sheets is shown in figure 17. The discharge is initially greater from the European ice sheet, but thereafter the Laurentide discharge dominates. The peak of Laurentide discharge occurs in the early Holocene, when the ice sheet is still very large and the climate has warmed considerably.

The discharge from storage and the rate of volumetric decay are not helpful indices of climate change, as a small net melting rate from a large ice sheet can give a larger discharge rate than a large net melting rate from a small ice sheet. The average rate of surface lowering for the two ice sheets can however be estimated from the rate of volume change and the contemporary area. This is shown in figure 18. In this case, the European ice sheet shows a greater surface melting rate than the Laurentide ice sheet until the former has receded into the Scandinavian mountains where cooling at higher elevations reduces the melting rate. The contrast between the two continents during deglaciation, of a colder North America and a warmer Europe, is the opposite of the current state and presumably reflects the greater atmospheric impact of the Laurentide ice sheet.

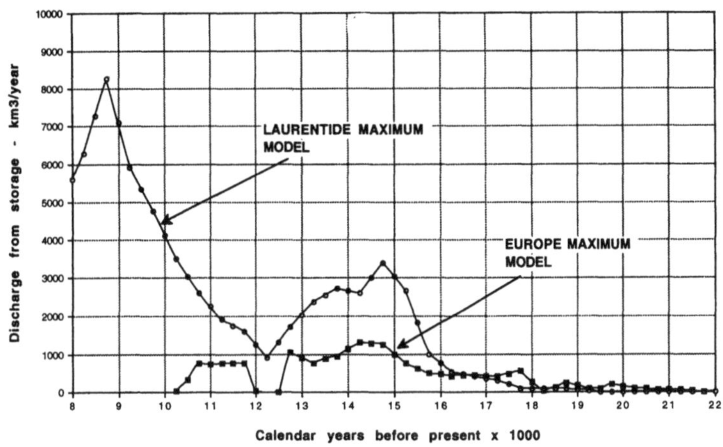

Figure 17 *Flux rate to the oceans of meltwater discharged from storage in the European and Laurentide ice sheets during the last deglaciation.*

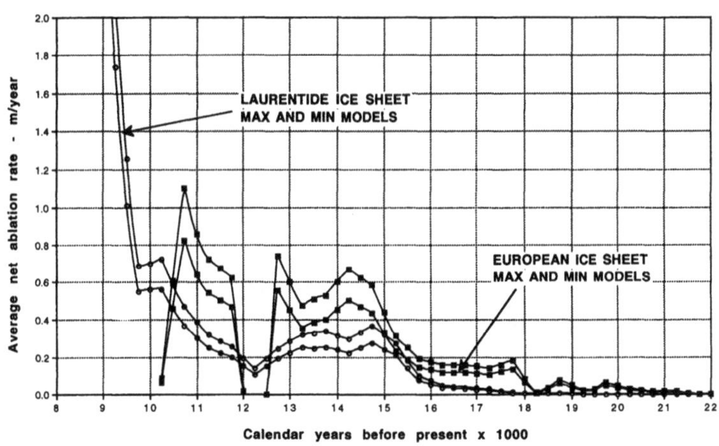

Figure 18 *Average rate of surface melting of the European and Laurentide ice sheets during the last deglaciation.for maximum and minimum models.*

8.0 INSTABILITY OF THE MARINE MARGINS OF ICE SHEETS

Dynamic instabilities in glaciers leading to surging behaviour have long been recognised. In general, it is thought that surging results from uncoupling between the glacier sole and its bed, because of build up of basal water pressures. This reduces friction at the bed, causing enhanced sliding, which cannot be sustained because of an inadequate snow and ice flux from accumulation on the glacier surface. Glaciers prone to this behaviour show cyclical phases of build-up and surging. Some glaciers, for instance Jakobshavn Isbrae, an outlet glacier of the west Greenland ice sheet, carry such a strong ice flux because of convergent flow into it, that it is able to sustain surging velocities (>1000m/year) in a approximately steady state.

The probability of surging behaviour at the southern margin of the Laurentide ice sheet has long been recognised (Wright, 1973) but this appears to have been relatively localised and not to have had a major impact on the rest of the climate system. However, observations of iceberg-dropped detritus from cores in the Atlantic (Grousset and others,1993) suggest that major surges of the marine margins of at least the Laurentide ice sheet have occurred, and that these have been sufficiently large to impact on other parts of the global climate system. Thus, although the landward margins of the mid-latitude ice sheets have fluctuated largely in response to external changes in the atmosphere and ocean, the marine margins of ice sheets may have suffered from large dynamic instabilities which, at certain periods, dominated their behaviour (figure 2)..

Such instabilities may arise from four causes:

a)- buoyancy of the marine margin;

b)- thermal instabilities which effect sliding;

c)- build-up of a deforming till layer at the sole;

d)- build-up of large water pressures at the glacier sole.

b) and c) will both be strongly effected if large water pressures build up at the glacier sole. It is suggested that all four mechanisms are likely to be more effective at the marine margins of the mid-latitude ice sheets; a) because of the marine position, and b) to d) because of the geology which happens to characterise those margins.

A perennial cause of instability at a marine margin is the buoyant force exerted as the ice sheet flows into the sea, which reduces friction in the terminal zone, producing enhanced flow which is able to draw down the ice sheet surface. Thinning of the glacier can lead to rapid grounding line retreat and positive feedback leading to catastrophic ice retreat (Hughes, 1975).

Isostasy tends to introduce a negative feedback, whereby an advancing glacier produces crustal depression, deeper water in the terminal zone, enhanced buoyancy, drawdown and a slowing down of advance or change to retreat. An example of the effect of this process on a specific ice sheet is given in 8.1, and its potential for cyclical oscillation is described in 8.2.

8.1 An ice sheet over the British Isles through a glacial cycle

The ice sheets which have covered much of the British Isles from time to time during recent glacial periods developed in an area with a major arm of the sea, the Irish Sea and the North Channel between Scotland and Ireland, in its core area. It has been suggested by Eyles and McCabe (1989) that the decay of the ice sheet in this central, Irish Sea sector was driven by rapid calving in a marine environment.

In seeking to simulate the behaviour of an ice sheet over such a topographically complex area, it has been necessary to use a very small mesh spacing of 20km. The ice sheet model was driven by a simple sinusoidal oscillation of the ELA but run in the quasi-inverse mode by attempting to fit the model to some aspects of glacial geology, and using the rate of ELA rise and fall, and the orientation of the ELA plane as driving parameters which are adjusted to ensure that the model fits the data.

The model was used to simulate the last advance and decay of the British ice sheet during isotope stage 2. Its rate of advance and decay during this period is poorly known, but it is known to have overrun sites at Bishopbriggs in central Scotland after 27,500 B.P, and Inchnadamph in N.W.Scotland after 24,600 BP (Boulton and others,1991; Gordon and Sutherland, 1993) and advanced over a site on the east coast of England by 18,500 (Penny and others, 1969). It is known to have receded from Glanllynau, North Wales by 14,700 BP (Coope and Brophy,1971), from the Isle of Man by 15,000, from the north-east coast of Scotland by 17,700 BP and from Mull on the west coast of Scotland by 13,100 BP (Gordon and Sutherland,1993). A period of strong climatic cooling commenced in early Stage 2 in the N.E.Atlantic at about 28,000 BP (Ruddiman et al, 1986). After the glacial maximum at about 20,000 BP, the warmest part of the succeeding warming phase occurred at about 12,000 BP, during which Sissons (1976) has suggested that the British ice sheet finally decayed in Scotland prior to its regrowth during the Younger Dryas. We therefore assume a sinusoidal cooling/warming cycle beginning at 28,000 BP, reaching its peak by 20,000 and ending by 12,000 BP.

The additional geological information which was used to constrain the model is the known extent of the ice sheet at the glacial maximum (Boulton et al, 1991), and the patterns of till lithology and erratic transport. Patterns of erratic transport in Scotland are summarised by Boulton and others (1991) and more generally over Britain by Charlesworth (1957). The constraint that we impose upon the model is that it should not produce erratic transport in directions for which there is no evidence of transport. We do not require it to simulate precise patterns of transport because of the problem of multiple reworking.

Local till sequences may show lithological changes which also reflect systematic changes in ice sheet flow patterns. For instance, Kirby (1969) has shown that in the area south of Edinburgh, a till rich in northern erratics from the Highlands reached the site before a till of the same last glaciation reached the site from the south-west (Southern Uplands), reflecting arrival of ice derived from the growing Highland ice dome before ice from the growing Southern Upland dome. Similar shifts in the derivation direction of till material can be recognised in northern England (Eyles and others, 1982) and northern Ireland (McCabe, 1983). The imposed constraint is that the model ice sheet must follow this pattern of successive occupancy of ice from different centres.

The British ice sheet was volumetrically insignificant in global terms. We therefore assume a global background sea level cycle around the British Isles as given by Fairbanks (1989) curve to 20,000 BP and the estimate of Shackleton (1987) between 20,000 and 28,000 BP. The model then generates a local isostatic component of crustal flexure which is added to the presumed global component to produce a history of relative sea level around the ice sheet margin. Modelled relative sea level changes in the Irish Sea are shown in figure 20.

Figure 19 shows the model which obeys the all the principal constraints imposed upon it. The till lithology constraint is one of the most important in controlling lowering of the ELA during ice sheet build-up. The ice sheet over Britain had a divide with a SW-NE trend located over the north-western area. To simulate this pattern requires an ELA surface dipping towards the NW. In the model, as this surface descends, ice first nucleates in the mountains of NW Scotland and NW Ireland. To ensure that Highland ice arrives in the Edinburgh region before ice from the nearer Southern Uplands ice centre, the rate of ELA lowering must be so slow that the Highlands ice mass grows to a substantial size before nucleation and growth begins in the Southern Uplands. A rapid snowline descent precludes this.

North-western Britain has a plethora of mountain masses, with the consequence that the constrained model reconstructs about 16 independent nucleation areas, with complex patterns of confluence during growth.

The ice in the Irish Sea area has a very distinctive pattern of behaviour. As global sea level descends, ice moves readily into the shallow water areas (water depths between 50 and 100m), with ice from eastern and western flanks of the basin eventually merging in the channel between South Wales and Ireland. However, isostatic depression of the lithosphere beneath and immediately beyond the advancing ice sheet causes a local increase in relative sea level, so that the ice sheet begins to retreat in the area of the deep water channel in the southern Irish Sea even before some other sectors of the ice sheet margin have reached their maximum extent. Retreat in this sector is extremely rapid. Within 1500 years of the

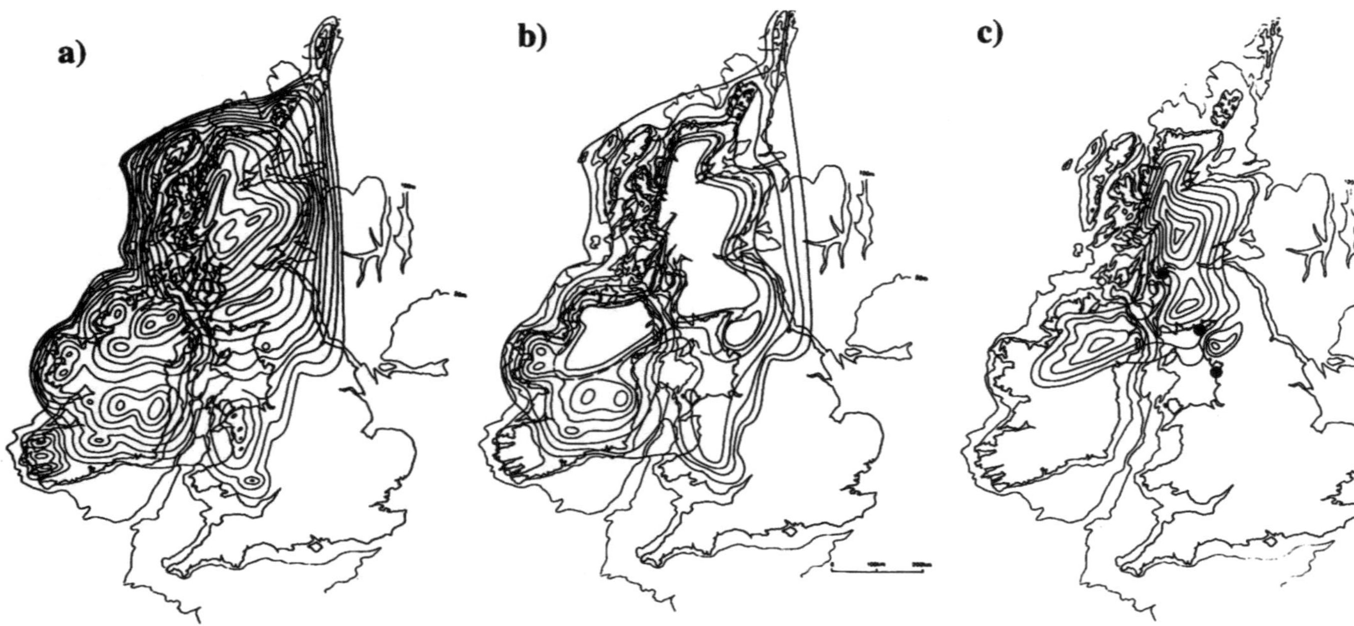

Figure 19a-c. *Modelled advance and decay of an ice sheet over the British Isles using a quasi-inverse approach. The maps show displacement of the ice sheet margin at 500 year intervals. a) Advance of the ice sheet between 28,000 BP and 20,000 BP. b) Retreat of the ice sheet between 20,000 and 17,000 BP. c) Retreat between 17,000 and final decay shortly after 13,000 BP.*

commencement of retreat, a large calving bay has developed in the central Irish Sea. We conclude therefore that the results of the modelling experiment are compatible with the conceptual model of deglaciation of the Irish Sea Basin advanced by Eyles and McCabe (1989).

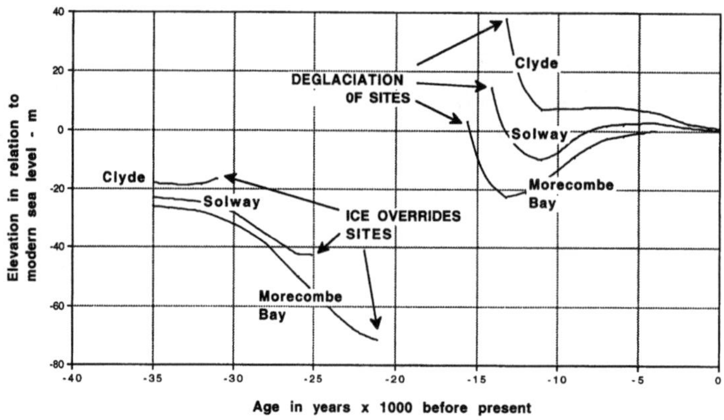

Figure 20 *Modelled relative sea level curves for a series of sites down the eastern shore of the Irish Sea . From north to south they are Cyde, Solway, Morecombe Bay. Their locations are shown in figure 19c.*

There is a potentially important and unexpected result of the model. Where the ice sheet withdraws from a sea area where there has been very active calving, such as on the south-western coast of Scotland, a readvance or stabilisation of the margin tends to occur. This is because strong calving at the marine margin produces strong drawdown and a large total ablation rate which leads to retreat even when the accumulation rate is high. When the ice sheet margin recedes into shallow water or onto land, reduction in the total ablation rate due to loss of the calving mechanism may be sufficient to cause a readvance or stabilisation of the margin.

8.2 Dynamic oscillations at the marine margin due to interactions between buoyancy and crustal flexure

The combination of grounding-line migration and the consequent ice sheet drawdown together with isostasy, can generate an oscillation at a marine margin. We consider an ice sheet which advances over a continental shelf which has been exposed by sea level fall due to global ice sheet build-up. The global component of sea level change is assumed constant at a level which would keep the unglaciated shelf dry. The ice sheet is driven by a prescribed mass balance function. We include the lithosphere flexure model and grounding line retreat model.

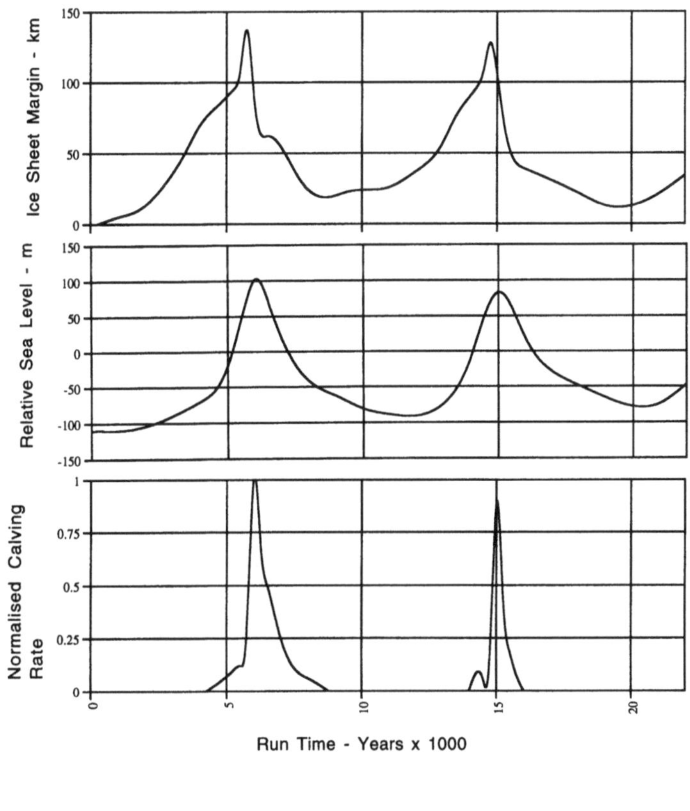

Figure 21 *Oscillations in the position of the marine margin of an ice sheet, of iceberg calving rate and of relative sea level as a consequence of the interaction between stability of the marine margin and relative sea level.*

As the ice sheet advances over the shelf, the crust is depressed beneath the ice sheet. If the lithosphere is given a relatively large flexural rigidity, considerable depression can take place beyond the ice sheet margin such that relative sea level rises significantly at the margin. This produces strong grounding line retreat, ice drawdown and, ultimately, retreat of the ice sheet margin from the shelf and from marine contact. Withdrawal of the ice sheet from the shelf permits isostatic uplift, falling relative sea level and retreat of the sea from the shelf. With the calving component of ablation gone, the ice sheet returns into a state of positive balance, and readvances onto the shelf. The cycle is then repeated (figure 21).

Oscillations are damped if the lithosphere is weak, if the diffusivity of the asthenosphere is low, if the shelf and hinterland have a strong seaward slope so that grounding line retreat is limited, if the ELA is very near sea level (which inhibits retreat) and if the ELA is very high or has a very small slope (which tends to produce unstable retreat).Oscillations are enhanced by the opposite trends

Under appropriate conditions we might thus expect periodic oscillations of an ice sheet with a marine margin. Use of reasonable parameter values for an ice sheet of width 2000km and a shelf of 150km produces a period of oscillation of between about 4,000 and 8,000 years. The Antarctic ice sheet does not have properties which lend itself to this process. The continental shelf is too deep and the ELA is effectively at sea level.

8.3 Thermal instabilities, sliding and the development of major ice sheet oscillations.

When basal ice reaches its melting point, it is assumed to slide over its bed. The rate of sliding will be a function of bed roughness but also of the effective pressure (ice pressure minus water pressure) at the sole (equation 4). The initiation of sliding fundamentally alters the velocity structure near to the bed of the ice sheet. Without sliding, this ice is virtually stagnant. Sliding introduces large horizontal velocities.

The distribution of stable melting and frozen bed zones is shown in figure 7 for the European ice sheet at the last glacial maximum. If the melting zone were to progressively extend towards the divide, it would strongly effect the vertical velocity field by producing an enhanced downward flow from the cold/frozen zone to the warm/freezing zone.

These changes in velocity structure dramatically alter the near-bed heat budget.and transport mechanisms. Without sliding, diffusion dominates, with sliding, advection dominates. In addition, a new heat source is introduced at the bed, friction generated by sliding. All this implies that the near-bed temperature gradient will be steeper when the ice is sliding because of enhanced cold ice advection.

We have modelled these changes (figure 22) from an initial state where the bed of the ice sheet is everywhere frozen and no sliding occurs. For simplicity, the ice sheet is assumed to extend to the ocean on both sides, with a zone of accumulation to its margin, and its expansion prevented by marine calving. Both margins are thus analogous to the marine margins of a northern hemisphere mid-latitude ice sheet. The modelling procedures are those described in section 3.0.

The basal ice nearest to the margin warms first to the melting point, because strain heating of ice is largest there. It starts to slide. The enhanced vertical velocity only acts to advect relatively warm ice to the bed, permitting the penetration of the melting zone towards the interior to continue. It does so until it extends further back towards the ice sheet divide, where the ice is much colder. Eventually, a point is reached where the enhanced vertical velocity advects ice which is cold enough to produce refreeezing. The melting zone then begins to shrink. The enhanced cold ice advection and steepening of the near-base temperature gradients produce a hysteresis in which the basal ice is now colder then before and enhanced vertical velocities associated with the melting/freezing interface can advect colder ice which leads to expansion of the freezing zone. The melting zone shrinks rapidly leading eventually to an ice sheet frozen to its bed throughout.

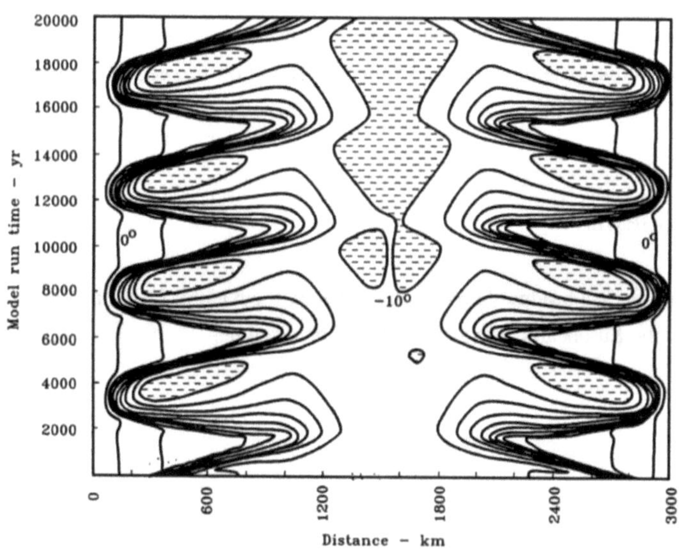

Figure 22 *Time-distance diagram showing a modelled thermal oscillation in an idealised ice sheet limited on both flanks by calving into the sea. The ice sheet margins extend from 0 to 1000km. At time zero, the bed is entirely frozen, but melting begins in the marginal zone and extends inwards. It has reached a first maximum extent by about 1800 years of model time. It is associated with enhanced sliding in the outer zone leading to drawdown and increased calving into the sea. As this phase is exhausted, the ice sheet cools and sliding is progressively inhibited. The maximum extension of the non-sliding cold zone occurs by about 3500 years of model time. The cycle then repeats itself. It is an entirely dynamic effect, unforced by external climate.*

A quiescent phase then follows, in which geothermal heat and strain heating gradually warm the ice sheet base so that the cycle can restart.The cycles shown in figure 22 have a period of about four thousand years, which is similar to the period of Heinrich events in the Atlantic. The cyclical periods of sliding will produce enhanced iceberg fluxes at the marine margins.

We would argue that the marine margins of the northern hemisphere ice sheets will be suscepible to this mechanism because of their particular geology. The beds of both the north-western margin of the European ice sheet and the north-eastern margin of the Laurentide ice sheet are mountainous and composed of old, hard, impermeable rocks. The southern margins overrode relatively young permeable aquifers. The contrast between the moutainous left hand and flat right hand margins of figure 2 exemplify this contrast. In the southern areas, permeable rocks will tend to draw down basal water pressures so that basal sliding is inhibited (equation 4), in the northern areas, water prssures will be high, enhancing sliding. Thus we expect the northern margins of the ice sheet to be more susceptible to the mechanism. Moreover, where there is a marine margin, ice advance, which will tend to dissipate the effect of cyclical warming of the bed, can be inhibited by calving.

ACKNOWLEDGEMENTS

Much of this article is a product of research funded by the Natural Environment Research Council and by the European Community EPOCH programme.

REFERENCES

Bard E , Arnold M , Fairbanks RG , Hamelin B (1993) ^{230}Th/^{234}U and ^{14}C ages obtained by mass spectrometry on corals. Radiocarbon 35:191-199

Berger A , Fichefet Th , Gallee H , Marsiat I , Tricot C , van Ypersele JP (1990) Physical interactions within acoupled climare model over the last glacial-interglacial cycle. Trans. Roy. Soc. Edin. 81:357-370

Boulton GS , Smith GD , Morland LW (1984) The reconstruction of former ice sheets and their mass balance characteristics using a non-linearly viscous flow model J Glaciol 30:140-152

Boulton GS, Smith GD , Jones AS , Newsome J (1985) Glacial geology and glaciology of the last mid-latitude ice sheets. Journal of the Geological Society 142: 447-474

Boulton GS , Peacock JD , Sutherland DG (1991) Quaternary. *In* Craig GY (ed) Geology of Scotland, Geological Society, London.

Boulton GS , Payne A (1992) Reconstructing past and predicting future regional components of global change: the case of glaciation in Europe. Waste disposal and Geology. International Geological Congress Proceedings , Workshop 21-51-134

Budd WF , Jenssen D (1975) Numerical modelling of glacier systems. IASH Publ 104:257-291
Budd WF Jenssen D Radok U (1971) Derived physical chatracteristics of the Antarctic ice sheet ANARE interim reports 120
Böse M (1990) Reconstruction of flow directions south of the Baltic sea during the Weichselian glaciations 19:217-226
Büdel J (1949) Die raümliche und zeitliche Gliederung des Eiszeitklimas. Die Naturwissenschaften 36:105-112
Charlesworth JK (1957) The Quaternary Era. Arnold London
Coope GR (1977) Fossil Coleopteran assemblages as sensitive indicators of climatic changes during the Devensian (Last) cold stage Phil Trans Roy Soc Lond B 280:313-340
Coope GR Brophy JA (1971) Late-glacial and environmental changes indicated by a coleopteran succession from North Wales Boreas 1:97-142
Dansgaard W (1964) Stable isotopes in precipitation Tellus 4:436-468
Eyles N , Sladen JA , Gilroy S (1982) A depositional model for stratigraphic complexes and facies superimposition in lodgement tills Boreas 11:317-333a
Eyles N , McCabe AM (1989) The Late Devensian Irish Sea Basin: the sedimentary record of a collapsed ice sheet margin Quat Sci Revs 8: 307-351
Fairbanks RG (1989) A 17,000 year glacio-eustatic sea level record: influence of glacial melting rates on the Younger dryas event and deep-ocean circulation Nature 342:637-642
Gordon JE , Sutherland DG 1993 Quaternary of Scotland. Chapman and Hall London
Glen JW (1955) JW (1955) The creep of polycrystalline ice. Proc Roy Soc Lond Ser A 228:519-538.
Grousset F , Labeyrie LD , Sinko JA , Bond G , Duprat J , Cortijo E , Huon S. (1993) Patterns of ice-rafted detritus in the Glacial North Atlantic (40-55°N). Palaeoceanography 8:175-192.
Hindmarsh RCA , Boulton GS , Hutter K (1989) Modes of operation of thermo-mechanically coupled ice sheets. Ann Glaciol 12:57-69
Hirvas H , Lagerbäck , R Makinen H , Minell H , Nenonen K , Olsen L , Riber K , Rodhe L , Sutinen R , Thoresen R (1986) Map of Quaternary Geology, ice flow indicators, Northern Scandinavia. Nordkalott Project, Sheet 3.
Hughes T (1975) The West Antarctic Ice Sheet: instability, disintegration and initiation of ice ages. Reviews of Geophysics and Space Physics 13:502-526
Huybrechts P (1986) A three-dimensional time-dependent numerical model for polar ice sheets: some basic testing with a stable and efficient finite different scheme. Report 86-1 Vrije Universiteit Belgium
Johnsen SJ (1977) Stable isotope homogenisation of polar ice and firn Isotopes and impurities in snow and ice IASH Publ 118 :210-2191977
Kirby RP (1969) Till fabric analyses from the Lothians, central Scotland Geogr Annlr 51A:48-60
Lundquist J (1986) Late Weichselian glaciation and deglaciation in Scandinavia Quat Sci Revs 5:269:292
Mahaffy MW (1976) A three-dimensional model of ice sheets: tests on the Barnes Ice Cap, Northwest Territories. J Geophys Res 81:1059-1066
McCabe AM (1983) Glacial Geomorphology The Quaternary History of Ireland Academic Press London
Mangerud J (1991) The Scandinavian ice sheet through the last interglacial/glacial cycle Klimageschichtliche Probleme der Letzten 130,000 Jahre Fischer Stuttgart
Penny LF , Coope GR , Catt JA (1969) Age and insect fauna of the Dimlington Silts, East Yorkshire. Nature 224:65-67et al 1969
Ritz C (1987) Time dependent boundary conditions for the calculation of temperature fields in ice sheets. The physical basis of ice sheet modelling IASH Pub 170:207-216

Ruddiman WF , Shackelton NJ , McIntyre A (1986) North Atlantic sea-surface temperatures for the last 1.1 million years. North Atlantic Palaeoceanography Geological Society Publication 21:155-173
Sissons JB (1976) Scotland. Methuen London
Shackleton NJ (1987) Oxygen isotopes, ice volume and sea level. Quat Sci Rev 6:183-190
Tushingham AM , Peltier WR (1991) Ice-3G: A new global model of late Pleistocene deglaciation based upon geophysical predictions of post-glacial relative sea-level change. Jour Geophys Res 96B:4497-4523.
Wright HE (1973) Tunnel valeys, glacial surges and subglacial hydrology of the Superior Lobe, Minnesota. Geol Soc America Mem 173:251-276
Zagwijn WH (1985) An outline of the Quaternary stratigraphy of the Netherlands. Geologie en Mijnbouw 64:17-24

THE CLIMATIC RECORD FROM ANTARCTIC ICE NOW EXTENDS BACK TO 220 kyr BP

Jouzel J.[1,2], Lorius C.[2], Petit J.R.[2,1], Ritz C.[2], Stievenard M.[1], Yiou P.[1]
Barkov N.I.[3], Kotlyakov V.M.[4], Lipenkov V.[3]

Laboratoire de Modélisation du Climat et de l'Environnement,
CEA/DSM CEN Saclay,
91191, Gif sur Yvette Cedex,
France

Abstract. The isotopic content of ancient waters (deuterium and oxygen 18) gives a key access to past climatic changes because of the essentially linear relationship that exists between the isotopic content of a precipitation and the temperature of the site (at least for middle and high latitudes). This approach has been extensively used for extracting high resolution climatic records from both Antarctic and Greenland ice that now extend beyond 200 kyr BP (Dansgaard et al., 1993 ; Jouzel et al., 1993). Here we present the Vostok isotope temperature record which has been extended back to 220 kyr BP and the associated glaciological timescale. We then briefly discuss differences and similarities between this Antarctic record and the new climatic records recently obtained in Central Greenland .

1. Introduction

Vostok drillings were initiated in 1970 in a very cold site (mean annual temperature, -55.5°C) particularly well suited for obtaining long and undisturbed records (altitude, 3490m asl, ice thickness, 3700 m and accumulation, 2.3 g.cm^{-2} a^{-1}). A large part of the published results were obtained on core 3Γ, the first to cover the

---

2) Laboratoire de Glaciologie et Géophysique de l'Environnement, CNRS, BP96, 38402, Saint Martin d'Hères Cedex, France
3) Arctic and Antarctic Research Institute, Beringa Street 38, 199226, St Petersburg, Russia
4) Institute of Geography, Staromonetny, per 29, 109017, Moscow, Russia

NATO ASI Series, Vol. I 22
Long-Term Climatic Variations
Edited by J.-C. Duplessy and M.-T. Spyridakis

glacial-interglacial cycle. The drilling of this hole was started in 1980 and reached 2083 m in 1982 (Fig. 1) with most of the Vostok analytical records extending down to this depth. In 1985, this core reached a final depth of 2202 m, with only limited additional results (Petit et al., 1990; Jouzel et al., 1990). Core 4Γ started in 1984 reached a final depth of 2546 m in 1990. A new core, 5Γ, is now underway at its current depth (November 1993) of 2746 m.

The climatic record is extended through the measurement of the ice deuterium content in the bottom part of core 4Γ down to 2546 m. This parameter (δD in versus SMOW) was measured continuously below 1920 m on 1 m ice increments with a more detailed sampling (half a meter) between 2100 and 2414 m. Those samples cover time spans ranging from ~70 to 150 years.

The new results are plotted vs depth in Fig. 2 along with data from the bottom part of core 3Γ (Jouzel et al., 1987a). The shallowest depth plotted, 1500 m, corresponds to the end of the Last Interglacial (LIG) and is dated at about 110 kyr BP. As shown in this figure the agreement between the 2 profiles is excellent over their common part. The 3Γ results below 2083 m (Jouzel et al., 1990) are not reported because this part of the profile is discontinuous. The comparison suggests no sizeable age difference between similar levels in the cores which could result from a different inclination of the two holes.

Beyond the extension of the temperature record obtained for the last climatic cycle, the focus of this article is on the establishment of a timescale for the bottom part of the Vostok core starting from the Last Interglacial downward. As strikingly illustrated by the Devils Hole continental record, the dating of climatic events over this period is crucial to get new insights on the mechanisms of glacial interglacial cycles (Winograd et al., 1992; Imbrie et al., 1993).

2. The Dating of the Vostok Ice Cores

Because the accumulation rate is too low for clear isotopic or other annual signals to form, obtaining an accurate time scale for ice cores on the East Antarctic Plateau proved more difficult than for cores such as Byrd (Hammer et al., 1985), Dye 3 (Hammer et al., 1986), GRIP (Johnsen et al., 1992) and GISP2 (Taylor et al., 1993) drilled in higher accumulation sites of West Antarctica and Greenland. Further dating methods are necessary and fall in two main categories. The glaciological approach combines an ice-flow model and an accumulation model; it has the advantage of providing an independent dating. The second type of approach essentially provides a relative chronology. It consists in correlating through

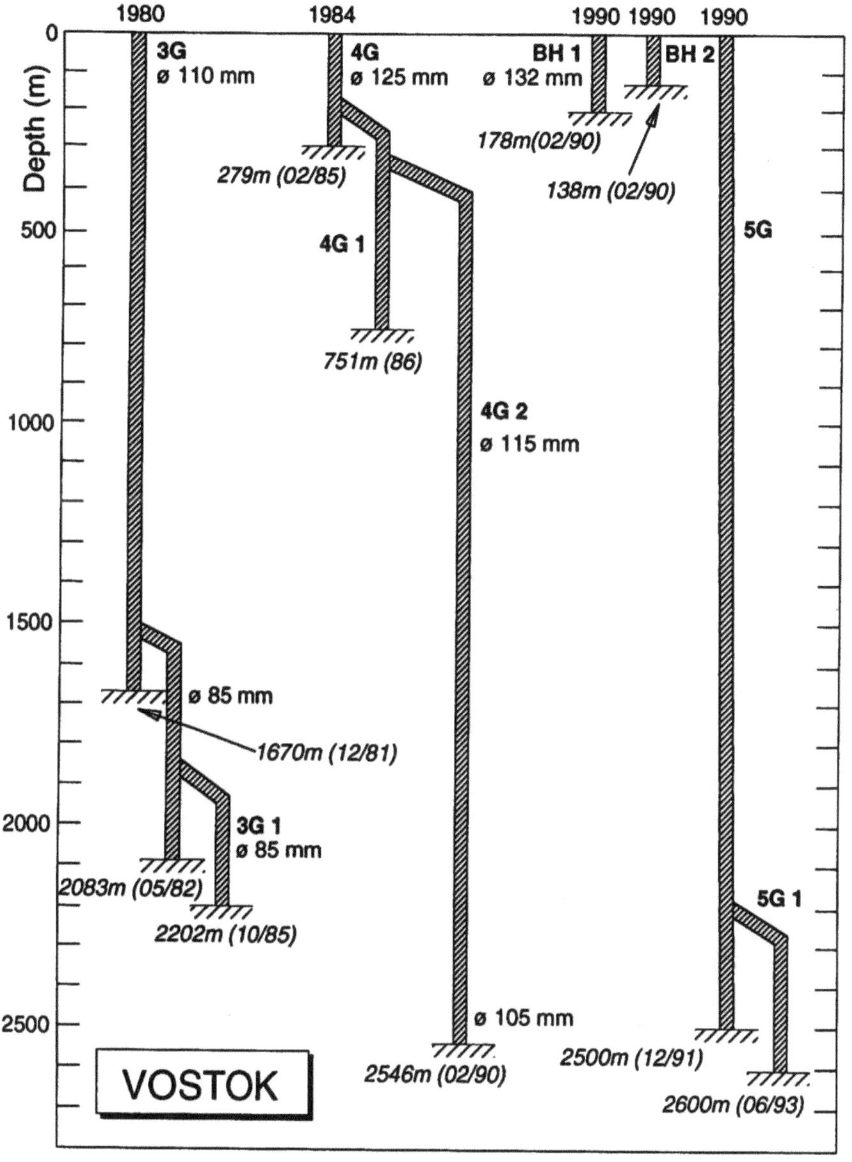

Figure 1 : Deep drillings performed by the RAE (Russian Antarctic Expedition) at Vostok since 1980. Cores 3 and 4 were stopped in 1985 and 1990 respectively. The ice thickness at Vostok is ~ 3700 m.

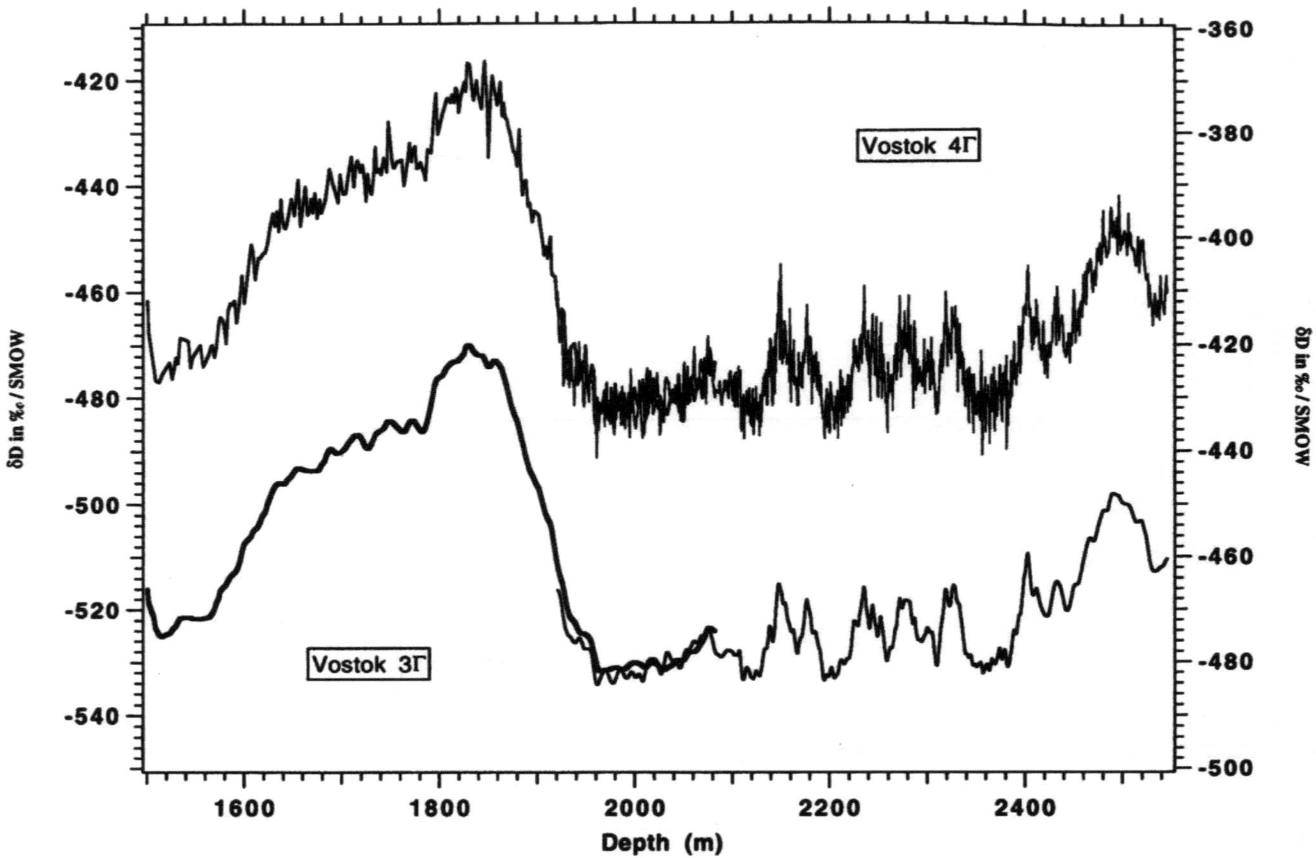

Figure 2 : Deuterium profile in cores 3Γ and 4Γ with respect to depth.The upper curves (left scale) are raw data whereas in the lower curves (right scale), a smoothing procedure has been applied

common climatic or stratigraphic indicators the ice record with other, and up to now deep-sea cores, paleorecords.

The strategy followed for the Vostok core was to develop a glaciological dating to be used as a reference chronology (Lorius et al., 1985) because we believe that the Vostok core, (and ice cores, in general), offers a unique source of information which will be most valuable if dated independently of the deep-sea cores record. In parallel, various ways of correlating the Vostok and the deep-sea core records were explored (Petit et al., 1990, Shackleton et al., 1992, Pichon et al, 1992; Sowers et al., in press) that, for example, have already brought important information on the timing between climate forcings and responses. We do however claim that none of these stratigraphic correlations provide a "revised" Vostok chronology as yet. This would require the marine chronology to be certain which is still debated and the "revised" chronology be unique which is only true to a certain extent as discussed in Sowers et al. (in press) for the last climatic cycle.

We shall keep the Lorius et al. strategy (1985) for the bottom part of the Vostok core with however, as an additional condition, the use of a time reference level that constrains the timescale but no longer keeps the Vostok record fully independent of other paleorecords. We extend this glaciological dating approach down to the bottom of the core, taking into account information on accumulation derived from a new core drilled at Dome B on the Antarctic Plateau (Vaikmae et al., 1993). Such an approach requires taking into account both past accumulation changes at the site where the ice originates and for the ice layer thinning with depth. At Vostok, the thinning is estimated from a two-dimensional glaciological model well suited to the Vostok area as fully described by Ritz (in press). The accumulation change is estimated directly from the temperature record (Lorius et al., 1985), which is itself deduced from the isotopic profile. The uncertainty on the accumulation rate dominates the overall uncertainty on the Vostok timescale. The accuracy was estimated to be no better than 10-15 Kyr at 164 Kyr BP (Lorius et al.,1985). Minor modifications were later introduced in particular the use of δD instead of $\delta^{18}O$ for estimating the temperature and thus the accumulation (Jouzel et al., 1989). The modification never exceeds ~2%, i.e., far less than the above uncertainty so that the timescale published by Lorius et al (1985) has been used since 1985 as the reference Vostok timescale for the last glacial-interglacial cycle.

One basic asssumption of this model is that the present-day accumulation is constant upstream of Vostok over all the region from where the Vostok ice found at depth originates. This was reasonable given the data then available but is now contradicted by the isotopic profile recently obtained along a 780 m core drilled at Dome B (Vaikmae et al., 1993). The beginning of the Holocene is found at 500 m depth instead of 300 m at Vostok which implies that the Dome B accumulation rate is 1.65 time the Vostok one. Ritz (in press) extended the model down to 2,400 m with various assumptions for the spatial variation of the present day accumulation rate between Vostok and Dome B. As nothing is known yet on this spatial variation, we adopt here the simplest assumption of a linear increase between Vostok and Dome B and extended this timescale down to 2,546 m. We take into account all the model improvements incorporated by Ritz (In press). They concern the use of the δD instead of the $\delta^{18}O$ profile, the use of a more accurate correction of the isotopic content of the ocean and numerous improvments concerning the ice-flow model. As expected, this additional unknown in the accumulation rate accentuates the timescale uncertainties. Taking the full range of possibilities the accuracy may be no better than ± 20 % at the bottom of the core (Ritz, in press).

This uncertainty is reduced if it is feasible to assign the age for at least one given level. We choose to keep fixed the 110 Kyr BP level (1,534 m) by adjusting the present day Vostok accumulation rate. This choice is done for several reasons : 1) it minimizes the difference with the Lorius et al. timescale down to 110 Kyr BP ; 2) stratigraphic correlations between Vostok and ocean records show no significant discrepancy for this 110 Kyr BP level, dated at 108 and 112 kyr BP in Pichon et al. (1992) and in Sowers et al. (1993) respectively, and 3) it leaves open the discussion for the LIG and earlier periods for which the dating and the timing of climatic events in various parts of the system is still debated. As in Lorius et al. (1985), we assume that the accumulation rate has varied in the past according to the following equation:

$$A(t) / A(t=0) = (\partial p_s / \partial T)(t) / (\partial p_s / \partial T)(t=0)$$

where A is the accumulation rate, t the age and p_s the saturation water vapor pressure. This equation invokes the fact that, over the Antarctic Plateau, the accumulation rate is largely governed by the derivative of the saturation pressure at the level where the precipitation forms, i.e. above the inversion layer (Lorius et al., 1985 ; Jouzel et al., 1987a). The saturation vapor pressure is directly derived from the temperature of snow formation (atmospheric temperature) reconstructed from the δD profile of the ice (Figure 8). Both the GRIP (Dahl-Jensen et al., 1992) and GISP 2 (Alley et al., 1993) Greenland cores show a strong correlation between the annual layer thickness and the isotopic temperature supporting the general validity of this method of estimating accumulation change in polar regions.

With these assumptions, the accumulation rate at Vostok today is calculated to be 1.98 g cm^{-2} yr^{-1}, 13% lower than the value of 2.3 g cm^{-2} yr^{-1} estimated for the very recent period (~ 10 yr) and used in Lorius et al. (1985). Our estimate falls at the lower end of the estimated range of 2 - 2.4 g cm^{-2} yr^{-1} (Ritz, in press) but we note that it well agrees with the value of 2.00 ± .04 g cm^{-2} yr^{-1} estimated over the past 170 yr using the Tambora eruption as a time marker (Legrand and Delmas, 1987). Our new chronology agrees well with that of Lorius et al. (1985) to 110 kyr BP (Fig. 3) ; the biggest discrepancy is 3.5 kyr around 70 kyr BP (Jouzel et al., 1993).

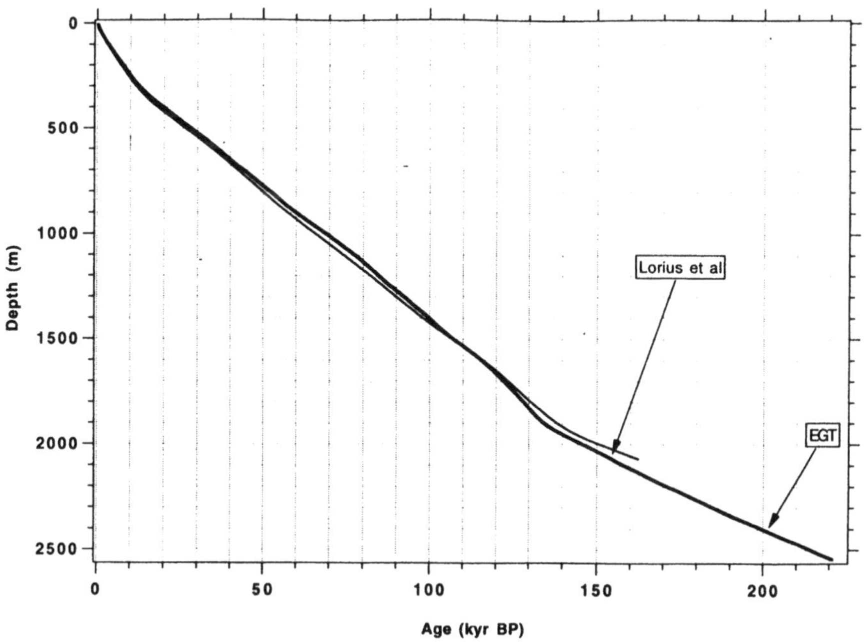

Figure 3 : Comparison of the Lorius et al. (1985) and of the EGT (this work) age-depth timescales.

With these two additional assumptions (linear increase of the present-day accumulation between Vostok and Dome B and imposed 110 kyr BP level), the bottom of the 4Γ core (2546 m) is dated at 220 ± 20 Kyr BP (Fig. 3). Still, the choice of the 110 Kyr BP level is somewhat arbitrary and any dating error there will directly reflect in the dating uncertainty. We will refer to this timescale as to the extended glaciological timescale (EGT).

3. Interpreting Isotopic Profiles in Polar Ice

We now examine in a relatively detailed manner how this isotopic record can be interpreted in a temperature record. Both hydrogen (^1H) and oxygen (^{16}O) have two isotopes ^2H and ^3H (deuterium, D and tritium, T), ^{17}O and ^{18}O. Among the isotopic forms of the water molecule, only HDO et $H_2{}^{18}O$ are of climatic interest. D/H and $^{18}O/^{16}O$ ratios are commonly expressed in δ notation with respect to V-SMOW (Vienna Standard Mean Ocean Water) with isotopic ratios R_{SMOW} of D/H = 155.76 10^{-6} and of $^{18}O/^{16}O$ = 2005.2 10^{-6}. δ values are given by : $\delta = (R_S/R_{SMOW}) - 1$, in which R_S is the isotopic ratio of the sample. δ values are expressed in per mill.

HDO et $H_2{}^{18}O$ saturation vapor pressures are slightly lower than that $H_2{}^{16}O$. As a result, the condensed phase (either liquid or solid) is, at equilibrium, isotopically enriched with respect to the vapor phase. The fractionation coefficients, α, then correspond to the ratios of D/H or $^{18}O/^{16}O$ in the condensed and in the vapor phases, α is practically equal to the ratio of the saturation vapor pressures of the corresponding molecules and depends only on the temperature and on the phase change considered (Table).

Temperature	Liquid/vapor equilibrium			Solid/vapor equilibrium		
	α^2H	$\alpha^{18}O$	$\frac{\alpha^2H-1}{\alpha^{18}O-1}$	α^2H	$\alpha^{18}O$	$\frac{\alpha^2H-1}{\alpha^{18}O-1}$
+ 20	1.0850	1.0098	8.7			
0	1.1123	1.0117	9.6	1.1330	1.0152	8.8
- 20	1.1492	1.0141	10.6	1.1744	1.0187	9.2

Table : HDO / $H_2{}^{16}O$ and $H_2{}^{18}O$ / $H_2{}^{16}O$ fractionation coefficients as a funtion of temperature and of phase change along with the ratio of the equilibrium and of the kinetic isotopic effects.

A second isotopic effect "the kinetic effect" is due to the fact that molecular diffusivities in air are lower for isotopic molecules than for $H_2{}^{16}O$ which induces an additional effect for non-equilibrium processes (evaporation and condensation). D_i/D ratios (D_i for isotopic molecules and D for $H_2{}^{16}O$) are independent of temperature and equal to .9755 et à .9723 for HDO and $H_2{}^{18}O$ respectively. The equilibrium isotopic effect (α-1 ratio) is 8 to 10 times higher for deuterium than for oxygen 18 (Table) whereas the kinetic effects are of the same order.This difference between the relative importance of equilibrium and kinetic isotopic effects implies that oxygen 18 is more sensitive than deuterium to non-equilibrium processes. In general, for that reason, it is interesting to analyze both isotopic species (Jouzel, 1986).

a

b

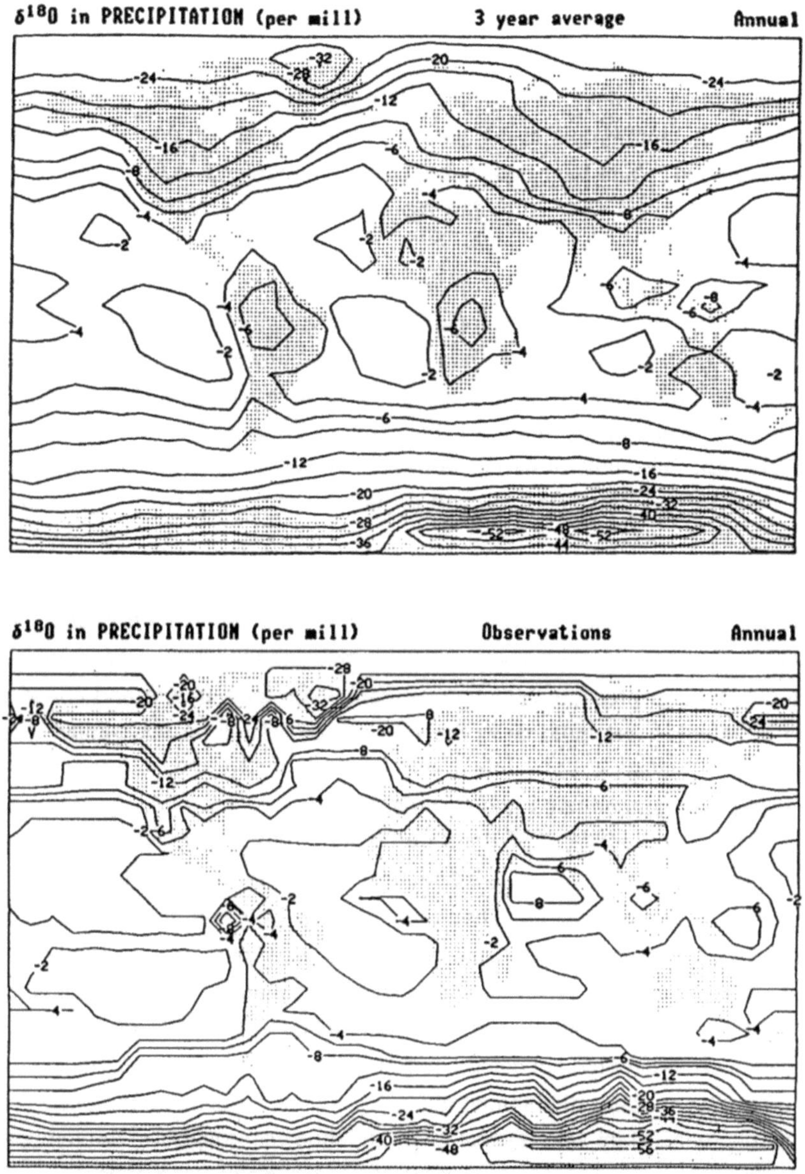

Figure 4 : Global distribution of $\delta^{18}O$ in precipitation for the present-day climate : the lower map is derived from observations, the upper map is produced from a 3-year simulation performed with the NASA/GISS model.

Global scale characteristics of δD and $\delta^{18}O$ distributions in precipitation are the following (Craig, 1961; Dansgaard, 1964) :

1) existence of a seasonal cycle (summer maximum and winter minimum) well marked for mid and high latitude continental precipitation,

2) general decrease of the mean annual isotopic contents from equatorial to polar regions (Figure 4b),

3) existence of a linear relationship between these annual means and the mean annual temperature at the site, Ts, for Ts <15°C (Figure 5b); this relationship is illustrated in Figure 6 for East Antarctica and Greenland, and

4) δD et $\delta^{18}O$ are distributed along the Meteoric Water Line (MWL) $\delta D = 8*\delta^{18}O + 10$.

Those characteristics are easily explained from a simple isotopic Rayleigh model (Dansgaard, 1964) in which isotopic fractionations occuring in an isolated air mass travelling from its oceanic source towards polar regions are considered. The condensed phase is assumed to form in isotopic equilibrium with the surrounding vapor and to leave the air mass just after it formed. The precipitation isotopic content, δ_p, then may be expressed as a function of the vapor isotopic content at the origin of the air mass, δ_0, and of its initial, m_0, and final, m_c, water vapor content:

$$\delta_p = \alpha_c (1+\delta_0) ({}^{m_c}/{}_{m_0})^{\bar{\alpha}-1} - 1$$

in which α_c is the fractionation coefficient when precipitation forms and $\bar{\alpha}$ its mean value over the air mass trajectory. The water vapor content is proportional to the saturation vapor pressure (which depends only on temperature and phase change) and is inversely proportional to the air pressure. Thus, in this simple model, the isotopic content of a precipitation only depends on δ_0 and on the initial and final condensation temperatures and air pressure.

Merlivat and Jouzel (1979) showed that δ_0 may be, under certain simplifying assumptions, expressed as a function of sea surface temperature, T_W, relative humidity, h, and windspeed. They extended the Rayleigh approach in accounting for the fact that part of the liquid droplets formed may be kept in the cloud and (Jouzel and Merlivat, 1984) for the kinetic fractionation which occurs when snow forms by inverse sublimation in a sursaturated environment. δ_p then essentially depends on the source conditions, h and T_W, on the proportion of liquid kept in the cloud on the condensation temperature, T_c. Figure 7 illustrates the observed global scale decrease of δ_p with respect to T_c and thus with respect to the temperature of the site which varies as T_c over most of the the globe. One exception is in polar regions where a strong temperature inversion is observed in particular over East

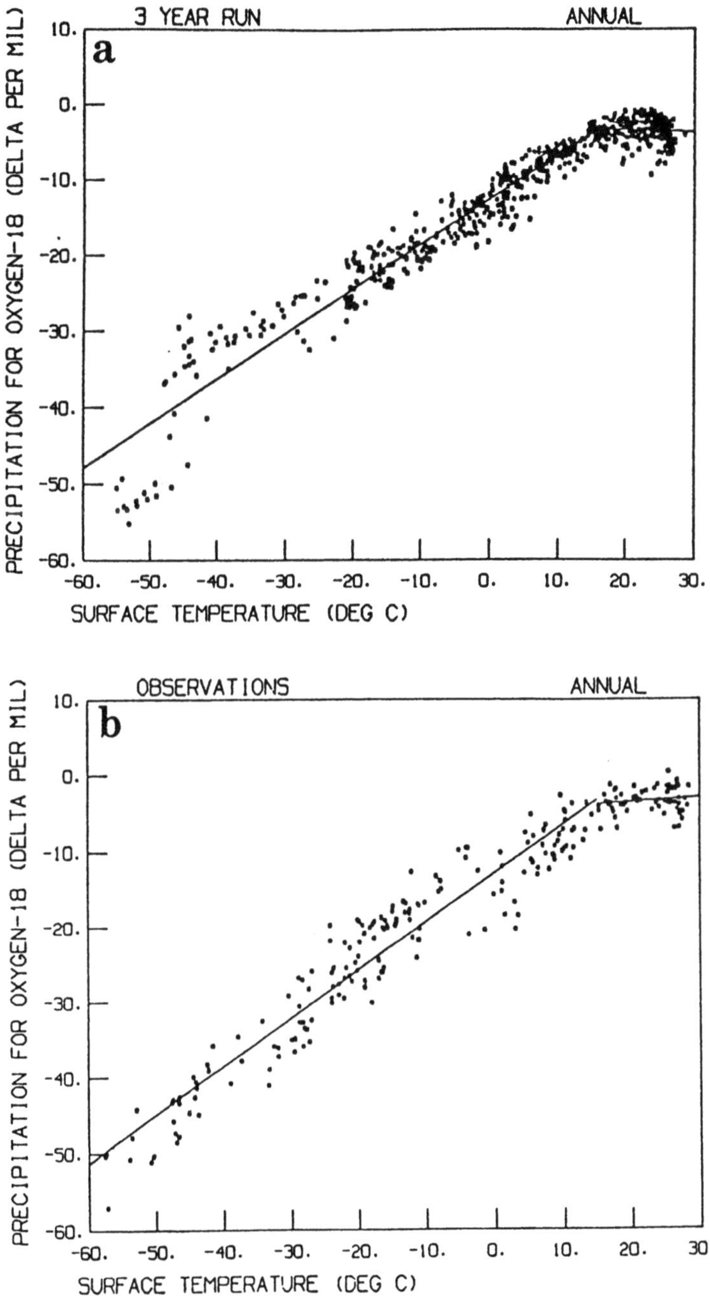

Figure 5 : $\delta^{18}O$ in precipitation versus local temperature (annual average): a) simulation performed with the NASA/GISS model, and b) observations.

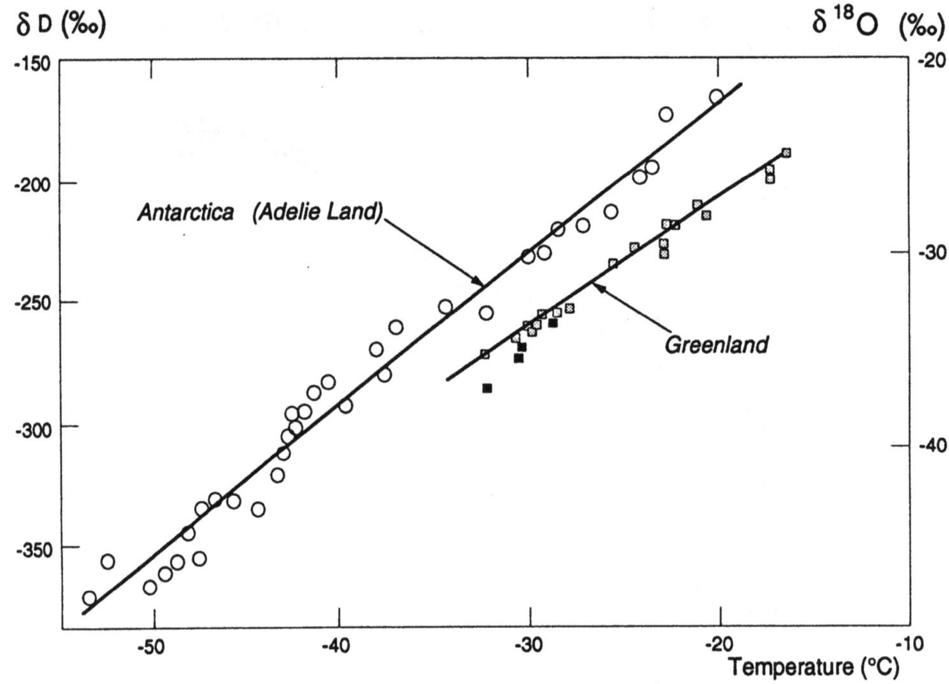

Figure 6 : Plot of the isotopic content of snow with respect to the local temperature (annual average). Antarctic data (δD, left scale) are from Lorius and Merlivat (1977) and Greenland data ($\delta^{18}O$, right scale) are from Johnsen et al. (1989).

Antarctica. The straight line representing the Antarctic data of Figure 6 (Dumont d'Urville - Dôme C axis) is reported both with respect to surface temperature, T_S, and to the temperature at the inversion level, T_i which is very close to the temperature of condensation, T_C. (Robin, 1977). The slope obtained in this latter case (1.12 ‰/°C for $\delta^{18}0$) is very close to that predicted by the Rayleigh type model assuming a source temperature of 20°C (1.1 - 1.2 ‰/°C). Provided an adequate choice of its parameters, this model thus allows a quite correct simulation of data observed over East Antarctica.

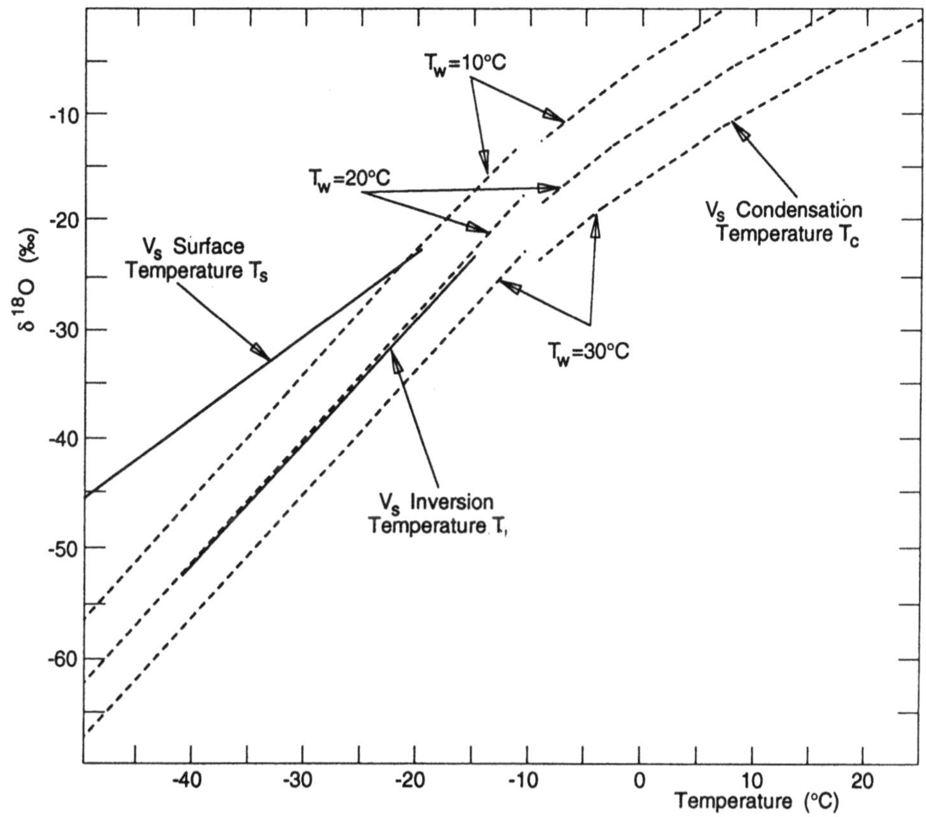

Figure 7 : $\delta^{18}O$ in precipitation as calculated from a Rayleigh model. The three sets of curves are for different Sea Surface Temperature (T_W). The curves are obtained in using the approaches followed by Merlivat and Jouzel (1979) for the liquid phase and by Jouzel and Merlivat (1984) for snow formation. The solid lines correspond to the East Antarctic data plotted with respect either to T_S (surface temperature) or to T_i (inversion temperature).

Figure 7 shows the influence of the sea surface temperature, T_W, on δ_p. As examined by Merlivat and Jouzel (1979), Jouzel and Merlivat (1982), Johnsen et al. (1989) and Petit et al. (1991), source conditions (temperature and humidity) also influence the deuterium / oxygen 18 relationship and thus the deuterium excess. For T_C>-20°C the precipitation points fall on a straight line $\delta D = s * \delta^{18}O + d$ with s and d depending essentially of h and T_W and being practically independent of atmospheric processes (Merlivat and Jouzel, 1979). For s=8 and d=10 (Meteoric Water Line) globally averaged estimates of 81% for h and of 26°C for T_W are inferred. This corresponds to reasonable values as the main water vaper source at a global scale is in subtropical regions. Extended to the solid phase (Jouzel and Merlivat, 1984), the model also accounts for deuterium excess values observed in

Greenland (Johnsen et al., 1989) and in Antarctica where δ becomes higher than 15 ‰ in central regions (Petit et al., 1991, Dahe et al., 1993). Those results as well as the isotope temperature relationships illustrated in Figure 7 are now well confirmed by Ciais and Jouzel (in press) who introduced the treatment of mixed clouds, (coexistence of liquid and solid phases), in this Rayleigh type model.

We will not further discuss the deuterium/oxygen 18 relationship and the deuterium excess in this article, but it is clear from the above results that simple isotopic models are appropriate to explain the main characteristics of δD and $\delta^{18}O$ in precipitation (although the existence of large drops which are out of equilibrium prevents to use them for convective systems). Indeed their ability to correctly simulate the present day temperature isotope relationships in polar regions is, up to now, the main justification of the standard practice of using the present day observed slope (spatial slope) to interpret the deep ice core isotopic profiles in terms of records of past temperature changes. However this slope is conceptually different from the "temporal $\delta^{18}O/T_S$ slope", i.e., the relationship describing the variation of $\delta^{18}O$ (or δD) with T_S when one stands at a single geographic location and goes back in time through several different climates, which is used by paleoclimatologists in this approach. Also, dynamically simple models are not adequate to account for the complexity of the processes involved in individual precipitation events and for the changes in boundary conditions and in atmospheric circulation that may occur during important climatic changes, such as the glacial-interglacial transition.

One way to deal with these aspects is to incorporate the isotopic cycle into a general circulation model (GCM). The modeling technique consists of initializing the ocean, atmosphere, and ground reservoirs with reasonable HDO and $H_2{}^{18}O$ concentrations in water, and, during a model simulation, letting the GCM transport the isotope masses through the atmosphere and ground via the same processes used to transport model water (advection, moist convection, re-evaporation of falling raindrops, evapotranspiration,etc...). Isotopic fractionation, including kinetic fractionation effects in subsaturated environments, is accounted for at every change of phase. Joussaume et al. (1984) pioneered this approach for present-day January climate with the GCM of the Laboratoire de Météorologie Dynamique (LMD). Jouzel et al. (1987b) generated a full annual cycle of isotope fields with the the GCM of the Goddard Institute for Space Studies (GISS). Like the LMD model, the GISS model successfully reproduces the important features of present-day global isotope distributions, (Fig 4a) including the spatial correlations between isotope concentration and temperature at mid to high latitudes (Fig 5a).

Both groups have now performed similar experiments for the climate of the Last Glacial maximum (18 kyr BP) with as a main objective to test the validity of the assumption, used by paleoclimatologists, that the temporal and the spatial slopes are identical. As far as Greenland and Antarctica are concerned they arrived to the conclusion that the spatial and temporal slopes are similar within 20 to 30 % (Joussaume and Jouzel, 1993; Jouzel et al., submitted). Although new experiments of this type are certainly necessary, this helps to justify the standard practice of applying the spatial slope to the interpretation of paleodata that we will use here. One must also note that this GCM approach accounts for such changes as the origin of precipitation or the seasonality regime that are often cited as limiting factors of the temperature interpretation (Robin, 1983; Fisher, 1992). This standard practice is now further supported by a study indicating that the long-term changes of isotopic composition of precipitation over mid and high-latitudes regions over the past three decades closely followed long-term changes of surface air temperature (Rozanski et al., 1992). Also, we note that applying it for the Dye 3 Greenland core led to a value quite comparable to that independently derived from ice and heat-flow modelling (Johnsen et al., 1992)

4. The Isotope Temperature Record

Using the EGT chronology, the temperature record reconstructed from the isotopic profile is shown on Figure 8b from the present back to 220 Kyr BP. We started from a composite deuterium profile constructed from published data (Jouzel et al.,1987a) down to 1,920m, by averaging 3Γ and 4Γ data down to 2,083m and from the new 4Γ data down to 2,546 m.

In order to correct for the concurrent change in the ocean isotopic content, we now take advantage of the fact that the $\delta^{18}O$ of trapped O_2 in the Vostok core is a first order proxy for the $\delta^{18}O$ of sea water which allows us to put the Vostok and marine records on a common temporal framework (Sowers et al., in press). We follow these authors in using the V19-30 $\delta^{18}O$ foraminiferal record, corrected for bottom water cooling during glacial periods, as a record of the changes in the isotopic composition of sea-water (multiplied by 8 for δD) and as a continental ice volume record. We extend it back to 250 kyr BP (end of marine glacial stage 8). The record, normalized so that the average Holocene value is 0 ‰, (curve 8c) is placed into the SPECMAP time scale constructed by orbital tuning (Martinson et al., 1987). The Vostok temperature record now entirely covers marine stage 6 which spans from 130 to 190 kyr BP and the later part of stage 7 characterized by 3 warm periods 7.1, 7.3, 7.5 separated by cold periods 7.2 and 7.4 centered around 200 and 225 kyr BP.

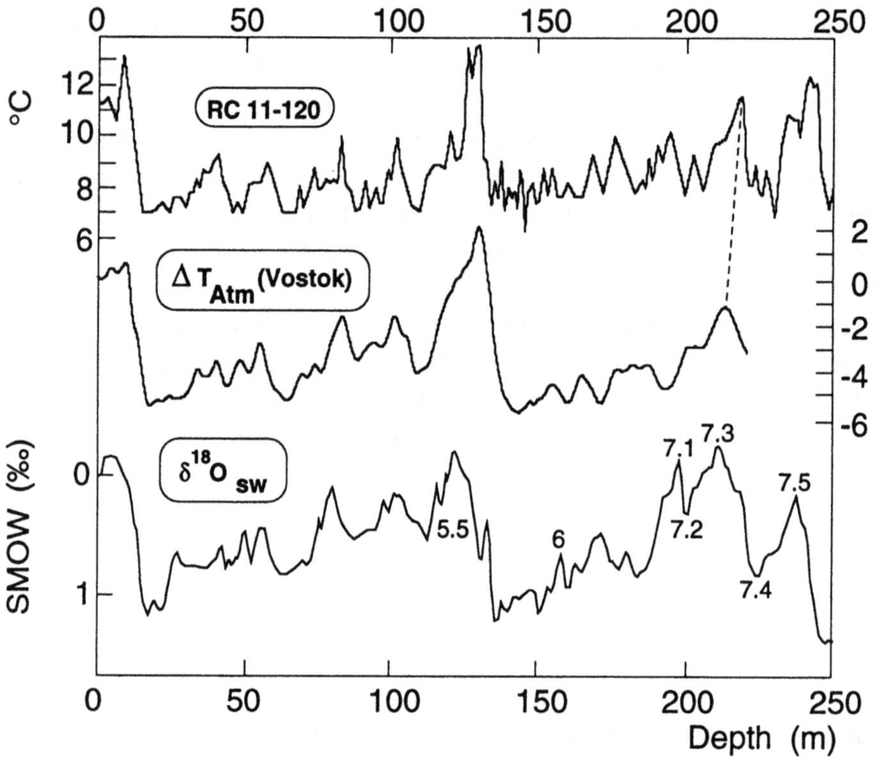

Figure 8 : Variations with time of :

a) Summer Sea-Surface temperature (SST) Indian ocean site RC11-120 (43°S),

b) The atmospheric temperature change at Vostok (above the inversion) as derived from the deuterium profile. We used published deuterium data down to 1920 m (Jouzel et al., 1987a) and extended this record down to 2083 m by averaging 3Γ and 4Γ data and to 2546 m using 4G data alone, and smoothed the entire record,

c) the change in the isotopic composition of sea-water derived from the V19-30 $\delta^{18}O$ foraminifera record and extended back to 250 kyr BP (Sowers et al., in press).

After applying a simple smoothing procedure, curve 8b is obtained with a δD-T gradient of 9‰/°C (equivalent to 1.12 ‰/°C for $\delta^{18}O$) which refers to the temperature change, ΔT_i, just above the inversion layer where the antarctic precipitation is formed. As discussed above, ΔT_i is directly accessible from the snow isotopic content and, also, this parameter is the more relevant for characterizing the global atmosphere as the existence of a strong inversion is restricted to central Antarctica. (In this approach, we do not account for possible changes in the altitude of the ice cap).

Back to 110 kyr BP the curve is essentially similar to that of Jouzel et al. (1987a) with the last ice age characterized by three minima, around 20, 60 and 110 kyr BP separated by well marked interstadials. The Last Interglacial is slightly shortened with a mid-peak duration of 18 ka (135-117 kyr BP) instead of 22 ka. Temperature starts to raise five years earlier (137 kyr BP) and the LIG peak is around 130 instead of 133 kyr BP. The new record shows a long cold period from ~140 to 200 kyr BP with, unlike the Last Ice Age, only minor oscillations. In central Antarctica all this period was practically as cold as the LGM (~6°C colder than the Holocene) but the two ice ages exhibit significantly different structures. The temperatures are warmer before ~200 kyr BP in our chronology with a well marked peak around 215 kyr BP suggesting that the bottom part of the core reached back into the end of the previous Interglacial Period. Still this peak is significantly colder than the Holocene (sligthly more than 1°C) and much colder (3°C) than the LIG Peak.

The Vostok temperature record compares well with sea surface temperature (SST) in various Indian ocean cores (Jouzel et al., 1987a, Pichon et al., 1992). For example, in the summer SST record of RC11-120 (curve 8a), the temperature peak corresponding to marine stage 7.1 (around 195 kyr BP) is less pronounced than 7.3 (around 218 kyr BP). Stages 7.1 and 7.3 are colder than the Holocene and much colder than the Last Interglacial. These features are in agreement with the Vostok temperature record (curve 8b). Correlating the Vostok temperature peak at 215 kyr BP with the RC11-120 peak at 218 kyr BP suggests that the bottom of the Vostok core corresponds to marine stage 7.4. The comparison with RC11-120 would thus indeed well support our glaciological timescale. However, there are limitations in the interpretation of those SST profiles located north of the Antarctic polar front and thus still under the strong influence of the subtropical/transitional waters (Pichon et al., 1992). Indeed, these authors showed that the comparison between Vostok and Southern Ocean temperatures is even better with MD 84-551 core which is located 12° closer to Antarctica (south of the polar front) but this temperature profile ends at 135 kyr BP. Correlating the Vostok and SPECMAP chronologies by matching temperature-depth curves will improve when we have a SST profile, covering two glacial-interglacial cycles, south of the polar front.

Jouzel et al. (1993) have extended back to the bottom of the core other ways of correlating the Vostok and the SPECMAP time scales respectively through the dust (Petit et al;, 1990) and the air $\delta^{18}O$ (Sowers et al., 1991; Sowers et al., in press) Vostok records. Various uncertainties restrict the comparison to about ± 10 kyr but within this uncertainty the EGT chronology is generally supported by correlations with SPECMAP.

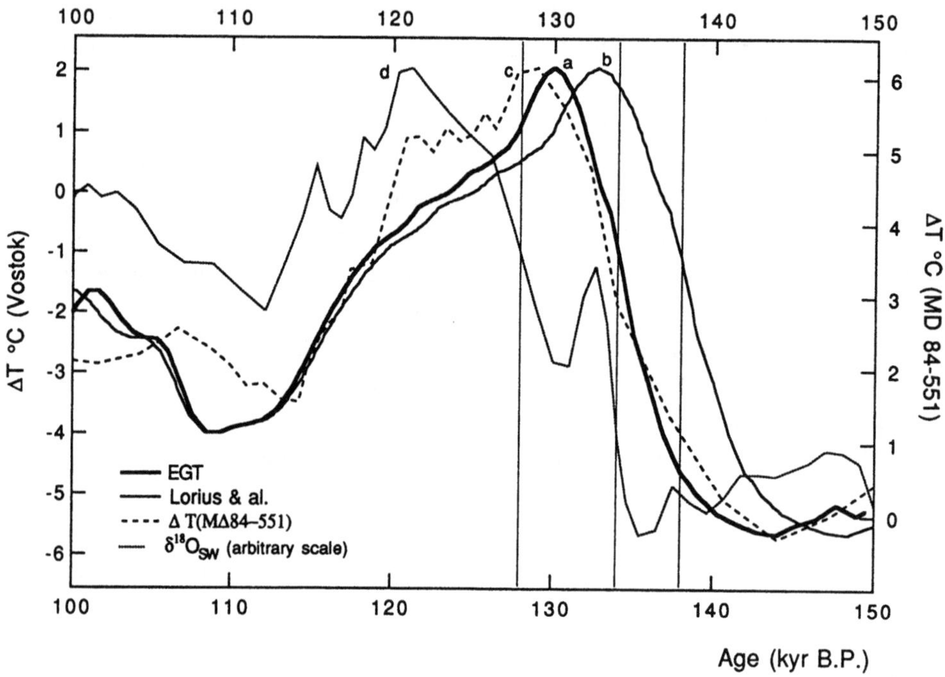

Figure 9 : The Vostok temperature record on various timescales for the 100-150 kyr BP period encompassing the LIG. Vostok temperature using either the EGT (a) or the Lorius et al. Timescales (left axis) c) the sea surface temperature estimated from core MD 84-551 (Pichon et al., 1992, right axis) and the seawater $\delta^{18}O$ (similar to 8c with an arbitrary scale chosen in such a way this curve is comparable to the temperature records). The 3 vertical lines correspond a) to mid -transition in seawater $\delta^{18}O$, proxy of sea-level and ice volume change, (128 kyr BP) b) to mid -transition in Vostok temperature using the Lorius et al. timescale (138 kyr BP) and c) using the EGT reference timescale (134 kyr BP).

The agreement between Vostok and SPECMAP chronologies now holds true even for the Last Interglacial (LIG). Information relevant to the LIG is summarized in Fig. 9. The figure illustrates the offset of ~ 10 kyr between the warming at Vostok in the Lorius et al. chronology and the sea-level change as indicated from the seawater $\delta^{18}O$ (Lorius et al., 1985; Jouzel et al., 1987a). The Vostok temperature record is plotted using the adopted EGT timescale along with the curve of sea surface temperature vs. age derived from core MD 84-551 (Pichon et al.,1992) . The correlation between those two curves (Pichon et al., 1992) indicates that the temperature at Vostok rose before (~ 5 kyr at mid - transition) sea level rose for this termination (termination 2) which is independently confirmed from the air $\delta^{18}O$

(Sowers eta l., 1991). This observation accounts for 5 kyr of the previous 10 kyr age difference between the temperature rise at Vostok and the deglaciation recorded in the seawater isotopic record (see above). The other part is accounted for by the fact the EGT timescale is, at mid-transition, ~ 4 kyr younger than the Lorius et al. (1985). The EGT and SPECMAP timescales now appear in fair agreement for the LIG. The EGT timescale supports the observation, now well documented, that the duration of the LIG is longer in the Vostok temperature and Southern ocean SST records than in the record of sea-level change. Our reconcilation of Vostok and SPECMAP chronologies is partly due to our decision to fix the age of Vostok at 1534 m at 110 kyr BP. On the other hand, the duration we calculate for the LIG is nearly independent of the age invoked for 1534 m depth : a 5 % change in age at 110 kyr BP will modify the duration of the LIG by only 1 kyr or so. Finally, we note that the agreement between the Vostok and Devils Hole records (Winograd et al., 1992) is less convincing with the EGT than with the the Lorius et al. timescale.

The link between the Vostok climate and the orbital forcing has been examined for the last climatic cycle through the spectral analysis of the temperature record (Jouzel et al., 1987a). We now extend this approach for the 220 kyr record. Following Yiou et al. (1991), we use the multitaper method (MTM) developed by Thomson (1982). Its high resolution makes the MTM method appropriate when the time series is short with respect to the periods we seek to examine. This is the case here for orbital periodicities (Berger, 1978) of 41 kyr (obliquity) and of 23 and 19 kyr (precession). The MTM method provides a statistical test for the validity of the location of each peak in the spectrum and for its amplitude. Fig 10 summarizes the results of the MTM spectral analysis for : a) for the entire 2546 m record (220 kyr in the EGT chronology), b) for the same series limited to 2,083 m using the EGT chronology and c) for the Lorius et al. time-scale. Using the EGT instead of the Lorius et al. chronology (Figure 10) causes i) a slight shift in the period of the obliquity peak (from 41.7 to 40. kyr), ii) retains the strong amplitude and significance of this peak and iii) makes the amplitude of the precessional peak (24.0 kyr) much stronger likely due to the shortening of the LIG but the second precessional peak near 19 kyr is no longer significant. Using the full time series and the EGT chronology leads to a decrease in the amplitude of the obliquity peak (41.0 kyr) which is even more significant but the precessional peak (24.6 kyr) is now weaker and less significant. This decrease in amplitude of orbital frequencies is the quantitative expression of uniformly cold temperatures throughout the period between 140-200 kyr BP. Over most of the last climatic cycle there is a strong relationship between local annual insolation which primarily reflects obliquity and temperature (Jouzel et al., 1987a). However this relationship is absent between

140-200 kyr BP : the insolation maximum at ~170 kyr BP has no clear temperature counterpart. Therefore, our new results confirm that apart the ~100 ka cycle, the Vostok record is dominated by a strong obliquity component and influenced only to a lesser degree by the precession (Jouzel et al., 1987a). They also draw attention on a possible difference in the link between high southern latitudes climate and astronomical forcing for the penultimate and the last ice ages.

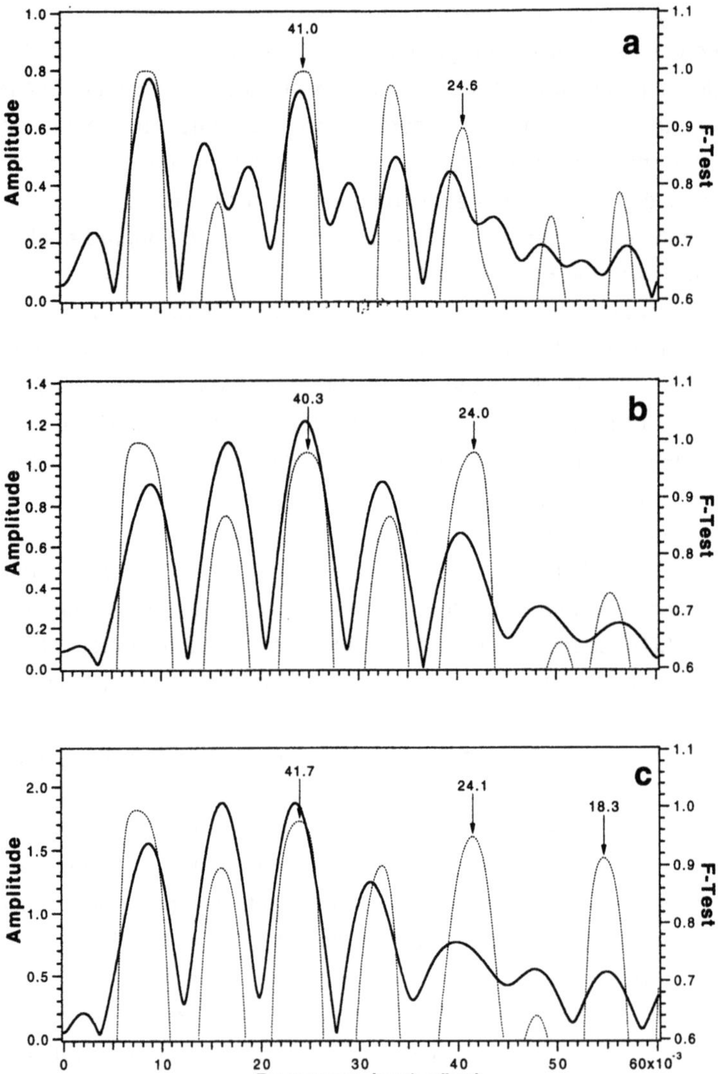

Figure 10 : Spectral analysis (MTM) of the Vostok time series. The thick curve represents the amplitude (left axis) and the thin curve represents the confidence value in percent (right axis). a) for the entire 2546 m record (220 kyr in the EGT chronology), b) for the same series limited to 2,083 m using the EGT chronology and c) for the Lorius et al. time-scale.

Conclusions

In this paper, we extend the Vostok climatic record to the bottom of the Vostok ice core and derive a new glaciological time scale for the entire core. The chronology gives an age of 220 ± 20 kyr for the bottom of the 4Γ core (2546 m below surface). This chronology also resolves the discrepancy, present in the glaciological chronology originally derived by Lorius et al. (1985), between the duration of marine stage 5e and the age of maximum warmth at Vostok. With the EGT chronology, the length of the LIG at Vostok is still 5 kyr longer than the length of stage 5e in deep-sea sediments. We believe that this important difference reflects the fact that, during termination 2, Antarctic temperatures rose ~ 5 kyr before the Northern Hemisphere ice sheets began to melt (Sowers et al., 1991; Pichon et al., 1992). The isotopic temperature record shows that there is a long cold period in East Antarctica which we date from about 140 to 200 kyr BP. The coldest isotopic temperatures measured during this period are similar to those measured during the last glacial maximum, but the duration of the cold period is much longer. Isotopic temperatures in the bottom of the Vostok 4Γ core corresponding to the end of marine stage 7.

We have now the opportunity to compare this new Antarctic climatic to the GRIP Greenland record (Summit) covering a comparable time-span; a continuous and very detailed $\delta^{18}O$ record is now available back to 250 kyr BP (Johnsen et al., 1992; Dansgaard et al., 1993 ; GRIP project members, 1993). We have drawn in Fig. 11 the Vostok record with each sample representing 200 years as in Dansgaard et al. (1993). Although dating methods are different they have in common to use a glaciological model with variable accumulation rate derived from the isotopic profiles and fixed points ; as in Vostok the characteristic correlation for marine stage 5d (110 kyr BP) is used for establishing the GRIP timescale (Dansgaard et al., 1993). As noted by these authors the violent δ shifts observed in Greenland cores are less pronounced in the Vostok records probably because the shifts in Greenland are connected to rapid ocean/atmosphere circulation changes in the North Atlantic region (Dansgaard et al., 1989, 1993 ; Bond et al., 1993). One example we have previously mentioned is that the Antarctic record shows a cold revearsal during the last deglaciation which may be the Southern Hemisphere counterpart of the Younger Dryas (Jouzel et al., 1992). In Fig. 11, the two events appear in phase within the uncertainty of the Vostok dating (~ 1000 years or slightly more at this level). The situation is different for the previous termination : whereas multiple "Younger-Dryas type" events are observed at GRIP between 135 and

125kyr BP (GRIP project members, 1993), the Antarctic temperature seems to increase in one single step during this deglaciation.

Dansgaard et al. (1993) also noted that the two records share common features : 1) there are similarities between the smoothed versions of the two records particularly during the last glaciation, 2) the Summit spectrum also shows the presence of a strong obliquity signal and of a considerably weaker precession signal and 3) as for Vostok, the last interglacial appears longer (nearly 20 kyr) than in the SPECMAP deep-sea core record. At last, we draw attention that although the isotopic glacial-interglacial changes are comparable between the two cores (the δ scales of Fig 11 are chosen in such a way that the isotopic signals are comparable) they may correspond to a significantly different temperature change i.e. 6°C (atmospheric level) at Vostok (Jouzel et al., 1987a ; Lorius et al. 1990) and 12°C at Summit (Johnsen et al., 1993). This is because the δ/temperature gradients used are different which is independently confirmed by the GCM simulations performed using the GISS model (Jouzel et al., submitted).

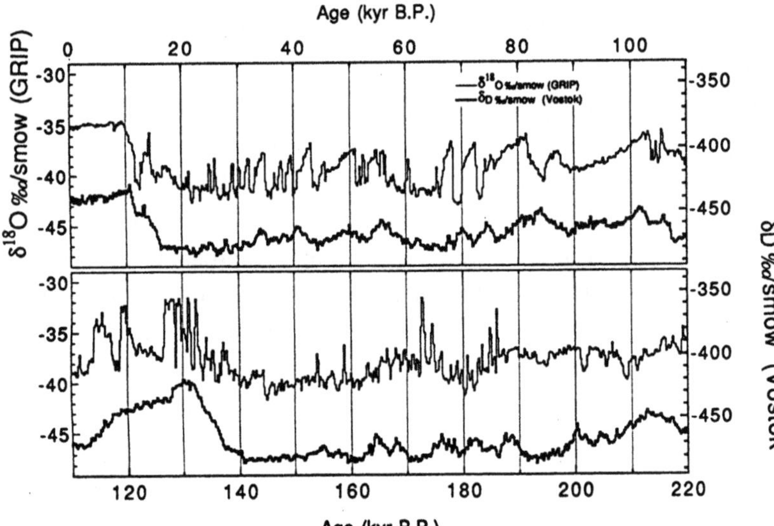

Figure 11 : Comparison of the GRIP (Dansgaard et al., 1993) and Vostok isotopic profiles over their common part (0-220 kyr BP) using 200 yr average values. The scales have been chosen in such a way the two curves have comparable amplitude.

There is still much more climatic information to extract from Vostok drillings : more than 1 km of ice below the level of 2546 m reached by core 4Γ (the ice thickness is ~ 3700 m) that may represent more than 500 kyr of time. There is a strong hope of reaching 2800 m depth by early 1994 (core 5Γ) a depth that may fully cover 2 climatic cycles back to 270 kyr BP.

Acknowledgements

Vostok is a joint project between Russia, France and USA. We thank all Russian, French and US participants in drilling, field work and ice sampling. We acknowlegdge the Russian Antarctic Expeditions (RAE), the Institut Français de Recherches et Technologies Polaires (IFRTP) and the Division of Polar Programs (NSF) for the logistic support. The project is supported in Russia, by the Russian Ministry of Sciences and in France, by PNEDC (Programme National d'études de la Dynamique du Climat), by Fondation de France and by the CEC (Commission of European Communities) Environment Programme. We thank J.M.Barnola, M.Bender, J.Chappellaz, C.Genthon, G.Raisbeck, D.Raynaud, T.Sowers and F.Yiou for discussions, S.J. Johnsen for GRIP isotopic data, J.C.Pugno for the figures and C.Alba, P.Doira and M.Delmotte for isotopic analyses.

References

Alley RB, Meese DA, Shuman CA., Gow AJ, Taylor KC, Grootes PM, White JWC, Ram M, Waddington ED, Mayewski PA, Zielinski GA (1993) Abrupt increase in Greenland snow accumulation at the end of the younger Dryas event. Nature 362 : 527 -529

Bond G, Broecker W, Johnsen SJ, McManus JC, Labeyrie L, Jouzel J. and Bonani G. (1993) Correlations between climate records from North Atlantic sediments and Greenland ice. Nature 365:143-147, 1993.

Craig H, Isotopic variations in meteoric waters (1961) Science 133: 1702 - 1703

Ciais P, and Jouzel J (1993) Deuterium and Oxygen 18 in precipitation : a n ew isotopic model including mixed cloud processes. J. Geophys. Res. (in press).

Dahe Q, Petit JR, Jouzel J, and Stievenard M, (1992) Distribution of stable isotope in Antartic surface snow along the route of the 1990 International Trans-Antarctica expedition. J. Glaciol., in press.

Dahl Jensen D, Johnsen SJ, Hammer CU, Clausen HB, and Jouzel J, (1992) Past accumulation rate. In: Ice Climate Interaction (NATO Editions), in press.

Dansgaard W (1964) Stable isotopes in precipitation. Tellus 16 : 436-468.

Dansgaard W, Johnsen SJ, Clausen HB, Dahl jensen D, Gunderstrup NS, Hammer CU, Steffensen JP, Sveinbjörnsdottir, Jouzel J and Bond G (1993) Ice core evidence of generally unstable climate. Nature 364: 407- 412 (1993).

Dansgaard W, White JWC and Johnsen SJ (1989) The abrupt termination of the Younger-Dryas climate event. Nature 339 : 532-534.

Fisher DA (1992) Stable isotope simulations using a regional stable isotope model coupled to a zonally averaged global model (1992) Cold regions Sci. and Tech., 21: 61 -77.

GRIP Project Members (1993) Climatic instability during the last interglacial revealed in the Greenland Summit ice - core. Nature, 364, 203 - 207, 1993.

Hammer CU, Clausen HB and Langway CC (1985) The Byrd ice core : Continuous acidity measurements and solid electrical measurements. Ann. Glaciol., 7, 214.

Hammer CU, Clausen HB and Tauber H (1986). Ice core dating of the Pleistocene/Holocene boundary applied to a calibration of the ^{14}C time scale. Radiocarbon 28: 284 - 291.

Imbrie J, Mix AC and Martinson DG (1993) Milankovitch theory viewed from Devils Hole. Nature 363: 531 - 533.
Johnsen SJ, Dansgaard W and White JWC (1989) The origin of Artic precipitation under present glacial conditions. Tellus 41B : 452-468.
Johnsen SJ, Clausen HB, Dansgaard W, Fuhrer K, Gundestrup N, Hammer CU, Iversen P, Jouzel J. and Stauffer B (1992) Irregular glacial interstadials recorded in a new Greenland core. Nature 359 : 311-313.
Joussaume S., Jouzel J., Sadourny R . (1984) Water isotope cycles in the atmosphere: First simulation using a general circulation model. Nature 311: 24-29
Joussaume S, Jouzel J (1993) Paleoclimatic tracers: An investigation using atmospheric general circulation model under ice age conditions; Part II; Water isotopes. J. Geophs. Res. 98: 2807-2830.
Jouzel J., Isotopes in cloud physics: Multiphase ans multistage condensation processes (1986) In: Handbook of Environmental Geochemistry, vol 2, The terrestrial environment, A, Fontes JC. and Fritz P Eds. (Amsterdam, Elsevier), 61-112.
Jouzel J, Barkov NI, Barnola JM, Bender M, Chappellaz J, Genthon C, Kotlyakov VM, Lipenkov V., Lorius C., Petit JR, Raynaud D, Raisbeck G, Ritz C, Sowers T, Stievenard M, Yiou F., Yiou P. (1993) Vostok ice cores : Extending the climatic record over the penultimate glacial period. Nature 364: 407 -412.
Jouzel J, Koster RD, Suozzo RJ and Russell GL (1993) Stable water isotope behavior during the Last Glacial Maximum : a GCM analysis, J. Geophys. Res., submitted.
Jouzel J, Lorius C, Petit JR, Genthon C, Barkov NI,. Kotlyakov V.M. and Petrov V.N (1987a) Vostok ice core : a continuous isotope temperature record over the last climatic cycle (160 000 years). Nature 329: 403-409.
Jouzel J., Merlivat L, and Lorius C (1982) Deuterium excess in an East Antarctic ice core suggests higher relative humidity at the oceanic surfaces during the last glacial maximum. Nature 299: 588-591.
Jouzel J., Merlivat L (1984) Deuterium and oxygen 18 in precipitation: modelling of the isotope effects during snow formation. J. Geophys. Res. 89:11749-11757.
Jouzel J, Petit JR., Raynaud D (1990) Paleoclimatic information from ice-cores : the Vostok records; The late cenoizoic ice age; Transaction of the royal Society of Edimburg : Earth Sciences 81 : 349-355.
Jouzel J, Petit JR, Barkov NI, Barnola JM, Chappellaz J, Ciais P, Kotlyakov VM, Lorius C, Petrov VN, Raynaud D, Ritz C (1992) The last deglaciation in Antarctica : further evidence of a younger Dryas type climatic event. In: The Last Deglaciation : Absolute and Radiocarbon Chronologies, Bard, W.S. Broecker Eds. (Springer Verlag, NATO/ASI Series), 229-266.
Jouzel J, Raisbeck G, Benoist JP, Yiou F, Lorius C, Raynaud D, Petit JR, Barkov NI, Korotkevitch YS and Kotlyakov VM (1989) The Antarctic climate over the late glacial period. Quat. Res. 312: 135-150.
Jouzel J, Russell GL, Suozzo RJ, Koster RD, White JWC and Broecker W.S. (1987b). Simulations of the HDO and $H_2{}^{18}O$ atmospheric cycle for present day conditions. J. Geophys. Res. 93: 14739-14760.
Legrand M, Delmas RJ (1987) A 220 - year continuous record of volcanic H2SO4 in the Antarctic ice-sheet. Nature 327 : 671-676.
Lorius C, Jouzel J, Raynaud D, Hansen J and Le Treut H (1990) The ice core record : climate sensitivity and future greenhouse warming. Nature 347: 139-145.
Lorius C, Jouzel J, Ritz C, Merlivat L, Barkov NI, Korotkevitch YS and Kotlyakov VM (1985) A 150 000 years climatic record from Antarctica. Nature 316: 591-596.

Lorius C and Merlivat L (1977) Distribution of mean surface isotope values in East Antarctica : Observed changes with depth in a coastal area.. Isotopes and Impurities in ice. IAHS Publ. 127-137.
Martinson DG, Pisias NG, Hays JD, Imbrie J, Moore TC and Shackleton NJ (1987) Age dating and the orbital theory of the ice ages : development of a high resolution 0 to 300 000 year chronostratigraphy. Quat. Res. 27: 1-30.
Merlivat L, and Jouzel J (1979) Global climatic interpretation of deuterium-oxygen 18 relationship for precipitation. J. Geophys. Res. 84: 5029-5033.
Petit JR, Mounier L, Jouzel J, Korotkevitch YS, Kotlyakov VI., Lorius C (1990) Paleoclimatological and chronological implications of the Vostok dust record. Nature 343: 156-158.
Petit J.R, White JWC, Young NW, Jouzel J, Korotkevitch YS (1991) Deuterium excess in recent Antarctic snow. J. Geophys. Res. 96: 5113-5122.
Pichon JJ, Labeyrie LD, Bareille G, Labracherie M, Duprat J and Jouzel J (1992) Surface water temperature changes in the high latitudes of the Southern Hemisphere over the last glacial-intergalical cycle. Paleoceanogr. 7: 289-318
Ritz C (in press) Chronology of the Vostok ice core based on precipitation and ice modeling. J. Glaciology.
Robin G. de Q. (1977) Ice cores and climatic changes. Phil.Trans.Royal Soc. London, B, 280: 143 -168.
Robin G. de Q. (1983) The climatic record from ice cores. In: The climatic record in polar ice sheets, G. de Q. Robin Ed.(Cambridge Univrsity Press, Cambridge UK), 180-195.
Rozanski K, Araguas- Araguas L and Gonfiantini R (1992) Relation between long - terrm trends of oxygen - 18 isotope composition of precipitation and climate. Science 258: 981-985
Shackleton NJ, Mix A and Hall (1992) Carbon isotope records from Pacific surface waters and atmospheric carbon dioxyde. Quat. Sci. Rev. 11: 387-400.
Sowers T, Bender M, Raynaud D, Korotkevich YS and Orchado J (1991) The $\delta^{18}O$ of atmospheric O_2 from air inclusions in the Vostok ice core. Paleoceanography 6: 679 - 696.
Sowers T, Bender M, Labeyrie L, Jouzel J, Raynaud D, Martinson D, Korotkevich YS. (1993) 135,000 year vostok, SPECMAP common temporal framework. Paleoceanography, in press.
Thomson DJ (1982) Spectral estimation and harmonic analysis. Proc. IEEE, 70: 1055-1096.
Vaikmae R, Jouzel J, Petit JR, Stievenard M and Toots M. (1993) A new 30000 year record from the Dome B ice core. Paper presented at VISAG, Cambridge UK.
Winograd IJ, Coplen TB, Landwher JM, Riggs AC, Ludwig KR, Szabo BJ, Kolesar PT., Revesz K.M. (1992) Continuous 500 000 - year calcite in Devils hole Nevada. Science 258: 255-260.
Yiou P, Genthon C, Ghil M, Jouzel J, Le Treut H, Barnola JM, Lorius C, Korotkevitch YN (1991) High-frequency paleo-variability in climate and in levels from Vostok ice-core records. J. Geophys. Res. 96: 20365-20378.

Lorius C and Merlivat L (1977) Distribution of mean surface isotope values in East Antarctica: Observed changes with depth in a coastal area. [illegible]

[illegible]

ORGANIC COMPOUNDS AS PROXY-INDICATORS OF SEA SURFACE PALAEOTEMPERATURE: THE U^{k}_{37} INDEX.

A.Rosell, J.Grimalt* and G.Eglinton
Environmental and Analytical Chemistry Section, School of Chemistry
University of Bristol, Cantock's Close
Bristol, BS8 1TS, U.K.

Abstract. The U^{k}_{37} index is a proxy-indicator of sea surface temperature (SST). It is derived from the relative abundance of two organic compounds (long-chain alkenones) which are commonly found in Quaternary marine sediments. These compounds are biosynthesized by some species of algae of the Class Prymnesiophyceae, which includes the coccolithophorids. U^{k}_{37} stratigraphy in ocean cores can provide time series records of local SST in the photic zone. Despite several biological and geochemical questions which constrain its applicability as a palaeothermometer, it has been successfully used in the study of temperature records over several glacial/interglacial transitions, and has increased the current understanding on the evolution of upwelling areas, el Niño events and rates of climatic change produced over hundreds, thousands or tens of thousands of years.

INTRODUCTION

The long-chain alkenones (see table 1) are a group of organic compounds that occur ubiquitously in marine sediments (Brassell, 1993). They are biosynthesized by a few species of phytoplankton and hence, in organic geochemical terms, they can be considered as biomarkers (Volkman *et al.*, 1980a; de Leeuw *et al.*, 1980); that is molecules whose occurrence can be related to inputs from a group of organisms, in this case the algae of the Class Prymnesiophyceae (Volkman *et al.*, 1980a; Marlowe *et al.*, 1984a). These algae are widespread dwellers of the present time oceans, and comprise both coccolith and non-coccolith bearing species. Large deposits of coccolithphorids tests are contained in the fossil record, for example in the Black Sea (Bukry, 1974). A predominant species is the coccolithophorid *Emiliania huxleyi* (McIntyre & Bé, 1967; Okada & Honjo, 1973; Okada & McIntyre, 1977).

----

* Departament de Química Ambiental, CID-CSIC, Jordi Girona 18-26, 08034 Barcelona, Catalonia.

NATO ASI Series, Vol. I 22
Long-Term Climatic Variations
Edited by J.-C. Duplessy and M.-T. Spyridakis

The first documentation of the presence of the alkenones in the marine environment was provided in the late 70's, after the analysis of sediments from Walvis Ridge (Boon *et al.*, 1978), off Southwestern Africa. In the following years, their structure (de Leeuw *et al.*, 1980) and source (Volkman *et al.*, 1980a,b) to the environment was determined, and their widespread incidence in marine sediments revealed. Their long occurrence in the geological record was shown when they were identified in Cenozoic sediments back to the Eocene (ca. 45Ma) (Volkman *et al.*,1980a; de Leeuw *et al.*, 1980; Marlowe *et al.*, 1984b), and in two Cretaceous black shales (Farrimond *et al.*, 1986). In the early 80's, it was demonstrated that there is a linear relationship between the relative abundances of the alkenones and the temperature of the waters where they were biosynthesized, either in laboratory cultures or the open sea (Marlowe, 1984). Further studies gave a numerical expression to this relationship with the definition of the U^k_{37} index (Brassell *et al.*, 1986a,b):

$$U^k_{37} = [37{:}2Me - 37{:}4Me]/[37{:}2Me + 37{:}3Me + 37{:}4Me]$$

Where U^k_{37} stands for **U**nsaturated **K**etones with **37** carbon atoms. 37:2Me stands for alkenone with 37 carbon atoms, a methyl side-chain and 2 double bonds or unsaturations.

The equation was later modified by removing the 37:4Me alkenone, because this compound is hardly ever found in open-sea marine sediments or water-column particulates. Thus (Brassell *et al.*, 1986a; Prahl & Wakeham, 1987; Prahl *et al.*, 1988):

$$U^k_{37} = [37{:}2Me]/[37{:}2Me + 37{:}3Me]$$

The calibration of the U^k_{37} index with water temperature was devised using laboratory cultures of *Emiliania huxleyi* (Marlowe, 1984; Prahl & Wakeham, 1987; Prahl *et al.*, 1988). Since the end of the 80's, the calibration curve is defined by the following expression (Prahl *et al.*, 1988):

$$T = [U^k_{37} - 0.039]/0.034$$

Besides temperature, other factors (e.g. light, nutrients) have not been found to have a significative influence on the U^k_{37} index (Marlowe, 1984; Prahl & Wakeham, 1987; ten Haven *et al.*, 1987; Sikes *et al.*, 1991). The physiological role of the alkenones in the algae and their biosynthetic pathway are still unknown. The fact that they are a major biochemical component in the plant cell and have a such response to growth temperature suggests that they serve some structural role and are localized in the cell membrane, maybe as fluidity stabilizers (Brassell *et al.*, 1986a,b; Prahl *et al.*, 1988) at different water temperatures.

The alkenones have also been reported in lacustrine sediments from both temperate (Cranwell, 1985 and 1988) and polar areas (Volkman *et al.*, 1988). They must have a different source than those from marine environments, as it suggests their different composition

(Cranwell, 1985 and 1988; Volkman *et al.*, 1988) and the fact that *E.huxleyi* is a restricted marine species. This lack of known biogenic precursors is a major drawback for the extension of the U^k_{37} index to lacustrine environments.

Table

Shorthand notation	Name	U^k_{37}	IPT
alkenones			
37:4Me	heptatriaconta-8,15,22,29-tetraen-2-one	*	+
37:3Me	heptatriaconta-8,15,22-trien-2-one	*	+
37:2Me	heptatriaconta-15,22-dien-2-one	*	+
38:3Et	octatriaconta-9,16,23-trien-3-one		
38:3Me	octatriaconta-9,16,23-trien-2-one		
38:2Et	octatriaconta-16,23-dien-3-one		
38:2Me	octatriaconta-16,23-dien-2-one		
alkyl alkenoates			
37:3OMe	methyl hexatriaconta-7,14,21-trienoate		+
37:2OMe	methyl hexatriaconta-14,21-dienoate		+
38:2OEt	ethyl hexatriaconta-14,21-dienoate		+

Names, and shorthand notation of the long-chain alkenones and alkyl alkenoates, biomarkers of the Prymnesiophyceae algae. The compounds marked with '*' are used in the U^k_{37} index. It has been suggested that in some circunstances another temperature index (IPT; Conte *et al.*, 1992) should be used (see text), which would include the compounds marked with '+'.

CHARACTERISTICS OF THE ALKENONES

The alkenones undergo less degradation in the environment than other compounds (e.g. sterols) biosynthesized by the same algae (Volkman *et al.*, 1980b and 1983). Their double bond configuration (*E*; not the biologically common *Z*), long chain length (C_{37}-C_{39}) and wide spacing between the double bonds are unusual characteristics for biolipids (Rechka & Maxwell, 1988); and may provide them with a special resistance to enzymatic attack. Hence, they are among the most abundant biomarkers in the sediment. However, their concentration in the water column and in the top surface sediment decreases through remineralization, microbial degradation, (Volkman *et al.*, 1983; Prahl & Wakeham, 1987; Prahl *et al.*, 1989; McCaffrey *et al.*, 1990; Conte *et al.*, 1992; Freeman & Wakeham, 1992), and, sometimes, by incorporation of sulphur (Sinninghe Damsté *et al.*, 1989). Nevertheless, both alkenones are altered proportionally to their abundance, and hence, the U^{k}_{37} index remains constant throughout, although recent research carried out in the Black Sea shows that U^{k}_{37} increases with water depth indicating that alkenones with more double bonds may be preferentially lost (Freeman & Wakeham, 1992). The alkenones are not affected by the conditions that cause dissolution of carbonate tests below the CCD (Volkamn *et al.*, 1980a; Brassell *et al.*, 1986a), and laboratory experiments support this evidence (Sikes *et al.*, 1991). Finally, modification of the ratio during the storage of cores at cool or room temperature for up to 4 years appears to be negligible (Sikes *et al.*, 1991).

METHODS OF ANALYSIS

It is not the purpose of this paper to discuss the full range of analytical methods available to determine the U^{k}_{37} ratio in sediments. However, the use of different procedures by the increasing number of research groups interested in carrying out their own determinations may result in discrepancies. We do not as yet know how much the value of the index in any sample can be affected by the use of different analytical methods. Evidence of this problem is being collected at the present time. Limited initiatives are under way internationally to carry out intercalibration exercises. The distribution of standardized sediment samples could facilitate this intercomparison between laboratories and methods, as would the general availability of synthetic standards of the alkenones (Rechka & Maxwell, 1988). Fortunately, their physico-chemical characteristics make them easily quantifiable by conventional gas chromatographic analysis and by mass spectrometry (Rechka & Maxwell, 1988; Rosell, 1994). Furthermore, once a routine method is set up, most of the analysis procedure can be automated using commercial robotic systems (Rosell, 1994).

Most of the sediment samples retrieved from the oceans in the last 20 years should be analysed with special caution, as the organic geochemical requirements for storage, manipulation and delivery are unlikely to have been taken into account. Analytical procedures based on mass spectrometric techniques are then particularly appropriate, as the targeted compounds can be quantified more selectively, avoiding additional fractionation steps that would otherwise lengthen the analysis time and create further problems (Rosell, 1994). Samples for U^{k}_{37} determination, and for the analysis of biomarkers, should be handled with care. They should be stored frozen in glass containers, as soon as they are retrieved. Their manipulation should be carried out using only metal, teflon or glass utensils, with special care to avoid organic contamination (e.g. from lubricant oils, plasticizers from bags and flasks), that could interfere with the analysis.

TEMPERATURE INFORMATION

The information which can be extracted from the U^{k}_{37}-stratigraphic analysis of deep-sea cores is in the form of time series of relative and absolute changes of sea surface temperature. However, in order to interpret correctly the U^{k}_{37}-temperature values, several considerations have to be made. These arise from the ecology of the Prymnesiophyceae, their fossil history and the limitations of the temperature calibration. Unfortunately, little is known of the ecology of *E.huxleyi* and other species of coccolithophorids.

First, the U^{k}_{37} measurements correspond to the temperature of the euphotic zone where the Prymnesiophyceae carry out their photosynthesis. The maximum depth of the photic layer may vary from a few centimetres to 150m depth, according to the intensity of the insolation and the translucence of the waters (McIntyre *et al.*, 1970). *E.huxleyi* has been found to live at depths of up to 200m. It contains a pigment which probably allows it to photosynthesize in the deepest parts of the photic layer (Prahl *et al.*, 1989; Freeman & Wakeham, 1992).

Second, the temperature information recorded in the sediments approximates to that of the time of the year when the maximum production of the algae occurred (Brassell *et al.*, 1986a; Sikes *et al.*, 1991; Conte *et al.*, 1992); this is when the productivity limiting factors were optimum for the algae to grow. In mid latitudes coccolithophorids bloom in spring (Conte *et al.*, 1992), summer and autumn, whereas at high latitudes only in summer (Okada & McIntyre, 1979; Smayda, 1980; Holligan *et al.*, 1983). In the eastern equatorial Atlantic, *E.huxleyi* is assumed to be more abundant in winter (Brassell *et al.*,, 1986a).

Third, the temperature calibration is based on a single strain of *Emiliania huxleyi* cultured at discrete temperature intervals between 8-25°C (Prahl & Wakeham, 1987; Prahl *et al.*,1988). Its applicability to the natural environment remains to be completely assessed. In some cases, the U^{k}_{37}-temperature values derived from the analysis of particulate material from the water column (Prahl & Wakeham, 1987; Conte *et al.*, 1992) and surficial sediments (Prahl *et al.*, 1988; Sikes *et al.*, 1991) from different oceanic locations are in agreement with the measured *in situ* temperatures of the overlying waters. Thus, the strain of *E.huxleyi* used for the calibration is presumably representative of the average natural population of *Prymnesiophyceae* for those areas studied. In other instances, they are not in agreement (Prahl *et al.*, 1988); the reason for this is likely to be due to changes in the physiology of the natural population of the algae producing the alkenones, or preferential degradation of one of the compounds (Prahl *et al.*, 1988 and 1989; Freeman & Wakeham, 1992). Different species and strains of Prymnesiophyceae co-exist in the oceans, perhaps synthesizing alkenones yielding different correlations with temperature. For instance, *Isochrysis Galbana* and *Geophyrocapsa Oceanica* have been found to produce a relationship with temperature which differs from that of *Emiliania huxleyi* (Marlowe, 1984; Volkman *et al.*, personal communication, 1993).

Fourth, the temperature boundaries of the U^k_{37} index are at the moment being established. The current calibration does not extend to water temperatures below 8°C or above 25°C, restricting its world wide application. However, recently, attempts have been made to extend it to waters just above zero degrees centigrade (Sikes & Volkman, 1992; Rosell, 1994).

Fifth, another question to be tackled is how far back in the past U^k_{37} can be applied. The coccolithophorid *Emiliania huxleyi*, likely main contributor of alkenones to contemporary oceanic sediments (Marlowe *et al.*, 1984b), is a rather recent species. It first appeared ca.250,000 years ago, in the late Pleistocene (McIntyre, 1970). Members of the family Gephyrocapsaceae, which includes the genus *Emiliania*, whether living or extinct, have or have had the capability to biosynthesize alkenones during the last 45Ma (Farrimond *et al.*, 1986). Nothing is known of the correlation of the U^k_{37} index from extinct species with temperature. At the moment, only from the information derived from their present day successors can we infer their past capabilities, following a molecular palaeoentological approach (Marlowe *et al.*, 1990); the joint analysis of nannofossil assemblages and U^k_{37} in sediment cores may help to shed some light. Finally, another approach is to compare U^k_{37} stratigraphic records with those derived from other palaeotemperature proxy-indicators (foraminifera and diatom transfer functions), and thereby assess the applicability of U^k_{37} in the geological past.

Finally, evidence of a poor correlation of the U^k_{37} index with temperature has been found during studies of particulate samples of the North Atlantic water column (Conte *et al.*, 1992), over a restricted range of water temperatures (10-16°C) during algal bloom episodes. A better correlation was found when other long-chain compounds (related to the alkenones; see table) were incorporated into a new calibration, IPT (Conte *et al.*, 1992). The discrepancies between these results and those from other studies (Prahl & Wakeham, 1987) may relate to differences in the biogeography of the source organisms of the alkenones (Conte *et al.*, 1992). Thus, temperature calibrations developed in one area may not be applicable to other regions (Conte *et al.*, 1992).

STRATIGRAPHIC INFORMATION

The stratigraphic distribution of the long-chain alkenones and the use of the U^k_{37} index was first assayed in a core from the Kane Gap, off northwest Africa (Brassell *et al.*, 1986a,b). The results revealed a parallelism between the U^k_{37} and the oxygen isotope record from the same core, over several glacial/interglacial transitions, showing that U^k_{37} was effective in registering climatic information on the same time scale as that of the Milankovitch forcing. The utility of this indicator has been further proven by showing, for instance, the correspondence of five U^k_{37} stacked records off the eastern equatorial Atlantic with the stacked oxygen isotope 'SPECMAP' curve (Poynter & Eglinton, 1991). Another example is the coincidence found between a yearly-laminated 20th century U^k_{37} record (Santa Barbara basin, California) and historic measurements of El Niño events (Kennedy & Brassell, 1992).

Whereas the widespread occurrence of the long-chain alkenones in marine surficial sediments has been documented for most areas of the oceans, their stratigraphic occurrence has been less comprehensively evaluated. Those sediment cores analyzed are mainly from low latitudes and upwelling areas. These locations are usually highly productive with high sedimentation rates and with sediments with a high organic content. Hence, they are easy to study by conventional organic geochemical methods. It is only recently that, some cores from mid and high latitudes in the North Atlantic have been examined (Rosell, 1994).

Stratigraphic studies of U^k_{37} in marine cores have been carried out in the upwelling areas off Mauritania (Ten Haven *et al.*, 1989; Poynter *et al.*, 1989; Eglinton *et al.*, 1992), Namibia (Rosell, 1994; Schneider *et al.*, personal comunication, 1992), Peru (McCaffrey *et al.*, 1990; Farrimond *et al.*, 1990), Oman (ten Haven & Kroon, 1991) and the central equatorial Pacific (Lyle *et al.*, 1992). High-sedimentation rate areas have been studied in the Santa Barbara basin (Kennedy & Brassell, 1992) (off southern California) and the Pigmy basin (Jasper & Gagosian, 1989) (northern Gulf of Mexico). Cores in the eastern equatorial Atlantic (Brassell *et al.*, 1986a,b; Poynter & Eglinton, 1991) have also been extensively surveyed, as well as some in the Indian Ocean (Poynter & Eglinton, 1991; Rostek *et al.*, 1992). Oligotrophic areas from the north Atlantic have been attempted (Rosell, 1994), as well as cores from the Norwegian Sea (Rosell, 1994). In summary, different climatic events have been explored, such as, for instance, glacial/interglacial variability (e.g. Brassell *et al.*, 1986a,b; Jasper & Gagosian, 1989; Poynter & Eglinton, 1991) abrupt climatic changes (Eglinton *et al.*, 1992), upwelling evolution (e.g. ten Haven & Kroon, 1991; Lyle *et al.*, 1992), 'El Niño' events (McCaffrey *et al.*, 1990; Kennedy & Brassell, 1992), and rates of climatic change on scales of hundreds, thousands or tens of thousands of years (e.g. Eglinton *et al.*, 1992; Lyle *et al.*, 1992). SST-U^k_{37} data have also been used to reconstruct variations in pCO_2 values (Jasper & Hayes, 1990).

To date, no exhaustive studies have been published comparing different palaeotemperature proxy-indicators: U^k_{37} and transfer functions derived from foraminiferal and diatomaceous assemblages. These indicators are bound to provide different SST information as they are derived from organisms with different life styles and cycles. U^k_{37} can be applied in areas where sediments are deposited below the carbonate lysocline, where transfer functions are unattainable. However, several questions remain yet unanswered about its reliability and applicability. In order to achieve further progress, a fuller understanding of the ecological behaviour and distribution of the Prymnesiophyceae algae is needed.

OTHER BIOMARKERS

Biomarkers may provide a wealth of information as yet hardly explored for palaeoclimatic assessment. Information on primary palaeoproductivity and palaeodisoxia can be obtained from the study of the abundances in sediments of phytoplankton biomarkers (e.g. photosynthetic pigments, long-chain alkenones and alkenoates, sterols) (Volkman *et al.*, 1983; Brassell *et al.*, 1986b; McCaffrey *et al.*, 1990; Poynter & Eglinton, 1991; Kennedy & Brassell, 1992; Repeta *et al.*, 1992). Knowledge of palaeowind, stress and direction can be gained from measurement of the average chain length, isotopic composition and abundances of n-alkanes and n-alkanols (Poynter *et al.*, 1989b; Poynter & Eglinton, 1991). pCO2 data can be derived from the measurement of the $\delta^{13}C$ values of individual long-chain alkenones and of the carbonate tests from co-occurring foraminifera (Jasper & Hayes, 1990). Finally, the ratio of two sterols, defined as STI (ten Haven *et al.*, 1987), showed a correlation with the U^k_{37} index values in sapropel samples from the eastern Mediterranean. However, no further studies, such as culture experiments or surveys of particulate matter and sediment samples from different temperature regimes, have been carried out. Thus, the relationship between STI and temperature remains to be verified.

REFERENCES

Boon J.J., van der Meer,F.W., Schuyl, P.J.W., de Leeuw, J.W., Schenck, P.A. and Burlingame, A.L. (1978) 1.Organic geochemical analyses of core samples from site 362, Walvis Ridge, DSDP Leg 40 Init.Reps.DSDP,38,39,40 and 41 Supplement pp. 627-637.

Brassell S (1993) Applications of biomarkers for delineating marine palaeoclimatic fluctuations during the Pleistocene. *In: Organic Geochemistry: Principles and Applications.* M.H.Engel and S.A.Macko (eds), Plenum Press New York pp. 699-738.

Brassell S.C., Eglinton G., Marlowe I.T., Pflaumann U. and Sarnthein M. (1986a) Molecular stratigraphy: a new tool for climatic assessment *Nature*, **320**, 129-133.

Brassell S.C., Brereton R.G., Eglinton G., Grimalt J., Liebezeit G., Marlowe I.T., Pflaumann U. and Sarnthein M. (1986b) Palaeoclimatic signals recognized by chemometric treatment of molecular stratigraphic data *Org.Geochem.*, **10**, 649-660.

Bukry,D. (1974) In: The Black Sea-Geology, Chemistry and Biology. Degens, E.T. and Ross, D.A., (eds.) Memoir 20, pp.353-363. American Association of Petroleum Geologists.

Conte M.H., Eglinton G. and Madureira L.A.S. (1992) Long-chain alkenones and alkyl alkenoates as palaeotemperature indicators: their production, flux and early sedimentary diagenesis in the Eastern North Atlantic *Org.Geochem.*, **19**, 287-298.

Cranwell P.A. (1985) Long-chain unsaturated ketones in recent lacustrine sediments *Geochim.Cosmochim.Acta*, **49**, 1545-1551.
Cranwell, P.A. (1988) Lipid geochemistry of late Pleistocene lacustrine sediments from Burland, Cheshire, U.K. *Chem.Geol.* **68** 181-197.
de Leeuw, J.W., van der Meer, F.W., Rijpstra W.I.C. and Schenck P.A. (1980) On the occurrence and sctrutural identification of long chain unsaturated ketones and hydrocarbons in sediments In: A.G.Douglas & J.R.Maxwell (eds.), Advances in Organic Geochemistry 1979, Pergamon, Oxford, pp 211-217.
Eglinton G., Bradshaw S.A., Rosell A., Sarnthein M., Pflaumann U. and Tiedemann R. (1992) Molecular record of secular sea surface temperature changes on 100-year timescales for glacial terminations I, II and IV *Nature*, **356**, 423-426.
Farrimond P., Eglinton, G. and Brassell, S.C. (1986) Alkenones in Cretaceous black shales, Blake-Bahama Basin, western North Atlantic. In: D.Leythaeuser and J.Rullköter (eds.), Advances in Organic Geochemistry 1985. Pergamon, Oxford, Org.Geochem., **10**, 897-903.
Farrimond P., Poynter J.G. and Eglinton G. (1990) A molecular stratigraphic study of Peru margin sediments, Hole 686B, Leg 112 Proc.Ocean Drill.Prgr., 112, 547-553.
Freeman K.H. and Wakeham S.G. (1992) Variations in the distributions and isotopic compositions of alkenones in Black Sea particles and sediments *Org.Geochem.*, **19**, 277-285.
Holligan, M.Vollier, D.S.Harbour, P.Camus and M.Champagne-Phillipe (1983) Satellite and ship studies of coccolithophore production along a cntinental shelf edge. *Nature* **304**, 339-342.
Jasper J.P. and Gagosian R.B. (1989) Alkenone molecular stratigraphy in an oceanic environment affected by glacial freshwater events *Paleoceanography*, **4**, 603-614.
Jasper J.P. and Hayes J.M. (1990) A carbon isotope record of CO2 levels during the late Quaternary *Nature*, **347**, 462-464.
Kennedy J.A. and Brassell S.C. (1992) Molecular stratigraphy of the Santa Barbara basin: comparison with historical records of annual climate change *Org.Geochem.*, **19**, 235-244.
Lyle M.W., Prahl F.G. and Sparrow M.A. (1992) Upwelling and productivity changes inferred from a temperature record in the central equatorial Pacific *Nature*, **355**, 812-815.
Marlowe I.T., Brassell S.C., Eglinton G. and Green J.C. (1984a) Long chain unsaturated ketones and esters in living algae and marine sediments *Org.Geochem.*, **6**, 135-141.
Marlowe I.T., Green J.C., Neal A.C., Brassell S.C., Eglinton G. and Course P.A. (1984b) Long chain (n-C37-C39) alkenones in the Prymnesiophyceae. Distribution of alkenones and other lipids and their taxonomic significance *Br.Phycol.*, **19**, 203-216.
Marlowe I.T., Brassell S.C., Eglinton G. and Green J.C. (1990) Long-chain alkenones and alkyl alkenoates and the fossil coccolith record of marine sediments *Chemical Geology*, **88**, 349-375.
Marlowe I.T. (1984) Lipids as palaeoclimatic indicators. Ph.D. thesis, University of Bristol, 273pp.
McCaffrey M.A., Farrington J.W. and Repeta D.J. (1990) The organic geochemistry of Peru margin surface sediments: I. A comparison of the C37 alkenone and historical El Niño records *Geochim.Cosmochim.Acta*, **54**, 1671-1682.
McIntyre A. and Bé A.W.H (1967) Modern coccoliths of the Atlantic Ocean, I. Placoliths and cyrtoliths. *Deep Sea Res.*, **14** 561-597.
McIntyre, A., Bé, A.W.H. and Roche, M.B.(1970) *Trans. N.Y. Acad. Sci. Ser. II* **32**, 720-731.
McIntyre,A. (1970) *Gephyrocapsa protohuxleyi* sp.m., a possible phyletic link and index fossil for the Pleistocene. Deep-Sea Res., **17** 187-190.
Okada H. and Honjo S.(1973) The distribution of oceanic coccolithophorids in the Pacific. *Deep Sea Res.*, **20** 355-374.
Okada H. and McIntyre A. (1977) Modern coccolithophores of the Pacific and North Atlantic Oceans. *Micropaleontol.*, **23** 1-55.
Okada, H., A.McIntyre (1979) Seasonal distribution of modern coccolithophores in the Western North Atlantic. *Marine Bio.* **54**, 319-328.

Poynter J.G., Farrimond P., Brassell S.C. and Eglinton G. (1989a) Molecular stratigraphic study of sediments from holes 658A and 660A, Leg 108 In: Proc.Ocean Drill.Prg., **108**, 387-394.

Poynter, J.G., Farrimond, P., Robinson, N. and Eglinton, G. (1989b) Aeolian-derived higher plants lipids in the marine sedimentary record: links with palaeoclimate. In: M.Leinen and M.Sarnthein (eds.), *Palaeoclimatology and Palaeometereology: Modern and Past Patterns of Global Atmospheric Transport*, Kluwer, Dordrecht, pp. 435-462.

Poynter J. and Eglinton G. (1990) Molecular composition of three sediments from Hole 717C: The Bengal fan Proc.Ocean Drill.Prgr., 116, 155-161.

Poynter J. and Eglinton G. (1991) The biomarker concept-strength and weaknesses Fresenius *J.Anal.Chem.*, **339**, 725-731.

Prahl F.G. and Wakeham S.G. (1987) Calibration of unsaturation patterns in long-chain ketone compositions for palaeotemperature assessment *Nature*, **320**, 367-369.

Prahl F.G., Muehlhausen L.A. and Zahnle D.L. (1988) Further evaluation of long-chain alkenones as indicators of paleoceanographic conditions *Geochim. Cosmochim. Acta*, **52**, 2303-2310.

Prahl F.G., de G.J., Lange , Lyle M. and Sparrow M.A. (1989) Post-depositional stability of long-chain alkenones under contrasting redox conditions *Nature*, **341**, 434-437.

Rechka J.A. and Maxwell J.R. (1988) Characterisation of alkenone temperature indicators in sediments and organisms *Org.Geochem.*, **13**, 727-734.

Repeta D.J., McCaffrey M.A. and Farrington J.W. (1992) Organic geochemistry as a tool to study upwelling systems: recent results from the Peru and Namibian shelves In: Upwelling systems: evolution since the early Miocene., Summerhayes C.P., Prell W.L. and Emeis K.C. (eds) : Geological Society Special Publication No 64, pp 257-272 abstract.

Rosell A. (1994) Ph.D. thesis, University of Bristol.

Rostek F., Ruhland G., Bassinot F.C., Muller P.J., Labeyrie L.F., Lancelot Y. and Bard E. (1993) Reconstructing sea-surface temperature and salinity using $\delta^{18}O$ and alkenone records. *Nature* **364** 319-321

Sikes E.L., Farrington J.W. and Keigwin L.D. (1991) Use of the alkenone unsaturation ratio Uk37 to determine past sea surface temperatures: core-top SST calibrations and methodology considerations *Earth and Planetary Science Letters*, **104**, 36-47.

Sinninghe Damsté, J.S., Rijpstra, W.I.C., Kock-van Dalen, A.C., de Leeuw, J.W. and Schenck, P.A. (1989) Quenching of labile functionalised lipids by inorganic sulphur species: Evidence for the formation of sedimentary organic sulphur compounds at the early stages of diagenesis. *Geochim.Cosmochim.Acta*, **53**, 1343-1355

Smayda T.(1980) Phytoplankton species succession, In: The Physiological Ecology of Phytoplankton, I. Morris, (ed)., University of California Press, Los Angeles, California.pp 493-570

ten Haven H.L., Baas M., Kroot M., de Leeuw J.W., Schenck P.A. and Ebbing J. (1987) Late Quaternary Mediterranean sapropels. III: Assessment of source of input and palaeotemperature as derived from biological markers *Geochim.Cosmochim.Acta*, **51**, 803-810.

ten Haven H.L., Rullköter J. and Stein R. (1989) Preliminary analysis of extractable lipids in sediments from the eastern North Atlantic (Leg 108): Comparison of a coastal upwelling area (Site 658) with a nonupwelling area (Site 659) In: Proc.Ocean Drilling Program, 108B, 351-360.

ten Haven H.L., and Kroon D. (1991) Late Pleistcene sea surface water temperature variations off Oman as revealed by the distribution of long-chain alkenones Proc.Ocean Drill.Prgr., Scientific results, 117, 445-452.

Volkman J.K., Eglinton G., Corner E.D.S. and Sargent J.R. (1980a) Novel unsaturated straight-chain C37-C39 methyl and ethyl ketones in marine sediments and coccolithophore Emiliana huxleyi In: A.G.Douglas and J.R.Maxwell (Editors), *Advances in Organic Geochemistry* 1979, Pergamon, Oxford, pp 219-227.

Volkman J.K., Eglinton G., Corner E.D.S. and Forsberg T.E.V. (1980b) Long-chain alkenes and alkenones in the marine coccolithophorid *Emiliania huxleyi. Phytochemistry*, **19**, 2619-2622.

Volkman J.K., Farrington J.W., Gagosian R.B. and Wakeham S.G. (1983) Lipid composition of coastal marine sediments from the Peru upwelling region, In: M.Bjoroy et al. (eds.) Advances in Organic Geochemistry 1981, pp. 228-240.

Volkman, J.K., Burton, H.R., Everitt, D.A. and Allen, D.I. (1988) Pigment and lipid compositions of algal and bacterial communities in Ace Lake, Vestfold Hills, Antartica. *Hydrobiologia* **165** 41-57.

EVIDENCE FOR A CHANGE IN ATMOSPHERIC CIRCULATION DURING THE YOUNGER DRYAS

N. A. S. Beveridge
Department of Earth Sciences,
University of Cambridge, U.K.

Abstract

Bulk major and trace element ratios from deep sea cores have been used to infer the intensity of aeolian transport over the past 30ka. Evidence from a high sedimentation rate core (ODP 658) suggests that wind speeds were more vigorous during the glacial. They also intensified during the brief Younger Dryas event, implying a short lived global reorganization in atmospheric circulation at this time. Cool sea surface temperatures resulting from increased trade wind driven upwelling along the N.W. African margin provide further evidence for increased wind vigour during the Younger Dryas.

Introduction

Stronger wind speeds during the glacial have been demonstrated using grain size analyses of the terrigenous fraction from sub-tropical deep-sea cores (Parkin, 1973; Sarnthein *et al.*, 1981). The most obvious explanation for a glacial increase in wind speed is the increased meridional temperature gradient and the contraction of temperate and sub-tropical climate belts associated with the southern advance of the polar front (Ruddiman and McIntyre, 1976; Sarnthein *et al.* 1981). However, detailed particle size analysis through short lived climatic events such as the Younger Dryas (Y.D.) has been difficult in most low latitude ocean sediment cores due to slow accumulation rates (<2cm/ka) coupled with the 'smoothing' effect of bioturbation. Ice cores exhibit much greater accumulation rates (50m/ka), and show substantial increases in dust content during the last glacial and Y.D. (Hammer 1978, figure 1). However dust content in ice cores is not exclusively related to wind strength, but is also a function of continental aridity.

One sub-tropical ocean core which exhibits sedimentation rates that are rapid enough to reflect short-lived climatic events is ODP 658A (Sarnthein and Tiedemann, 1990). High accumulation rates in this core (20cm/ka) are a result of its central position in the N.W. African

NATO ASI Series, Vol. I 22
Long-Term Climatic Variations
Edited by J.-C. Duplessy and M.-T. Spyridakis

upwelling cell and proximity to the Saharan coast where major dust outbreaks occur (Jaenicke and Schutz, 1978).

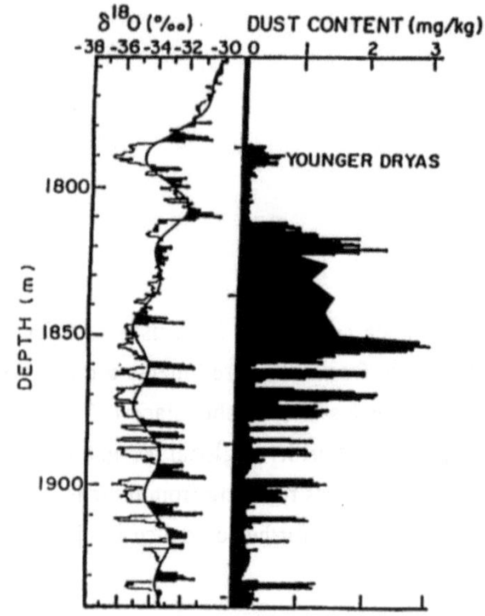

Figure 1: Oxygen isotopes and dust records for Greenland ice core Dye 3, for 45ka to 8ka (Hammer *et al.*, 1985). Note the large increase in dust content at the last glacial maximum and during the Younger Dryas.

The main dust supply to this region comes from high altitude, easterly zonal winds which are related to the intertropical convergence zone (ITCZ). Dust clouds associated with these zonal winds migrate in a westerly direction before circulating anticlockwise around a high altitude, high pressure system ('H', figure 2) once they reach the coast . Additional dust is supplied from the N.E. Trade winds which flow beneath the zonal wind layer, contributing a heavy dust load to a limited region, up to 200km offshore (Parkin, 1973;Sarnthein *et al.*, 1981). In this high resolution study, bulk major and trace element ratios from ODP core 658A off N.W. Africa are used to infer the relative intensity of wind speed over the past 30ka.

Approach

The mean grain size of terrigenous dust is a useful indicator of wind vigour. If we assume that fluvial output and current winnowing off the N.W. African margin plays a minimal role in shaping the grain size distribution as supported by Sarnthein *et al* (1981), then as wind speed increases, the mean grain size of dust deposited at any given distance from the dust source must also increase.

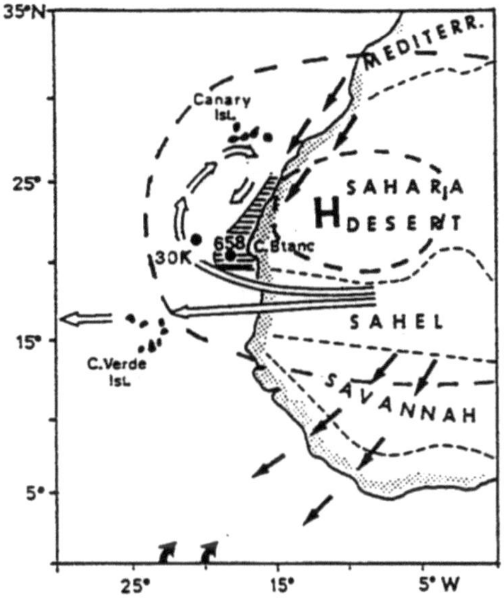

Figure 2:
Location map showing cores ODP 658 and 30K with major climate zones across Africa outlined. Solid arrows represent meridional trade winds; open arrows represent mid-tropospheric zonal winds (Easterly Jet). Dashed line indicates position of upper air high pressure system, centred on H. Hachured area indicates position of perennial coastal upwelling. (Diagram modified from Tiedemann *et al.*, 1978).

In this study, the mean size of the 63-10μm terrigenous dust fraction is used as an indicator of wind stress. Grain size analysis with a Sedigraph (5000ET) particle size analyser revealed that Saharan dust has a distribution with a modal peak around 24μm. 10μm was chosen as a lower cut off because grains finer than this tend to be scavenged out of the atmosphere by an aggregation process during rainstorms, rather than being sorted by aeolian processes (Parkin, 1973;McCave *et al.*, submitted).

One problem in grain size analysis with a SediGraph particle size analyser is that a large sample is required (4g dry weight). This is particularly unfortunate when a high chronological resolution is required, demanding sample intervals of the order of 2cm intervals from ODP cores. Average grain size may be determined indirectly by measuring the ratio of metals restricted mainly to the coarse and/or heavy mineral fraction to those which occur in the very fine clay fraction. This provides an estimate of the coarseness of the sediment, which can be used to infer the vigour of atmospheric circulation through time. Major and trace element analyses were performed on a Phillips PW1250 X-ray spectrometer and an optical inductively-coupled plasma spectrometer. All ages are given in 'calendar years' after correction of ^{14}C ages to a U/Th timescale according to Bard *et al.* (1990).

Aluminium (Al) is principally derived from clay minerals and is hence an indicator of fine clay detritus. Titanium (Ti) and Zirconium (Zr) on the other hand, are strongly partitioned into 'heavy' minerals such as ilmenite, rutile, titanomagnetite, augite, and zircon (Schmitz, 1987). Potassium (K) may occur within the clay mineral illite or its unweathered (and hence coarser) parent such as alkali feldspar or mica flakes, which are a known constituent of Saharan dust (Glaccum and Prospero, 1980). Since illite has a K/Al ratio of 0.2 whereas alkali feldspar and mica which have K/Al ratios of 1.0 and 0.3 respectively, we can interpret K/Al ratios above 0.2 as due to an increase in the abundance of relatively coarser K-bearing minerals. Hence it seems reasonable that K/Al ratios may also be used as a proxy for sediment coarseness. Ti, K, Al and Zr are also useful as tracers of terrigenous input in that they take no part in the biological cycle (Buat-Menard and Chesselet 1979).

Figures 3a and 3b show the strong correlation between mean grain size and Ti/Al ratios for core 30K. This supports the use of Ti/Al ratios as a measure of wind stress, provided that source area and mineralogy of the source area has remained constant over the past 30ka. Ti/Al ratios were first used by Boyle (1983) as a measure of the intensity of aeolian transport off the Peru margin, and have also been employed by Shimmield and Mowbray (1991) in similar studies on the Oman margin.

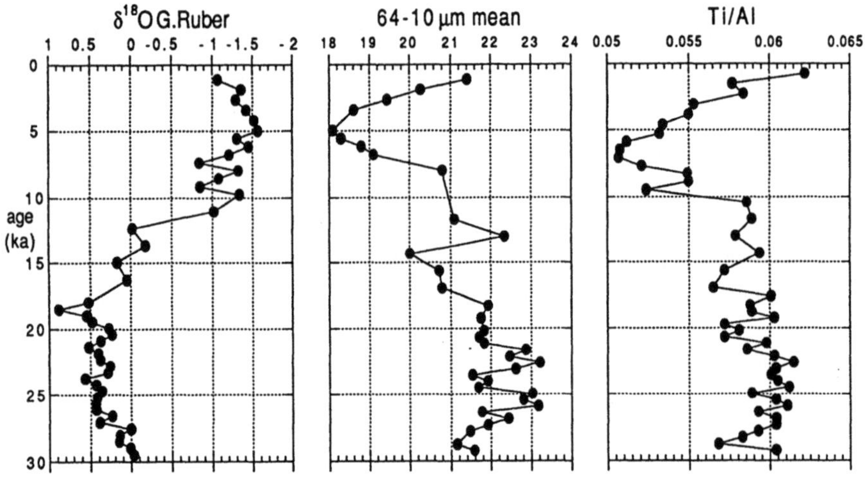

Figure 3a: Time series profiles of $\delta^{18}O$ (*G.ruber*), mean grain size of the 64-10µm non-carbonate fraction and bulk sediment Ti/Al ratios from a typical low sedimentation rate (3cm/ka) low latitude core, 30K (see figure 2 for location). Ages given in calendar years.

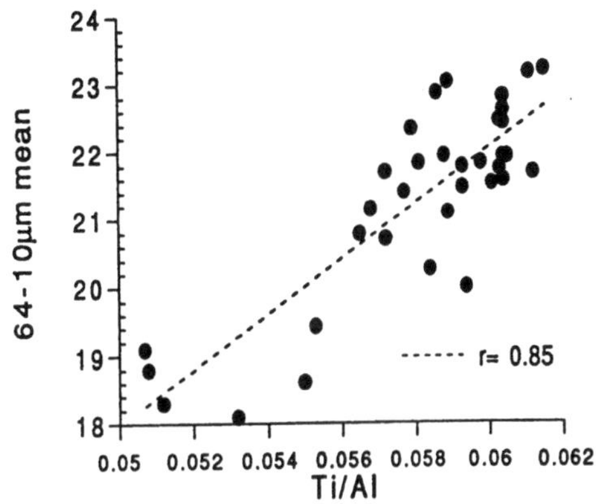

Figure 3b:
The relationship between Ti/Al ratio and the mean grain size of the 64-10mm non-carbonate fraction in core 30K.The strong correlation (r=0.85) supports the use of Ti/Al ratios as a proxy indicator of grain size.

Evidence in favour of this assumption is given in figure 4 which shows the remarkable similarity between a typical Holocene and glacial particle size distribution. Lake level data (Street and Grove, 1979) indicate that aridity and therefore chemical weathering over N.W. Africa during the last glacial was similar to the present day. This coupled with the fact that soil profiles take tens of thousands of years to develop, suggests that the mineralogy of the source area during the glacial would not have been substantially different to the present day.

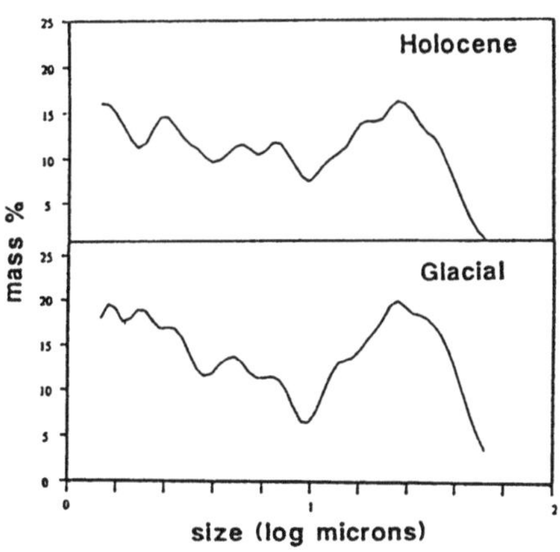

Figure 4:
Typical non-carbonate grain size distributions for the late Holocene (0-5ka) and glacial maximum from core 30K. Note the strong modal dust peak at 25μm. The similarity in grain size distributions indicates that the source must be similar for the present day and the glacial.

Results

Figure 5 shows a benthic $\delta^{18}O$ profile for core ODP 658 and the strong similarity between Ti/Al, Zr/Al and K/Al elemental ratios over the past 30ka. Ti/Al, Zr/Al and K/Al profiles all

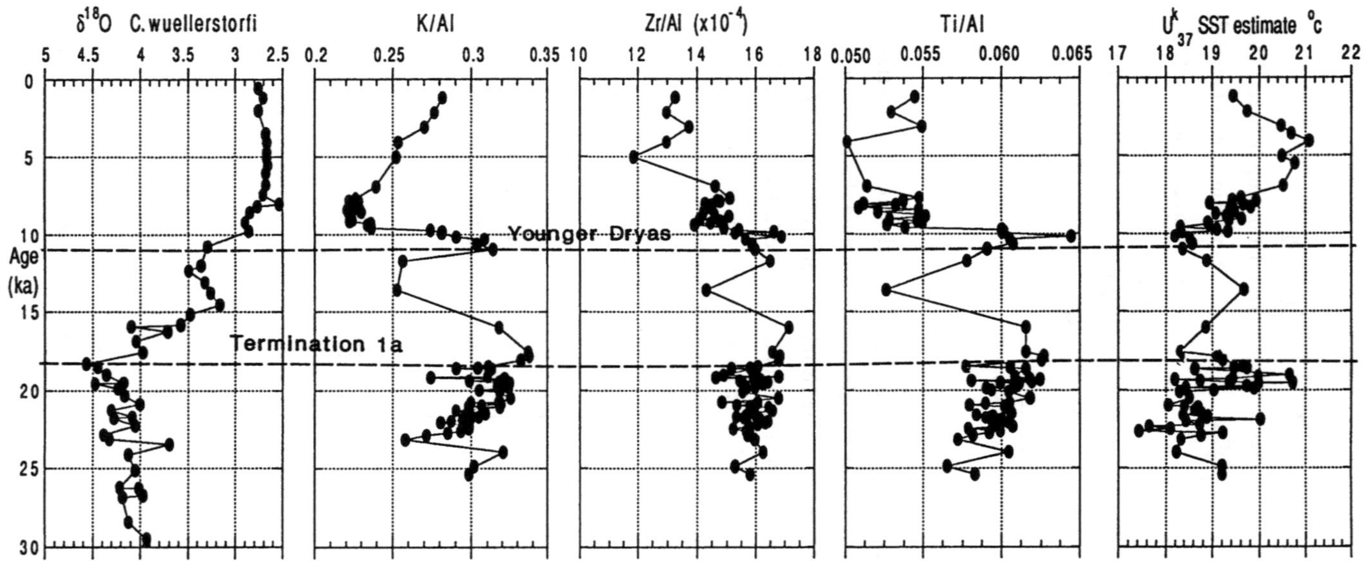

Figure 5: Major and trace element ratios for core OPD-658A compared to $\delta^{18}O$ *C.wuellerstorfi* (Sarnthein and Tiedemann, 1990) and U^{k}_{37} ratio estimates of SST (Eglinton *et al.* 1992). The lower dashed line repesents the start of Termination 1a and the upper dashed line represents the Younger Dryas. Strong wind speeds and strong upwelling are inferred during the last glacial and Younger Dryas from high Ti/Al, K/Al and Zr/Al ratios and cooler SSTs. Also note the significant lag of 2-3ka in trace metal ratios and SSTs after the start of Termination 1a.

show similar features, with high ratios indicating strong wind speeds during the last glacial and Younger Dryas, and low ratios (weak wind speeds) during the climatic optimum, rising up to higher values towards the present day. The pronounced Y.D. spike (10-12ka) indicates a rapid 'flip' in atmospheric circulation towards glacial conditions (stronger winds). Figure 5 also shows the strong inverse relationship between Ti/Al ratios and SSTs derived from U^{k}_{37} ratios (Eglinton *et al.*, 1992) for core 658A. Upwelling along the N.W. African margin is driven by the N.E. Trade winds which blow obliquely to the coast and so periods of strong upwelling, driven by strong wind speeds should be accompanied by cooler SSTs. During the Younger Dryas, more vigorous atmospheric circulation must have increased coastal upwelling and resulted in the lower sea-surface temperatures indicated by U^{k}_{37} ratios.

A further interesting point is the significant 2-3ka lag between the decline in Ti/Al, Zr/Al and K/Al ratios (hence wind speed) and the start of Termination 1a. This is also observed in U^{k}_{37} ratios which show that SSTs remained low for 2-3ka after the termination. Taken together with the trace element data, this provides convincing evidence that tradewind driven upwelling remained strong after Termination 1a. A pulse of meltwater related to the influx of icebergs into the N. Atlantic during Heinrich event 1 (Broecker *et al.*, 1992, Bond *et al.*, 1992) may provide a possible explanation for the persistence of the sharp latitudinal temperature gradients after the termination, which helped sustain vigorous atmospheric circulation.

Conclusions

1. The major and trace element ratios Ti/Al, Zr/Al and K/Al provide a useful indicator of sediment coarseness which may be related to wind speed.

2. Data from a high sedimentation rate core verify stronger glacial wind speeds but also indicate a rapid change in atmospheric circulation during the Younger Dryas. Wind speeds during the climatic optimum were much slower, but have risen sharply to the present day.

3. SSTs and wind speed are strongly linked through trade wind driven upwelling, and appear to indicate that upwelling remained strong for 2-3ka after the termination.

Acknowledgements

I thank Nigel Higgs and John Thomson for running samples at IOS Deacon Laboratory. I am also grateful to G.Eglinton, M.Zhou and A. Rosell for getting me involved with the ODP658 project, providing samples and for useful discussions. This paper benefitted from comments by T.J. van Andel, B.Manighetti and I.N. McCave. Support for this study was provided by a NERC studentship GT4/91/GS/14.

References

Bard, E., B. Hamelin, R. G. Fairbanks and A. Zindler (1990) Calibration of the ^{14}C timescale over the past 30,000 years using mass spectrometric U-Th ages from Barbados corals. Nature 345: 405-410

Buat-Ménard P. and R. Chesselet (1979) Variable influence of the Atmospheric flux on the trace metal chemistry of oceanic suspended matter. Earth Planet. Sci. Lett. 42: 399-411

Bond, G., H. Heinrich, W. Broecker, L. Labeyrie, J. McManus, J. Andrews, S. Huon, R. Jantschik, S. Clasen, C. Simet, K. Tedesco, M. Klas, G. Bonani and S. Ivy (1992) Evidence for massive discharges of icebergs into the North Atlantic ocean during the last glacial period. Nature 360: 245-249

Boyle, E. A. (1983) Chemical accumulation variations under the Peru current during the past 130,000 years. J. Geophys. Res. 88: 7667-7680

Broecker, W., G. Bond, M. Klas, E. Clark and J. McManus (1992) Origin of the northern Atlantic's Heinrich events. Clim. Dyn. 6: 265-273

Eglington, G., S. Bradshaw, A. Rosell, M. Sarnthein, U. Pflaumann and R. Tiedemann (1992) Molecular record of secular sea surface temperature changes on 100-year timescales for glacial terminations I,II, and IV. Nature 356: 213-216

Glaccum, R. A. and J. M. Prospero (1980) Saharan aerosols over the tropical North Atlantic - Minralogy. Mar. Geol. 37: 295-321

Hammer, C. U., H. B. Claussen, W. Dansgaard, N. Gunderstrup, S. J. Johnsen and S. J. Reeh (1978) Dating of Greenland ice cores by flow models, isotopes, volcanic debris, and continental dust. J. Glaciol. 20: 3-26

Jaenicke, R. and L. Schutz (1978) Comprehensive study of physical and chemical properties of the surface aerosols in the Cape Verde island region. J. Geophys. Res. 83: 3585-3599

McCave, I. N., B. Manighetti and S. C. Robinson Sortable silt and fine sediment size/composition slicing: Parameters for palaeoceanography. submitted to Paleoceanography

Parkin, D. W. (1974) Trade winds during the glacial cycles. Proc. R. Soc. Lond. 337: 73-100

Ruddiman, W. F. and A. McIntyre (1976) Northeast Atlantic paleoclimatic changes over the past 600,000 years. Geol. Soc. Am. Mem. 145: 111-146

TEMPERATURE OF THE LAST INTERGLACIAL BASED ON $\delta^{18}O$ IN MOLLUSKS

E. Bard, S. Cornu, J. Pätzold*
Géosciences de l'Environnement
JE DRED 192, Faculté de St Jérôme
Université d'Aix-Marseille III
13397 Marseille cedex 20
France

INTRODUCTION

Strombus bubonius is a thermophilous marine gastropod which today lives only in the shallow waters of the Gulf of Guinea. This marine snail requires an annual average temperature of 25°C to grow normally, but it also tolerates temperatures as low as 15 to 20°C in Senegal and as high as 31°C on the island of Bioko (Meco 1967, 1977). However, during the last interglacial period (LIP; between 115,000 and 135,000 yr BP, Bard et al. 1992) the habitat of *Strombus bubonius* extended much farther to the north, and well preserved specimens can commonly be found in Pleistocene beaches from the Canary islands and Mediterranean coastline.

High resolution records of $\delta^{18}O$ in mollusks can be used to reconstruct paleotemperatures by means of the calibration equation established by Epstein et al. (1953). In this work we present $\delta^{18}O$ and SST data obtained on *Strombus bubonius* from Mediterranean fossil beaches corresponding to the LIP. The fossil specimens were found in two classical Mediterranean sites: Palma Nova (4 specimens), which is a Pleistocene fossil deposit on the island of Majorca (Pomar Goma & Cuerda Barcelo, 1979; Cuerda Barcelo 1989) and Monastir in Tunisia (3 specimens), where a prominent fossil beach is situated 30 m above present sea level (Paskoff and Sanlaville, 1980; Sorel and Kamoun, 1980; Oueslati et al., 1982). Both terraces have been assigned to the

* Fachbereich Geowissenschaften, Universität Bremen, D 2800 Bremen 33, Germany

NATO ASI Series, Vol. I 22
Long-Term Climatic Variations
Edited by J.-C. Duplessy and M.-T. Spyridakis

LIP by stratigraphic correlation with terraces dated by U-Th (Stearns & Thurber 1965, Bernat et al. 1985). High precision dating by thermal ionization mass spectrometry will be performed on these shells and will be presented elsewhere. For the discussion of the stable isotope signal it is sufficient to attribute these shells to the LIP.

METHODS

In order to obtain time series of stable isotope analyses, we sampled the shells following the spiral in the direction of growth. The different $\delta^{18}O_c$ measurements obtained on one particular shell represent different times of calcification. Since our sampling scheme followed precisely the growth spiral, it can logically be expected that the successive $\delta^{18}O_c$ measurements compose a time series with an unknown and variable time step. As has been shown for other species of *Strombus* (Wefer & Killingley 1980) which live much longer than *Strombus bubonius* (only 3-5 years, Meco 1977), the $\delta^{18}O_c$ time series generally exhibit a cyclic behavior which can be attributed to the seasonal SST cycle. In addition, the shape of the $\delta^{18}O_c$ cycles is often asymmetric since the growth rate of the snail varies during the year, being higher during the warm season than during the cold season. In other words, the time step between measurements varies from the beginning (juvenile phase) to the end of the shell's growth (adult stage), as well as between the seasons, with a growth rate maximum during the caloric summer.

We used the correlation program ANALYSERIES (Paillard & Labeyrie, in prep.) in order to convert the whorl number into time. Extremes of the raw $\delta^{18}O_c$ series were correlated with the extremes of a regular sine wave. This procedure is illustrated in Figure 1 which represents measurements from a fossil shell from Monastir (E20). One byproduct of this mapping procedure is the determination of a mean annual growth rate curve expressed as months versus whorl number. The mean annual growth rate decreases regularly during the life of the *Strombus* as illustrated in Fig. 1c. Our results suggest that the mean annual growth-rate is usually proportional to the log of time (Fig. 1d). However the "instantaneous" growth rate appears to vary widely during a seasonal cycle. This results in asymmetric seasonal $\delta^{18}O_c$ curves (Fig.1b) constructed by our mapping technique, which employs only the correlation of extremes. The results of a simple model suggest that *Strombus bubonius* exhibits a seasonal growth-rate variation between 40 and 80% of the mean annual value (Cornu et al. 1993).

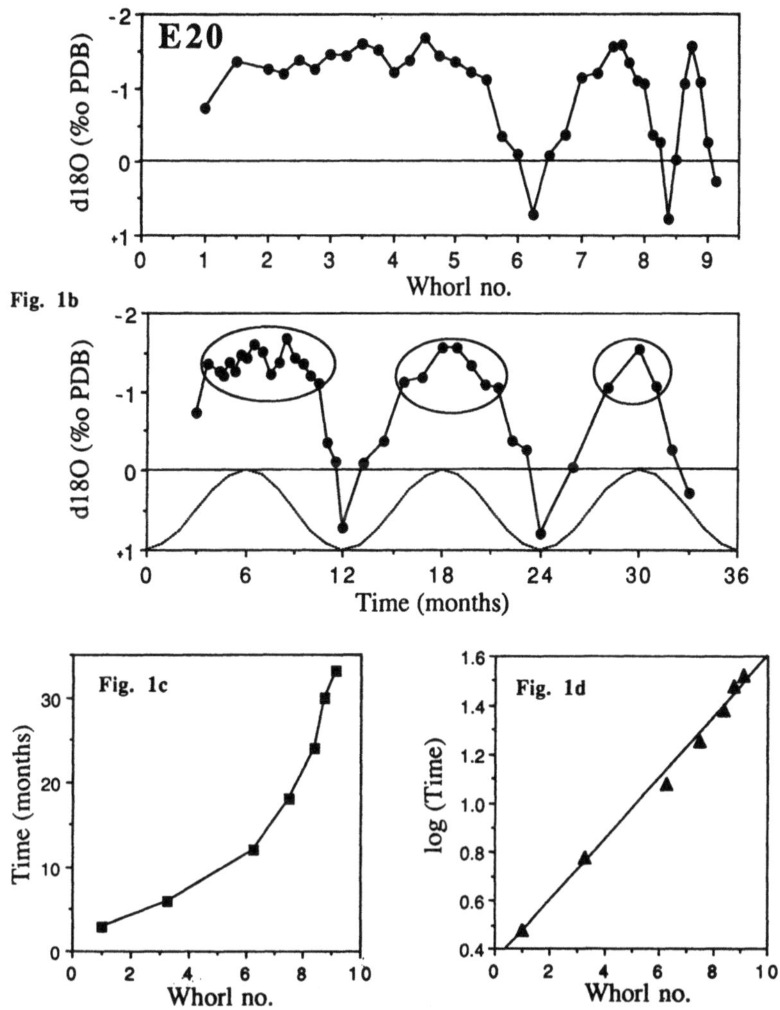

Figure 1: Example of $\delta^{18}O$ record obtained for a *Strombus* shell (E20 from Monastir).
1a: $\delta^{18}O$ versus whorl number starting from the umbilicus.
1b: $\delta^{18}O$ versus time in months. This time series has been obtained by correlation of the raw record shown in Fig. 1a with a regular sine wave. Between the $\delta^{18}O$ extremes, a constant growth rate was assumed. Consequently, this mapping procedure takes into account the slow variation in growth rate during the gastropod life but does not correct for seasonal growth rate changes. The mapping of the raw record with the sine wave was performed with the correlation program ANALYSERIES (Paillard & Labeyrie in prep.).
1c: Mapping function obtained through the correlation procedure. This mapping function describes the variation of the mean annual growth rate versus time expressed in months.
1d: Same as Fig. 1c but the y-axis is logarithmic. This graph shows that the mean annual growth rate decreases logarithmically through the life of the mollusk.

Isotopic SST's were estimated for the extremes of the $\delta^{18}O_C$ record, both warm and cold seasons. Since shell growth is relatively rapid during the warm season, $\delta^{18}O_C$ values were averaged. The warm-season $\delta^{18}O_C$ values are usually based on numerous individual $\delta^{18}O_C$ measurements, whereas the cold-season estimates are usually based on a single $\delta^{18}O_C$ datum.

DISCUSSION

In order to calculate paleotemperatures we assumed (hypothesis 1) that during the LIP the Mediterranean Sea was characterized by about the same $\delta^{18}O_W$ distribution as today (Pierre et al. 1986). In fact, the LIP sea level was about 0 to 10 meters higher than at present, which was responsible for a global lowering of the $\delta^{18}O_W$ on the order of 0.1°/oo (Shackleton, 1987; Bard et al.1992; Lambeck & Nakada, 1992). This value is negligible when compared with other sources of errors in our approach, and has therefore been neglected.

In Fig. 2 the mean annual isotopic temperature obtained for each seven fossil *Strombus* specimen is compared with the modern SST obtained from the Levitus program. The main result is that the paleotemperatures are much higher than the modern ones by about 7°C (closed circles). The mean isotopic temperature is on the order of 26°C which is near the SST observed for the modern habitat in the Gulf of Guinea (open circles).

Nevertheless, the temperature difference reconstructed here for the LIP is greater than the 1-3°C obtained from other proxy data such as pollens (Guiot et al. 1989), foraminifera (Sancetta et al. 1973, Hays et al. 1976, CLIMAP 1984), mollusks (Keen et al. 1981) and isotopes in polar ice (Jouzel et al. 1987, GRIP 1993). Moreover, other data obtained from *Strombus* shells from Fuerteventura (Canary islands) also indicate a small difference, on the order of 2-3°C (Bard, Pätzold, Meco, Petit-Maire in prep.).

A likely explanation for this discrepancy lies in the assumption that the salinity during the LIP was the same as the present-day salinity. Mediterranean hydrological conditions have varied widely during the last million years, with important and abrupt changes of the evaporation-precipitation balance. Thunell & Williams (1989) have even suggested that the antiestuarine circulation, which characterizes the modern Mediterranean, was reversed to an estuarine mode during recurrent periods. These reversals of the thermohaline circulation were associated with the deposition of sapropels in the eastern Mediterranean and reduced hemipelagic sediments in the western Mediterranean. Large variations of salinity and $\delta^{18}O_W$ occurred during these

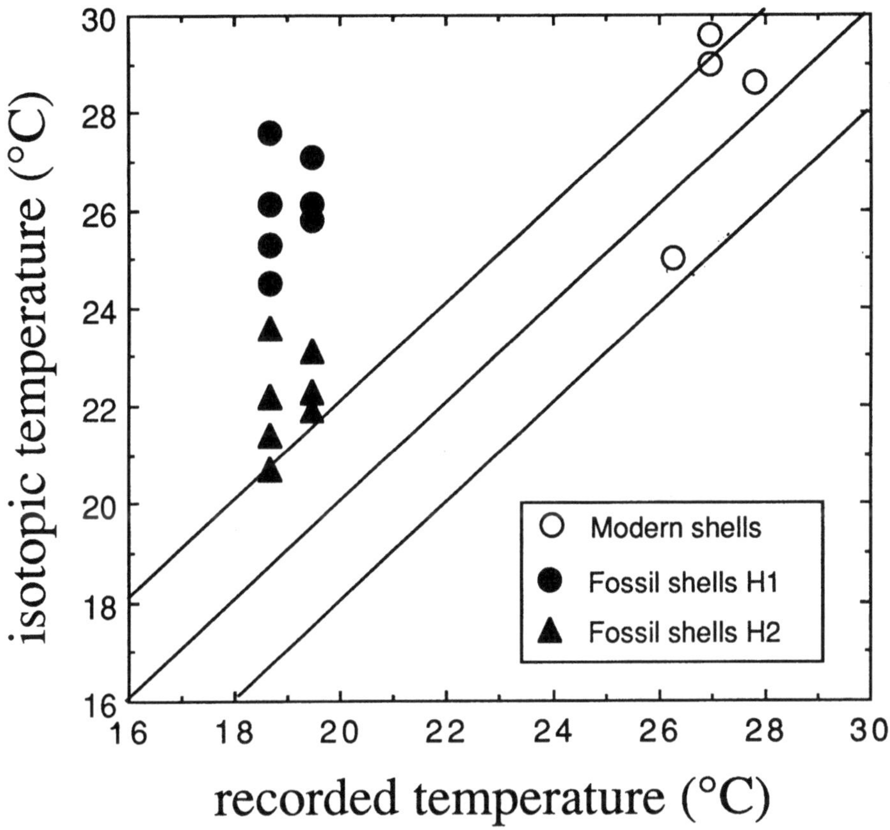

Figure 2: Mean annual isotopic temperature versus modern SST obtained by using the Levitus atlas (1982).
Open circles: data obtained on modern *Strombus* shells from three sites in the Gulf of Guinea (Cornu et al. 1993).
Closed symbols: data obtained on the fossil *Strombus* from Majorca (4 specimens) and Monastir (3 specimens). The isotopic paleotemperatures were obtained by assuming that the Mediterranean $\delta^{18}O_w$ composition during the LIP was the same as today (hypothesis 1).
Closed triangles: data obtained on the fossils but in this case the isotopic paleotemperatures were obtained by assuming that the Mediterranean $\delta^{18}O_w$ distribution during the LIP was different from the modern and characteristic of a period of sapropel deposition (hypothesis 2).

drastic changes in Mediterranean hydrology. In addition, these periods were also characterized by an intensification of African monsoons and increased Nile discharge, which amplified the salinity-$\delta^{18}O_W$ variations due to the reversal of the thermohaline circulation (Rossignol-Strick et al. 1982).

Such oceanographic changes also took place during the early part of the Holocene and during the LIP. These periods are referred as S1 and S5, respectively, following the sapropel numbering scheme introduced by Ryan (1972). More recently, Fontugne & Calvert (1992) published records established from a piston core from the eastern Mediterranean Sea which clearly confirms the contemporaneity of sapropel 5 with the LIP. In addition, their results indicate a large difference of about 2.3 $^{0}/oo$ for $\delta^{18}O_W$ between the LIP and modern period which, of course, cannot be attributed to a temperature difference but rather to a strong freshening of the eastern Mediterranean surface waters.

Our second working hypothesis is that the *Strombus* fossils lived during the period of lowered salinity which corresponds to S5 deposition. For the most recent sapropel (S1), Thunell & Williams (1989) reconstructed $\delta^{18}O_W$ changes on the order of -0.8 and -1.2 $^{0}/oo$ for the western and the eastern Mediterranean basins respectively. We will assume that the value of -0.8 $^{0}/oo$ was also valid during the deposition of S5 which occurred during the LIP (hypothesis 2). This value corresponds approximately to a difference of about 2 $^{0}/oo$ of the SSS in the western Mediterranean. Changing the $\delta^{18}O_W$ in equation 1 by about -0.8 $^{0}/oo$ leads to a significant decrease of the paleotemperature estimates by about 3.5°C (Fig. 2, closed triangles). In this case, we derived a mean temperature change on the order of 3°C in rather good agreement with other paleotemperature estimates obtained for the LIP period (Sancetta et al. 1973, Hays et al. 1976, Keen et al. 1981, CLIMAP 1984, Jouzel et al. 1987, Guiot et al. 1989, GRIP 1993).

Acknowledgements

We thank J. Meco, N. Petit-Maire, J.C. Duplessy and G. Wefer for fruitful discussions. We are grateful to M. Bernat, J. Cuerda-Barcelo for providing *Strombus* shells. This work is coordinated and supported by CLIP, PNEDC and PRCO.

REFERENCES

Bard, E., Fairbanks, R. G., Hamelin, B.(1992) How accurate are the U-Th ages obtained by mass spectrometry on coral terraces. *In* Kukla G & Went E, eds: Start of a Glacial, NATO ASI serie I, vol. 3., Springer-Verlag: 15-21

Bernat, M., Paskoff, R., Sanlaville, P. (1985) Datation de terrasses marines de la côte est de la Tunisie: méthode Io-U appliquée aux mollusques fossiles, un exemple de contamination subactuelle. *Revue de Géologie Dynamique et de Géographie Physique*, 26 (3): 157-161

CLIMAP Project members (1984) The Last Interglacial Ocean. *Quat. Res.* 21: 123-224.

Cornu S., Pätzold J., Bard E., Meco J., Cuerda-Barcelo J.(1993) Paleotemperature of the last interglacial period based on $\delta^{18}O$ of *Strombus bubonius* from the western Mediterranean Sea. *Paleoclim. Paleogeogr. Paleoenv.* 103, 1-20.

Cuerda Barcelo, J. (1989) Los tiempos cuaternarios en Baleares. éd. Conselleria de Cultura, Educacio i Esports Govern Balear, chap. 2.3: 105-155.

Epstein, S., Buchsbaum, R., Lowenstam, H. A. and Urey, H.C. (1953) Revised carbonate-water isotopic temperature scale. *Bull. Geol. Soc. Am.*, 64: 1315-1325.

Fontugne, M. R. and Calvert, S. E. (1992) Late Pleistocene variability of the carbon isotopic composition of organic matter in the eastern Mediterranean: monitor of changes in carbon sources and atmospheric CO_2 concentrations. *Paleoceanography*, 7: 1 -20.

GRIP members (1993) Climate instability during the last interglacial period recorded in the GRIP ice core. *Nature* 364, 203-207.

Guiot, J., Pons, A., de Beaulieu, J. L., and Reille, M. (1989) A 140,000-year continental climate reconstruction from two European pollen records. *Nature* 338: 309-313.

Hays, J. D, Imbrie, J., and Shackleton, N. J. (1976) Variations in the Earth's orbit: Pacemaker of the ice ages. *Science*, 194: 1121-1132.

Jouzel, J., Lorius, C., Petit, J.R., Genthon, C., Barkov, N.I. Kotlyakov, V.M., and Petrov, V.M. (1987) Vostok ice core: a continuous isotope temperature record over the last climatic cycle (160,000 years). *Nature* 329: 403-406.

Keen, D. H., Harmon, R. S., and Andrews, J. T. (1981) U series and amino acid dates from Jersey. *Nature* 289: 162-164.

Lambeck, K. and Nakada, M. (1992) Constraints on the age and duration of the last interglacial period and on sea-level variations. *Nature* 357: 125-128.

Levitus, S. (1982) Climatological atlas of the world ocean. *NOAA Professional Paper*, 13. Rockville. Maryland USA.

Meco, J., (1967) Données actuelles pour l'étude paléontologique du *Strombus bubonius* Lamarck. *in* Actes du VI Congrès Panafricain de Préhistoire et des Etudes du Quaternaire. Dakar 1967. ed. H. J. Hugot. Chambery Imprimeries Reunies.

Meco, J., (1977) Los strombus Neogenos y Cuaternarios del Atlantico Euroafricano (Taxonomia, biostratigrafia y paleoecologia). Doctoral thesis Universidad Complutense de Madrid. published in *Paleontologia de Canarias*, 1. Edicion del Excmo. Cabildo insular de Gran Canaria.

Oueslati, A., Paskoff, R., and Sanlaville, P., (1982) Le Tyrrhénien de Tunisie: essai de synthèse. *Bull. Soc. géol. France*, XXIV, 2: 173-178.

Paskoff, R. and Sanlaville, P. (1980) Le Tyrrhénien de la Tunisie: essai de stratigraphie. *C. R. Acad. Sc. Paris*, 290: 393-396.

Pierre, C., Vergnaud Grazzini, C., Thouron, D. and Saliège, J.-F. (1986) Compositions isotopiques de l'oxygène et du carbone des masses d'eau en Méditerranée. *Mem. Soc. Geol. It.*, 36: 165-174.

Pomar Gomá, L. and Cuerda Barceló, J. (1979) Los depósitos marinos Pleistocénicos en Mallorca. *Acta Geologica Hispanica*, 14: 505-513.

Rossignol-Strick, M., Nesteroff, V., Olive, P., and Vergnaud Grazzini, C. (1982) After the deluge: Mediterranean stagnation and sapropel formation. *Nature* 295: 105-110.

Ryan, W.B.F. (1972) Stratigraphy of late Quaternary sediments in the eastern Mediterranean.*in* The Mediterranean Sea. ed D.J. Stanley, Dowden, Hutchinson & Ross, Inc., Stroudsburg, PA.: 149 -169.

Sancetta, C., Imbrie, J., and Kipp, N. G. (1973) Climatic record of the past 130,000 years in North Atlantic deep sea core V23-82:Correlation with the terrestrial record. *Quat. Res.* 3: 110-116.

Shackleton, N.J. (1987) Oxygen isotopes, ice volume and sea level. *Quat.Sci.Rev.* 6: 183-190.

Sorel, D. and Kamoun, Y. (1980) Sur la stratigraphie des dépôts quaternaires marins de la région de Monastir (Tunisie orientale). *C. R. Acad. Sc. Paris*, 291: 1019-1022.

Stearns, C. E., and Thurber, D. L. (1965) Th^{230}-U^{234} dates of late Pleistocene marine fossils from the Mediterranean and Moroccan Littorals. *Quaternaria*: 29 - 42.

Thunell, R.C. and Williams, D. F. (1989) Glacial-Holocene salinity changes in the Mediterranean Sea: hydrographic and depositional effects. *Nature*, 338: 493 - 496.

Wefer, G. and Killingley, J. S. (1980) Growth histories of strombid snails from Bermuda recorded in their O-18 and C-13 profiles. *Mar. Bio.*, 60: 129-135.

RECONSTRUCTION OF CLIMATIC ZONALITY IN THE INDIAN OCEAN DURING THE NEOGENE

Elena V. Ivanova
P.P. Shirshov Institute of Oceanology
Russian Academy of Sciences
Moscow 1192
Russia

1. INTRODUCTION

Several attempts were made to reconstruct Neogene biogeographic zonality of oceans, based on the distribution of planktonic foraminifera in sediment cores from DSDP sites (Kennett et al., 1985 ; Sancetta, 1978 among others). In 1982, my colleagues and I made our first reconstructions of climatic zonality in three oceans for the Pliocene (Ivanova and Barash, 1982 ; Oskina et al., 1982) and later for the Miocene (Ivanova, 1987 ; Barash, 1989). This paper summarizes my earlier publications on Neogene climatic zonality in the Indian Ocean. The new data (Huber, 1991 ; Krasheninnikov, 1986 ; Vincent and Toumarkine, 1990) essentially refined my previous results.

Six time intervals were selected for this study: the earliest Miocene (22,,5 -21 m.y., Zone N 4), the late early Miocene (17 - 15 m.y., Zone N 8), the late middle Miocene (11,7 - 10,5 m.y., Zone N 14), the latest Miocene (6 -5 m.y., Zone N 18), the earliest Pliocene (5 - 4 m.y., Zone N 19), and the latest Pliocene (3- 1,8 m.y., Zone N 21). Biostratigraphic correlations for different parts of the Indian Ocean and time-scale assignments were made by Barash (1989). Climatic zonality was plotted on palinspastic reconstructions by Zonenshine et al. (1984). Individual sites have been accordingly backtracked.

The main questions raised in this study are : Is it possible to reconstruct paleoclimatic zonality on the basis of semi-quantitative and qualitative data on foraminiferal occurrence pusblished in Deep-Sea Drilling reports (Huber, 1991 ; DSDP reports 22-29, 1974-1975 ; Vincent and Toumarkine, 1990) ? If so, were the climatic zones stable during the Neogene or did they migrate ?

Along with DSDP sites (DSDP reports 22-29, 1974-1975 ; Loutit, 1981 ; Srinivasan and Kennett, 1981), we used ODP sites from legs 115 and 119 (Huber, 1991 ; Vincent and Toumarkine, 1990) as well as several sediment

NATO ASI Series, Vol. I 22
Long-Term Climatic Variations
Edited by J.-C. Duplessy and M.-T. Spyridakis

cores (Kennett and Srinivasan, 1975) and onshore sections from Australia (Lindsay, 1969 ; Quilty, 1978 ; Wright, 1977), the Bay of Bengal (Krasheninnikov and Basov, 1986) and Java (Saint-Marc and Suminta, 1979). I met the same difficulties as other investigators (Kennett et al., 1985 ; Sancetta, 1978) : poor core recovery, small number of cores with suitable age assignments, in some cases poor preservation due to dissolution. Other difficulties arise from differences in taxonomy, various size fractions analyzed, non-exhaustive lists of species. However, in spite of these difficulties, we could recognizegeneral trends of climatic changes.

2. METHOD

It is well established that planktonic foraminifera are sensitive indicators of Sea Surface temperature (SST) and thus of climate (Bé, 1977; Ivanova, 1983 ; Kennett et al., 1985 ; Barash, 1989 among many others). Mapping the distribution of modern species in the Indian Ocean surface sediments, I grouped the various species into five different groups (equatorial, tropical, subtropical, temperate and polar) applying the criteria proposed earlier by Barash (1989) : latitudinal position of each species area, location of maximum abundance and gradient of abundance toward area boundaries. Taking into account the relative proportions of aech group in surface-sediment assemblages (Ivanova, 1983), I recognized five climatic zones in the Indian Ocean (Fig.1). Similar findings from planktonic foraminiferal assemblages in surface sediments were previously obtained by Bé and Tolderlund (1971) and Bé and Hutson (1976).

Using all available data, my colleagues and I compiled distribution maps for each species in the World Ocean for selected Neogene time intervals. Then, we assumed that a given species or its ancestors did not essentially change their preferable environmental conditions (mainly SST). Based on this modern analogue approach, we compared for every time-slice the distribution of extinct species with that of their extant descendant species and we were able to place extinct species within the same five climatic groups as those used for modern species. We went down step by step from the late Pliocene to the earliest Miocene. In order to estimate the relative proportions of each of the five groups in the assemblages at the various sites for every time-interval, I had to interpret published semi-quantitative data on species abundance (Huber, 1991 ; DSDP reports 22-29, 1974-1975 ; Vincent and Toumarkine,

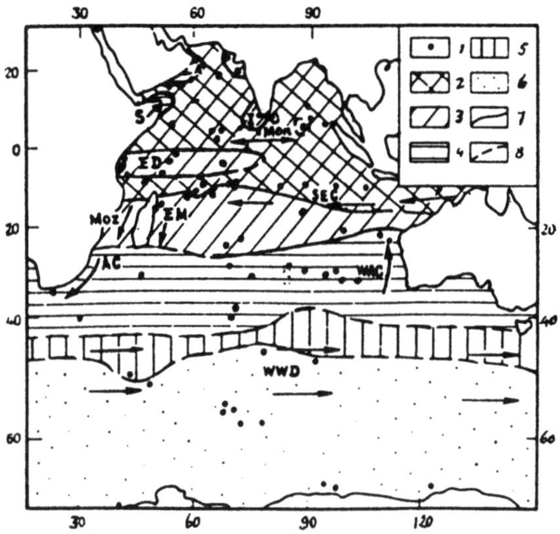

Figure 1. Recent climatic zonality of the Indian Ocean based on planktonic foraminiferal assemblages distributyion and major elements of surface circulation.

Stations :	1
Climatic zones :	2 - Equatorial-Tropical ; 3 - Tropical ; 4 - Substropical ; 5 - Temperate ; 6 - Subpolar-polar.
Boundaries :	7 - Established ; 8 - Assumed ; ED - Equatorial Divergence.
Upwellings :	A - Arabian ; S - Somalian ; I - Indian.
Currents :	Mon - Monsoon ; AC - Agulhas ; WAC - West Australia; WWD - West Wind Drift.

1990) and transform them into usable numbers. Samples in selected stratigraphic intervals for each site were averaged to obtain representative foraminiferal assemblages. In most DSDP/ODP reports, no actual foraminiferal counts are done, because such time-consuming studies are beyond the scope of "Initial Reports". The relative abundances of species, however, are usually visually estimated and reported using four or five categories of abundance which are defined by different authors according to various criteria. Usually, these categories are designated by using a numbering system or a letter system (A: abundant ; C: Common ; F: Frequent ; R : Rare) according to an estimate of the number of specimens present in the sample examined or to percentage classes. I used these so-called semi-quantitative data to assign rank numbers (Table 1) . The data were used to reconstruct the type of assemblage at every site and map the corresponding climatic zones (Ivanova, 1983). The migration of climatic zones during the Neogene can be inferred from the comparison of the maps (Ivanova, 1988). When semi-quantitative

TABLE 1

Berggren et al., leg 22 Kennett, leg 28			Vincent et al., leg. 24 Boltovskoy, leg 26			Fleisher, leg 23 Huber, leg 119		Vincent and Toumarkine leg 115	
Abundance	No. of tests	rank	Abundance	No. of tests	rank	Abundance	rank	Abundance	rank
1	1-10	1	R Very R	< 10	1	R	1	/(R)	1
2	11-25	3				F	3		
3	26-50	7	C	10-100	11	C	11	x(C?)	11
4	51-100	15							
5	> 100	30	A	> 100	30				

Transformation of semiquantitative data on species abundances into ranks

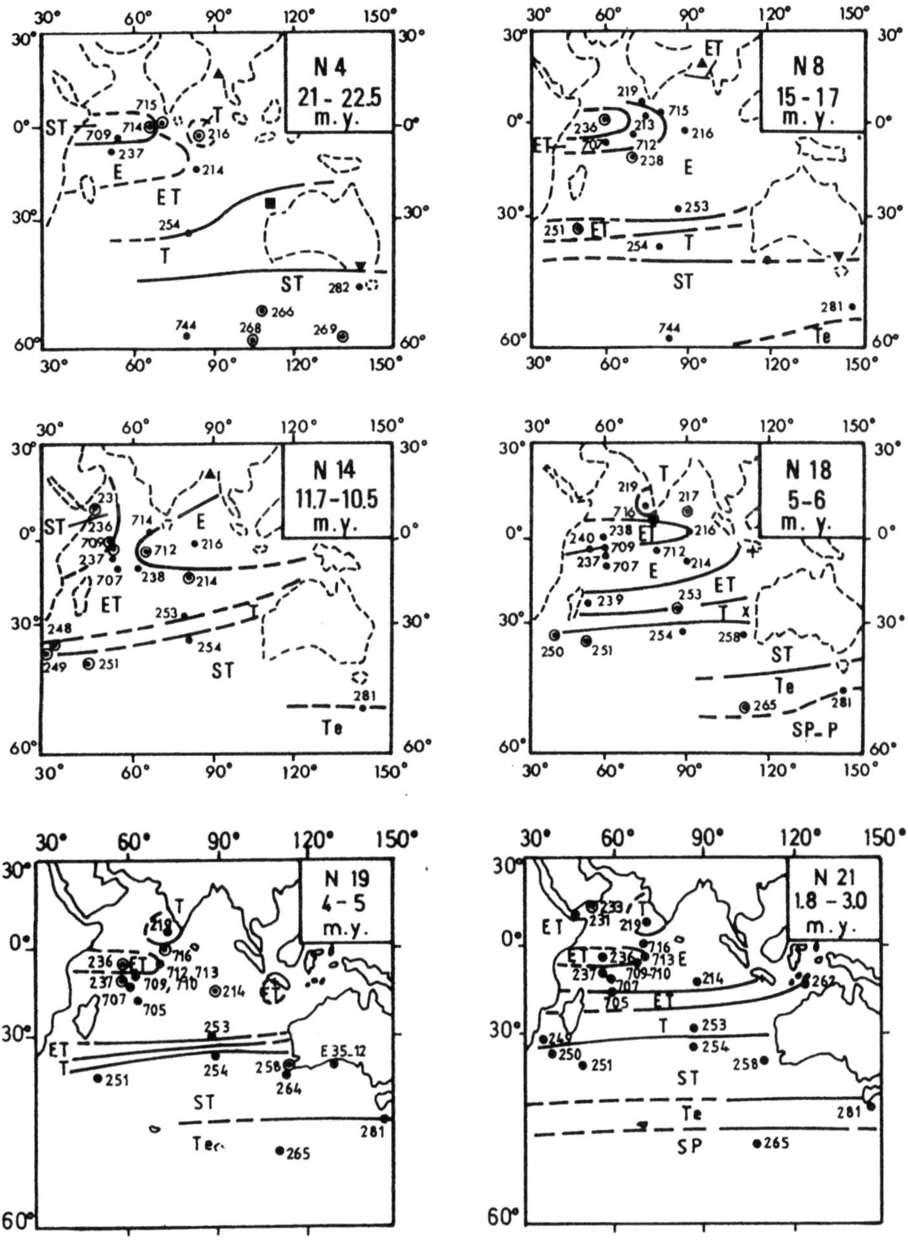

Figure 2. Neogene climatic zonality of the Indian Ocean based on planktonic foraminiferal assemblages distribution and plotted on palinspastic reconstructions by Zonenshine et al. with backtracking of individual sites.

Climatic zones :	E : Equatorial ; ET : Equatorial-Tropical ; T : Tropical ; ST : Substropical ; Te : Temperate ; SP : Subpolar.
Symbols :	DSDP and ODP sites : • with good planktonic foraminifera preservation ; ⊙ with poor preservation ;
other data :	▲ (Banerjie) , + (Saint-Marc & Suminta, 1979), ▼ (Lindsay, 1969), ■ (Quilty, 1978), x (Wright, 1977), core E 35-12 (Kennett & Srinivasan, 1975)
Boundaries :	Strong line - established, dotted line, assumed.

data were not available (as in legs 25, 27, 29) and species occurrences were reported only as present or absent, we used as an index the ratio of the number of species reported as present within each climatic group versus the total number of species defining that group and assumed that this index characterizes the climatic zone in the same way as abundance rank or test percentages (Oskina et al., 1982).

3. RESULTS

The position of equatorial, equatorial-tropical, tropical, subtropical, temperate and subpolar zones was reconstructed for six selected time intervals and compared to each other as well as to the position of these zones in Recent time (Fig. 1, 2).

The comparison of maps allows to infer the following trends of climatic changes : the progressive cooling of high latitudes during the Neogene (except for the early Pliocene) was accompanied by equatorward migration of cold-water climatic zones (Ivanova, 1988), while at low latitudes, climatic oscillations occured without identifiable trends. The warmest conditions with the most extended equatorial zone existed at the end of the early Miocene, in part due to the closure of Tethys (Berggren and Hollister, 1974), and during the early Pliocene. The earliest Miocene and the end of the middle Miocene were relatively cold : the warmest equatorial zone contracted, while equatorial-tropical zones widened. Later, during the Quaternary, the equatorial zone almost disappeared, and today this zone exists only in the southern Red Sea (Barash, 1989).

Boundaries between climatic zones shifted during the Neogene with a latitudinal amplitude of 10° to 30° (for example, expansion of the equatorial zone in the Eastern Indian Ocean 17 - 15 Ma, equatorward shifting of subtropical, temperate and subpolar zones).

During the early-middle Miocene, the climatic conditions in the western and eastern sectors of the Northern Indian Ocean were different from each other. By late Miocene, this difference largely disappeared and latitudinal zonality became more distinct. The latter probably points to strengthening of gyral systems after closure of the Indonesian Seaway, as noted by Kennett et al (1985). During the coolings, the equatorial divergence intensified dramatically, so that subtropical foraminiferal assemblages appeared, while, during warm episodes, this area was occupied by tropical and equatorial-tropical assemblages.

The Arabian-Somalian coastal upwelling can be traced beginning at least from the end of the middle Miocene, when it was rather strong as shown by the appearance of cold-water subtropical assemblages. That means that the Northern Indian Ocean has been affected by monsoonal circulation at least since 12-10 Ma.

4. DISCUSSION

We assumed that each planktonic foraminiferal species belonged to the same climatic group as its ancestors during the whole evolutionary time span. So, we believe that similar environmental conditions controlled species distribution patterns, and that a given species or its ancestors have not changed their preferable temperature range (climatic zone) with time. Therefore, one can deal with the same climatic zones, characterized by similar oceanographic parameters (water masses) and reflected by approximately the same proportions of species groups during the Neogene, in spite of evolutionary changes in individual species. Area shifts of species groups are thus interpreted as reflecting the migration of climatic zones and oceanographic boundaries, but not as a change in environmental preference of the species. This suggestion is supported also by CENOP data based on quantitative data analysis(Kennett et al., 1985).

Our approach allowed us to reconstruct the climatic zonality for several selected Neogene time-intervals. The maps (Fig. 2) show that the latitudinal position of zones changed significantly during the Neogene, reflecting both climatic oscillations and regional tectonic events : the closure of the Tethys and Indonesian seaway resulted in the development of gyral (Kennett et al., 1985) and monsoonal circulations.

One can also notice the development of the equatorial divergence as well as coastal monsoon upwellings in the Northern Indian Ocean, reflected by relatively cold-water assemblages.

5. CONCLUSION

1). Using semi-quantitative and qualitative data on planktonic foraminifera from DSDP/ODP volumes and from other sources, climatic zonality for six Neogene time intervals in the Indian Ocean was reconstructed.

2). Comparison of the climatic zonality maps revealed some latitudinal shifts of the zones and corresponding water masses. Migration trends of the zones were different at high and low latitudes : progressive cooling and equatorward shift for the former contrasts with oscillations for the latter.

3). Evolution of the major surface circulation patterns can be inferred from the data obtained including current systems, gyres, equatorial divergence and coastal upwellings.

Acknowledgments. I thank my colleagues NS Oskina and NS Blyum for fruitful cooperation during this work. I am also grateful to E Vincent, C Sancetta, A Boersma and JP Kennett for valuable discussions.

REFERENCES

Banerjie RK, Baruah RM, Gupta SC (1981) Micropaleontology of marine Miocene subsurface sequence of Galsi area, Bengal Basin. J. Geol. Soc. India 22: 55-68.

Barash MS (ed.) (1989) Neogene-Quaternary paleoceanology based on micropaleontological data. Nauka, Moscow (in Russian).

Bé AWH (1977) An ecological, zoogeographic and taxonomic review of Recent planktonic foraminifera. In Oceanic Micropaleontology, 1, Ramsay ATS (Ed), Academic Press, London, pp. 1-106.

Bé AWH and WH Hutson (1977) Ecology of planktonic foraminifera and biogeographic patterns of life and fossil assemblages in the Indian Ocean. Micropaleontology, 23, 369-414.

Bé AWH and DS Tolderlund (1971) Distribution and ecology of living planktonic foraminifera in surface waters of the Atlantic and Indian oceans. In The micropaleontology of Oceans, BM Funnell and WR Riedel (Eds), Cambridge University Press, Cambridge, UK, pp. 105-149

Berggren WA, Hollister CD (1974) Paleobiogeography and history of circulation in the Atlantid Ocean. In "Studies in Paleoceanography" Hay WW (ed). Soc. Econ. Paleont. Mineral., Spec. Publ. No. 20 : 126-186.

Deep-Sea Drilling Project Initial Reports (1974-1975) Washington US Government Printing Office. vol. 22-29.

Huber BT (1991) Paleogene and early Neogene planktonic foraminifer biostratigraphy of sites 738 and 744, Kerguelen plateau (Southern Indian Ocean). In "Proceedings of Ocean Drilling Programme, Scientific Results" Barron J., et al. (ed). Washington US Government Printing Office, Vol. 119 : 427-450.

Ivanova EV (1983) On paleotemperature analysis on planktonic foraminifera. Okeanologia 23 : 456-463.

Ivanova EV (1987) Miocene climatic zonality of the Indian Ocean.Okeanologia 27 : 951-957.

Ivanova EV (1988) Migration of climatic zones in the Indian Ocean during the Neogene. Dokl. Akad. Nauk. SSSR 302 : 922-925 (In Russian).

Ivanova EV, Barash MS (1982) Experience in evalutating climatic zonality in the Indian Ocean for Pliocene from the deep-sea drilling data. Okeanologia 22 : 71-79 (In Russian).

Kennett JP, Keller G, Srinivasan MS (1985) Miocene planktonic foraminiferal biogeography and paleoceanographic development of the Indo-Pacific region. In "The Miocene Ocean: Paleoceanography and Biogeography" Kennett JP (ed). Geol. Soc. Amer. Mem. No. 163: 197-236.

Kennett JP, Srinivasan MS (1975) Stratigraphic occurences of the Miocene planktonic foraminifer *Globoquadrina dehiscens* in Early Pliocene sediments of the Indian Ocean. Rev. Espan. Micropal. 7 : 5-14.

Krasheninnikov VA, Basov IA (eds) (1986) Cenozoic stratigraphy of the Southern Ocean. Nauka (Moskva), 206 p (in Russian).

Lindsay JM (1969) Cenozoic foraminifera and stratigraphy of the Adelaide Plains Subbasin, South Australia. Bull. Geol. Surv.South Australia, 42 : 60 p.

Loutit TS (1981) Late Miocene paleoclimatology : Subantarctic water mass, Southwest Pacific. Mar. Micropal. 6 : 1-28.

Oskina NV, Ivanova EV, Blyum NS (1982) Pliocene climatic zonality of the Atlantic, Indian and Pacific oceans. Dokl. Akad. Nauk SSSR 264 : 400-407 (In Russian).

Quilty PG (1978) The late Cretaceous-Tertiary section in Challenger No.1 (Perth Basin- details and implications. Bureau of Mineral Resources (Canberra), Geol. Geophys. Bull. 192 : 109-135.

Sancetta CA (1978) Neogene Pacific microfossils and paleoceanography. Mar. Micropal. 3 : 347-376.

Saint-Marc P, Suminta (1979) Biostratigraphy of late Miocene and Pliocene deep water sediments of eastern Java (Indonesia). J. Foram. Res. 9 :106-118.

Srinivasan MS, Kennett JP (1981) Neogene planktonic foraminiferal biostratigraphy and evolution : equatorial to subantarctic, South Pacific. Mar. Micropal., 6 : 499-533.

Vincent E, Toumarkine M (1990) Neogene planktonic foraminifers from the western tropical Indian Ocean, leg 115. In "Proceedings of the Ocean Drilling Programme, Scientific Results" Duncan RA et al. (eds). Washington US Governement Printing Office. vol. 115 : 795-838.

Wrigth CA (1977) Distribution of Cenozoic foraminifera in the Scott Reef N 1 well, Western Australia. J. Geol. Soc. Australia, 24 : 269-277.

Zonenshine LP, Savostin LA, Sedov AP (1984) Global paleogeodynamic reconstructions for the last 160 m.y. Geotektonika 3 : 3-16 (In Russian).

SURFACE AND DEEP WATER CIRCULATION CHANGES DURING THE LAST CLIMATIC CYCLE

Jean-Claude Duplessy and Laurent D. Labeyrie
Centre des Faibles Radioactivités
Laboratoire mixte CNRS-CEA
91198 Gif Sur Yvette Cédex, FRANCE

The global oceanic circulation experienced major variations during the last climatic cycle as a result of changes at the sea surface. Micropaleontological and isotopic analysis of deep sea sediments allows to reconstruct these variations and to relate them to major changes of the Earth's climate system.

1. INTRODUCTION

Ocean circulation and climate are strongly interconnected. The waters of the ocean are continually moving and this motion is driven by energy from the sun and the rotation of the Earth. The sun drives the ocean circulation through the circulation of the atmosphere (winds) and by causing variations in the temperature and salinity of sea water which in turn control its density. Deep water is formed at high latitudes, where the atmosphere extracts heat from the sea surface, and where a salinity increase results from sea ice formation during winter conditions. Both processes make the surface water sufficiently dense to sink to great depth and flow toward the equator. Above the deep and bottom water masses of the ocean lie the intermediate water masses, of which sinking is associated with the subpolar convergences in both hemispheres.

The global oceanic circulation has changed dramatically during the last million years, because conditions prevailing in the high latitudes of both hemispheres have varied considerably. For instance, the Norwegian Sea was essentially ice-covered during most of the last glaciation. It did not form any bottom water until the last

NATO ASI Series, Vol. I 22
Long-Term Climatic Variations
Edited by J.-C. Duplessy and M.-T. Spyridakis

deglaciation (Duplessy et al., 1988a). Similarly, the location of intermediate water formation sites experienced large climate-related changes. As a consequence the fluxes and characteristics of these water masses also varied. In this paper, we shall review the link between ocean circulation and climate, using the example of the last climatic cycle (the last 150,000 years), which is best documented and can be simulated using general circulation models.

2. GEOCHEMICAL AND MICROPALEONTOLOGICAL METHODS OF PALEOCEANOGRAPHY

Cores from the ocean floor contain material deposited at a fairly constant rate, generally in the order of a few centimeters per thousand years. In order to derive a useful information on past climates, sediment must first be dated. Then, its fossil content (foraminifera, coccoliths, ...) and its isotopic composition are translated into quantitative estimates of those climatic variables (temperature, salinity, geochemical tracers) that physical models can use and simulate.

2.1 Isotopic and geochemical tracers

Most elements occur in nature as a mixture of stable isotopes. These isotopes of an element differ slightly in their chemical and physical properties, as a consequence of the differences in their thermodynamic properties. Therefore, many natural processes entail slight separation of stable isotopes. For example, the evaporation and precipitation of water give rise to variations in the $^{18}O/^{16}O$ ratio (the $H_2{}^{16}O$ vapor pressure is larger by about 1% than $H_2{}^{18}O$ vapor pressure at 15°C). During the transport of the atmospheric water vapor from the areas of high evaporation (low latitudes) towards high latitudes, the successive rain episodes correspond to a progressive isotopic distillation of the water vapor which drives to extreme in the difference between the isotopic composition of ice stored in the polar ice sheets, and that of water in the ocean (Dansgaard, 1964 ; Broecker, 1989). Chemical processes also fractionate the isotopes, so that the $^{18}O/^{16}O$ ratio in calcite shells laid down by organisms is slightly greater (by about 4%) than that of the water they inhabit. This fractionation factor is particularly interesting, because it is temperature dependent. This means that it may be possible to estimate past ocean temperatures by measuring the $^{18}O/^{16}O$ ratio of calcitic fossils (Emiliani, 1955 ; Shackleton, 1974; Duplessy, 1978).

As these isotopic variations are small, they are generally expressed in the δ unit defined by the relationship :

$$\delta = [(R_{sample} / R_{standard}) - 1] \times 1000$$

in which the reference standard is PDB for carbonate samples and SMOW for water samples.

A precise interpretation of the variations of the $^{18}O/^{16}O$ ratio of the calcitic shells of foraminifera in deep sea sediments is however difficult, because both the sea water temperature and $^{18}O/^{16}O$ ratio vary with the climate: First, since continental ice sheets are depleted in ^{18}O, as the amount of ice present on the continent increases during a glaciation, the water remaining in the ocean becomes richer in ^{18}O and its $^{18}O/^{16}O$ ratio increases. This isotopic enrichment is entirely reflected in the calcitic test of foraminifera. Second, as the temperature at which the shell formed decreases, the $^{18}O/^{16}O$ ratio of the calcite increases (Epstein et al., 1953 ; Shackleton, 1974). Thus high $\delta^{18}O$ values indicate cold climate. Conversely, temperature increases and ice cap melting will be reflected by low $\delta^{18}O$ values. Special effort must therefore be made to estimate separately the variations of the two variables, sea water temperature and $\delta^{18}O$.

Labeyrie et al. (1987) and Shackleton (1987) have independently derived a $\delta^{18}O$ record which shows only the mean sea water $\delta^{18}O$ variations during the last 135,000 years due to continental ice volume changes. Both estimates exhibit a fair agreement. As these sea water $\delta^{18}O$ variations constitute 60 to 100% of the benthic records of most sediment cores (depending on their location and depth), the $\delta^{18}O$ variations in a deep sea core can be used to determine unambiguously the sediment deposited during the major events of the last climatic cycle. This is the basis of the isotope stratigraphy (Shackleton and Opdyke, 1973).

Carbon isotopes constitute also an excellent tracer for both the geochemical cycle of carbon and deep water circulation. Kroopnick (1985) has demonstrated that the distribution of $\delta^{13}C$ of the total CO_2 (ΣCO_2) dissolved in sea water delineates the general distribution of water masses and that modern $\delta^{13}C$ gradients in the deep ocean follow the net flow of the deep waters. The distribution of the $\delta^{13}C$ value of the ΣCO_2 dissolved in intermediate, deep and bottom waters of the ocean can be used as a tracer of the abyssal circulation under the following conditions: (i) within a deep water mass, the $\delta^{13}C$ value decreases with increasing oxidation of organic matter, such that the $\delta^{13}C$ decrease is related both to the surface productivity and the time elapsed since the water mass was isolated from the atmosphere; (ii) when two water masses are compared, the $\delta^{13}C$ value in each depends not only on their residence time at depth, but also on the ratio between the CO_2 produced from organic matter oxidation and the bicarbonate and carbonate ions derived from carbonate dissolution (note that this ratio is poorly controlled under past conditions); (iii) mixing between two water masses results in $\delta^{13}C$ values intermediate between the two original components (see a detailed discussion in Duplessy and Shackleton, 1985). As a consequence, high $\delta^{13}C$ values are found in deep waters only in the vicinity of well-ventilated areas, where surface waters has sunk to great depth. Then $\delta^{13}C$ decreases along the flow lines of a well-identified deep water mass (Fig. 1). Some species of benthic foraminifera (Genus *Cibicides*) closely record the modern $\delta^{13}C$ distribution in the world ocean and provide a good proxy-indicator for the reconstruction of the past $\delta^{13}C$ gradients in the ocean (Duplessy et al., 1984).

The interpretation of carbon isotope data is however complicated because the average oceanic $\delta^{13}C$ has been altered in the past by a transfer of carbon from the continental biosphere (reduced during Glacials) or continental shelves into the ocean (Shackleton, 1977 ; Broecker,. 1982). In view of these difficulties, Boyle (1981) developed a new tracer of deep ocean circulation, the Cd content of benthic foraminiferal tests. This method rests on the observation that the distribution of Cd in the modern ocean is similar to that of Phosphorous. Since North Atlantic Deep Water (NADW) forms from nutrient-depleted near surface waters of subtropical origin, variations in the nutrient content of deep waters reflect the intensity of the decomposition at depth of organic matter falling from the surface. This provides a useful tracer of thermohaline circulation patterns of the deep ocean. Since Cd is incorporated into the shells of benthic foraminifera in proportion to its concentration in bottom waters overlaying the sediment, Cd can be used as a proxy for nutrient distribution in the past ocean. Extensive comparisons of informations provided by Cd and $\delta^{13}C$ in the same deep sea cores have proved the general reliability of both proxy-indicators and evidenced a few discrepancies, which are still under discussion (Boyle, 1992).

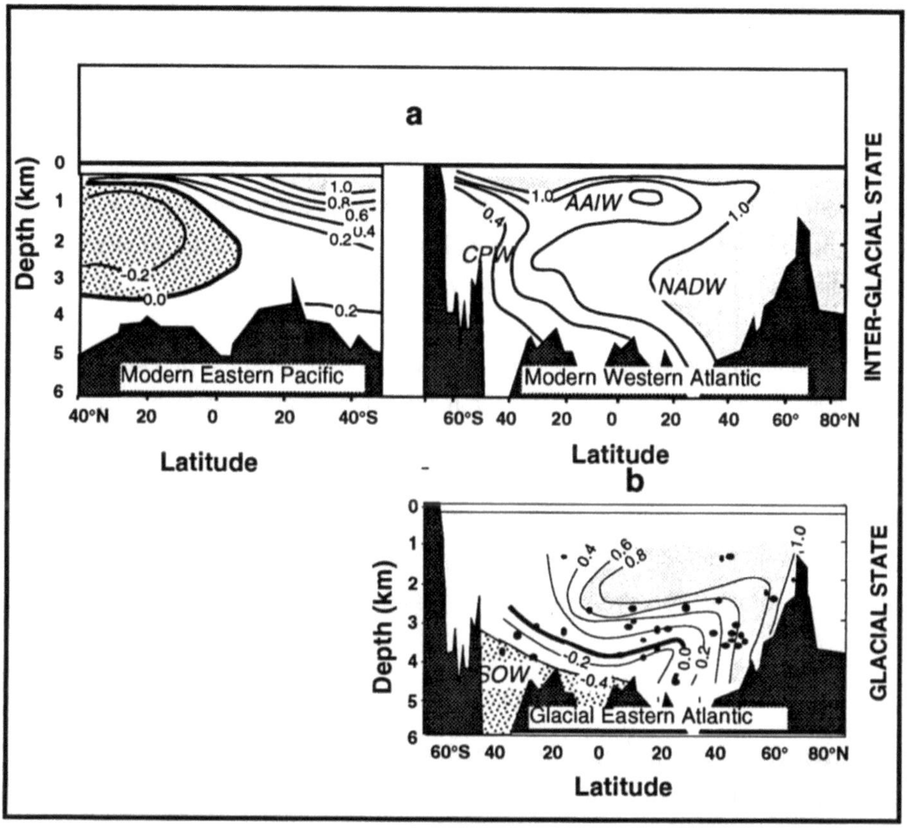

Figure 1: reconstruction of the changes in $\delta^{13}C$ across the Atlantic Ocean for the modern period and the Last Glacial Maximum (from Labeyrie et al., 1992).

2.2 Transfer functions

Transfer functions are empirically derived equations for calculating quantitative estimates of past atmospheric and oceanic conditions from paleontological data. Ideally, this problem requires the formulation of a model that explains the biological responses in terms of the multitude of factors that comprise the observed ecological system. The basic assumptions are: (i) No significant changes occur to the interactions among organisms over the time scale studied and within the ecosystems sampled. (ii) The modern observations provide sufficient information for analyzing the fossil observations, and further that a snapshot of modern spatial patterns is a sufficient basis for interpreting changes through time. (iii) The biological responses are systematically related to the physical attributes of the biotic environment. This assumption implies that the relationships can be represented by mathematical expressions linking species or assemblages to environmental parameters such as temperature or salinity.

Transfer functions techniques, which use multivariate statistical procedures, were first applied to marine plankton data (foraminifera) by Imbrie and Kipp (1971). They provide typically sea surface temperature (SST) estimates with an accuracy of ± 1.5°C. Other statistical techniques have been used in recent years, such as the modern-analogue technique (Prell, 1985 ; Guiot, this volume). These attempts basically confirm that SST estimates for fossil samples are correct if the data base used for calibration contains modern samples similar to those present in the past sediment.

3. GLACIAL-INTERGLACIAL CONTRASTS IN INTERMEDIATE AND DEEP WATERS

3.1 Circulation changes

The reconstruction of the $\delta^{13}C$ distribution of the total CO_2 dissolved in the deep ocean during the last interglaciation (isotopic substage 5e, about 127,000 to 115,000 years ago) showed that deep water sources were active in both hemispheres, according to a pattern similar to present-day conditions. However, the production of Antarctic Bottom water (AABW) was stronger than that of NADW, resulting in an east-west $\delta^{13}C$ gradient stronger than the modern one (Duplessy et al., 1984). During the phase of rapid ice-growth which followed the last interglaciation, the eastern subpolar North Atlantic Ocean maintained warm sea surface temperatures comparable to those of today and the Norwegian Sea continued to be an active source of deep waters. However, the relative importance of the AABW flux compared to that of NADW increased, a trend which was further enhanced during glacial conditions (Duplessy and Shackleton, 1985).

The peak of oxygen isotope stage 2 was considered as a typical example of full glacial conditions. In the very high latitude of the North Atlantic, a Glacial North Atlantic Deep Water mass, with high $\delta^{13}C$ values is found. This water mass extends toward the south mainly at intermediate depth (Fig. 1). Below 3000 m, most of the deep eastern Atlantic was filled with bottom water originating from the Southern Ocean so that the relative contribution of water from the southern hemisphere as compared to that from the northern hemisphere in the deep Atlantic was much greater than today (Duplessy et al., 1988b). This is in agreement with the Cd data of Boyle and Keigwin (1982 ; 1985), the $\delta^{13}C$ data of Oppo and Fairbanks (1987), and the distribution of benthic foraminifera described by Streeter and Shackleton (1979).

In the Indian Ocean, a very strong and sharp discontinuity separated the high $\delta^{13}C$ values of Intermediate Water to a depth of about 2000 m from negative $\delta^{13}C$ values deeper than 2300 m (Kallel et al., 1988). This glacial pattern is in sharp contrast with the modern situation where the vertical $\delta^{13}C$ gradient in the North Indian Ocean is lower than 0.1 per mil in this depth range. This change in the vertical profile originates from a decrease of the $\delta^{13}C$ values of Bengal Gulf Deep Water by about 1‰ during last glacial maximum and indicates a very weak ventilation of the deep Indian Ocean during that period. By contrast, Intermediate Water was more ventilated than today. Because of a lowering of sea level during glacial conditions, Red Sea and Arabo-Persian Gulf contributions to Intermediate Water were probably very reduced or absent. Three sources are possible for the glacial Intermediate Water in the Northern Indian Ocean: (i) dense water formed in the Northern Arabian Sea by the increased evaporation/precipitation ratio and the possible decreased temperatures of the surface water described by CLIMAP (1981) and Duplessy (1982), (ii) expansion to the North of 10°S of the Antarctic Intermediate Water (AAIW) in the absence of a northern local source and (iii) increased flow of intermediate water from the Pacific Ocean through the Indonesian Archipelago. A more complete coverage is required to determine unambiguously the origin of the glacial North Indian Intermediate Water.

Under present conditions, most areas of the Pacific Ocean lie below the carbonate compensation depth, and foraminifera are absent in the sediment. However, it has long been observed that, in contrast with the deep north Atlantic Ocean, dissolution was less intense in most of the Pacific Ocean during isotope stage 2, with a deepening of a few hundred meters of the carbonate compensation depth (Broecker, 1982). This indicates a significant change in the chemistry of the deep waters, with a drop in the water total CO_2 content which may possibly be related to an increase in ventilation at intermediate depth. By contrast with modern conditions, at the last glacial maximum, benthic foraminifera are often present in North Pacific sediments shallower than 4000 m. The glacial profile for the Pacific Ocean exhibits $\delta^{13}C$ values which are identical to the modern values for the 1000–2600 m depth range. Deeper than 3000 m, $\delta^{13}C$ values are lower than modern ones by 0.3‰ (Duplessy et al., 1988b). As the global mean oceanic $\delta^{13}C$ also decreased by 0.3‰ during the last glacial maximum, the $\delta^{13}C$ profile shows that Pacific Deep Waters were as ventilated as today while intermediate waters above 2600 m were more ventilated than today during the last glaciation. The origin of the ventilated water cannot be traced because the core coverage is insufficient.

3.2 Ice volume, sea-level and global $\delta^{18}O$ variations

The sea-level sinks when glacier ice piles up on the continents. It rises when continental ice melts. Sea-level variations are recorded along shorelines in the form of marine erosional benches and depositional terraces. Radioactive dating of the coraline material found at the terraces shows when the high sea stands. However, the absolute positions of sea stands during these times is difficult to establish, because land areas are generally not tectonically stable (Bloom et al., 1974). The sea-level record of the last deglaciation is the only one which has been reconstructed in great details by a set of cores drilled in several submerged terraces surrounding Barbados Island (Fairbanks, 1989).

Indirect methods must therefore be used to reconstruct sea-level variations during older periods. They usually rely on stable isotope variations in benthic foraminifera. ^{13}C values of benthic foraminifera from the Norwegian sea show that this area was an active source of deep water sinking during isotope stages 4 and 5, i. e. from 125,000 to 65,000 years Before Present (BP) (Duplessy et al., 1988a,b). The temperature of the deep Norwegian basin was therefore close to -1°C during this whole period and the $\delta^{18}O$ record of benthic foraminifera reflects only global ocean variations due to ice volume changes (Labeyrie et al., 1987). The disappearance of benthic foraminifera genus *Cibicides* in Norwegian Sea sediment during the glaciation (isotope stages 2-4) prevent us to estimate the deep water temperature in the Norwegian basin associated with glacial climate. However, during this period, the Pacific deep waters were very cold and, at a first approximation, did not experience any significant temperature change. Labeyrie et al. (1987) therefore derived the mean global ocean water $\delta^{18}O$ record by piecing together the Norwegian sea $\delta^{18}O$ record for the interglacial periods and the $\delta^{18}O$ record of Pacific core V19-30 for the glaciation.

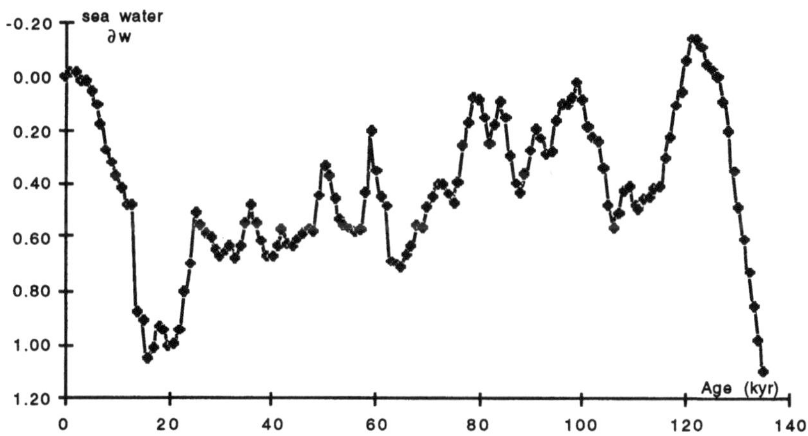

Figure 2: Variations of the $^{18}O/^{16}O$ ratio of the global mean oceanic water during the last 135,000 years. The value 0 indicates that the isotopic composition of the ocean is identical to that of the standard, which is the mean modern sea water isotopic composition (SMOW). Positive ∂w values (reported downward) indicate that the $^{18}O/^{16}O$ ratio of the global mean oceanic water was higher than the modern one, in relation with the waxing of continental ice sheets strongly depleted in ^{18}O.

Figure 2 shows the variations of the global mean $\delta^{18}O$ of the ocean during the last 135,000 years. After a rapid deglaciation which occurred about 128,000 years ago, the Earth's climate was characterized by a volume of continental ice smaller than the present one. As a consequence, the sea level was higher than today by about 6 meters. This interglacial period, called Eemian, lasted for about 11,000 years and was followed by a first phase of moderate glaciation, which had culminated by about 107,000 years BP. This ice advance was followed by several phases of ice disintegration, but the volume of ice was always larger than today. The sea-level was estimated to be about -18 meters during the two high levels around 80,000 and 99,000 years BP. Several phases of continental ice accumulation developed further until the peak of the glaciation 18,000 years ago, when the sea-level was 120 meters below the present one. At that time, about 50,000,000 km^3 of ice were frozen, mainly over Canada and northern Europe. The final deglaciation began about 15,000 years ago and the ice melted in two major separate phases, each lasting about 2,000 years (Fairbanks, 1989). During these phases, the sea level increased at a rate which reached up to 2-3 meters per century.

3.3 Deep water temperatures

Changes in deep water temperature in the major oceanic basins during the last climatic cycle can be reconstructed by subtracting the global mean ocean water $\delta^{18}O$ record from each benthic foraminiferal record. Results of such calculations are reported in Figure 3. At the end of the last interglacial, in the Pacific and Southern oceans, the cooling to glacial temperatures around or below 0°C occurred within a few thousand years between 115,000 and 110,000 years BP. This also dates the first major change in North Atlantic hydrology. However, in the Atlantic Ocean, a second cooling phase was associated with the stage 5a-4 transition, so that the deep water temperature was close to 0°C in all oceanic basins during isotope stages 2-4. By contrast, the temperature of intermediate waters did not change significantly in the Indian Ocean, so that the sharp $\delta^{13}C$ gradient separating intermediate from deep water was associated to a strong temperature gradient of about 2.5°C in a few hundred meters around a water depth of 2000 meters.

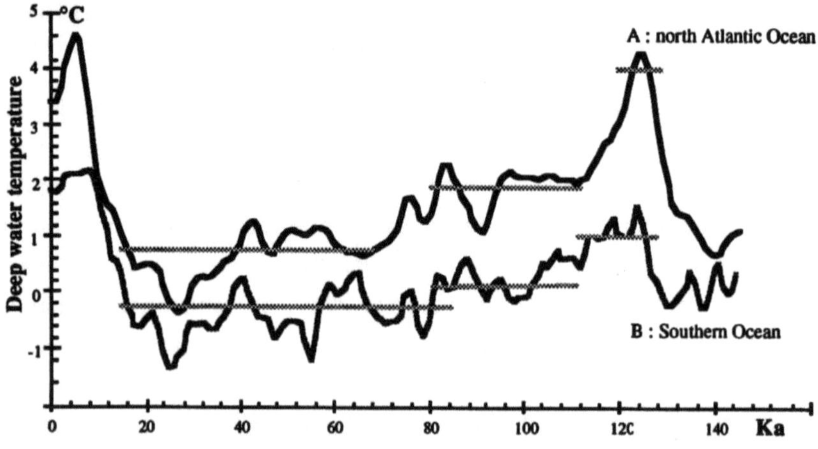

Figure 3: Reconstruction of the temperature of the deep water (at 3 km depth) in the North Atlantic and in the Southern Ocean during the last climatic cycle.

4. RECONSTRUCTION OF SURFACE WATER SALINITY : A NEW METHOD

In order to obtain a complete reconstruction of the ocean in the geological past, it is necessary to estimate both the temperature and the salinity of the various water masses. Sea-surface temperature can be estimated by planktonic fauna analysis (Imbrie and Kipp, 1971), but a similar method cannot be used to reconstruct surface salinities in the high latitude ocean, because temperature and salinity are too closely correlated in these areas.

Another possible approach to estimate past salinity relies on oxygen isotope analysis. Polar water has lower $^{18}O/^{16}O$ ratios than does tropical and temperate waters. These differences are closely tied to the ocean surface salinity distribution (Craig and Gordon, 1965) ; the higher the salinity, the higher the $^{18}O/^{16}O$ ratio. Excess evaporation concentrates the less volatile $H_2{}^{18}O$ along with salt, while precipitation, more depleted in $H_2{}^{18}O$ than ocean water, decreases both salinity and $^{18}O/^{16}O$ ratio of the sea surface water. The $^{18}O/^{16}O$ ratio of planktonic foraminifera records these variations since it depends on both the $^{18}O/^{16}O$ ratio and the temperature of the sea water in which foraminifera have grown.

More precisely, the $\delta^{18}O$ of foraminiferal $CaCO_3$ directly reflects variations of the $\delta^{18}O$ of the ambient water and the oxygen isotope fractionation between $CaCO_3$ and water increases by about 0.25 ‰ for each degree the water is cooled (Epstein et al., 1953; Shackleton, 1974). This is expressed by the paleotemperature equation:

$$T = 16.9 - 4.38 \times (\delta^{18}O_{carbonate} - \delta^{18}O_{water}) + 0.10 \times (\delta^{18}O_{carbonate} - \delta^{18}O_{water})^2$$

which is valid in a wide temperature range (-1°C to 27°C).

If the temperature of deposition of foraminiferal calcite is known (for instance through a transfer function estimate), it becomes feasible to calculate the $^{18}O/^{16}O$ of the water in which the foraminifera have grown. Using a large set of cores from the North Atlantic and the Southern oceans, Duplessy et al. (1991) have statistically established the relationship linking the $^{18}O/^{16}O$ ratios of *Neogloboquadrina pachyderma* and *Globigerina bulloides,* two species common in the North Atlantic, to the summer sea-surface water temperature (SST) and the ambient $^{18}O/^{16}O$ ratio.

The comparison of the Levitus summer SST with that derived from the isotopic measurements on *G. bulloides* throughout the North Atlantic Ocean and the Southern Ocean. shows that the $\delta^{18}O$ values of *G. bulloides* is a linear function of the summer SST only in the range 7 to 22°C (Fig. 4). We call this temperature range the "optimal temperature range" for this species. Figure 4 also shows that the isotopic temperature of *G. bulloides* is generally lower than the summer SST. A statistical analysis performed on core tops raised from locations where the summer SST is in the range 7 to 22°C shows that the isotopic temperature given by *G. bulloides* is related to the observed summer temperature by the simple relationship $T_{isotope}$ = summer SST minus 1°C.

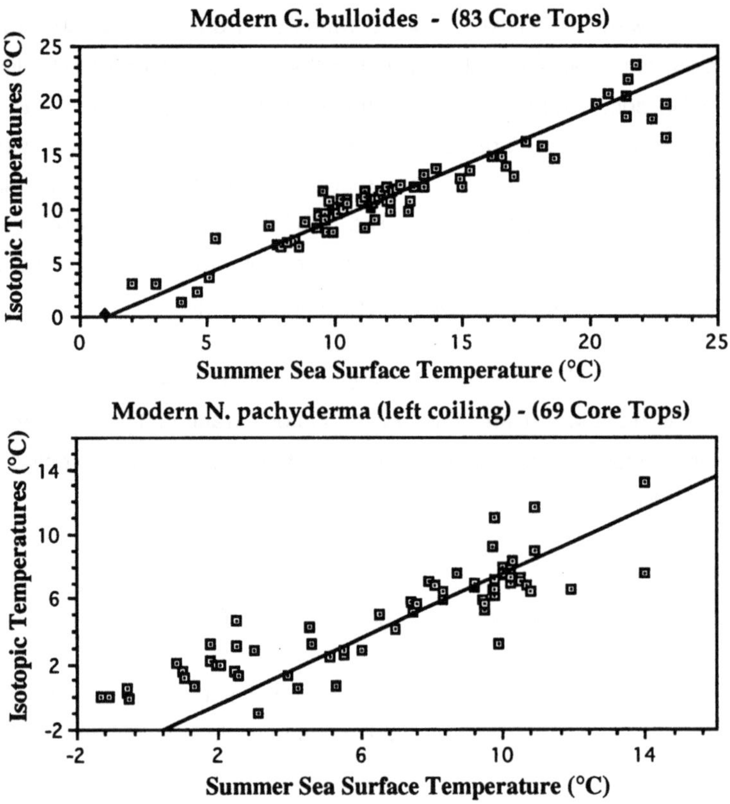

Figure 4: Comparison of the isotopic temperature of recent *G. bulloides* and *N. pachyderma* with the modern August sea surface temperature. The straight lines have for equation isotopic Temperature = summer SST - K, where K=1 for *G. bulloides* and K=2.5 for *N. pachyderma.*

Similarly, the isotopic temperature given by *N. pachyderma* was compared to the summer SST (Fig. 4). Within the temperature range -1.5°C to 3°C, the isotopic temperatures given by *N. pachyderma* are close to 0°C, and are often warmer than the atlas mean summer values. This is true even when a permanent ice cover is generally present at the surface. This indicates that, in very cold water, *N. pachyderma* develops only when it can find a water mass of which the temperature is close to 0°C. A statistical analysis performed on core tops raised from locations where the summer SST is in the range +3 to + 10°C shows that the isotopic temperature given by *N. pachyderma* (left coiling) is related to the observed summer temperature by the simple relationship $T_{isotope}$ = summer SST minus 2.5°C.

Duplessy et al. (1991) have therefore defined a parameter, which they called T*, and which represents a statistical estimate of the mean calcification temperature for each species : T* = Summer SST -1°C for *G. bulloides* and T* = Summer SST - 2.5°C for *N. pachyderma.* Taking all the *N. pachyderma* $\delta^{18}O$ values from sites where the summer SST is higher than 3°C and lower than 10°C and all the *G. bulloides* $\delta^{18}O$ values from sites where the summer SST is higher than 7°C and lower than 22°C, Duplessy et al. (1991) found that the isotopic temperatures calculated from the $\delta^{18}O$ values of these species are not significantly different from the temperature T* (Fig. 5). They are highly correlated (r^2 = 0.918) and the slope of the regression line (0.959 with s = 0.04) is not significantly different from 1 at the 2 s level. We shall assume in the following section that this statistical relationship is also valid under different climatic conditions.

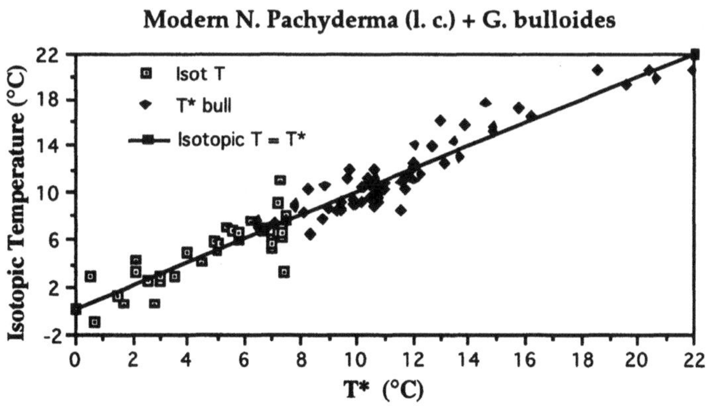

Figure 5: Isotopic temperature versus effective temperature of calcification T*.

5. THE SURFACE TO DEEP WATER CONNECTION DURING THE LAST GLACIAL MAXIMUM

5.1 North Atlantic surface salinity reconstruction

The previous calibration of *G. bulloides* and *N. pachyderma* (left coiling) was used to reconstruct surface water $\delta^{18}O$ and salinity in the North Atlantic Ocean during the last glacial maximum (Fig. 6). Low salinity values were found in the Bay of Biscay as a consequence of the outflow of European rivers (which carried meltwater from snow and from the southern margin of the ice sheets) and in the Norwegian-Greenland sea as a consequence of the disappearance of the North Atlantic Drift during glacial conditions (Kellogg, 1976 ; Ruddiman and McIntyre, 1973). The presence of the Labrador current along the eastern coast of North America is marked by salinity values noticeably lower than those found in the central Atlantic at about 50-55°N. In contrast, high salinities are found north of the thermal polar front, from 45°N to 55°N and from 30°W to 40°W. This reconstruction shows that the surface salinity distribution during the last glacial period was noticeably different from the present one (Fig. 6). This was due to both the disappearance of the North Atlantic drift and the input of freshwater resulting from local precipitation, ice melting and river outflow. The presence of a salty surface water mass (comparatively saltier than the modern one in the central North Atlantic from 45°N to 55°N) indicates that high latitudes of the glacial North Atlantic still received a significant flux of salt, which was required to permit the formation of deep and intermediate waters.

5.2 Surface density

Using this reconstruction of sea surface salinity and the temperature reconstruction obtained by the CLIMAP group (1981), the density field can be computed using the formula of Cox et al. (1970). Labeyrie et al (1992) showed that first order estimates of deep water density may be derived from the $\delta^{18}O$ of benthic foraminifera, the salinity being more homogeneous in deep than surface waters. These authors have demonstrated that the area located from 45°N to 55°N and from 30°W to 40°W was the only place where winter cooling was strong enough to increase the density to a value similar to that of deep waters (Fig. 7). They concluded that this area was the main location for deep water formation during the last glacial maximum. This pattern is consistent with the presence of a small well-ventilated deep water mass which

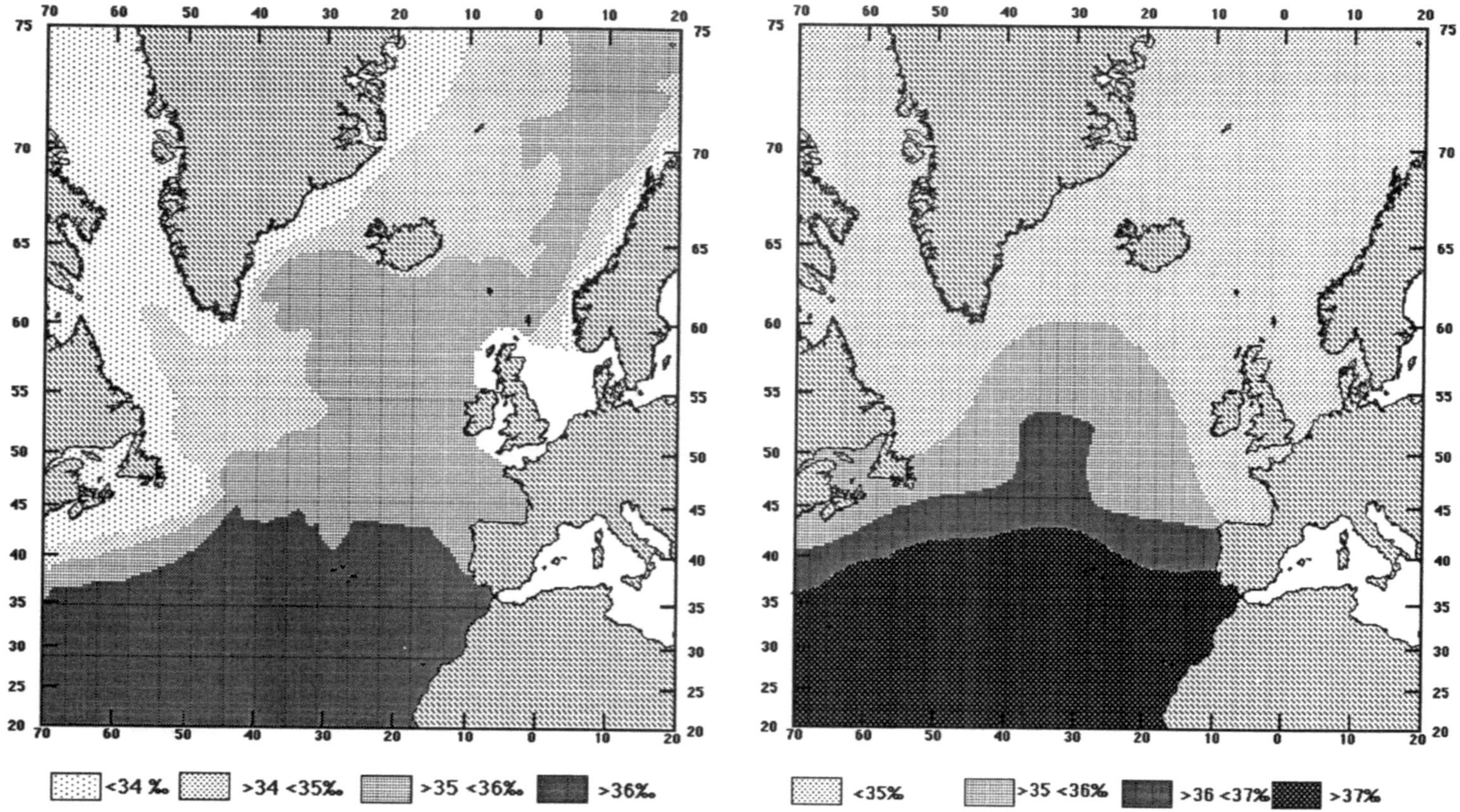

Figure 6: Surface water salinity distribution in the North Atlantic Ocean during present time (left) and during the Last Glacial Maximum (right).

was present at the bottom of the glacial North Atlantic Ocean, north of 45°N (Duplessy et al., 1988b). During glacial conditions, the flux of salt from low to high latitude surface waters was, like today, an essential component of the global ocean circulation.

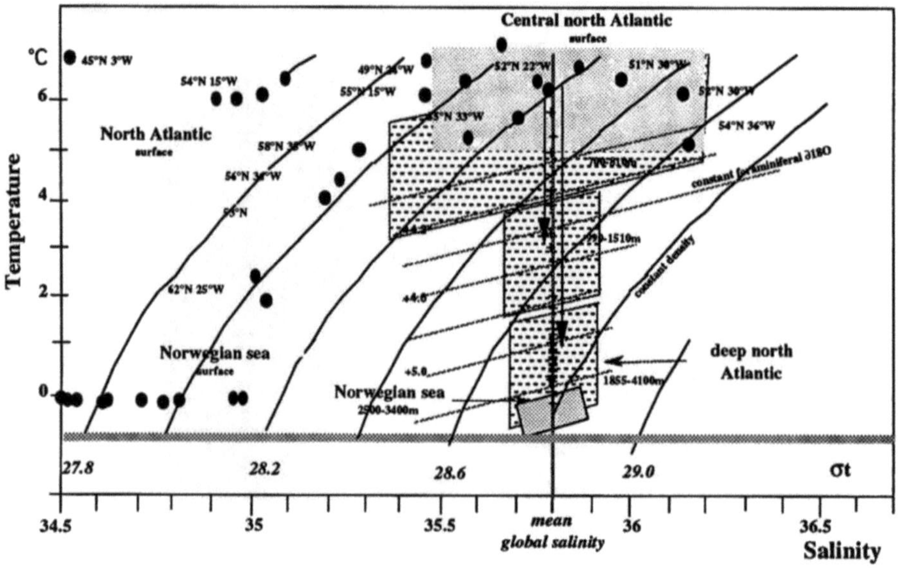

Figure 7: Distribution of the different water masses in the North Atlantic and Norwegian Sea during the Last Glacial Maximum on a T, S, σt diagram. Benthic foraminifera $\delta^{18}O$ are plotted at the intersection between the corresponding equilibrium isotopic fractionation lines and the estimated mean North Atlantic salinity for that period. Surface Sea water temperature are derived from Transfer Functions and salinity calculated using the planktic foraminifera $\delta^{18}O$ and temperature estimates . The potential areas of deep water formation areassociated with the vertical arrows (which simulate winter cooling). Crosses represent the benthic foraminifera $\delta^{18}O$; plain circles are for planktic foraminifera:.

6. THE RAPID CHANGES DURING THE LAST GLACIAL PERIOD.

Detailed analysis of sediment cores from the North Atlantic Ocean by sedimentological, micropaleontological, or isotopic methods (at resolution of 1-2 cm) show a rapid and large amplitude variability, especially during the glacial periods. This is attributed to a succession of large ice surge events, the so-called "Heinrich events" (Heinrichs 1988, Grousset et al (1993). Each of these events corresponds to a massive invasion of icebergs which melt along the northern boundary of the Polar front, between 40 and 55°N. This decreases abruptly the surface water temperature and salinity. Both phenomena affect in opposite direction the planktic foraminifera $\delta^{18}O$, but the resulting effect is a reduction by 1 to 2 ‰ of the foraminiferal $\delta^{18}O$, the salinity decrease being preponderant (Bond et al 1992). The spikes of low $\delta^{18}O$ water may be used as a tool to vizualize the mixing betwen surface water and deep water. Figure 8 shows the high resolution $\delta^{18}O$ analysis of planktic and benthic foraminifera along core SU 90-08 (43°31'N, 30°25'W, 3080 m), located within North Atlantic Deep Water. The age scale has been built by AMS ^{14}C dating of the planktic foraminifera. The location of the Heinrich layers (HL) in the core are indicated by large increases in magnetic susceptibility. The ice-rafted detritus HL3 peak is not present in that core, as the icebergs were probably following a more northern track during that period (Grousset et al 1993). However, each magnetic susceptibility peak is associated with a negative planktic $\delta^{18}O$ anomaly, and followed by a similar anomaly in the benthic foraminifera $\delta^{18}O$, after 2-3 kyrs. This is also true for the event which occurred during the LGM (HL2). Therefore, although mixing times were significantly larger than for the modern ocean, deep convection did occur during the last glacial period, even during events which lowered the surface water salinity.

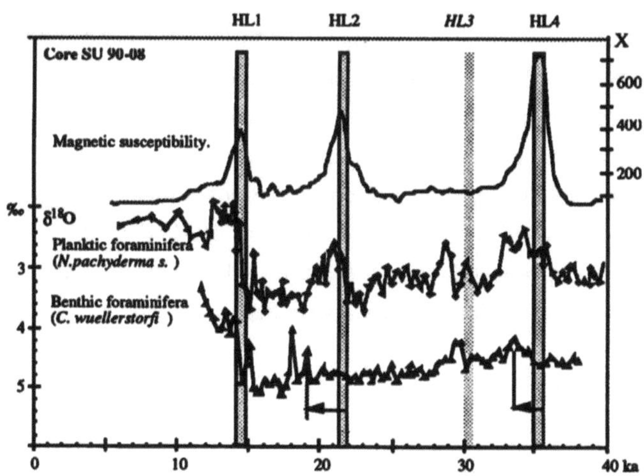

Figure 8: $\delta^{18}O$ analysis of planktic and benthic foraminifera along core SU 90-08 (43°31'N, 30°25'W, 3080 m). The age scale has been built by AMS ^{14}C dating of the planktic foraminifera. The location of the Heinrich layers (HL) in the core are indicated by large increases in magnetic susceptibility.

7. THE LAST DEGLACIATION

Micropaleontological and stable isotope records of planktonic foraminifera in two deep sea cores from the North Atlantic were used by Duplessy et al (1992) to estimate sea surface temperature and salinity change over the past 18,000 years, during the melting phase of the Laurentide and European ice sheets. Both core have been accurately dated by Accelerator Mass Spectrometric measurements of Carbon 14 in planktonic foraminifera.

The reconstructed records show that both sea surface temperature and salinity of the northern North Atlantic experienced large variations during the last 18,000 years (figure 9). A primary forcing factor for surface salinity was the injection of meltwater in the northern North Atlantic, which led to water stratification and decreasing SST. Both effects led to decreased evaporation, increased precipitation and reduced poleward advection of salty subtropical water. These feedbacks amplified the initial freshwater signal, resulting in a significant density decrease for surface water (Figure 9), a shutdown or a weaker flux of NADW (Boyle and Keigwin, 1987), a reduced heat transport to high latitudes and cold conditions over the northern North Atlantic and the neighboring European continent. Such coupled oceanic and atmospheric phenomena occurred during both the Oldest and Younger Dryas cold events. In disagreement with model simulations (Maier-Reimer and Mikolajewicz, 1989 ; Stocker and Wright, 1991), paleoclimatic reconstructions show that sea surface temperature and salinity were not affected in the northern North Atlantic when the freshwater input occurred at low latitude. Here the amplifying feedbacks did not operate. During the principal phases of ice-sheet melting, meltwater was transported by the Mississippi River to the Gulf of Mexico and, therefore, warm climatic conditions prevailed over the North Atlantic while active thermohaline circulation favored the dilution of meltwater in the world ocean.

8. CONCLUSION

We have now in hand the micropaleontological and geochemical methods to fully reconstruct the response of the surface and deep ocean to major climatic changes. Results obtained today demonstrate that the global ocean circulation experienced dramatic changes during the last climatic cycle and is very sensitive to climate and

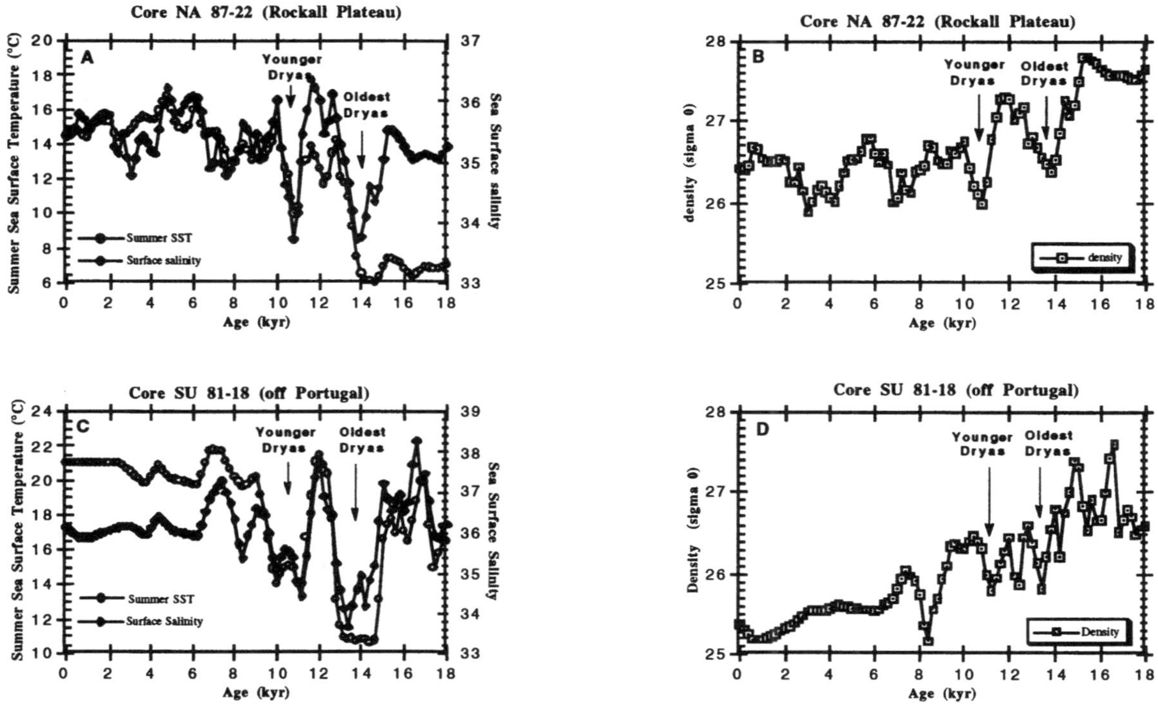

Figure 9: Sea surface salinity and temperature estimates for:
A Core NA87-22, off Ireland
C Core SU 81-18, off Portugal

and the associated surface water density changes, respectively **B** and **D**. All values have been interpolated every 200 years. Densities are expressed as sigma values defined by $\sigma = (d - 1) \times 1000$

conditions prevailing at the sea surface. Abrupt and short climate changes, such as the Younger Dryas, may result from rapid changes in sea surface conditions leading to a decreasing heat transfer to the high latitudes of the northern hemisphere.

The oceanic circulation, often viewed as a "conveyor belt" (Broecker, 1987 ; Broecker and Denton, 1989) is only one of many elements which together constitute the Earth's climate system. It stands out because of its dramatic impact on the climate of the North Atlantic and western Europe, but we still have to understand the complete sequence of feedbacks which may lead to the abrupt climate cooling and warming depicted by the paleoclimatic record.

REFERENCES

Bloom AL, WS Broecker, JMA Chappell, RK Matthews, and KJ Mesolella (1974) Quaternary sea level fluctuations on a tectonic coast : New $^{230}Th/^{234}U$ dates from the Huon Peninsula, , Quat.Res. 4: 185-205.

Bond G, H Heinrich, WS Broecker, L Labeyrie, J McManus, J Andrews, S Huon, R Jantschik, C Clasen, C Simet, K Tedesco, M Klas, and G Bonani (1992) Evidence for massive discharges of icebergs into the glacial North Atlantic, Nature 360: 245-249.

Boyle EA (1981) Cadmium, zinc, copper and barium in foraminifera tests. Earth. Planet. Sci. Lett.,: 11-35.

Boyle EA (1992) Cadmium and $\delta^{13}C$ paleochemical ocean distribution during the stage 2 glacial maximum, An. Rev. Earth Planet. Sci. 20: 245-287.

Boyle EA and LD Keigwin (1982) Deep circulation of the North Atlantic over the last 200,000 years: Geochemical evidence, Science 218: 784–787.

Boyle EA and LD Keigwin (1985) Comparison of Atlantic and Pacific paleochemical records for the last 215,000 years: Changes in deep ocean circulation and chemical inventories, Earth Planet. Sci. Lett. 76: 135–150.

Boyle EA and LD Keigwin (1987) North Atlantic thermohaline circulation during the last 20,000 years: Link to high latitude surface temperature, Nature 330: 35–40.

Broecker WS (1982) Glacial to interglacial changes in ocean chemistry, Prog. Oceanogr. 11: 151–197.

Broecker WS (1982) The biggest chill, Natural History Magazine, 74-82, October issue.

Broecker WS (1989) The salinity contrast between the Atlantic and Pacific Oceans during glacial times, Paleoceanography 4: 207-212.

Broecker WS and GH Denton (1989) The role of ocean-atmosphere reorganizations in glacial cycles, Geochim. Cosmochim. Acta 53: 2465-2501.

CLIMAP Project Members (1981) Seasonal reconstructions of the earth's surface at the last glacial maximum, Geol. Soc. Am. Map and Chart Series, MC-36.

Cox RA, MJ McCartney, and F Culkin (1970) The specific gravity /salinity/ temperature relationship in natural sea water, Deep-Sea Res. 17: 679-689.

Craig H and Gordon LI (1965) Deuterium and oxygen 18 variations in the ocean and the marine atmosphere, in: *Stable isotopes in oceanographic studies and paleotemperatures,* Ed. E Tongiorgi, 9-130, CNR, Pisa.

Dansgaard W (1964) Stable isotopes in precipitation, Tellus, 16: 436-468.

Duplessy JC (1978) Isotope Studies. In:*Climatic Change,* Ed. by J Gribbin, Cambridge University Press, 46-67.

Duplessy JC (1982) Glacial to interglacial contrast in the Northern Indian Ocean. Nature 295: 494-498.

Duplessy JC, and NJ Shackleton (1985) Response of global deep-water circulation to the Earth's climatic change 135,000–107,000 years ago, Nature 316: 500–507,.

Duplessy JC, NJ Shackleton, RK Matthews, W Prell, WF Ruddiman, M Caralp, and CH Hendy (1984) ^{13}C record of benthic foraminifera in the last interglacial ocean: Implications for the carbon cycle and the global deep water circulation, Quat. Res. 21*:*225–243.

Duplessy JC, LD Labeyrie, and PL Blanc (1988a) Norwegian Sea deep water variations over the last climatic cycle: Paleooceano-graphical implications, In: *Lectures Notes in Earth Sciences,* 16: 83–116, edited by H Wanner and U Siegenthaler, Springer Verlag , New York.

Duplessy JC, Shackleton NJ, Fairbanks RG, Labeyrie L, Oppo D and Kallel N (1988b) Deep water source variations during the last climatic cyle and their impact on the global deep water circulation, Palaeoceanography 3: 343-360.

Duplessy JC, LD Labeyrie, A Juillet-Leclerc, F Maitre, J Duprat, and M Sarnthein (1991) Surface salinity reconstruction of the north Atlantic ocean during the last glacial maximum, Oceanologica Acta 14: 311-324.

Duplessy JC, LD Labeyrie, M Arnold, M Paterne, J Duprat and TCE Van Weering , North Atlantic Sea Surface Salinity and Abrupt Climatic Changes, Nature 358: 485-488, 1992.

Emiliani C (1955) Pleistocene temperatures, J . Geol. 63: 538-578.

Epstein S, Buchsbaum R, Lowenstam HA and Urey HC, 1953, Revised carbonate-water isotopic temperature scale, Geol. Soc. Am. Bull. 64: 1315-1325.

Fairbanks RG (1989) A 17,000-year glacio-eustatic sea level record : influence of glacial melting rates on the Younger Dryas event and deep-ocean circulation, Nature 342: 637-642.

Grousset F, L Labeyrie, J Sinko, G Bond, J Duprat, M Cremer, and S Huon (1993) Patterns of ice rafted detritus in the Glacial North Atlantic (40-55°N), Paleoceanography 8: 175-192.

Guiot J (1993) Statistical analyses of biospherical variability, in *Long Term Climatic Variations: Data and modelling,* JC Duplessy and M Spyridakis Eds, Springer Verlag, this volume.

Heinrich H (1988) Origin and consequences of cyclic ice rafting in the Northeast Atlantic Ocean during the past 130,000 years, Quat. Res. 29: 142-152.

Imbrie J and NG Kipp (1971) A new micropaleontological method for quantitative paleoclimatology: Application to a late pleistocene Caribbean core. In: *The Late Cenozoic Glacial Ages,* KK Turekian ed., 71-181, Yale University Press.

Kallel N, LD Labeyrie, A Juillet-Leclerc, and JC Dupless (1988) A deep hydrological front between intermediate and deep-water masses in the glacial Indian Ocean, Nature 333: 651-655.

Kellogg TB (1976) Late Quaternary climatic changes : Evidence from deep sea cores from Norwegian and greenland Seas, In: *Investigation of Late Quaternary paleoceanography and paleoclimatology,* RM Cline and JD Hays Eds, Geol. Soc. Am. Mem. 145: 77-110.

Kroopnick P (1985) The distribution of ^{13}C of ΣCO_2 in the world oceans, Deep Sea Res. 32: 57–84.

Labeyrie LD, JC Duplessy, and PL Blanc (1987) Variations in mode of formation and temperature of oceanic deep waters over the past 125,000 years, Nature 327: 477–482.

Labeyrie D, JC Duplessy, J Duprat, A Juillet-Leclerc, J Moyes, E Michel, N Kallel and NJ Shackleton (1992) Changes in the vertical structure of the North Atlantic Ocean between glacial and modern times, Quat. Sci. Rev. 11: 401-413.

Maier-Reimer E, and U Mikolajewicz (1989) Experiments with an OGCM on the cause of the Younger Dryas, In: *Oceanography,* Ed. by A Ayala-Castanares, W Wooster and A Yanez-Arancibia, UNAM Press, 87-100.

Oppo DW, and RG Fairbanks (1987) Variability in the deep and intermediate water circulation of the Atlantic during the past 25,000 years: Northern hemisphere modulation of the Southern Ocean, Earth Planet. Sci. Lett. 86: 1–15,.

Prell WL (1985) The stability of low-latitude sea surface temperature: An evaluation of the CLIMAP reconstruction with emphasis on the positive SST anomalies, Rep. TR 025, DOE/ER/60167-1, US Dept. of Energy, Washington DC.

Ruddiman WF and A McIntyre (1973) Time-transgressive deglacial retreat of polar waters from the North Atlantic, Quat. Res. 3: 117-130.

Shackleton NJ (1977) Carbon-13 in Uvigerina: Tropical rainforest history and the equatorial Pacific carbonate dissolution cycle, In: *The Fate of Fossil Fuel CO_2 in the Oceans,* Ed. by N R Anderson and A Malahoff, pp. 401–428, Plenum, New York.

Shackleton NJ (1974) Attainment of isotopic equilibrium between ocean water and the benthonic foraminifera genus *Uvigerina* : Isotopic changes in the ocean during the last glacial, Colloque CNRS N° 219, 203-210, CNRS, Paris.

Shackleton NJ (1987) Oxygen isotopes, ice volume and sea-level, Quat. Sci. Rev. 6: 183-190.

Shackleton NJ and ND Opdyke (1973) Oxygen isotope and paleomagnetic stratigraphy of of equatorial Pacific core V28-238: oxygen isotope temperatures and ice volumes on a 10^5 and 10^6 year scale, Quat. Res. 3: 39-55.

Stocker T F and DG Wright (1991) Rapid transitions of the ocean's deep circulation induced by changes in surface water fluxes Nature 351: 729-732.

Streeter SS and NJ Shackleton (1979) Paleocirculation of the deep North Atlantic: 150,000 year record of benthic foraminifera and oxygen-18, Science 203: 168–171.

STATISTICAL ANALYSES OF BIOSPHERICAL VARIABILITY

J. GUIOT
UA CNRS 1152
Fac. de St-Jérôme
F-13397 Marseille cedex 13
France

1. INTRODUCTION

Climate and vegetation are interdependent. Climate types have been often used to define vegetation types and to draw maps of potential vegetation (Walter and Box, 1976; Mathews, 1983). Certain vegetation features have also often been used to recognize climate types: for example, the olive tree defines the northern limit for the mediterranean climate (Emberger, 1955).

On world maps, the boundaries of the natural vegetation zones, soil types, and climatic regions roughly coincide (Emanuel *et al.*, 1985). As plant productivity is also related to climate (Lieth, 1972) and climate is an important soil-forming factor (Post *et al.*, 1982), almost all systems for classifying climate incorporate relationships between vegetation distribution and climate (Holdridge, 1947; Matthews, 1983).

Plant distributions respond to changes in summer warmth, winter cold and moisture balance (Woodward, 1987; Prentice *et al.*, 1992). Climatic changes during the Late Quaternary have produced continual changes in the distribution of vegetation types. The patterns of these changes can be reconstructed from continental-scale syntheses of pollen data (*e.g.* Huntley, 1988, 1990a; Prentice *et al.*, 1991).

In long-settled regions the vegetation patterns have been modified by humans, and changes attributed to human impact are widely registered in European pollen diagrams (Behre, 1988). This does not prevent the use of pollen data to reconstruct past climate changes at a continental scale. The importance of climate at this scale is indicated by the fact that the European distributions of major taxa are as spatially coherent as they are in other continents. Furthermore many European species' distribution limits have been related to specific climatic controls on physiological processes (*e.g.* Piggott and Huntley, 1981; Woodward, 1988).

NATO ASI Series, Vol. I 22
Long-Term Climatic Variations
Edited by J.-C. Duplessy and M.-T. Spyridakis

The abundance pattern of some European taxa can even be predicted from the climatic relationships of their North American equivalents. So Huntley *et al.*, (1989) have established the relationship between the *Fagus* pollen distribution and climate independently on both sides of the Atlantic Ocean. The reconstruction of these distributions using the relationship calibrated over the sea agrees with the observations (Figure 1). This would not be possible unless climate were the major control in both continents. The spatial coherence of past changes in vegetation patterns, reconstructed from pollen data, also strongly argues for the primacy of climatic control in Europe (Huntley, 1988 ; 1990 a,b) as has been demonstrated for eastern North America (Prentice *et al.*, 1991).

Two approaches are possible to test the relationships between climate and biosphere. The forward approach consists to develop models of vegetation, to simulate the distribution of the vegetation according to a certain state of the climate, and to compare the simulated vegetation with the paleodata. This approach is particularly useful to understand the complex effect of climatic changes on the vegetation. The second approach is the inverse approach, including the well known transfer function. It consists to calculate non-causal statistical equations between climate and modern proxy data and to reconstruct past climate by applying these equations to paleodata. This approach has largely improved our knowledge of the past.

2. THE INVERSE APPROACH

Past variations of the climate are recorded by biological microorganisms trapped in the sediments. These archives of the past are also called "proxies" that it is necessary to decode properly. Apart some isotopic or chemical parameters measured on them, their composition is in itself a record of the climate. It is the reason why paleobiologists identify the remains of these organisms. They count either the proportion of each taxon present either the presence/absence of specific taxa. Each taxon can live in a particular range of particular climatic factors according to Figure 2.

The problem for the paleobiologist is to know very well the response of plants or animals to climate and other biotic factors. The tolerance of some taxa can be large and then the climatic information that we can extract from it is restricted. It can be refined by considering other taxa with a different tolerance, as shown in Figure 3. We have then to consider appropriate assemblage of taxa which are able to restrict the climatic range and then to refine the climatic information contained in it.

To deal with the information contained in these assemblages it has been necessary to use appropriate statistical tools called "transfer functions".

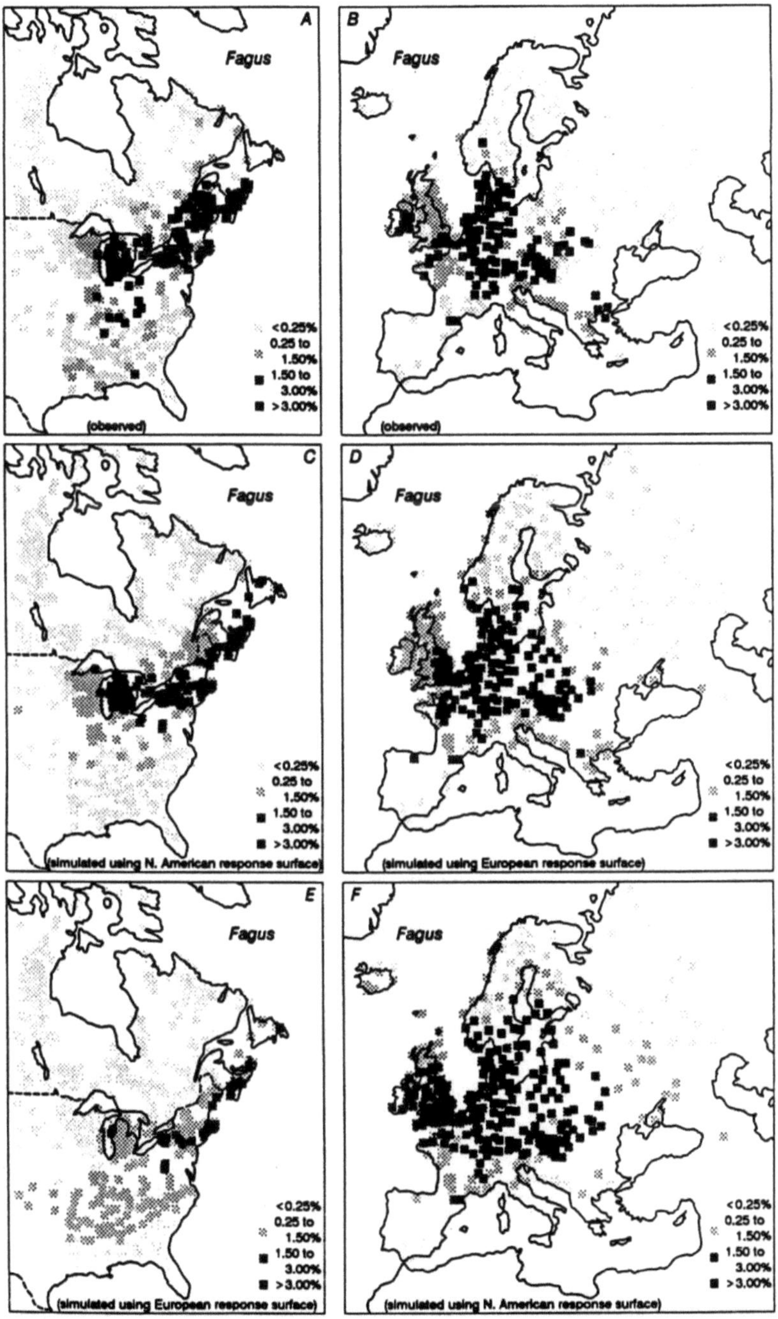

FIGURE 1. Pollen-abundance values. (A) Observed values for North American surface samples; (B) interpolated values fro European meteorological stations; (C) values simulated for NA using response surface; (D) values simulated for Europe using the European response surface; (E) values simulated for NA using the European response surface; (F) values simlulated for Europe using the NA response surface.

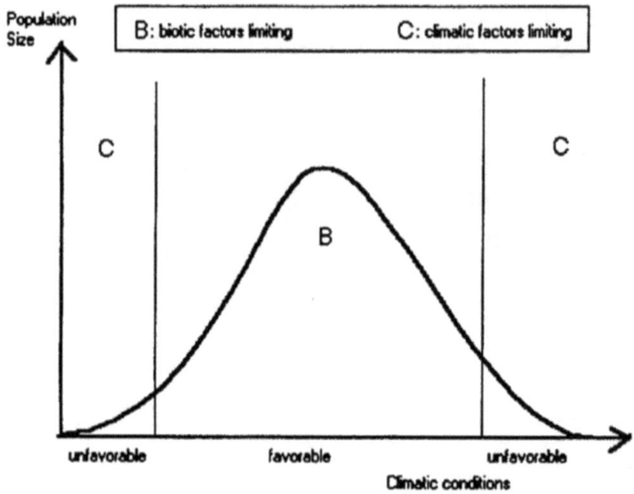

FIGURE 2. The relative numbers of some taxa are shown schematically against a background of climatic conditions. For extreme climatic environments, population density is low and climate can be considered as a limiting factor. Where climatic factors are favourable, other factors, such as soil, competition ..., are more important in determining population size.

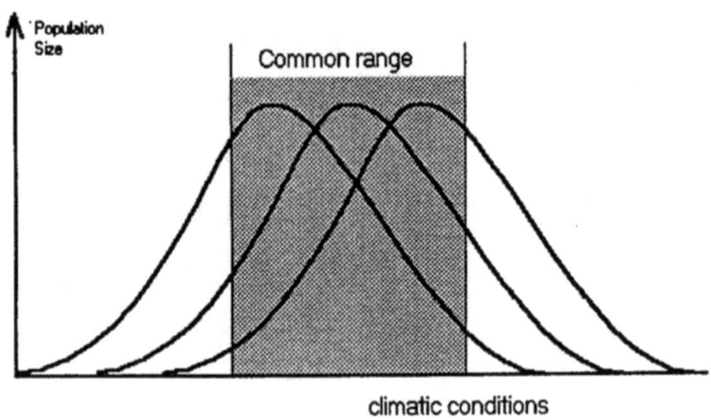

Figure 3. Optimum climatic range of several taxa.

2.1. Biological proxy data

2.1.1. Pollen data

The vegetation history is widespreadly known from pollen data. The pollen grains are dispersed each year and accumulated in sediments. The durable wall of the pollen makes that it is well preserved during hundreds of thousands and even millions of years. Its particular form makes that the palynologist is able to recognize the plant sometimes to the species, more frequently to the genus. The technique has been introduced in 1916 by the swedish botanist von Post. Sediments samples are chemically prepared to destroy all what is not pollen and mounted on thin plates. Using microscopes, the palynologist counts all the taxa identified on the plate, stopping usually after 300-500 grains in temperate vegetation by after 1000-3000 grains in tropical vegetations. The percentages of each taxon counted provides what is called a *pollen spectrum*. The succession of pollen spectra through a sediment core provides what is called a *pollen diagram*. Such a pollen diagram shows the variations in the composition of the vegetation in function of the time. By qualitative comparison with modem pollen spectra, it is possible to deduce roughly climatic changes. The use of transfer functions is now generalized. Figure 4 gives an example of partial pollen diagram at la Grande Pile in France accompanyed with the corresponding temperature and precipitation reconstruction (Guiot *et al.*, 1992).

An important problem for such data is that the pollen is sometimes dispersed far away from the source (it is particularly true in open vegetation). Macro remains (needles, fruits ...) found in the sediment then help to determine if the plant was really local. Another problem comes from the fact the the palynologist does not always recognize the species. A same taxon can recover very different species with very different ecological behaviour. The study of assemblages instead of isolated taxa may help to precise indirectly the species.

All the European and North-American continents are extensively covered by pollen diagrams, providing a mine of information for the past. These data are now collected in two pollen data bases: the European Pollen Database (EPD) in Arles (France) and the North American Pollen Database (NAPD) in Springfield (Illinois, USA). Other continents are also covered by such data, especially Africa and South America, but the density is still to be improved.

2.1.2. Beetles

Animal migrations are very rapid when compared to the plant migrations. It is especially true for the beetles populations - the largest order of the insects (more than 200 000 known species). Because they can leave an area rapidly when the

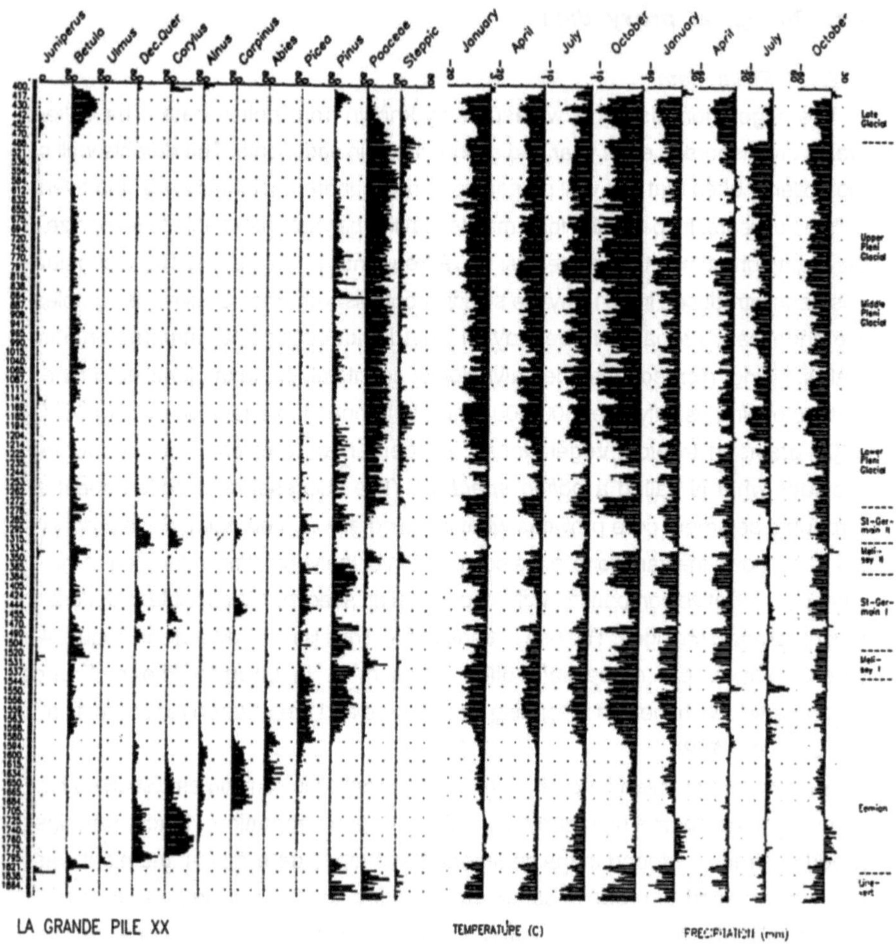

Figure 4. Variations in % of some typical taxa of the pollen diagram of la Grande Pile and corresponding reconstruction of monthly temperature and precipitation anomalies (deviations from modern climate) for 4 months representative of the 4 seasons. Pollen frequencies are given in %, temperatures in degrees C and precipitations in mm. The vertical axis gives the depths in cm. After Guiot et al., 1992.

climatic conditions no longer suit them, beetles do not necessarily have to perish or adapt. Of course, climate is not the only factor controlling the habitat of a certain species of beetle: competition, predators, available vegetation ... are others. An adequate selection of species and the study of assemblages are able to reduce the complexity of these interactions.

The British entomologist G.R. Coope, is a pioneer in this domain. He established a method to identify the beetle species from their remains in the sediment. He proved that the ecology of these beetles was the same in Late Quaternary than in the Present so that they are good climatic indicators (Coope, 1977; 1986).

They established also a very well diversified modern data base for about 200 different species and he determined their climatic tolerance, especially related to the temperature of the warmest and coldest month. A particular transfer function based on the presence/absence of the species in relation with the climate has enabled him to reconstruct the temperature variations of the British Isles for the last 50,000 years (Atkinson et al., 1986; 1987). Figure 5 shows very well the rapidity of response of beetles to climatic changes at about 13,000 yr BP. as the assemblages modified abruptly from glacial to temperate composition.

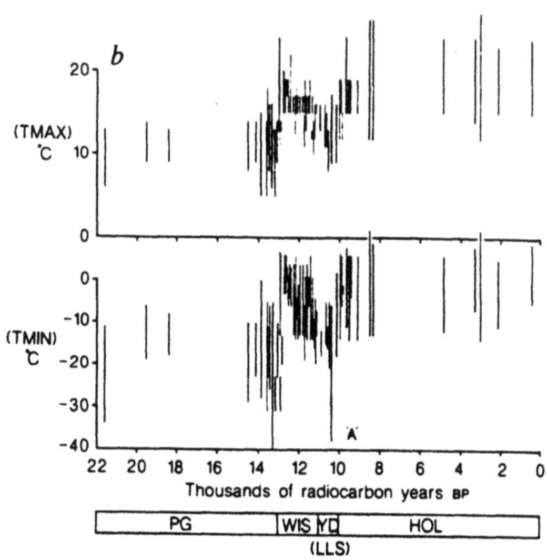

Figure 5. Reconstruction of January and July temperatures from beetles in the British Isles using mutual climatic range method (Atkinson et al., 1987).

2.1.3. Molluscs

Mollusc remains are frequently well preserved in loess sequences so that specific identifications are possible (Lozek, 1964; Puisségur, 1976). Sparks (1961) divided molluscs into four groups on the basis of their distribution and from the development of faunal assemblages through time made climatic conclusions. The molluscs can be identified to the species level and they do not suffer from distording effects like transport and reworking. That is two advantages over pollen data. Another advantage is a rapid response to climate change. But they are not found in lacustrine or peat bogs cores, so that these proxy data are not so widespread than pollen sequences. Loess soil being less frequent in warm period, the information extracted from these molluscs are often limited to cold period (Rousseau, 1987). Contrarily to beetles assemblages, numerical methods can be applied to abundances and not only presence/absence. Rousseau compiled modern data from several authors (including himself) to establish the basis of a transfer function (Rousseau, 1991) applied to the Achenheim sequence in Alsace, France (Figure 6).

2.1.4. Tree-rings

Tree-growth provides a variety of measurements which can be used as proxy climate records. As records which may be assigned accurately to a particular year and place, tree-growth occupies a special position in the catalogue of proxy climate information alongside layered lake sediments, ice cores and historical data. Moreover trees are present in a great number of places on the Earth. The key is to be able to find annual rings marked by succession of light spring wood and dense summer wood. Tropical trees provide rarely clear annual chronologies because of seasons not enough marked. The pioneer of dendroclimatology is A.E. Douglass (1914) and the pioneer of modern dendroclimatology is H.C. Fritts (Fritts *et al.*, 1971; Fritts, 1976).

Tree-rings have been analysed at date by statistical techniques based often on more or less complicated conceptual models (Fritts, 1976). The traditionnal treeTring analysis involves two steps: (1) the tree-growth modelling (including the detrending and the response function), (2) the calibration of the tree-ring data into climate (also called transfer function).

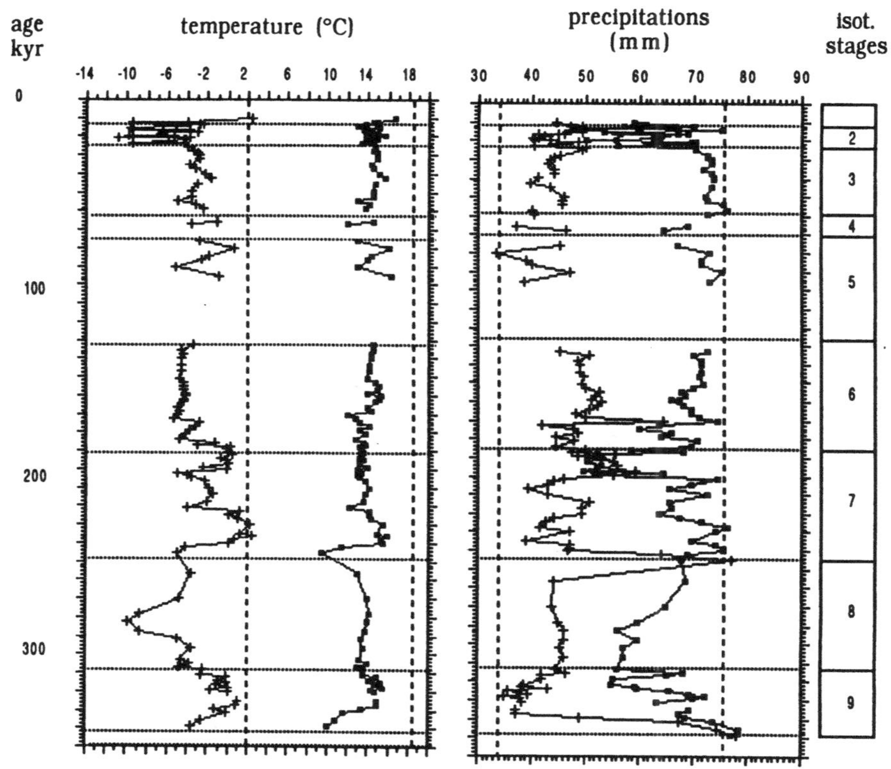

Figure 6. Reconstructions of the mean February (crosses) and August (squares) temperature and precipitation in the Achenheim sequence. Dotted vertical lines indicate the present conditions. After Rousseau (1991).

We can use a conceptual linear aggregate model to represent the the growth (after Cook and Kairiukstis, 1990):

$$\mathbf{R_t = A_t + C_t + \delta_t\, D1_t + \delta\, D2_t + E_t}$$

$\mathbf{R_t}$: observed ring-width series

$\mathbf{A_t}$: age-size-related trend in ring width - geometrical trend (negative exponential) often modified by competition

$\mathbf{C_t}$: climatically related environmental signal - precipitation, temperature, evapotranspiration

δ_t : binary indicator (0 or 1)

$\mathbf{D1_t}$: pulse caused by a local endogenous disturbance - forest management

$\mathbf{D2_t}$: pulse caused by a standwide exogenous disturbance - insect, fire, desease, pollution ..

$\mathbf{E_t}$: the largely unexplained year-to-year variability not related to the other signals - microclimate, soil, errors ....

The trend to estimate (G_t) is usually a function of A_t , $\delta D1_t$ and $\delta D2_t$. Nevertheless, C_t is also affected by a trend (ex. global warming) and a part of $\delta D2_t$ may be the signal of interest (ex. pollution)

$\mathbf{G_t}$ is often estimated as a linear trend, a negative exponential (more or less sophisticated), a polynomial, a low-pass filter, ARMA processes.

The estimation of the common signal intends to reduce E_t and the part of $D1_t$ badly removed: this is done by computing a mean chronology of several trees/cores.

The response of the growth to the climate is estimated by **response functions**, i.e. multiple linear regression between the tree-ring indices mean series and the climatic variables assumed to have a potential influence on the growth.

Adequate modelling enables to study pollution effects and forest decline (Becker, 1987; Tessier *et al.*, 1991). Figure 7 presents several tree-growth series of larchs and firs from the valley of Maurienne in the french Alps. In the second part of the series (after 1940), the valley were much more polluted by fluor than before 1940. A statistical relationship has been calibrated on the "unpolluted" period, which has enabled to provide an estimated tree-growth in "quasi-natural" conditions. The estimates are usually higher during the second part than the actual growth for the firs but not for the larches. The latter loss their needles in the winter and are then unable to accumate the pollution, at the contrary of the firs. This method illustrates well the effect of the pollution on the tree-growth.

Studies by Parker and Henoch (1971) and Schweingrüber *et al.* (1978) have shown that the maximum density of several species of conifer growing on cold and wet sites generally contain information on summer temperature.

The efficiency of the density as compared to that of the ring width is well illustrated by the reconstructions of Schweingrüber (1988): the yearly distribution of ring density over western Europe matches fairly with the yearly distribution of summer temperature, and that without any calibration.

Several studies showed that stable isotopes measured in the wood (Hydrogen, Oxygen, Carbon) could provide another information source on the temperature and wetness. As shown by Becker et al. (1991) for Late Glacial pine trees in Germany, this technique is really promising, but the problem is not simple. Indeed, the isotopical composition (O18/O16, D/H) of the precipitation is dependent upon the temperature at the location of the precipitations. The tree which absorbs this water incorporates this composition, but it is modified at several levels: time decay induced by stored water, D and ^{18}O enrichment in the stomats. The place where are sampled the trees is then crucial to facilitate the interpretation.

2.2. The transfer function

For a long time Quaternary paleoecologists have used intuitive methods to reconstruct paleoclimates from biological data. The commonest approach has been to compare the present-day distributions of selected species with the corresponding distribution of climate variables thought to be critical in determining the distributions of each species. This approach, called *indicator-species*, is univariate because the species are analysed separately and with respect to one climatic variable at a time. This limitation can be the source of a number of problems, reviewed by Birks (1981). In particular, the species respond to combinations of climatic variables and are limited by different climatic factors in different part of their ranges. Different climatic factors are intercorrelated and these correlations are a function of the climatic regimes in which the species live. The relationship of a species to climate may change within its ecological range due to differentiation of physiological ecotypes. The climatic influence on the species is complex because of competitive effects of other species present.

Multivariate indicator-species approaches are preferable, as they avoid these problems to some extent by considering the simultaneous relationship of several species' distributions to more than one climatic variable. Such methods have also been used by paleoecologists in an intuitive, non-quantitative way for many decades and even in a quantitative way with a limited number of taxa and climatic variables (*e.g.* Iversen, 1944). But these methods still disregard much climatically relevant information: on other species and on the species' abundances.

Multivariate statistical methods developed during the last two decades provide the most powerful approach. They are still necessarily based on a few hypotheses, namely (Birks, 1981): (1) that climate is the ultimate cause of changes in the paleobiological data; (2) that the ecology of the species considered has not varied between the period analysed and the present time, and consequently that the relationship between the species and the climate was constant in time; (3) that the modern observations contain all the necessary information for interpreting the fossil data. One major consequence of these hypotheses is that the species' present distribution must be at least approximately in equilibrium with climate. Because the response time of the biota to climatic changes is generally uncertain, it is worthwile to attempt reconstructions using various kinds of data (*e.g.* pollen, insects, molluscs) with potentially different response times, and to compare the results.

The term *transfer funtion* was apparently used for the first time in this context by Webb and Bryson (1972). According to Sachs *et al.* (1977), the transfer function procedure is defined as a method which produces calibrated quantitative estimates of some parameter of a past environment, such as seasonal or monthly air or ocean surface temperature, from proxy data such as biological assemblages as biological assemblages.

Transfer function equations are derived by multivariate techniques from an adequate "calibration" sample of modern distributional data, and applied to older samples to estimate environmental parameters for past times. The transfer function model is thus defined on the basis of a *response funtion* $\mathbf{X} = \mathbf{R(C,D)}$, which relates the set of biological responses X to a set of climatic factors C and nonclimatic ones D. If climate is the major factor, it is possible to approximate the response function by $\mathbf{R_C(C)}$. The inverse of this function response enables one to calculate climatic parameters from known biological responses: $\mathbf{C} = \mathbf{T(X)}$. T is called the transfer function. The methods are presented in a more ecological context in Ter Braak and Prentice (1988) and Prentice (1980, 1986).

The transfer function needs to be calibrated on data. If the data have an annual resolution (tree-rings), it is possible to calibrate on time series. This is the most favourable situation. For low resolution data (pollen, molluscs ...), we need spatial data for calibration. This is the cause of some problems such as the effect of the history of the data and the lack of analogues.

In spite of the numerous works devoted to the *transfer function* problem and the many different suggestions that have been made to improve transfer function methods, the methods can be in fact classified into a very few categories.

The first category contains the extrapolation methods, usually based on regression. The method of Imbrie and Kipp (1971) for deep sea cores or the method used by Huntley and Prentice (1988) for pollen or that of Roux *et al.,* (1991) for diatoms are included in this category. These method can provide reconstructions completely outside the range of the calibration data, which is an advantage when they can be controlled but they can be really unrealistic in some cases.

The second category contains interpolation methods, of which many are based on the use of a similarity index between fossil and modern biological assemblages to determine the modern analogues of the fossil data. The recontruction is given by a weighted average of the climate values for the modern analogues. The techniques differ in the way they compute the similarities. They include the response surfaces method (Bartlein and Prentice, 1986; Webb et al., 1987), the mutual climatic range method (Atkinson et al., 1987) and the paleobioclimatic analogues method (Guiot et al., 1989). The mutual climatic range method is slightly different from the rest as the similarities between assemblages do not depend upon the abundances of the species but only on their distribution (presence/absence).

Because extrapolation methods can provide estimates beyond the limits of the calibration data, there is a great risk of obtaining unrealistic estimates. This problem does not arise with interpolation methods which are unable, in return, to guess what happens outside the reference domain and for this reason interpolation methods are strongly recommended over extrapolation methods.

In the deep sea cores domain, methods are also shifting from the standard method introduced by Imbrie and Kipp (1971) which, nevertheless, seems still appropriate for the planktonic foraminifera (for which it has been done) or diatoms (Pichon *et al.*, 1992) to interpolation methods more powerful for other organisms like dinoflagelates (De Vernal et al., submitted).

Some methods recontruct one climatic variable at a time (*e.g.* regression). These methods are unable to take into account the interaction of the climate components and *a fortiori* to isolate the best climatic variable to reconstruct from the others that influence the data. Such methods reproduce perfectly the correlations between the climatic variables in the reference data set (Elovicheva and Bogdel; 1987):

It is generally important (a) to collect diversified modern data where the various climate components are as little correlated as possible and (b) to use a method that can separate the effects of the different components, even in the presence of some intercorrelations among components and intractions among the effects of different components. Interpolation methods satisfy this requirement, being not based on correlations between variables but rather between samples. They have the advantage that they consider the climate (at least the variables that are available) as a multivariate whole.

Particular problems arise when the defects of the univariate approach are amplified by the problem of a restricted reference data set. These are the methods based on the distributions of species along a modern altitudinal gradient (Adam and West, 1983). The present-day gradients of temperature and precipitation are known so it is easy to obtain climatic estimates. But because precipitation and temperature are both assumed to be perfectly correlated with elevation, Adam and West (1983), for example, have reconstructed perfectly parallel precipitation and temperature curves, whereas, over the same pariod (140,000 years), Guiot *et al.* (1989) have shown that the two variables have a complex and distinct history.

2.3. Characteristics of the biological proxy data

Biological proxy data can be characterized:

(1) By the *type of variable* measured (*e.g.* species abundance, species distribution, ring-width, density, isotopic composition).

(2) By the *sample resolution*, *i.e.* the minimum time interval between two samples, from one year to several millennia.

(3) By the length of the record, from a fraction of the Holocene to a much longer period which is nevertheless limited by the maximum period over which the ecology of the species can be assumed not to change.

(4) By the whether the reference data are distributed in the time or the space domain. The first case is the most favourable, but practicable only for data with a yearly (or quasi-yearly) minimum sampling interval, so that a large range of climatic situations is available in the instrumental meteorological series. In the second case, the references diversity must be found in the space domain. Modern samples are collected together with climatic *normals* usually computed from instrumental data averaged over about 30 years.

(5) By the response time, *i.e.* the time necessary to be in equilibrium with the climate. The response of the biological system to environmental changes may vary from a few years (1 to 3 years for the tree-growth) to several centuries (vegetation and its pollen record).

Independently of the indirect human influence on climate, humans influence biological system by pollution, local forest clearances, extension of agriculture and so, disturbing the proxy records to a greater or lesser degree. The effect of these anthropogenic factors are function of the response time of the biological system.

Table 1 shows that the proxy data for terrestrial climates may be grouped in two major categories. The first concerns only tree-rings, which may handled using the most powerful statistical methods because of their annual resolution and because they cover the recent period. The instrumental meteorological series usually provides sufficient diversity for a reliable calibration. Time-series analyses are then applicable, allowing response time be estimated and the climatic signal to be separated from the anthropogenic (see several studies in Cook and Kairiukstis, 1990).

TABLE 1

DATA	VARIABLE	RESOLUTION	LENGTH	RESPONSE TIME
tree-rings	width	1 yr	$10\text{-}10^4$ yr	a few years
	density			
pollen	abundances	10-1000 yr	$10^3\text{-}10^5$	10-1000
insects	distribution	10-1000 yr	$10^3\text{-}10^5$	short
	abundance			
molluscs	abundances	10-1000 yr	$10^3\text{-}10^5$	short
micro-	abundances	10-1000 yr	$10^3\text{-}10^5$	short
mammals				
diatoms	abundances	1-100 yr	$10^2\text{-}10^4$	short

Structural characteristics of biological proxy series for reconstructing past climates

The second category of proxy data includes species assemblage data measured by species abundance and/or qualitative distribution. The data typically cover a much longer time period but with a poorer resolution. They potentially represent a very large range of climatic changes for which analogues can only be found by sampling across a very large space domain. Consequently it is more difficult to use time-series analyses, to infer response times and to separate human influence from the climatic forcing. The comparative study of various types of data with different response times can however give a partial solution to this problem.

2.4. The combination of several proxy data

It is more and more clear that the combination of several types of proxy data is very promising to deconvoluate the various climatic components found in the biological data or to solve the problem of lack of modern analogues to some past environments. This is true as well at the historical time scale than at the Late Quaternary time-scale. We want here illustrate this approach with several examples.

At the Late-Quaternary time-scale, a first approach has previously been adopted in order to improve the temperature estimates from pollen data during glaciall periods in La Grande Pile using organic carbon measurements and to correct precipitation estimates using mineral clay data (Seret *et al.*, 1992). But the method has been refined in the two next examples

2.4.1. Pollen and lake level data

Pollen data have often been used to reconstruct temperature and/or precipitation patterns (*e.g.* Guiot et al., 1989), but we are not always sure that the two signals are reconstructed independently. As shown by A. Street-Perott in this volume, lakes provide an alternative source of information about precipitation changes. Their area and depth respond to the balance between precipitation + runoff versus evaporation + discharge. These changes are recorded as shorelines or in various stratigraphic indicators in the lake sediments themselves.

A combined approach is explored in Guiot *et al.* (1993). Relationships between modern vegetation and climate were explored by multiple regressions between climate variables and the log-transformed frequencies of the 28 pollen types in 675 modern pollen spectra.

The temperature relationships of the vegetation are adequately represented by the mean January, mean July, and mean annual temperatures (r^2=0.62 to 0.66). The moisture relationships proved more complicated. Annual precipitation has a low r^2 (0.53). The difference between annual precipitation and actual evapotranspiration, P-E, which approximates runoff and is considered to be a good predictor of lake-level changes in humid regions (Harrison *et al.*, 1993), has a still lower r^2 (0.42). The difference between annual precipitation and potential evapotranspiration, P-PET, which can be considered as a general climatic moisture index, however achieved a higher r^2 (0.66), similar to those obtained for the temperature variables. This analysis illustrates the fact that vegetation and lakes respond to different aspects of the moisture budget. The climatic moisture index P-PET, will be used to compare the moisture signal in pollen and lake-level data, as a compromise between these two aspects.

As shown in a previous section, we do not use multiple regression to reconstruct climate from pollen data, but rather the paleobioclimatic analogues method. For each fossil spectrum, the few closest modern spectra are determined and their value of the P-PET variable is compared to the lake-level status of the surrounding region. For that, an interpolated moisture index is calculated from the lake data. The modern pollen analogues are rejected if their P-PET are not consistent with the lake moisture index, i.e. if they are not either both greater, either both less or both close to the modern value at the same location (using an arbitrary threshold).

The lake-derived moisture index map for 6000 yr B.P. (Figure 7) shows clear similarities with the (mean) pollen-based reconstruction of P-PET anomalies. However, there is a region of disagreement extending from southern Sweden to the northern Balkans. Here, the mean pollen-based reconstruction shows conditions better than present while the lakes were lower than present. The pollen analogues in this region were poor (relative distances > 20%) and the reconstructed P-PET anomalies have a large error bar. This suggests that the lake-level data could provide a useful constraint.

The use of the lake-derived moisture index to constrain the selection of modern pollen analogues accentuated the large-scale regional contrasts in reconstructed P (Figure 8). The constraint generally increased the distances between fossil pollen spectra and their analogues by less than 15%. This result implies that almost equally-close analogues are associated with a wide range of P-PET values, leading to ambiguity in the reconstructions based on pollen data alone. The lake-level constraint reduced this ambiguity, producing a more spatially coherent map.

The reconstructed precipitation maps for 6000 yr B.P., with and without the lake-level constraint (Figure 8), closely resemble the corresponding P-PET maps (Figure 7). In contrast, reconstructions of annual mean temperature anomalies were hardly altered by the constraint (Figure 9). This reflects the fact that the modern analogues for a given pollen spectrum usually span only a narrow range of mean annual temperature, in contrast with the situation for annual precipitation and P-PET.

2.4.2. Beetles and pollen

The pollen sequence studied here is located at La Taphanel, french Massif Central, and cover the Late Glacial and the Holocene. Ponel and Coope (1990) have studied the beetles approximately from 16 to 8.5 Ka BP. The mutual climatic range method (see section 2.1.2) has produced estimates of the July temperature. From these reconstructions, we retain only the fact that the July temperature is > or < 12°C.

For each fossil pollen spectrum, the distance with modern spectra are sorted in ascending order. Only the few analogues with a July temperature in agreement with the conditions of Table 2 (> or < 12°C) are kept. The Older and Younger Dryas July temperature reconstructed from pollen alone disagrees completely from the beetles-based one. If we select the analogues according to beetles, we increase the distance of the best analogues by less than 2%.

For each fossil pollen spectrum, the distance with modern spectra are sorted in ascending order. Only the few analogues with a July temperature in agreement with the conditions of Table 2 (> or < 12°C) are kept. The Older and Younger Dryas July temperature reconstructed from pollen alone disagrees completely from the beetles-based one. If we select the analogues according to beetles, we increase the distance of the best analogues by less than 2%.

6000 BP : PRECIPITATION ANOMALIES (from pollen)

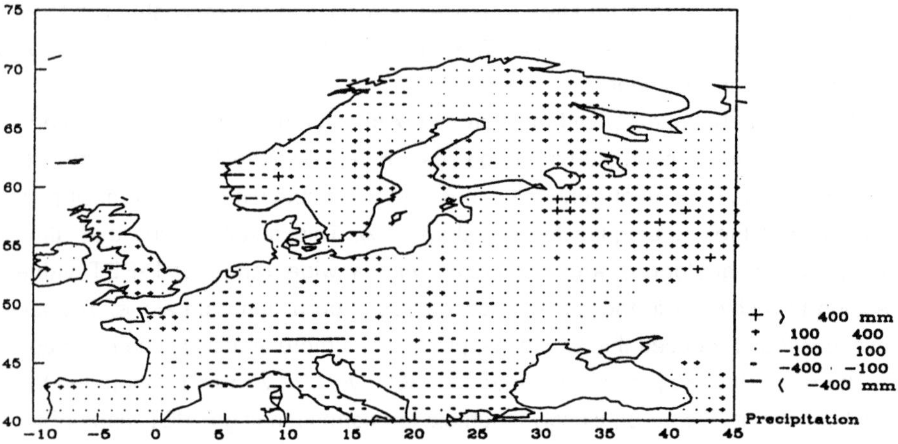

6000 BP : PRECIPITATION ANOMALIES (Pollen constrained by lakes)

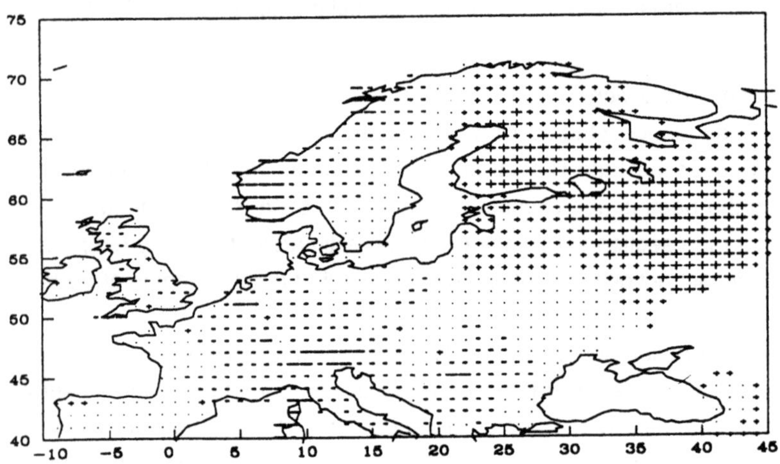

Figure 7. Comparison, for 6000 years BP., of interpolated moisture index derived from lake-level data and the P-PET variable derived from pollen.

6000 BP : MOISTURE INDEX ANOMALIES (from lakes)

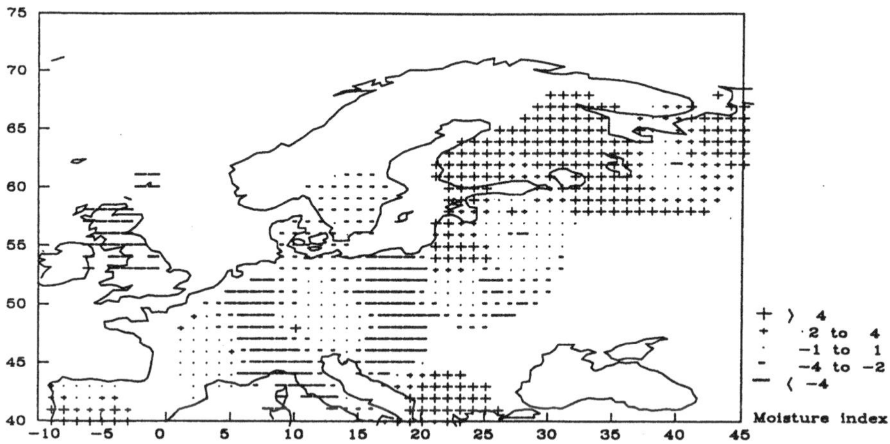

6000 BP : P-PET ANOMALIES (Pollen constrained by lakes)

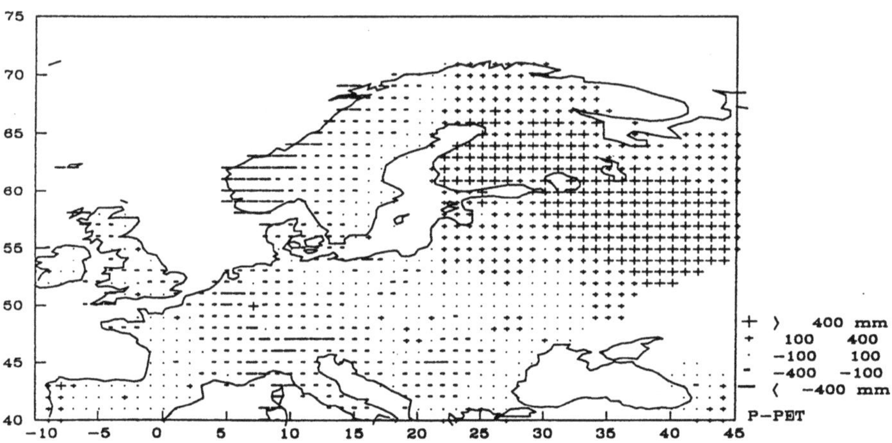

Figure 8. Precipitation patterns deduced from pollen alone and from pollen constrained by lakes.

6000 BP : TEMPERATURE ANOMALIES (from pollen)

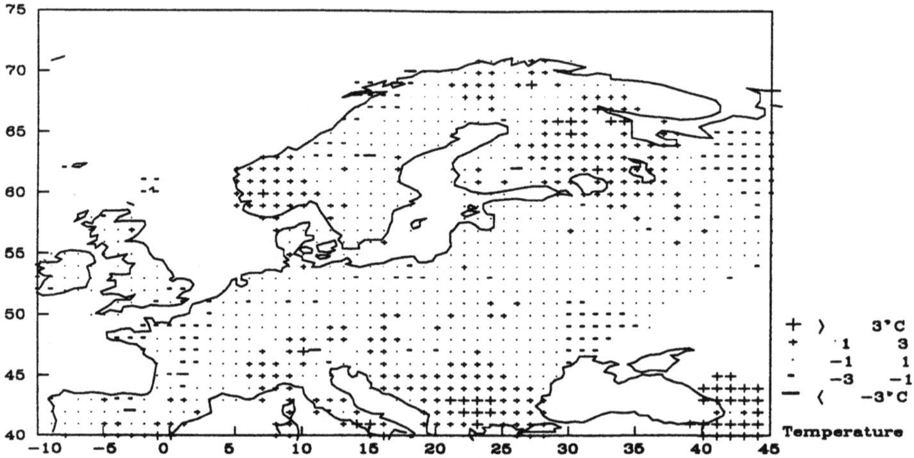

6000 BP : TEMPERATURE ANOMALIES (Pollen constrained by lakes)

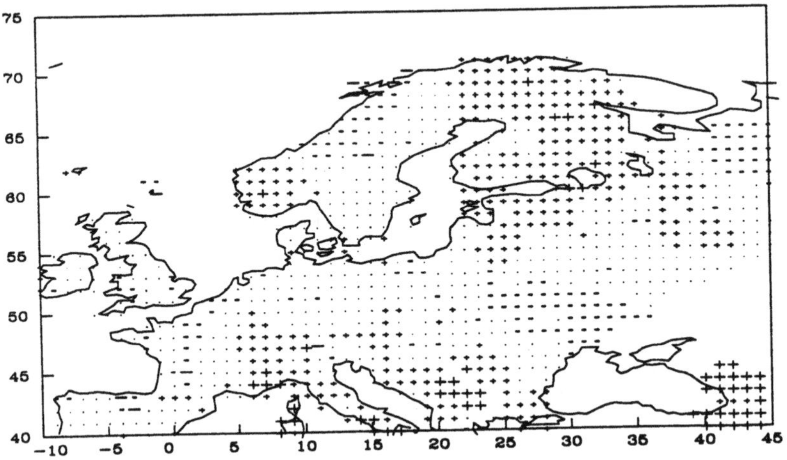

Figure 9. Temperature patterns deduced from pollen alone and from pollen constrained by lakes.

The method is controlled by comparing the January temperature reconstructed using the pollen analogues constrained by beetles with those obtained by beetles alone. The latter have a large range due to the great tolerance of the beetles to the winter conditions. They encompass then the temperatures obtained from pollen+beetles, which indicates an equivalent degree of continentality.

TABLE 2

Period/sample	Tjul(b)	Tjul(p)	Tjul(p+b)	Dist(p)	Dist(p+b)	Tjan(b)	Tjan(p+b)
Preboreal (20)	>12	14 to 15	14.4	1.3	1.3	-15 to 5	2.5
Younger Dryas (15)	<12	12 to 15	11	6.3	7.8	-24 to 0	-7.5
Alleröd (11)	>12	12 to 13	12.6	1.3	1.3	-15 to 2	-11
Older Dryas (8)	<12	12 to 15	13	1.9	2.1	-13 to 4	-8.8
Bolling (6b)	>12	12.5 to 19	20	5.4	5.4	-17 to -2	2.3
Oldest Dryas (3)	<12	11.7 to 13	10.7	1.9	1.9	-25 to 0	-6.7

Main steps of the climatic reconstruction from pollen constrained by beetles at la Taphanel. Six samples are selected to illustrate the method. Tjul(b) gives the beetles-based contraints (> or < 12°C); Tjul(p) gives the reconstructed July temperature from pollen alone and Tjul(p+b) from pollen constrained by beetles. Dist(p) gives the minimum distance from pollen alone (in % of a threshold calculated on 100 pairs of spectra randomly taken) and Dist(p+b) gives the minumum distance if we take into account the beetles constraints. Tjan(b) gives the climatic range of January temperature calculated from beetles and Tjan(p+b) gives the corresponding value obtained from the same analogues selected for Tjul(p+b).

2.4.3. Tree-rings and historical archives

Proxy data originating in historical documents are usually acute indicators of the past climate. They are archives, *i.e.* written records of environment clues very abundant from the 11th century (Alexandre, 1987) or phenonological records, *i.e.* data on the timing of recurrent biological phenomena, such as grapes harvest, wheat harvest, blossoming of plants, animal migration. These data have the disadvantage to be discontinuous in time and in space.

Biological proxy-data have the advantage, particularly in Europe and North-America, to be abundant and continuous. They are essentially tree-ring series, *i.e.* the annual growth of some arboreal species. Mostly ring widths are considered, but ring densities are often better correlated with the climate of more specific periods of the year. Dense networks of such data exist at least since the beginning of the 16th century (Bradley and Jones, 1992) and a good deal of long series are available for the last millennia. Consequently the strong interest of these data is a good space and time representability. Their weak point - less true for the density data (Schweingrüber *et al.*, 1979) - lies in the difficulty to separate the various components (*e.g.* temperature and precipitation) of the climatic signal (Serre-Bachet, 1989).

For regions where no historical or tree-ring data are available, ice cores isotopic data offer advantage of an annual time-resolution for the historical period (Dansgaard *et al.*, 1971).

All these records are generally qualitatively presented in the literature and sometimes converted into climatic scale. A technique enabling to take advantage of the complementarity of these data is presented here. More details can be found in Guiot (1992a, 1992b).

The aim is to estimate meteorological data from a various nuimber of proxy series. The annual series gridded by Jones *et al.* (1985) between latitudes 35°N and 55°N (by steps of 5°) and between longitudes 10°W and 20°E by steps of 10° provide 20 predictands series from 1851 to 1984 with missing data mainly before A.D.1900.

The set of the proxy series (predictors) used in this study is far from being exhaustive. A part of them is made up of 11 tree-ring chronologies of various species from various sites. Another group of proxy series is made up of data extracted from historical archives (7 series). A third category of data is provided by three ^{18}O series in the Arctic ice which are indicators of temperature. To these data are added the first three principal components of the 17 longest cedar ring series in Morocco (Till and Guiot, 1990). The period A.D.1068-1979 is retained as a period where a sufficient information is available.

The predictand and the predictor matrices contain missing data. The first step is to estimate them independently within the two sets.

The second step is the calibration of the proxy series on the annual temperature series on the basis of the common observations. The method used is a little complex because of the spectral difference of various series. This is the spectral canonical regression (Guiot, 1985). Bootstrapping enables us to provide realistic confidence intervals to the estimates (Guiot, 1989).

An indicator of the reliability of the reconstructions is obtained from the comparison with independent reconstructions. Briffa *et al.* (1988) has reconstructed the summer (April to September) temperature in Western Europe. In particular, it appears from both works that the year 1816 was one of the coldest year in the millennium, mainly on the Atlantic coast, France, and Corsica.

Figure 10 presents the reconstructed annual temperature in Europe in four points. There exists for cold periods a clear disconnection, (Serre-Bachet *et al.*, 1992), between Western Mediterranean coast (40°N, 0°W) and the others points

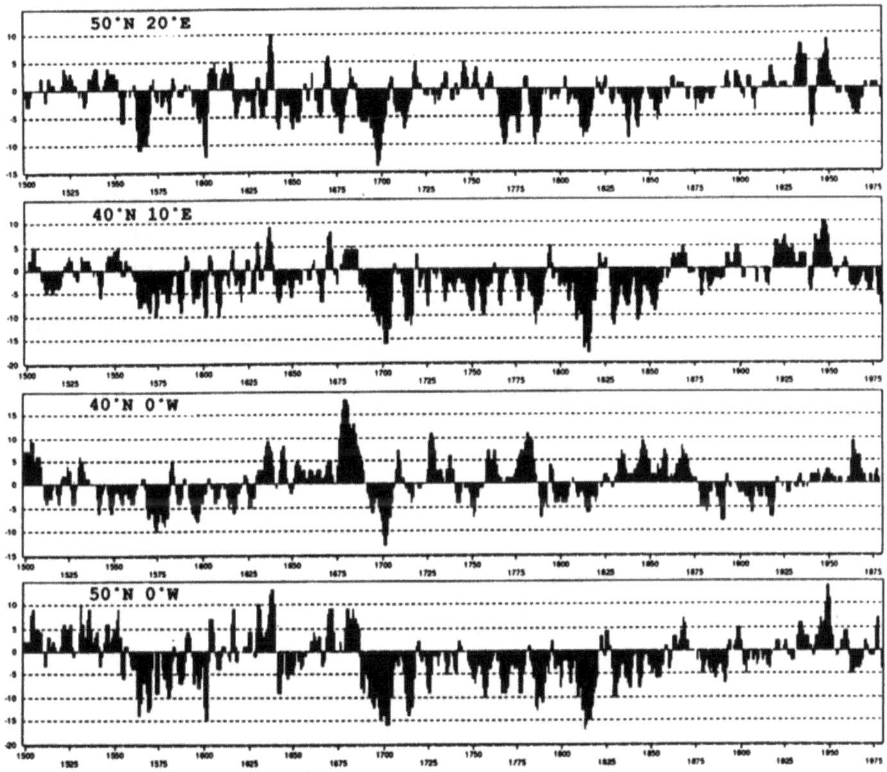

Figure 10. Annual temperature (in °C x 10) at four gridpoints of the Jones et al. (1985) network expressed as departures from the mean of the 1951-1970 period et reconstructed from tree-rings, historical archives data and ^{18}O isotopic data (Serre *et al.*, 1992).

including central Italy (40°N, 10°E). The central Mediterranean region is then more correlated with western Europe than with Morocco and Spain.

In the so-called "Little Ice Age", we remark two periods really colds: A.D.1550-1620 and A.D.1690-1860. From A.D.1860, we observe a clear warming trend until nowadays. The southwest has apparently not experienced these cold periods and consequently there is no significant warming during the 20th century. The precipitation reconstructions in Morocco (Till and Guiot, 1990) indicates that the previous centuries were dry, although extreme years, as 1816, knew an important moisture.

Even if a small part of the data used could be now improved, this work proves that the combination of tree-ring series and historical archives data provide reconstructions with a increased reliability. The precision obtained (about 0.3°C) - which can be improved by using more data and of better quality - is suffiently acute to analyse climatic changes as such occurred during the last millenium and particularly the recent human impact on the climate. Further improvements will be possible thank to the EURO-CLIMHIST database of Schwarz-Zanetti et al. (1992) coupled with the tree ring database of the NGDC (NOAA, Boulder).

3. THE FORWARD APPROACH

Several types of models exist for the biosphere. A first approach of the relationships between vegetation et climate is given by static "Global biome models" (Prentice *et al.*, 1992). These models are based on a more or less coarse grid on which biomes are deduced from constraints to plant growth (monthly temperature, precipitation, soil texture, water capacity of the soil, fertility, topography, solar radiation ...) (Prentice *et al.*, 1989). These models are unable to simulate the transient response of the vegetation to changing environment. Dynamic models are necessary for that, such as the forest succession model of Shugart (1984) "that simulates interactions among individual trees through their effects on and responses to changes in the physical environment of a *circa* 1000 m^2 "(Prentice *et al.*, 1989, p2). These models are powerful tools for regional studies, but too detailed for global studies. A recent version (Prentice *et al.*, in press) has been validated for typical forest ecosystems such as boreal and deciduous forests. Pollen data are proved also to be adequate for validating such models (Lotter and Kienast, 1992).

We will focus on the biome model which is being soon coupled to AGCM to test sensitivity of vegetation to climate changes. It has been also used to simulate the LGM vegetation and the carbon stored in it (Prentice et al., submitted).

The Biome model (Prentice et al., 1992) is based on the climatic tolerances (upper and/or lower bounds) of plant types. So for a given climate, it is possible to determine which plant is able to live. A dominance hierarchy is then applied to reduce the list to potential dominants. The biomes emerge through the interaction of constituent plant and are not *a priori* set. Biomes which are not anymore existing nowadays can so emerge from Quaternary climates with no modern analogues. The model of Box (1981) is of this type such as these of Woodward (1987) or Prentice *et al.* (1992), but while the climatic tolerances in the first type are determined statistically, in the second type, they are physiologically.

The model of Prentice was build on a small number of plant functional types for which the environmental constraints are well known and the selection of driving variables was restricted by the need to derive values of these variables from climatic data set available at sufficient resolution for the whole globe. They are described in Table 3. This table is essentially empty to avoid to take not well known constraints which could give a too modernist point of view.

TABLE 3

Plant functional type	Tc	GDD5	GDD0	Tw	E/PE	D
Trees						
tropical evergreen	>15.5				>0.80	1
tropical raingreen	>15.5				0.45-0.95	1
warm-temperate evergreen	>5				>0.65	2
temperate summergreen	-15/15.5	1200			>0.65	3
cool-temperate conifer	-19/5	900			>0.65	3
boreal conifer	-35/-2	350			>0.75	3
boreal summergreen	<5	350			>0.65	3
Non-trees						
sclerophyll/succulent	>5				>0.28	4
warm grass/shrub				>22	>0.18	5
cool grass/shrub		>500			>0.33	6
cold grass/shrub			>100		>0.33	6
hot desert shrub				>22		7
cold desert shrub			>100			8

Environmental constraints: Tc (mean temperature of the coldest month), GDD5 (growing degree-days on 0°C base), GDD0 (growing degree-days on 5°C base), Tw (mean temperature of the warmest month), E/PE (Priestley-Taylor coefficient, i.e. the ratio of actual evapotranspiration E to potential evapotranspiration PE), D (dominance class for each plant type in the model) (from Prentice et al., 1992).

The environmental used take into account: (1) the low-temperature tolerance of the plant (related to Tc), (2) the chilling requirements for rapid budburst in the following spring (related to GDD5), (3) period with temperatures sufficient for growth (related to GDD0 or GDD5), (4) moisture requirements (related to E/PE). The soil moisture algorithm is a "bucket model", i.e. the soil is treated as a single moisture store whose size depends on the soil texture. Evapotranspiration is computed on the basis of the site latitude, orbital parameters, soil water-storage capacity, monthly means or totals of the temperature, precipitation and sunshine (as percentages of the possible sunshine hours). Monthly data (from Leemans and Cramer, 1990) are interpolated to yield quasi-daily values. The grid size is 0.5°. The soil data have been digitized by Zobler (1986), also on a 0.5° grid, from the FAO soils map (FAO, 1974).

After applying the environmental sieve to determine which plant types can occur, the dominance class value D of these types are examined and only those in the highest class (lowest D) present are retained in the code indicating the biome type. The efficiency of the model has been tested using the vegetation data of Olson *et al.* (1985).

4. THE CONTINENTAL CARBON STORAGE PROBLEM : TWO APPROACHES

Estimates of the variations of exchangeable carbon stored on the continents in relation to glacial-interglacial oscillations should help to clarify the role of the terrestrial biosphere in the global cycle of carbon, since this role in modulating changes in atmospheric CO_2 concentrations (Barnola *et al.*, 1987) is still a matter of debate. Duplessy *et al.* (1988) have estimated that the mean $\delta^{13}C$ content of the global ocean during the last glacial maximum (LGM) was 0.032% lower than today, an increase of 0.03-0.04 % in the global ocean $\delta^{13}C$ being usually attributed to a post-LGM increase in storage of isotopically light terrestrial organic carbon (Shackleton, 1977).

Several estimates of glacial-interglacial variations in terrestrial carbon storage have been made (Prentice and Fung, 1990; Adams *et al.*, 1990; Friedlingstein et al., 1992; Prentice et al., submitted; Van Campo et al., submitted). Prentice and Fung (1990), using the GISS atmospheric general circulation model (AGCM) and a bioclimatic scheme, calculated the sizes of the carbon reservoir in biomass and soils for the LGM and for the current climate, and concluded that there was almost no change. Prentice *et al.* (submitted) used a biome model driven by modern climate data and an AGCM simulation to generate vegetation patterns for the LGM and for today. Their results suggest a 200-400 Pg C increase in biomass and soils after the LGM.

Only Adams *et al.* (1990) have attempted to estimate the change based on terrestrial geological evidence. They used available continental data to produce maps of vegetation for the two periods, and suggested a 1350 Pg C increase in the terrestrial reservoir after the LGM.

This data-based estimate is consistent neither with the $\delta^{13}C$ shift in the ocean, nor with the model-based estimates, nor with the new data-based estimate of Van Campo et al. (submitted). They compared the distribution of terrestrial ecosystems today and at the LGM, using CLIMAP (1981) reconstructions of LGM earth's surface. Although CLIMAP reconstructions do not use all of the information used to produce the more recent Adams *et al.* reconstruction, there are still large areas with few data, e.g. Siberia (Velichko et al., 1991) or central Amazonia (Van der Hammen, 1991), where any reconstruction of LGM vegetation patterns based on the available data must be speculative. In consequence, the large difference between Adams *et al.* estimate and the CLIMAP-based estimate of Van campo et al. (submitted) cannot be accounted for by significant differences in the data but must be attributed to a divergent interpretation of the data.

In this study, as in the other ones, an equilibrium state between the vegetation, the soil and the climate is assumed, owing to the fact that the last glaciation lasted several millenia. Such an equilibrium will not hold under conditions of rapid climatic change. Another limitation comes from unknown physiological effects of low glacial atmospheric CO_2 concentration on the vegetation (Prentice et al., submitted).

4.1. Using geological evidences (inverse approach)

Adams *et al.* (1990) used published palynological, pedological and sedimentological data to produce global maps of vegetation and ice cover at 18 kyr, which they compared with present natural vegetation. Vegetation was classified in 24 biomes on the basis of the categories of Olson et al. (1985). Man influenced types were disgarded. Each biome is attributed a value of carbon density for vegetation (Olson *et al.*, 1985) and for soil (Zinke *et al.*, 1984).

In Van campo et al. (submitted), the approach is based on CLIMAP (1981) which reconstructed the earth's surface characteristics at the LGM including continental ice caps, continent elevation, bathymetry, albedo and sea-surface temperature (SST), on a 2° x 2° grid. In order to compare the distribution of the ecosystems today and at the LGM, six predictors were interpolated to a 1°x1° grid: vegetation type, soil type, annual SST, SST amplitude, elevation, and distance to the sea. The mean and the standard deviation of each predictor were calculated for each ecosystem. Vegetation was subdivided according to its structure.

Olson's ecosystem classification was assimilated to the CLIMAP database by establishing structural correspondences between the CLIMAP vegetation codes and Olson's ecosystem types. Soils were subdivided according to their reflectivity. The soil classification were used only to separate tundra (dark) and steppe (light), and to distinguish bogs (black) storing carbon in peat under cool to cold climates. Macroscale relief was helpful in defining the montane vegetation types (tropical montane forest, high shrubhland), and to identify alpine tundra. The SST estimates were aslo taken into consideration, since SST has a major influence on the temperature and precipitation of adjacent continents. The SST at the closest marine point was attributed to each land gridpoint. Distance to the sea was used to characterize the continentality. Finally 26 ecosystems were retained to describe the present vegetation of the world. The area of these ecosystems for Present time and LGM are calculated. Using the usual densities, the total storage of carbon is deduced.

These reconstructions show a peak in the modern carbon stored in the vegetation in the equatorial latitudes in relation with the high proportion of the global biomass constituted by the equatorial forests. A lower but larger second peak is observed at middle-high latitudes, corresponding to temperate and boreal forests biomass. At the LGM, the first peak is much reduced, and the second disappears entirely, due to the contraction of the forested areas. In the soil there is a continuous increase of carbon from 20°N up to 60-65°N, where the bogs clearly dominate.

The greatest glacial-interglacial total carbon increase per degree of latitude is clearly registered around 60°N, while ice or tundra were replaced by boreal forests (Van Campo *et al.*, submitted).

Keeping in mind the limitations mentionned above, these results based on CLIMAP land and sea surface reconstructions show that the terrestrial biosphere contained substantially less carbon at the LGM than it does today (about - 715 Pg). The change in biomass and soil carbon is thus confirmed as a factor contributing to the difference in $\delta^{13}C$ in atmospheric CO_2 from ice age to now (Marino *et al.*, 1992; Leuenberger *et al.*, 1992). The magnitude of this change is broadly consistent with the estimated 0.032% change in $\delta^{13}C$ of the global ocean (Duplessy *et al.*, 1988).

4.2. Using models (forward approach)

Prentice and Fung (1990) used an atmospheric general circulation model (GISS model: Hansen et al. 1984) simulation at 18,000 yr BP. and Present to predict the biome distributions. This distribution was essentially based on the UNESCO classification scheme (Mathews, 1983), which makes the approach strongly dependent on the modern relationship between biomes and climate (man action excepted). Assigning modern mean carbon densities to each biome, they concluded that carbon lost to the terrestrial biosphere by inundation of the continental shelves approximately balanced carbon gained in other regions with former ice sheets.

Prentice *et al.* (submitted) developped this approach by using their biome model running on climate simulated by the MPI AGCM (Lautenschlager and Herterich, 1990), which enables biomes not existing today. The distribution of biomes at 18,000 yr BP. showed encouraging similarities with paleodata, so that this approach seems reasonable. As in the previous attempts, they assigned modern mean carbon densities to each biome and assumed these to have been the same at the LGM. They calculated a modern total carbon storage of 785 PgC in biomass and 1337 PgC in soils, which represents a total increase of 313 PgC (with a range of 215-400 Pg C). Taking into account the additional post-LGM increase of carbon in peatlands (essentially in the taiga biome), the model results suggest a range of 400-700 Pg C for the total increase, quite similar to the 400-500 Pg C increase estimate inferred from the ocean (Duplessy et al., 1988).

The difference between Prentice and Fung (1990) and Prentice et al. (1992) is partly due to the model used (GISS for the first and Hamburg model for the second) and partly due to the biome prediction schemes used.

4.3. Comparison of the results

In order to compare the different available estimates, the area anomalies (LGM minus Modern) expressed in $10^6 km^2$ have been calculated for each major ecosystem type (Figure 11).

Although the definitions of the ecosystem types can be slightly different among the authors, a broad comparison is possible. The results of Friedlingstein *et al.* (1992) have not been included in this comparison because their ecosystem classes are too broad and do not match the other ones.

In the two model simulations analysed (Prentice and Fung, 1990; Prentice et al., submitted), tropical forests are shown to be slightly more extensive than present, but the two data-based methods suggest they were about $8x10^6 km^2$ less extensive.

All four studies show major reductions of boreal and temperate forests. All four also show increases in xerophytic woodlands and open vegetation types, but to very different extents; the largest change being about $32x10^6 km^2$ as estimated by Van Campo et al. The changes in desert area are quite contradictory between Prentice and Fung, who produce an unrealistically small desert area at the LGM ($-12x10^6 km^2$), and in Adams *et al.* ($12x10^6 km^2$).

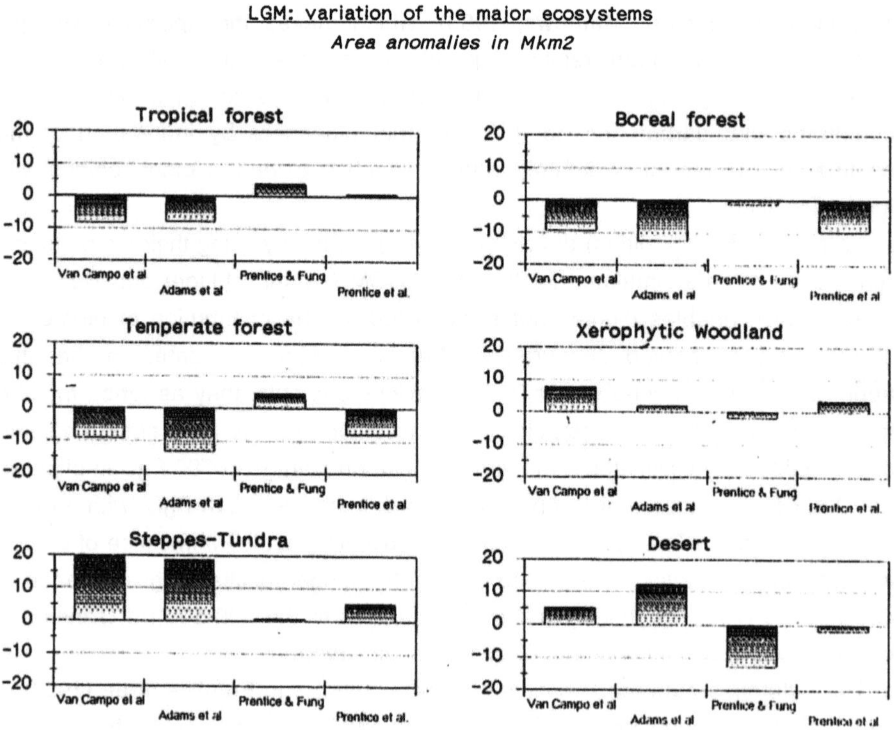

Figure 11. Area anomalies (in Mkm2) of 6 major ecosystems for the Last Glacial Maximum (LGM-Modern) reconstructed by 4 authors: Van Campo *et al.* (submitted), Adams *et al.*, 1990, Prentice and Fung (1990) and Prentice *et al.* (submitted); after Van Campo *et al.* (submitted).

The Van Campo et al. results agree with the model simulation of Prentice *et al.* on a relative stability in total desert area. Our estimates of changes in ecosystem areas are also most similar to those of Prentice *et al.* for four out of the six main groups of ecosystems. In terms of the global terrestrial carbon budget, the Van Campo et al. estimate corresponds to a 715 Pg C (34%) increase, about halfway between the no change estimate of Prentice and Fung, and the 1350 Pg (58%) estimate of Adams *et al.*.

Half of the discrepancy, i.e. 300 Pg C, between Adams *et al.* and Van Campo et al. may be found in the low carbon storage of their northern steppe-tundra, which is considered by Van Campo et al. partly as a carbon-rich tundra , according to the darkness of the associated soil in the LGM CLIMAP reconstruction. The two most recent estimates using AGCMs, in spite of the more realistic vegetation model of Prentice *et al.*, converge to about a 15% decrease (without considering peat bogs). Considering the peat bogs (1/3 taiga area), Prentice et al. estimates (15 to 35%) are very close to the estimates of Van Campo et al. (20% to 44%).

The model approach enables global studies. The estimates will be improved in the future by using more sophisticated vegetation and CO_2 exchange models. The data coverages will also improve thanks to the global database. These data have mainly a role of validating the results of models. Adequate statistical methods of extracting the information contained in these data are necessary.

REFERENCES

Adam, D.P., and West, G.J. (1983) Temperature and precipitation estimates through the last glacial cycle from Clear Lake, California, pollen data, Science, 219: 168-170.

Adams, J.M, Faure, H., Faure-Denard, L., McGlade, J.M., Woodward, F.I. (1990) Increases in terrestrial carbon storage from the Last Glacial Maximum to the Present. Nature 348: 711-714.

Alexandre, P. (1987) "Le climat en Europe au Moyen-Age". Ecole des Hautes Etudes en Sciences Sociales, Paris, 825p.

Atkinson, T.C., Briffa, K.R., Coope, G.R., Joachim, M.J., and Perry, D.W. (1986) Climatic calibration of coleopteran data. In BERGLUND, B.E. (ed): Handbook of Holocene Palaeoecology and Palaeohydrology, Wiley, Chichester.

Atkinson, T. C. Briffa, K. R. Coope, G. R. (1987) Seasonal temperatures in Britain during the past 22,000 years, reconstructed using beetle remains. Nature, 325: 587-592.

Barnola, D. Raynaud, Y.S. Korotkevitch, and C. Lorius (1987). Vostok ice core: a 160,000 year record of the atmospheric CO2. Nature 329: 408-414.

Bartlein, P.J., Prentice, I.C. (1986) Climatic response surfaces from pollen data for some eastern North American taxa. Journal of Biogeography, 13: 35-57.

Behre, K.-E. (1988) The role of man in European vegetation history. In "Vegetation History", B. Huntley and T. Webb III, (Eds.) Kluwer, Dordrecht, 633-672.

Becker, B., Kromer, B., and Trimborn, P. (1991) A stable-isotope tree-ring timescale of the Late Glacial/Holocene boundary. Nature, 353: 647-649.

Becker, M. (1987) Bilan de santé actuel et rétrospectif du sapin dans les Vosges. Etude écologique et dendrochronologique. Ann. Sci. Forestières, 43: 379-402.

Birks, H.J.B. (1981) The use of pollen analysis in the reconstruction of past climates: a review. in Wigley T. et al. (eds): Climate and History. Cambridge University Press, Cambridge, 111-138.

Bradley, R. and Jones, P.D., eds (1992) "The climate since 1500". Routledge, London, 679 p.

Briffa, K.R., Jones, P.D. and Schweingruber, F.H. (1988) Summer temperature patterns over Europe: a reconstruction from 1750 A.D. based on maximum latewood density indices of conifers. Quat. Res. 30: 36-52.

Box, E.O. (1981) Macroclimate and plant forms: an introduction to predictive modelling in phytogeography. Junk, The Hague.

CLIMAP Projects Members (1981). Geol. Soc. Am. Map Chart Ser. MC-36.

Cook, E.R. and Kairiukstis, L.A. (1990) Methods of Dendrochronology: Application in the Environmental Sciences, Kluwer Academic Press and IIASA, Dordrecht.

Coope, G.R. (1977) Fossil coleopteran assemblges as sensitive indicators of climatic changes during the Devensian (Last) cold stage. Proeedings of the Philosophical Transactions of the Royal Society of London, B 280: 313-340.

Coope, G.R. (1986) Coleoptera analysis. Handbook of Holocene Palaeecology and Palaeohydrology, Berglund (ed), Wiley, Chichester, 775-794.

Dansgaard, W., Johnsen, S.J., Clausen, H.B. and Langway, C.C. (1971) Climatic record revealed by the Camp Century ice core. In: Turekian K.K. (ed): The Late Cenozoïc Glacial Ages, New Haven, Yale University Press.

De Vernal, A., Guiot, J., and Turon, J.L., submitted. Late and Postglacial paleoenvironments of the Gulf of St.Lawrence: marine and terrestrial palynological evidence. Géographie Physique et Quaternaire.

Douglas, A.E. (1914) A method of estimating rainfall by the growth of trees. In "Climatic factor", L. Hunlington (ed), Carnegie Inst. of Washington Publ., 192: 101-122.

Duplessy, J.C. et al. (1988) Paleoceanography 3: 343-360.

Elovicheva, Ya., Bogdel, I. (1987) Reconstruction of paleoclimate and vegetation of the Byelorussian holocene using bog and lake deposit data, Paleohydrology of the temperate zone III, Mines and Lakes, Ac. Sc. Estonian SSR, Inst. Geology, 152-166.

Emanuel, W.R., Shugart, H.H., and Stevenson, M.P. (1985) Climatic change and the broad-scale distribution of terrestrial ecosystem complexes. Climatic Change, 7: 29-43.

Emberger, L. (1955) Une classification biogéographique des climats. Rev. Trav. Lab. Géo. Zoo., Fac. Sci. Montpellier, sér. bot. 7: 3-43.

FAO/UNESCO (1974) Soil map of the world 1:5,000,000. FAO, Paris.

Friedlingstein, C. Delire, J.F. Müller, J.C. Gérard (1992). The climate induced variation of the continental biosphere: a model simulation of the Last Glacial Maximum. Geophys. Res. Lett. 19: 897-900.

Fritts, H.C. Blasing, T.J., Hayden, B.P. and Kutzbach, J.E. (1971) Multivariate techniques for specifying tree-growth and climate relationships and for reconstructing anomalies in paleoclimate. J. Appl. Meteorology, 10: 845-864.

Fritts, H.C. (1976) Tree Rings and Climate. Academic Press, London, 567pp.

Guiot, J. (1985) The extrapolation of recent climatological series with spectral canonical regression. J. Climatology, 5: 325-335.

Guiot, J. (1989) Method of calibration and comparison of methods. In: Cook E.R. and Kairiukstis L.A. (eds) Methods of Dendrochronology, Kluwer Academic Pub. & IIASA, 165-178 and 185-193.

Guiot, J., Beaulieu, J.L. de, Pons, A., and Reille, M. (1989) A 140,000-year climatic reconstruction from two European pollen records. Nature, 338: 309-313.

Guiot, J., Reille, M., Beaulieu, J.L. de, and Pons, A. (1992) Calibration of the climatic signal in a new pollen sequence from La Grande Pile. Climate Dynamics, 6: 259-264.

Guiot, J. (1992a) The combination of historical documents and biological data in the reconstruction of climate variations in space and in time. Paläoklimaforschung / Paleoclimate Research, 7, Special Issue: ESF Project "European Paleoclimate and Man" 2, B. Frenzel, C. Pfister & B. Gläser (eds).

Guiot, J. (1992b) The climat of central Canada and Southwestern Europa reconstructed by contining various types of proxy data : a detailed analysis of the 1810-1820 period. The year without a summer ? world climate in 1816, C.R. Harington (ed), Canadian Museum of Nature, Ottawa, 291-308.Hansen J. et al., 1984. Geophys. Monogr. ser. (M. Ewing Symp. 5), 29: 130-163.

Guiot, J., Sandy P., Harrison, S.P. and Prentice, I.C. (1993) Reconstruction of Holocene precipitation patterns in Europe using pollen and lake-level data. Quat. Res. 40: 139-149.

Hansen, J., et al. (1984) Geophysical Monographies series (M. Ewing Symp. 5) 29: 130-163.

Harrison, S.P., Prentice, I.C., and Guiot, J. (1993) Climatic controls of Holocene lake-level changes in Europe. Climate Dynamics (in press)

Holdridge, L.R. (1947) Determination of World plant formations from simple climatic data. Science, 105: 367-368.

Huntley, B. (1988) Glacial and Holocene vegetation history: Europe. In "Vegetation History" (B. Huntley and T. Webb III, Eds.), pp. 341-384. Kluwer, Dordrecht.

Huntley, B. (1990a): European vegetation history: palaeovegetation maps from pollen data - 13 000 yr B.P. to present. J. Quat. Sci. 5: 103-122.

Huntley, B. (1990b) Dissimilarity mapping between fossil and contemporary pollen spectra in Europe for the past 13,000 years. Quat. Res. 33: 360-376.

Huntley, B., and Prentice, I.C. (1988): July temperatures in Europe from pollen data, 6000 years before present. Science 241: 687-690.

Huntley, B., Bartlein, P.J., and I.C. Prentice (1989) Climatic control of the distribution and abundance of beech in Europe and North America. J. Biogeography 16: 551-560.

Imbrie, J., and Kipp, N.G. (1971) A new micropaleontological method for quantitative paleoclimatology: application to a late Pleistocene Caribbean core. In K.K. Turekian (ed): The Late Cenozoïc Glacial Ages, Yale University Press, New Haven, 71-181.

Iversen, J. (1944) Viscum, Hedera and Ilex as climate indicators. A contribution to the study of the post-glacial temperature climate. Geologiska Föreningens i Stockholm Förhandlingar, 66: 463-483.

Jones, P.D., Raper, S.C.B., Santer, B., Cherry, B.S.G., Goodess, C., Kelly, P.M., Wigley, T.M.L., Bradley, R.S. and Diaz, H.F. (1985) A gridpoint surface air temperature data set for the northern hemisphere. U.S. Dept. of Energy, Washington, TRO22.

Lautenschlager, M. and Herterich, K. (1990) J. Geophys. Res., 95: 547.

Leemans, R., and Cramer, W. (1990) The IIASA database for mean monthly values of temperature, precipitation and cloudiness of a global terrestrial grid. WP-90-41, IIASA, Laxenburg.

Lieth, H. (1972) Modelling the primary productivity of the world. Nature and Resource, 8: 5-10.

Leuenberger, M., U. Siegenthaler, C. Langway, (1992) Carbon isotope composition of atmospheric CO2 during the last ice age from an Antarctic ice core. Nature 357, 488-490.

Lotter, A.F. and Kienast, F., 1992. Validation of a forest succession model by means of annually laminated sediments. Geol. Surv. Findland, Spec. pap.14: 25-31.

Lozek, V. (1964) Quatärmollusken der Tschechoslowakei. Rozpr. Ustred. ustavu. Geol. 31: 1-374.

Marino, M.B. McElroy, R.J. Salawitch, and W.G. Spaulding (1992) Nature, 357: 461-465.

Matthews, E. (1983) Global vegetation and land use: new high-resolution data bases for climate studies. J. Climate Applied meteorology, 22: 474-487.

Olson, J.S., J.A. Watts, and L.J. Allison (1985) Major world ecosystem complexes ranked by carbon in live vegetation: A Database. NPD-017, Carbon Dioxide Information Center, Oak Ridge National Laboratory, Oak Ridge, Tennessee.

Parker, M.L., Hennoch, W.S.E. (1971) The use of Engelman Spruce late-wood density for dendrochronological purposes. Can. J. Forest Research, 1: 90-98.

Pichon, J.J., Labeyrie, L.D., Bareille, G., Labracherie, M., Duprat, J., and Jouzel, J. (1992) Surface water temperature changes in the high latitudes of the southern hemisphere over the last glacial-interglacial cycle. Paleooceanography, 7: 289-318.

Piggott, C.D. and Huntley, J.P. (1981). Factors controlling the distribution of Tilia cordata at the northern limits of its geographical range. I. Distribution in north-west England. New Phytologist 87: 817-839.

Post, W.M., Emanuel, W.R., Zinke, P.J., and Sangenberger, A.G. (1982) Soil carbon pool and World life zones. Nature, 298: 156-159.

Ponel, P. and Coope, G.R., 1990. Late glacial and early Flandrian Coleoptera from La Taphanel, Massif central, France: climatic and ecological implications. J. Quat. Sci., 5: 235-249.

Prentice, I.C. (1986) Multivariate methods for data analysis. Handbook of Holocene Palaeecology and Palaeohydrology. In Berglund (ed), Wiley, Chichester, 775-794.

Prentice, I.C. (1980) Multidimensional scaling as a research tool in quaternary palynology, a review of theory and methods. Rev. Paleobotany Palynology, 31: 71-104.

Prentice, I.C., et al. (1989) Developing a global vegetation dynamics model: results of an IIASA summer workshop. IIASA, Laxenburg, RR-89-7.

Prentice, I.C., Sykes, M.T., Lautenschlager, M., Harrison, S.P., Denissenko, O., and Bartlein, P.J. (submitted). Modelling global vegetation patterns and terrestrial carbon storage at the last glacial maximum. Global and Planetary Changes.

Prentice, I.C., Bartlein, P.J., and Webb, T. III (1991) Vegetation and climate change in eastern North America since the last glacial maximum. Ecology 72: 2038-2056.

Prentice, I.C., Cramer, W., Harrison, S.P., Leemans, R., Monserud, R.A., and Solomon, A.M. (1992) A global biome model based on plant physiology and dominance, soil properties and climate. Journal of Biogeography 19: 117-134.

Prentice, I.C., Sykes, M.T., and Cramer, W. (in press). A simulation model for the transient effects of climate change on forest landscapes. Ecological Modelling.

Prentice, K.C., and I.Y. Fung (1990) The sensitivity of terrestrial carbon storage to climate change. Nature, 346: 48-51.

Puissegur, J.J. (1976) Mollusques continentaux quaternaires de Bourgogne. Significations stratigraphiques et climatiques. Rapports avec d'autres faunes boréales de France. Mémoire géol. Univ. Dijon, 3: 1-241.

Rousseau, D.D. (1987) Paleoclimatology of the Achenheim series (middle and upper Pleistocene, Alsace, France): a malacological analysis. Palaeogeography, Palaeoclimatology, Palaeoecology. 59: 293-314.

Rousseau, D.D. (1991) Climatic transfer function from Quaternay molluscs in European loess deposits. Quaternary Research 36: 195-209.

Roux, M., Servant-Vildary, S., and Servant M. (1991) Inferred ionic composition and salinity of a Bolivian Quaternary lake, as estimated from fossil diatoms in the sediments. Hydrobiologia, 1: 1-18.

Sachs, H.M., Webb III, T., Clark, D.R. (1977) Paleoecological transfer functions. Ann. Rev. Earth Planet. Sci., 5: 159-178.

Schwarz-Zanetti, W., Pfister, C., Schwarz-Zanetti, G., and Schüle, H. (1992) The EURO-CLIMHIST database - a tool for reconstructing the climate of Europe in the pre-instrumental period from high resolution proxy data. Paläoklimaforschung / Paleoclimate Research, 7, Special Issue: ESF Project "European Paleoclimate and Man" 2, B. Frenzel, C. Pfister & B. Gläser (eds).
Schweingrüber, F.H., Fritts, H.C., Bräker, O.U., Drew, L.G., and Schaer, E. (1978) The X-ray technique as applied to dendroclimatology. Tree-Ring Bulletin, 38: 61-91.
Schweingrüber, F.H., Braker, O.U. and Schär, E. (1979) Dendroclimatic studies on conifers from Central Europe and Great Britain. Boreas 8: 427-452.
Schweingrüber, F.H. (1988) Climatic information for the past hundred years in width and density of conifer growth rings. Lecture Notes in Earth Sciences 16, Springler-Verlag, Berlin, 35-55.
Seret, G., Guiot, J., Wansard, G., Beaulieu, J.L. de and Reille, M. (1992) Tentative paleoclimatic reconstruction linking pollen and sedimentology in La Grande Pile (Vosges, France). Quat. Sci. Rev., 11: 425-430.
Serre-Bachet, F. (1989) Tree-rings in the Mediterranean area. Paläoklimaforschung / Paleoclimate Research, 7, Special Issue: ESF Project "European Paleoclimate and Man" 2, B. Frenzel, C. Pfister & B. Gläser (eds).
Serre-Bachet, F., Guiot, J., and Tessier, L. (1992): Dendroclimatic evidence from SW Europe and N.W Africa. In: Bradley and Jones (eds): Climate since 1500 A.D., London, Routledge, 349-365..
Shackleton, N.J. (1977) Carbon-13 in Uvigerina: tropical rainforest history and the Equatorial Pacific carbonate dissolution cycles. In The Fate of Fossil Fuel CO2 in the Oceans, N.R. Anderson, A. Malahoff, Eds , New York, Plenum, 401-428.
Shugart, H.H. (1984) A theory of forest dynamics, Springer-Verlag, New-York.
Sparks B.W. (1961) The ecological interpretation of Quaternary non-marine mollusca. proc. Linnean Soc. London 172: 71-80.
Ter Braak, C.J.F., and Prentice, I.C. (1988) A theory of gradient analysis. Advances in Ecological Research 18, Academic Press, London, 271-317.
Tessier, L., Serre-Bachet, F., et Guiot, J. (1990) Pollution fluorée et croissance raiale des conifères en Maurienne (Savoie, France). Ann. Sci. Forestières, 47: 309-323.
Till, C. and Guiot, J. (1990) Reconstruction of precipitation in Morocco since AD.1100 based on Cedrus Atlantica tree-ring widths. Quat. Res., 33: 337-351.
Velichko, A.A., B. Frenzel, M. Pecsi, Eds, Atlas of Paleoclimates an Paleoenvirnments of the Northern Hemisphere (Gustav Fischer Verlag, Stuttgart, 1991).
Van Campo, E., Guiot, J., and Peng Changhui (submitted) The database reappraisal of the terrestrial carbon budget at the Last Glacial Maximum. Global and Planetary Changes.
Van der Hammen (1991) Climatic Change 19: 37-47.
Walter, H. and Box, E.O. (1976) Global classification of natural terrestrial ecosystems. Vegetatio, 32: 76-81.
Webb III, T., and Bryson, R.A. (1972) Late- and Postglacial climatic change in the northern Midwest, USA: quantitative estimates derived from fossil pollen spectra by multivariate statistical analysis. Quat. Res., 2: 70-115.
Webb, T., Bartlein, P.J. and Kutzbach, J.E. (1987) Post-glacial climatic and vegetational changes in eastern North America since 18 Ka: comparison of the pollen record and climate model simulation. In W.F. Ruddiman & H.E. Wright (eds) North America and Adjacent Oceans during the last deglaciation. Decade of North American Geology, vol K-3. Geol. Soc. America, Boulder, Colorado.
Woodward, F.I. (1987) "Climate and Plant Distribution". Cambridge University Press, Cambridge.

Woodward, F.I. (1988) Temperature and the distribution of plant species. Symposia of the Society for Experimental Biology 42: 59-75.
Zinke, P.J., Stangenberger, A.G., Post, W.M., Emanuel, W.R., Olson, J.S., 1984: Worldwide organic soil carbon and nitrogen data, ORNL/TM-8857, Oak Ridge National Laboratory, Oak Ridge.
Zobler, (1986) A world soil file for global climate modeling. NASA technical memorandum 87802. 32p.

LACUSTRINE DIATOMS FOR RECONSTRUCTING PAST HYDROLOGY AND CLIMATE.

Françoise Gasse
Laboratoire d'Hydrologie et de Géochimie isotopique
URA-CNRS723, Bâtiment 504
Université Paris-Sud
91405 Orsay Cedex
France

1. INTRODUCTION

Diatoms (Bacillariophyceae) are microscopic algae which colonize all types of waterbodies. Diatoms have a external skeleton (a frustule, composed of two valves), made of hydrated silica and which ensures their preservation when sinking through the water column and in the sediments. Diatoms are one of the most sensitive biological indicators of continental and marine environments. The close relationships exhibited between contemporary diatom commmunities and environmental parameters enables the numerical estimation of palaeo-environmental conditions from the sedimentary records by using transfer functions. During the last decade, diatoms have thus emerged as a powerful tool for environmental monitoring and palaeoenvironmental reconstruction.

Over the continents, one of the most spectacular application concerns the assessment of lake acidification trends in North America and Europe, associated with increased atmospheric acid loading (Charles et al., 1989 ; Charles and Smol, 1990 ; Birks et al., 1990). Palaeoecological studies offered the only means of obtaining direct evidence of past changes from lakes that lack historical data over the last century. Such historical perspectives allow a better predict how lakes will response in the future.

Recent concern about climate changes has also stimulated interest in the diatom records of closed lakes which may fluctuate in water level and water chemistry in response to seasonal, interannual or longer-term climatic fluctuations, especially in arid and semi-arid regions. In this paper, the complex relationships which link lake level (or area) and lake salinity with climatic parameters are summarized. On the basis of some case studies, methods applied for reconstructing water depth and water chemistry by using diatoms are presented. The limits of such approaches for palaeoclimatic research are discussed. We show

NATO ASI Series, Vol. I 22
Long-Term Climatic Variations
Edited by J.-C. Duplessy and M.-T. Spyridakis

that, in favourable cases, diatom-inferred palaeohydrological variables can be integrated in models of palaeolake systems and may help, indirectly, to estimate palaeoclimate parameters.

In the oceanic domaine where chemical conditions are relatively stable compared with that of continental waterbodies, diatom assemblages appear to be closely correlated with Sea Surface Temperature (SST)(Pichon et al., 1987). Although SST estimates have been firstly derived from transfer functions based on Foraminifera (CLIMAP Members, 1976), diatom-inferred SST are of utmost interest in high latitudes where Foraminifera are rare. Reconstruction of SST changes in the Southern Ocean (Labracherie et al., 1989 ; Pichon et al., 1992) allows interhemispheric comparison and correlations with Antarctic ice cores. Continental diatoms contained in marine sediments can also be used for palaeoclimatic reconstruction. These diatoms may be either windblown from the continents, or river transported. They may inform on changes in atmospheric circulation and/or river discharge. This topic is discussed here with the example of freshwater diatom influx in intertropical Atlantic during the Late Quaternary period.

2. LAKE SYSTEMS AT STEADY STATE. RELATIONSHIPS BETWEEN WATER DEPTH, WATER CHEMISTRY AND CLIMATIC CONDITIONS

Fluctuations in the extent and depth of closed lakes have been long recognized as a sensitive indicator of change in Precipitation/Evaporation (Street and Grove, 1976; Street-Perrott and Roberts, 1983). Fluctuations in the water balance may also result in changes in the concentration of dissolved salts. However, the lake water balance is the end product of interactions between climate, vegetation, runoff and groundwater storage (Fig. 1). Lake water-level and concentration thus depend largely on the water cycle in the catchment. This complexity may be responsible for apparent discrepancies between pollen evidence and lake level records. This was for instance discussed to explain high lake levels in the northern Mediterranean region during the Last Glacial Maximum (Prentice et al., 1992), or for Lake Sumxi (western Tibet) where the maximum in lake volume deduced from diatoms lags ≈3000 yr behind the major wet pulse at 10,000 yr BP inferred from the pollen record (Van Campo and Gasse, 1993).

Any lake can be regarded as a system of transfer, recycling and storage of water, dissolved and solid material, and energy (thermic, photic and cinetic) (Fig. 2). At steady state, a lake basin is characterized by its morphometry and hydrological parameters which are site specific, and is in equilibrium with climatic conditions over the catchment. This equilibrium can be written as water, energy, and salt balances.

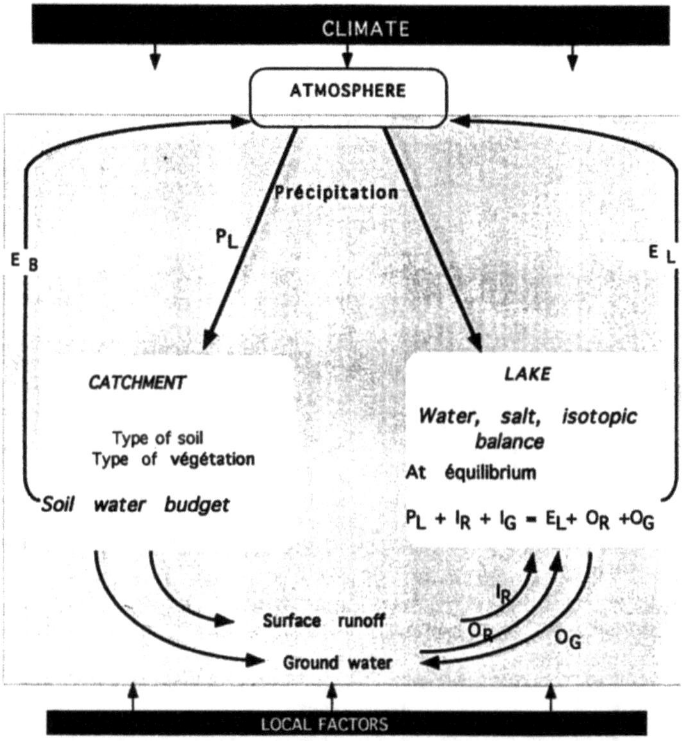

Figure 1. Schematic diagram illustrating interactions between climate, soil water budget, runoff and groundwater storage in the catchment area of a lake, and water and salt balances of the lake. See text for explanation of the symbols.

2.1. Water balance

The water balance of a lake is given by the equation :

$$dH/dt = \Phi I - \Phi o \qquad (1)$$

where dH is change in lake level,

dt is time

Φ = flux

I = inflow

O = outflow.

At steady state, $\Phi I = \Phi o$, or

$$P_L + I_R + I_G = E_L + O_R + O_G \qquad (2)$$

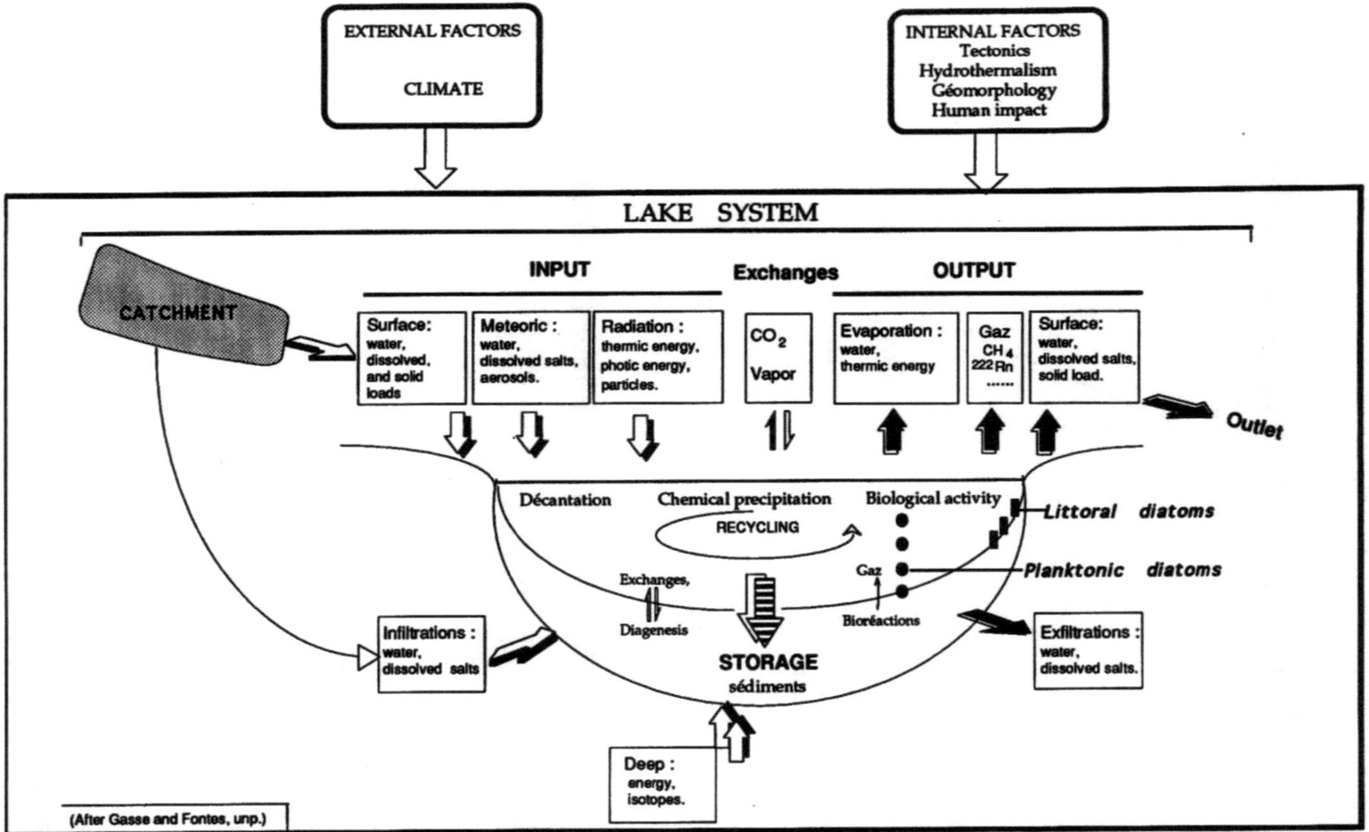

Fig. 1 : **Terms of water, salt and energy budgets** : Input; : Output; storage. (sedimentary profile)

Figure 2 - Functionnal diagram of a lake system.

where the parameters, expressed in water thickness, are :

P_L is precipitation on the lake surface

E_L is evaporation from the lake surface

I_R is the runoff from the catchment

I_G is the groundwater inflow

O_R is the surface discharge from the lake

O_G is the groundwater outflow.

The runoff from the catchment, I_R, is given by the equation :

$$I_R = (S_B - S_L) * P_B * C_R / S_L \qquad (3)$$

where :

S_B is the drainage catchment area,

S_L is the lake surface,

P_B is the precipitation on the catchment area,

C_R is the runoff coefficient on the catchment area, that is the ratio between the volume of water entering the lake by rivers and the volume of precipitation on the catchment. C_R depends on the rainfall regime, the types of soils, and the vegetation covers over the catchment.

Considering the favourable case of a closed lake (O_R=0) with negligible groundwater flow (I_G and O_G = 0), and with a relatively restricted drainage area (P_B = P_L = P), the lake surface, S_L (and thus the water level, H) is related to P and E_L as followed :

$$\frac{(S_B - S_L) * C_R}{S_L} = \frac{E_L - P}{P} \qquad (4)$$

2.2. Hydrological and energy balance

The equations presented below are derived from a hydrological and energy balance model established by Kutzbach (1980), and applied to Lake Chad (Kutzbach, 1980) and some tropical lakes (Kutzbach and Hastenrath, 1986).

At steady state :

$$P * (+S_L) = (E_B * S_B) + (E_L * S_L) \qquad (5)$$

where E_B is the evaporation on the drainage catchment.

The mean evaporation term on the land and lake surface is estimated from the mean annual energy budget :

$$E = R / [L * (1+B)] \qquad (6)$$

where :

R is the net radiation for the land or lake surface, and results from the surface radiative balance between adsorbed solar radiation and emitted and adsorbed longwave radiation,

L is the latent heat of evaporation,

B is the Bowen ratio of the land or lake surface, the Bowen ratio $B = Q / (L * E)$ expresses the partionning of energy between the sensible (Q) and evaporative ($L * E$) heat fluxes and depends mainly on the surface moisture, temperature, and other local conditions.

Equation (6) of lake and land evaporation rate may be used in equation (4) in terms of net radiation and Bowen ratio, in order to obtain a diagnostic equation for precipitation on the basin, as follows :

$$L * P * (S_B + S_L) = R_B * S_B / (1 + B_B) + R_L * S_L / (1 + B_L) \qquad (7)$$

Equation (7) expresses the simultaneous constraints on the hydrological and on the energy balance. It clearly shows that the characteristics of the catchment area have to be considered for reliable estimates of past precipitation from lake-level record.

2.3. Salt balance

Attention may focuss on surrogate indicators of water level, especially salinity, which shows close relationship with lake volume (Fig. 3).

We consider the variables :

Φ = flux

V=Volume

C= concentration

E= evaporation (mean annual value)

I = inflow

O = surface + groundwater outflow

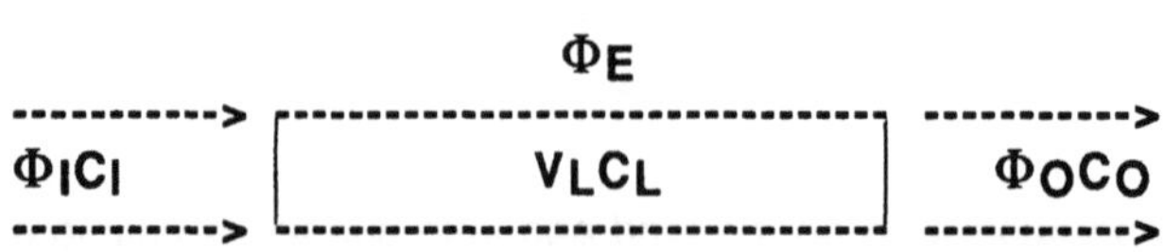

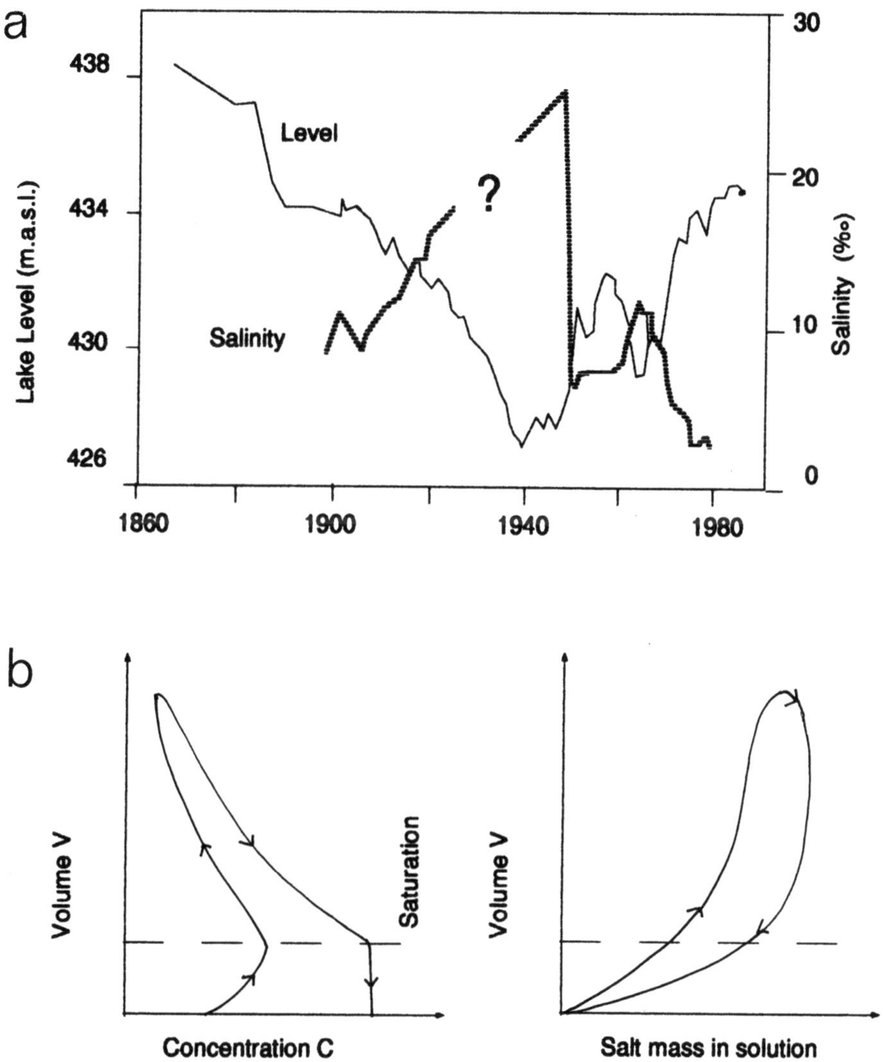

Figure 3

a) Lake-level and salinity measurements for Devils Lake (North Dakota, USA). No salinity measurements were made between 1923 and 1948, hence the dotted line and question mark between these dates. After Fritz, 1990.

b) Relationships between lake volume vs salt concentration, and lake volume vs salt mass in solution.

We consider the favourable cases of :

- 1) lakes with good mixing of the water : $C_L = C_O$
- 2) salts neither volatile ($C_E = 0$) nor precipitable (e.g. concentration in conservative ions such as Cl^-) :

$$\frac{dV_L}{dt} = \Phi_I - \Phi_E - \Phi_O \tag{8}$$

The salt balance is given by equation :

$$\frac{d\,(V_L C_L)}{dt} = \Phi_I C_I - \Phi_E C_E - \Phi_E C_O \tag{9}$$

At steady state ($dV_L = 0$), the water balance is : $\Phi_I = \Phi_E + \Phi_O$

Assuming that, at time t=0 (initial conditions), $C_L = C_I$, and C_I is constant through time :

$$C_L = \frac{\Phi_I C_I - [C_I - (\Phi_{II} - \Phi_E)\, e^{-(\Phi_I O/E)\, t}]}{\Phi_{IE}} \tag{10}$$

Equation (10) shows that the concentration of the lake water tends towards an equilibrium, or steady state S :

$$C_L \rightarrow C^S_L \approx \frac{\Phi_I C_I}{\Phi_O} = C_I\left(1 + \frac{\Phi_E}{\Phi_O}\right) \tag{11}$$

depending on climate (Φ_E) and hydrology (Φ_O).

Salt concentration in lakes can thus be regarded as a proxy indicator of climate change, although it does not allow a direct estimate of palaeoclimate variables.

3. DIATOM-INFERRED CHANGES IN LAKE LEVEL. PALAEOCLIMATIC INTERPRETATION

3.1. Lacustrine diatoms life forms and habitats

Diatom species can be classified according to three habitat categories :
(i) euplanktonic, (ii) facultative planktonic, and (iii) periphytic or benthic.

Euplanktonic forms are floating in open water. They may represent up to 100% of the taphocenose in the central part of deep, permanently stratified (meromictic) lakes. Facultative planktonic species spend part of their life cycle in the plankton and the other part as fixed on a substrate (meroplanktonic forms), or live indifferently as planktonic or benthic forms (tychoplanktonic). Benthic (or periphytic) species are attached on submerged substrates, e.g. macrophytes (epiphytic), mud (epipelic), rocks (epilithic)... Benthic forms are commonly called "littoral" because they characterize marginal areas lying within the photic zone, although they may develop on the entire bottom surface in large shallow lakes.

3.2. The Planktonic/Benthic ratio, an index of water depth fluctuations

The composition of the diatom assemblages in surface bottom sediments is usually closely related to water depth , and the planktonic/benthic ratio is a index commonly used as a rough depth indicator. The large predominance of benthic forms indicates that the site was in the photic zone, which may varies in thickness from tens of centimeters to 30-40 m depending on the water transparency. Planktonic percentages close to 100% suggest "deep" water, although depth cannot be estimated as planktonic flora is indifferent to the height of the water column below the surface.

In lakes of moderate size, attempts were made for quantitative estimates of water depth from the Planktonic/Benthic ratio. In Aguelmane Sidi Ali, a small lake (2 km long, 0.5 km wide) from the Middle Atlas in Morocco (Barker et al., 1993) , diatoms from surface mud were sampled (n=6) along a transect across the lake bed (0-15 m in depth). A logarithmic relationship was found between depth and the planktonic/benthic ratio, with a regression coefficient, r^2, of 0.92 (facultative planktonic species, which are most important in the samples from intermediate water depth, are excluded from the equation as they have no specific habitats requirements). The equation, which is site specific, was used to reconstruct lake level fluctuations in Sidi Ali over the last 6.5 kyr. However, the Planktonic/Benthic ratio is sensitive to several factors besides variations in lake morphometry, such as changes in mixing regime, changes in water transparency due to fluctuations in suspended solid load or planktonic productivity, etc... (Gasse, 1987). Large uncertainties thus affect diatom-inferred depth fluctuations, especially for high water level episodes.

Nevertheless, major regressions in very large deep lakes can be deduced from changes in diatom spectra with a relatively good precision. The occurrence of *in situ* littoral diatom assemblages at depth tens or hundred of meters below the present-day level undoubetly reflect a marked deficit in the water balance, which may be interpreted in terms of palaeoclimate.

3.3. Case study : water-level fluctuations of Lake Tanganyika during the last glaciation and deglaciation

Lake Tanganyika (3°20'-8°45'S, 31°20'E) is the deepest (1470 m) and the largest lake (32,600 km^2) lying in the East African Rift Valley lake (Fig. 4). The catchment area extends over 231,000 km^2 . The precipitation on the lake surface (P= 1000 mm yr^{-1}; see equations 1 to 7 for used symbols) and surface runoff (I_r = 780 mm yr^{-1} represents ≈ 63 % and 37 % of the water input, respectively. Presently, Evaporation (E_L=1700 mm yr^{-1}) is the major loss of water (94%). Surface outflow by the Lukuga river (O_R = 80 mm yr^{-1}) controls the maximum (and present) lake level. A minor drop of water level may lead to the closure of the lake (Degens and Hecky, 1973 ; Haberyan and Hecky, 1987).

Diatom evidence for a -350m, -400m water level during the Last Glacial Maximum

We refer here to the diatom study of a 10,13 m core, MPU XII, collected in the southern basin where the lake is 422 m deep (Figs. 4 and 5 ; Gasse et al., 1989). The core is ^{14}C-dated to 25,600 yr BP at its base. The sediments consist mainly of laminated or homogeneous diatomite.

From ≈25,600 to 21,700 yr BP, the lake was shallower than today. A depth of 100-200 m at the core site is compatible with the diatom spectra (*Cyclotella* sp. aff. *ocellata* dominant) and with sedimentology. The interval 21,700-12,700 yr BP corresponds to generally very shallow conditions. Littoral diatoms (such as *Rhopalodia* spp.) became progressively abundant. They are associated wiith sponge spicules and phytoliths which also reflect shallow environments. The minimum water depth is recorded by a peak of *Gomphonema clevei*, typically periphytic in modern East Africa (Gasse, 1986 a). This peak is ^{14}C-dated at 18,030±240 yr BP. The presence of intact frustules is a further indication that species was growing *in situ* at the core site, which was thus located in the photic zone. This reveals a drop in water level of at least 350 m, more probably of ≈400 m. A generally positive lake water balance was established from ≈ 16,000 yr BP, and the planktonic flora indicates that deep conditions were definitely restored to the lake shortly before 12,700 yr BP.

The 18,000 yr BP lowstand at -350 m corresponds to a lake surface and to a lake volume that are -42% (18,900 km^2) and -50% of the present values, respectively.

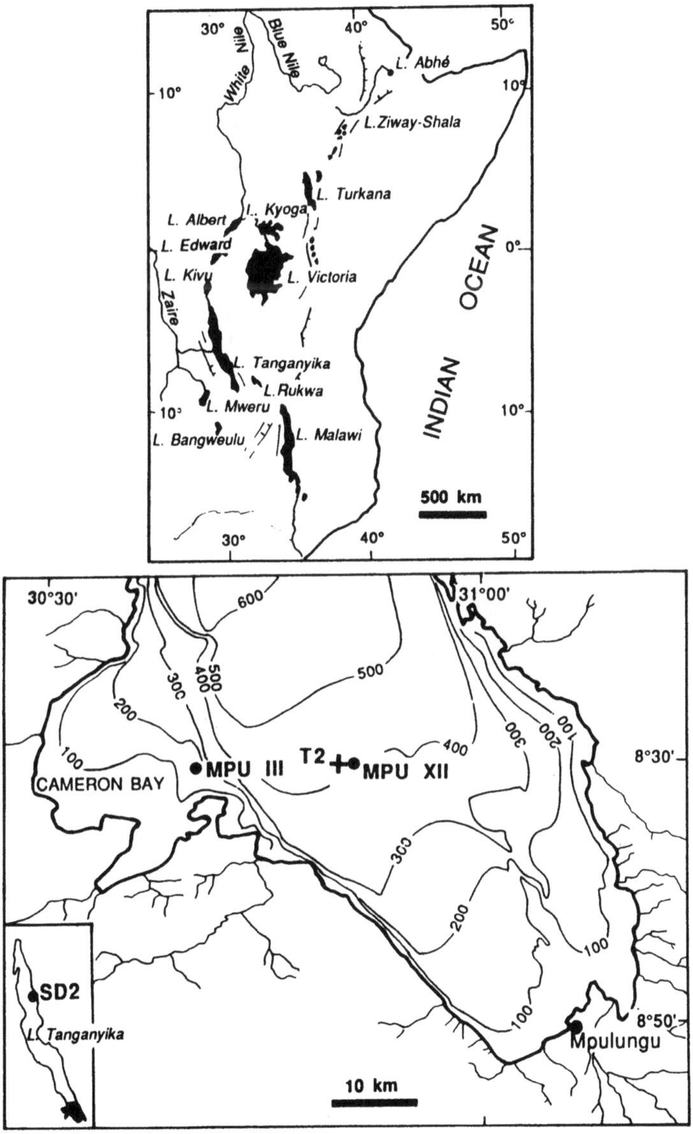

Figure 4 - Location of core MPU XII in Lake Tanganyika, East Africa.

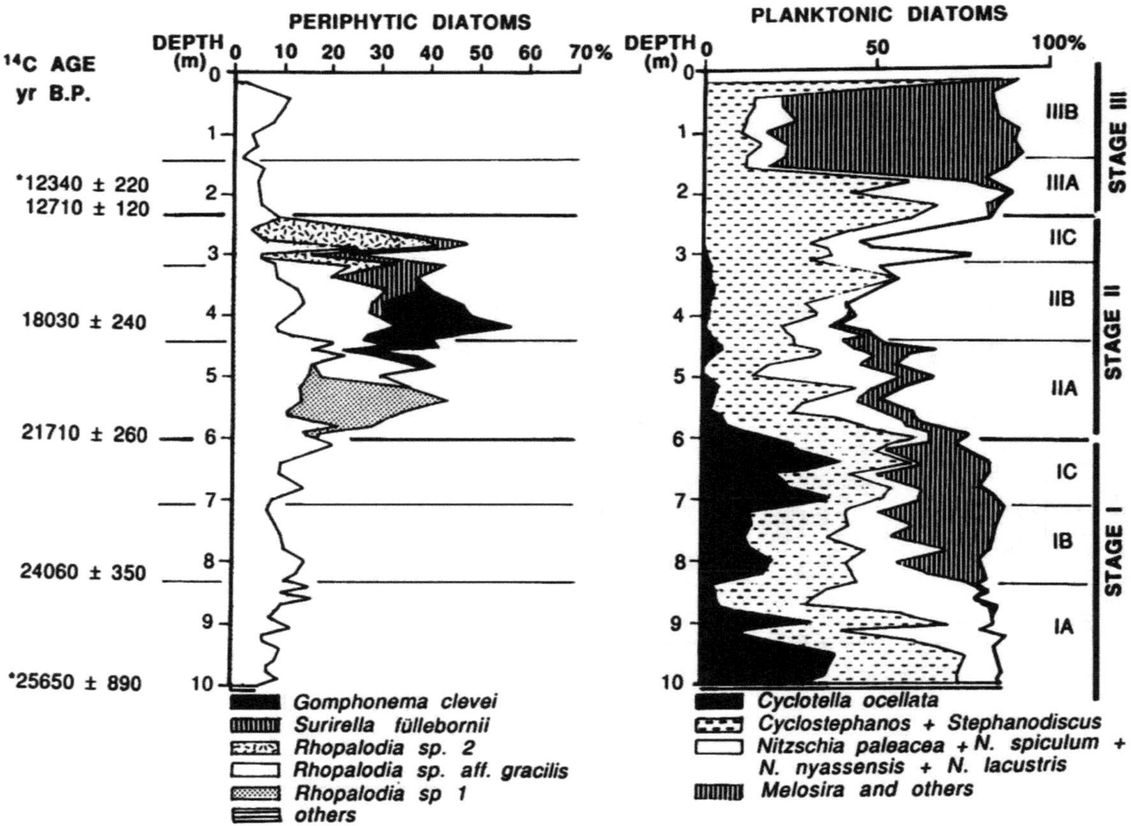

Figure 5 - Distribution of periphytic (= benthic), and planktonic diatoms along core MPU XII, Lake Tanganyika. After Gasse et al., 1989.

Estimates of past precipitation and evaporation

An hydrological and energy balance model derived from the Kutzbach's (1980) model was used to investigate the climatic conditions which could account for by a

-350 m level in lake Tanganyika during the Last Glacial Maximum (Bergonzini et al., 1993). Geological evidence indicates that Lake Tanganyika was not connected with Lake Kivu through the Rusizi river at that time (Stoffers and Kecky, 1978). The catchment area was reduced to 221,600 km^2, the S_B value considered in the following calcultations.

Table 1 presents the present-day budget. The terms are based on measured or estimated climatic and hydrological data (Griffiths, 1972 ; Bultot, 1993). Sensitivity experiments show that a 10 % decrease of land surface albedo and Bowen ratio implies an increase of 30 mm yr^{-1} and 77 mm yr^{-1} in precipitation, respectively.

The model was applied to the Last Glacial Maximum, considering that the palaeo-insolation at the top of the atmosphere for this latitude at 21,000 cal. yr (= 18,000 ^{14}C yr BP ; 420 W m^{-2}). Assuming that the atmospheric transmission coefficient has not changed, the palaeo-global radiation through clear sky is estimated at 327 W m^{-2}. A first experiment attempts to estimate Precipitation (P) changes associated with changes in lake surface only. A lake area -42 % compared to that of today is asociated with changes in P of ≈ -5 %. Changes in temperature deduced from pollen transfer functions (Bonnefille et al., 1992 ; Chalié, 1992) is estimated at -4.4°C between 18,000 and 16,000 yr BP. A temperature of -4.4°C and a lake surface of 18,900 km^2 correspond to a decrease in Precipitation of 11 %, and a decrease in evaporation from the lake surface (E_L) of ≈ 6 %. The decrease in rainfall amount reduces to only -8 % if the effect of a -10 % in land albedo, an albedo value proposed by CLIMAP Members (1976), is included in the model. Decrease in P is slightly slower than that deduced from the pollen study (-13.5 %). The difference may be due to the assumptions made in this model, which consider modern values for cloudiness, surface emissivity and Bowen Ratio. It is certainly unrealistic to use constant cloud cover for different precipitation rate, and a constant Bowen ratio for the drainage catchment area with different mean temperatures. The model could be refined by taking into account for changes in soil moisture over the catchment area, as they can be deduced from the characteristics of modern soils and changes in vegetation inferred from pollen records.

Although large uncertainties still remain on these results, the example of Lake Tanganyika shows that identification of low lake level by using diatoms represents a substantial help for past Precipitation and Evaporation estimates

TABLE 1

			LAKE			REDUCED DRAINAGE BASIN					
								Busumbura	Tabora	Kalemie	mountain
	Symbol	unit	calculated	estimated	observed	calculated	estimated	observed	observed	observed	observed
Radiation budget											
insolation	I	W/m2	410			410					
atm. transmission coef.	x	-	0.780		0,8	0.78					
global radiation (clear sky)	G°	W/m2		320	320		320	320	320	320	320
fractional cloud cover	C	-		0,4			0,5	0,5	0,5	0,5	0,625
global radiation (with clouds)	G	W/m2	236,8			216		207	235		216
surface albedo	α	-		0,06	0,06		0,2	0,2	0,2		0,2
net shortwave radiation	SW	W/m2	0222,59			172,8					
surface emissivity	em	-		0,96			0,92				
surface temperature	T	K°		298	298		296	296,9	302	296	289,3
relative humidity	U%	-		80%	80%		70%				
surface vapor pressure	e	mb	25.3			19.7		20,5	1635%		14,5
net longwave radiation	LW	W/m2	63,05			67.74		59			64
net radiation	R	W/m2	159.4			105.1		100		111	103
Bowen ratio evaporation											
Bowen ratio	B	-		0,21	0,21	0.53		0,59		0,41	0.29
evaporation	LE	W/m2	132		132	69		63	67	79	80
Hydrological budget											
evaporation	E	mm/yr	1690		1696	881		804	863	1013	1028
precipitation	P	mm/yr		1000	1000		1000	848	892	1153	1833
drainage	R	mm/yr	690		696	119		44	29	140	805

Present hydrological and energy budgets of Lake Tanganyika and its reduced drainage area. Calculations are based on the Kutzbach's model (1990). Observed data are after Griffiths (1972) and Bultot (1993).

through palaeohydrological modelling, provided the regional hydrology is well understood and provided the water budget of the catchment is considered.

4. DIATOM-INFERRED CHANGES IN IONIC CONCENTRATION. PALAEOHYDROLOGICAL, PALAEOCLIMATIC SIGNIFICANCE

4.1. Statistically derived relationships between modern diatom assemblages and lake water chemistry

In regions with a negative water balance, the composition of diatom assemblages is primarily controlled by the ionic concentration of the water. This is for instance demonstrated by the statistical treatment of a large data set (modern diatom communities and associated physico-chemical data) from East Africa (Gasse et al., 1983 ; Gasse and Tekaia, 1983 ; Gasse, 1986 a). The region considered contains thousands of lakes, spanning a gradient from fresh to hypersaline water and ranging from alpine peat bogs, hot springs or ephemeral playas to deep, permanently stratified lakes. The study of 214 samples, ranging from 4000 m.a.s.l. to -155 m.b.s.l. in elevation, 40 to 150,000 $\mu S\ cm^{-1}$ in conductivity, and 5 to +70°C, was performed. Good relationships were found between diatom communities, pH (r^2=0.89), alkalinity and total salinity (r^2=0.72), three parameters closely correlated together in waters of the bicarbonate-carbonate type, and ionic ratios (Alkaline/Alkaline earths, and $HCO_3^- + CO_3^{2-} / Cl^- + SO_4^{2-}$). Little information about thermal conditions could be stated (r^2=0.38). Similar conclusions were deduced from the other large data set currently available for the northern Great Plains of North America (Fritz, 1990). Transfer functions were thus established in order to infer chemical variables from diatoms in arid and semi-arid regions, and applied to the past (Gasse, 1986 b ; Fritz et al. , 1991).

Case study : the northern Great Plains region of North America

Surface-sediment and water samples were collected from 66 lakes with salinity ranging between 0.7 ‰ to 270 ‰. Fig. 6 illsutrates the distribution of selected taxa as a function of salinity. Detailed results are given in Fritz (1990) and Fritz et al. (1991). Statistical analysis of the data on water chemistry and diatoms were carried out using principal-component analysis, detrended correspondance analysis and canonical correspondence analysis. In all ordinations the first axis was strongly related to total salinity and many of its associated ionic components, and the second axis related to differences in brine type, in particular the difference between systems dominated by CO_3-HCO_3 and those dominated by SO_4 . (Fig. 7).

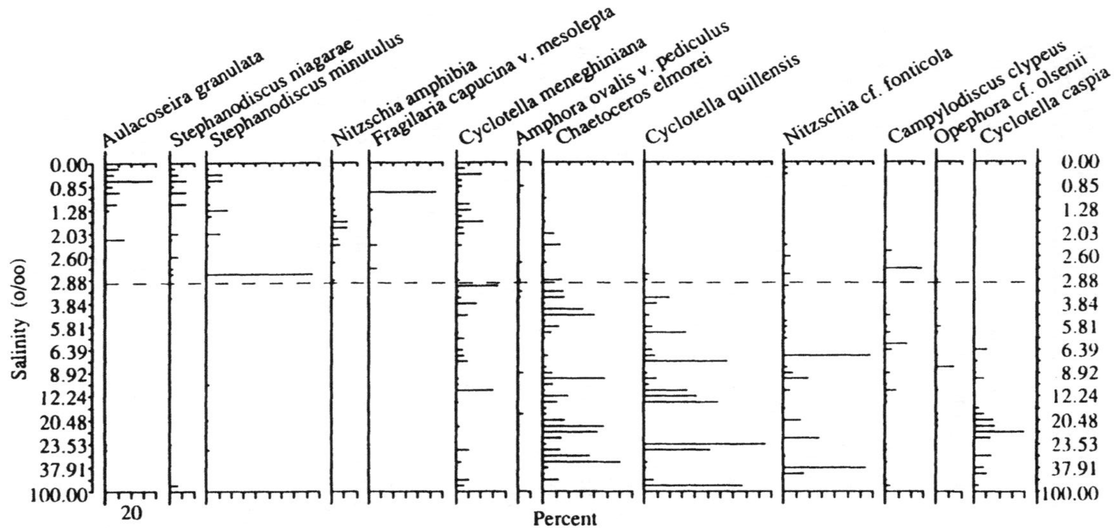

Figure 6. The relative abundance of selected diatom taxa in 55 sediments samples from lakes located in the Great Plains of North America. After Fritz, 1990.

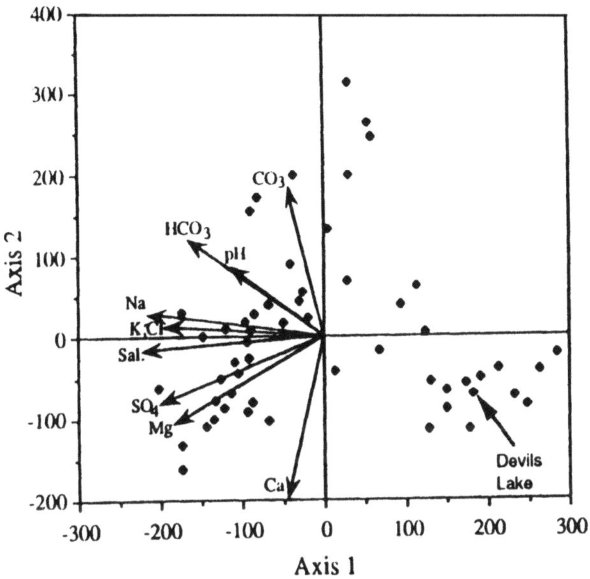

Figure 7. Canonical Correspondence Analysis ordination of 55 surface-sample sites in the Northern Great Plains, based on sedimentary diatom assemblages. The coordinates of samples are indicated. Arrows show the direction of maximal variation of measured environmental variables. After Fritz, 1990.

The weighted-averaging regression of the log-transformed salinity data was used to estimate the salinity optima and tolerances of all taxa that occurred with at least 2% abundance in any one sample. Estimated optima (abundance-weighted means) and tolerance (abundance-weighted standard deviations) of taxa are given by the equations :

$$\hat{u}_k = \sum_{i=1}^{m} y_{ik}\, x_i \;/\; \sum_{i=1}^{n} y_{ik} \qquad (12)$$

$$\hat{t}_k = \left[\sum_{i=1}^{n} y_{ik}\, (x_i - \hat{u}_k)^2 \;/\; \sum_{i=1}^{n} y_{ik} \right]^{1/2} \qquad (13)$$

where y_{ik} is the abundance of taxon k in sample i, x_i is the observed salinity of sample i, and $\hat{u}_k$, $\hat{t}_k$ are the optimum and tolerance of species k, respectively.

A weighted-average calibration with an inverse deshrinking regression was subsequently used to estimate the diatom-inferred modern salinity for each site in the data set. A good agreement was observed between inferred and measured salinity (Fig. 8):

$$\hat{x}_i = \sum_{k=1}^{m} y_{ik}\,\hat{u}_k \Big/ \sum_{k=1}^{m} y_{ik} \qquad (14)$$

$$\hat{x}_i = a = b\,\hat{x}_i \qquad (15)$$

where $\hat{x}_i$ is the inferred salinity, and a and b are the coefficients of the deshrinking regression of observed salinity x_i on $\hat{x}_i$.

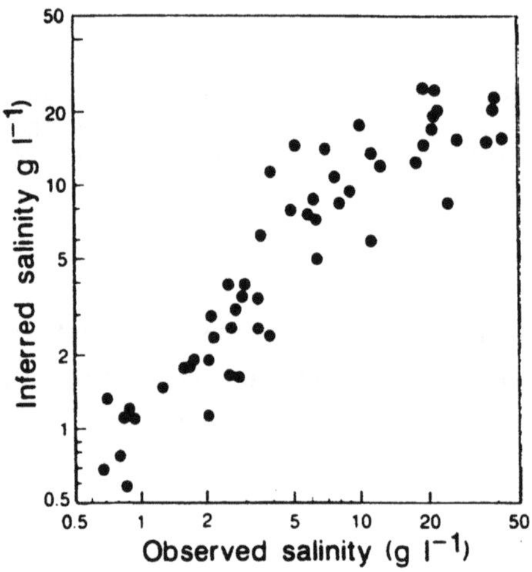

Figure 8. Observed salinity against diatom-inferred salinity (derived from weighted-averaging regression) for the surface-sample sites plotted on Fig. 7 (r = 0.91). After Fritz et al., 1991.

Two short cores from one of the investigated lakes, Devils Lake, were taken to evaluate the use of transfer function, because historical records show a strong hydrological response to climate change in the past century (Fig. 9). Salinity history is accurately reproduced for periods of fresh water and low salinity (<10 %) and high-salinity intervals can be clearly distinguished from those of low to moderate salinity. Discrepancies between the measured and diatom-inferred salinity during high salinity episodes are attributed to problems with the sedimentary record, such as poor diatom preservation, sediment mixing and reworking, or dating inaccuracies.

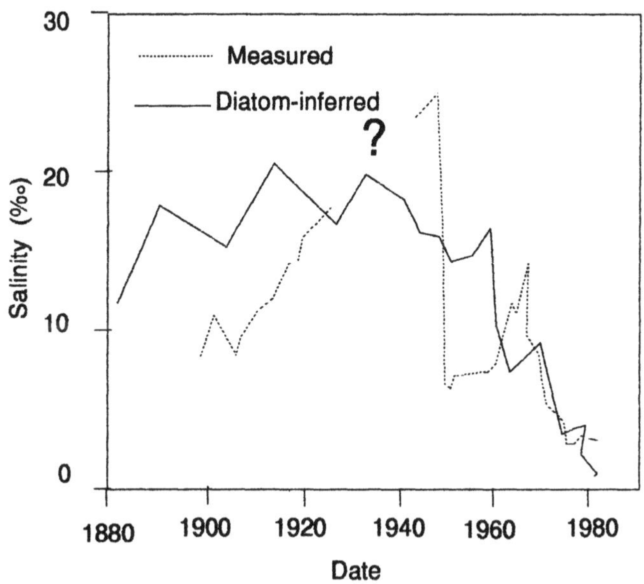

Figure 9. A comparison of the measured with the diatom-inferred salinity for the period of historic record in Devils Lake (North Dakota). No salinity measurements were made between 1923 and 1948, hence the dotted line and question mark between these dates. After Fritz, 1990.

The transfer function was applied to a 24-m long core from Devils lake and covering the Late Glacial and Holocene period (Fig. 10). Salinities in the lake fluctuated from ≈1 to 40‰. Results show that the Late-Wisconsin/early Holocene was characterized by freshwater taxa, abruptly replaced ≈8,000 yr BP by saline species indicating a phase of low water level and high salinity that persisted until ≈7,000 yr BP. Oscillations in salinity may indicate that maximum aridity occurred ≈8,000 yr BP, and that effective moisture fluctuated cyclicly throughout the remainder of the Holocene.

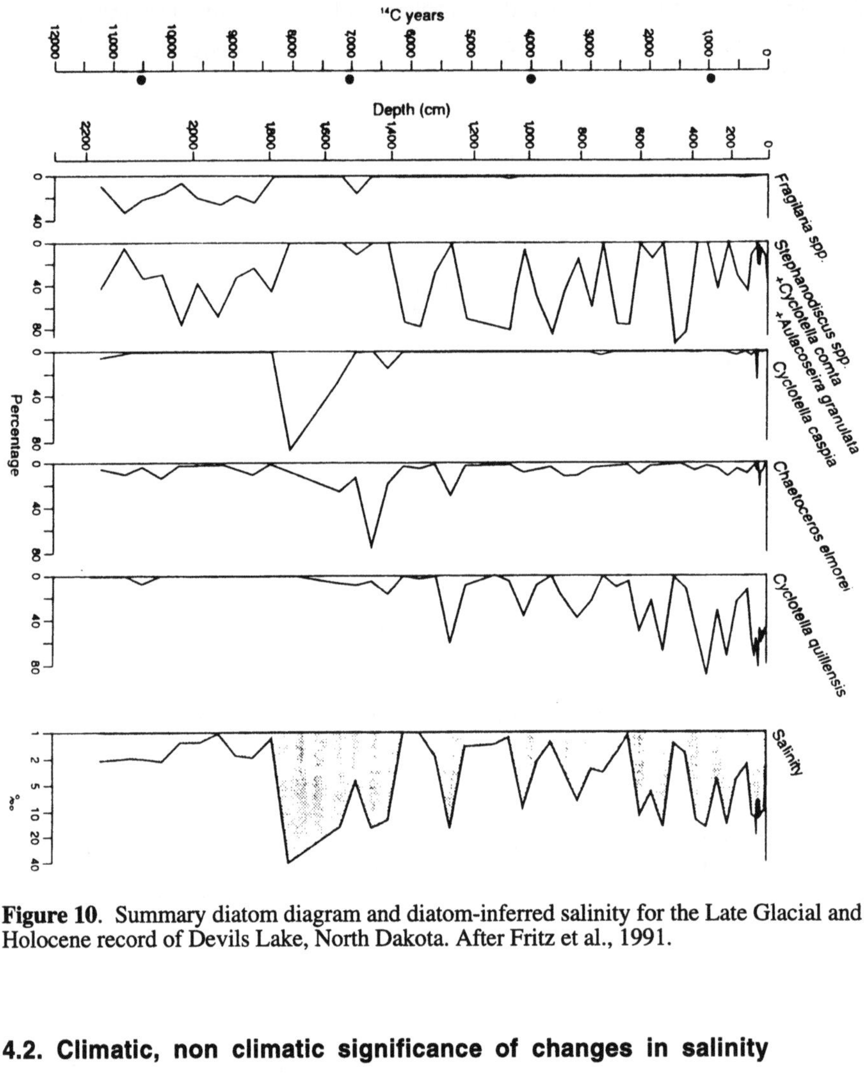

Figure 10. Summary diatom diagram and diatom-inferred salinity for the Late Glacial and Holocene record of Devils Lake, North Dakota. After Fritz et al., 1991.

4.2. Climatic, non climatic significance of changes in salinity

The interpretation of changes in salinity for palaeoclimatic purposes requires an understanding of the hydrological mechanisms responsible for salinity trends. One method to resolve the problem is the comparison of changes in diatom-inferred salinity and of ^{18}O content of authigenic inorganic carbonates associated with diatoms in the sediments. In seasonally dry climates, the enrichment of the lake water in heavy isotopes by evaporation is the major factor controlling the ^{18}O content (see discussion in Gasse and Fontes, 1992). Figure 11, Case 1, shows a parallel increase in diatom inferred-salinity and ^{18}O content, which can be attributed to increase evaporation. Paradoxically, increased precipitation and runoff may first result in increased salt concentration in the lake water by dissolution of salt crusts accumulated on the lake catchment during previous arid phases (Fig. 11, Case 2). This may occur at different time scales, from the season to 10^3 years. Conversely, climatic changes may be damped in lakes with permeable bottom where seepage of brines regulates the salinity which remains low whatever the climate (Fig. 11, Case 3). As is Lake Chad in the Sahel, which is a freshwater lake despite its negative water balance. These different cases are observed in sedimentary records.

Case study : the Bougdouma record (southern Niger)

Bougdouma is a small interdunal depression of the Manga plateau, west of Lake Chad (13°19'N, 11°40'E, 337 m.a.s.l.) which is occupied by an hyperakaline playa with a trona crust. It lies in the Sahelian zone where most of the precipitation is due to summer squalls (75 %) and monsoon rainfal (mean rainfal : 250 mm yr^{-1}, mean evaporation 2,000 mm yr^{-1}). It is supplied by an unconfined aquifer exclusively recharged by local rainfall. The post Glacial climate evolution is deduced from a 458 cm core (Gasse et al., 1990, and Gasse, unpublished). Figure 12 shows the fluctuations of $^{18}0$ content in carbonates and of diatom-inferred conductivity. Negative peaks in $\delta^{18}O$ reflect groundwater inputs. High $\delta^{18}O$ values indicate high evaporation rate and arid conditions. During the late Glacial period (>12,000-≈10,000 yr BP), $\delta^{18}O$ and salt concentration values are not corrrelated. The late Glacial corresponds to generally arid conditions, interrupted by sudden groundwater influxes before 12, 000 yr BP, at ≈ 12,000 and at ≈10,800-10,600 yr BP. The isotope concentration then increased and reached the maximum value observed in the section at ≈10,300 yr BP. Diatoms show that generally dilute conditions persisted up to that date, however, probably because of seepage of heavy solutions through the permeable lake bottom (aeolian sand) which may have regulated the salinity. At ≈10,000 yr BP, ≈300 years later than the maximum isotope concentration, diatoms indicate a brief episode of very saline-alkaline conditions. This is attributed to the leaching, by local rainfall or groundwater influx, of a highly soluble Na^+ carbonate crust deposited during a short but marked dry phase between ≈10,300 and 10,000 yr BP. Until 10,000 yr BP, salinity cannot be regarded as a climatic recorder. From ≈9,700 yr BP, the establishment of an early Holocene freshwater lake is registered both by salt dilution and by a sudden stable isotope decrease, with minimum at ≈9,200 yr BP. Clayish diatomite accumulated. The top of the sequence records several hydrological fluctuations. Changes in salinity, which correlate very well with changes in ^{18}O content, reflect the fluctuations of the freshwater influx - evaporation balance and is thus a climatic signal.

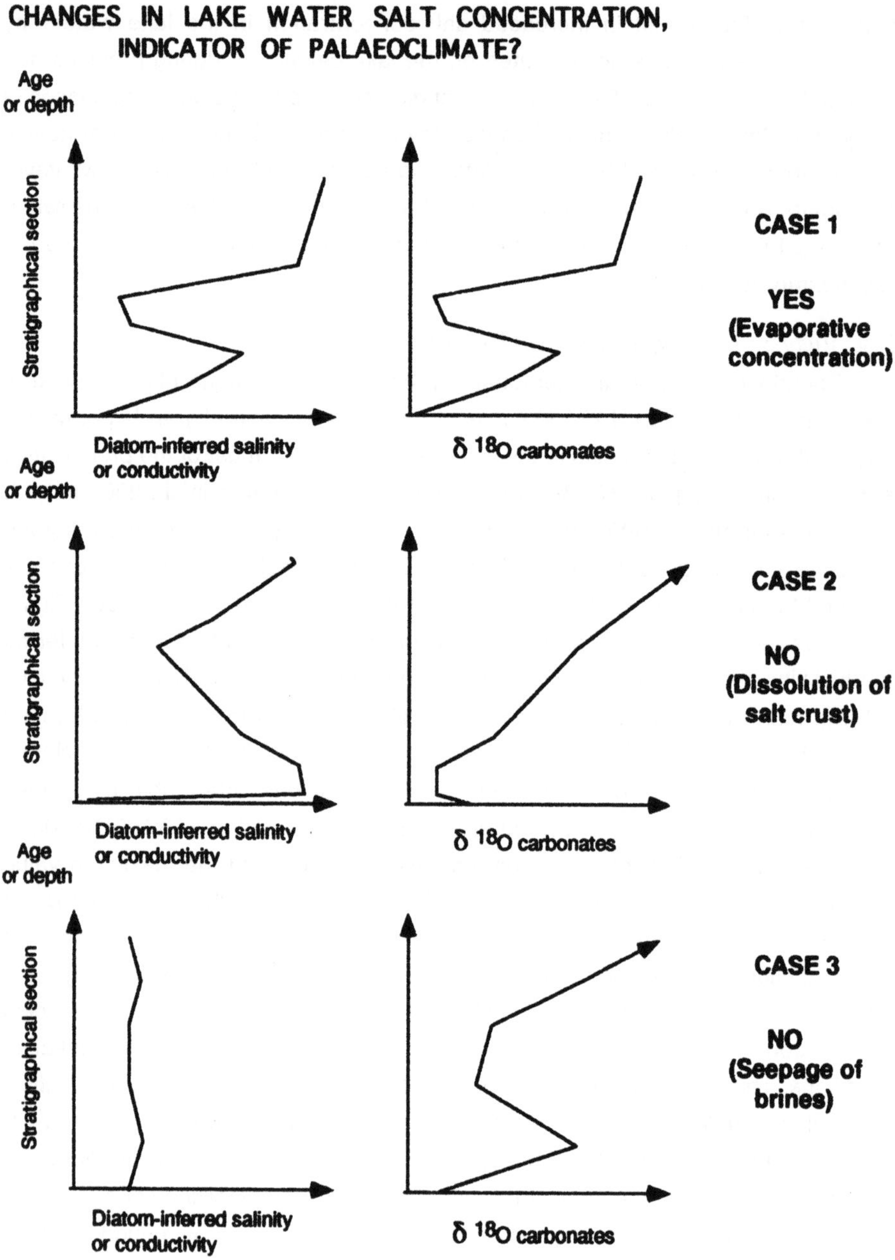

Figure 11. Palaeoclimatic, non-palaeoclimatic significance of lake salinity. Schematic diagram comparing theoretically diatom-inferred changes in salinity, and in ^{18}O content of associated inorganic carbonate.

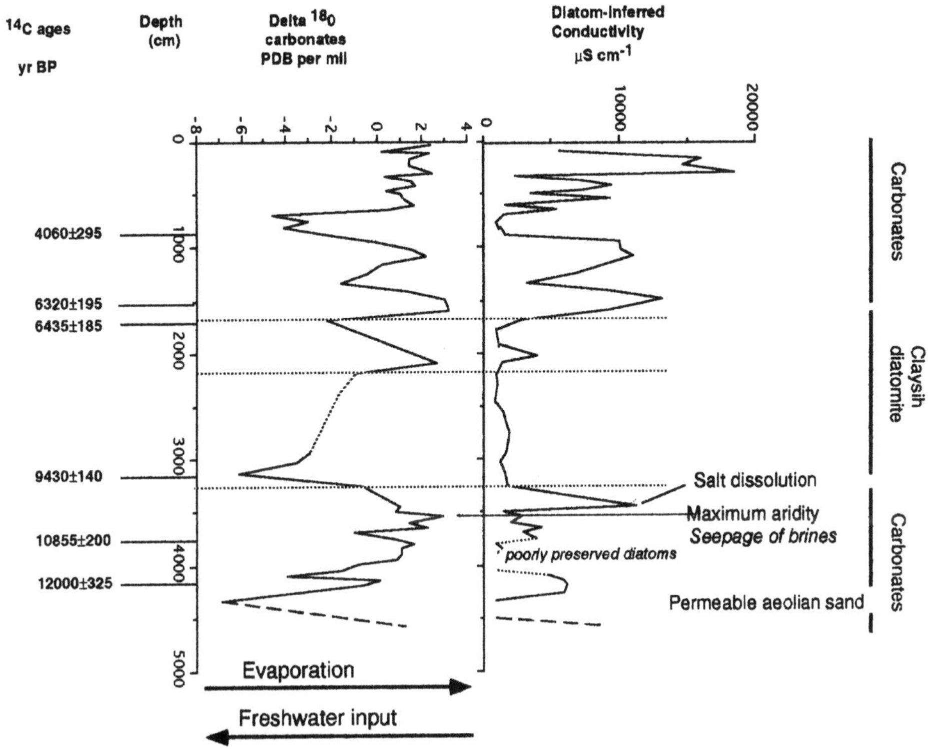

Figure 12. Content in ^{18}O of authigenic inorganic carbonate, and diatom-inferred conductivity for the Late Glacial and Holocene record of Boudgouma, Southern Niger. After Gasse et al., 1990 and Gasse, unpublished.Figure 12 . Content in ^{18}O of authigenic inorganic carbonate, and diatom-inferred conductivity for the Late Glacial and Holocene record of Boudgouma, Southern Niger. After Gasse et al., 1990 and Gasse, unpublished.

Although total salinity is correlated with the input and output terms of the water balance (see equations [12] to [15]), no model has been developed so far to infer climatic parameters from diatom-derived salinity. This would imply to take into account the effects of chemical precipitation over the salt budget, as illustrated by Figure 3b. A method for keeping out of this difficulty would be to consider the most conservative ions only, such as Cl^-, rather than total salinity.

4.3. Taphonomic problems : diatom preservation in saline waters

Distortions and biases to the fossil diatom record can arise in all environments from spatial and temporal mixing of communities and breakage of frustules, but in saline waters, an additional and more severe problem is caused by the chemical dissolution and diagenesis of diatom frustule silica. These phenomena occur during sedimentation in saline lakes and from contact between fossil diatoms and intersticial brines. Silica dissolution has long been recognized in the marine context where it is acute and can lead to erroneous interpretations (Shemesh et al., 1989), and where quantification of biogenic silica dissolution has recently been attempted (Pichon et al., 1992). If diatom based palaeoenvironmental reconstruction in continental waterbodies is to reach its full potential, the causes and implications of assemblage modification by dissolution need to be more thoroughly understood, with attention paid to the role of water ionic composition and concentration.

Dissolution experiments were performed on material from Lake Geneva, SE France. Dissolution of diatom silica from sediments were measured over a 92 day period at room temperature in six 3M solutions of different salts, in 0.6 M NaCl and in distilled water (Fig. 13 a). Methods, results and theorical considerations are thoroughly discussed in Barker et al. (1994). Results (Fig. 13 a) show different dissolution rate values, and also different saturation silica content. Dissolution is greater where high pH leads to ionic dissociation. In contrast, weakly acid solutions produce less dissolution than in distilled water. Similar patterns of silica solubility (rate and plateau) are observed for dissolution of living diatoms, monitored over 46 days in 3M solutions of Na_2CO_3 and NaCl, and distilled water (Fig. 13 b). An unknown inhibiting effect occurs, however, when the organic coating of frustules is preserved in contact with alkaline solutions. These results indicate that the amount of dissolved silica in saline solutions varies for different electrolytes of the same concentration and even of the ionic strenght.

Assuming that saturation is reached in our experiments, the differences occurring in the diatom assemblage composition after immersion in NaCl and Na_2CO_3 is also significant. Relatively little change is shown in the diatom composition of the samples dissolved in distilled water and NaCl, although total valve abundance declines gradually over 46 days. The most pronounced changes in the assemblages occurred after immersion in Na_2CO_3 where the diatom composition altered dramatically as a result of dissolution. For instance, the proportion of *Stephanodiscus neoastraea* increases in proportion from 6 to 80 % of the assemblage. The pattern of differential dissolution could be explained by differences in the surface area (SA) and volume (V) of individual diatom species.

The selective dissolution of diatoms in natural environments is thus a site specific phenomena, as the role of individual electrolytes differs by several orders of magnitude, and as changes in the diatom assemblage depend on the specific surface area of individual diatom taxa occurring in the original assemblage.

5. FRESHWATER DIATOM INFLUX IN INTERTROPICAL ATLANTIC : RELATIONSHIPS WITH CONTINENTAL RECORDS FROM AFRICA

During the last few years, several attempts were made to reconstruct atmospheric circulation and climate on the African continent using freshwater diatoms in Atlantic cores. Most of the authors (e.g. Parmenter and Folger, 1984, Pokras and Mix, 1985; Stabell, 1986 ; Pokras, 1987) assume that these freshwater diatoms are all windblown from Africa. During northern hemisphere summer, dust originating from the middle of the Sahara is transported westwards and concentrated along the continental margin from 10°N to 25°S (Prospero, 1981 ; Sarnthein et al., 1981) (Fig.14). In the northern winter, dust originates mainly along the southern edge of the Sahara, and the focus of activity is situated between 5°N and 15°N along the coast of Africa (Prospero, 1981). Over the continent, correlations between the late Quaternary glacial times and intertropical aridity, interpreted in terms of global climatic changes is clear (Kutzbach and Street-Perrott, 1985). Windblown freshwater diatoms in Atlantic sediments may thus be used as an index of enhanced winds and intertropical aridity. This is probably true for all fine sediments from midocean sites., and Pokras and Mix (1985) and Pokras (1987) used peaks of *Aulacoseira* (synonym with *Melosira*) in tropical Atlantic cores are a good indicator of aridity on the continent.

However, several authors also suggest that freshwater diatoms in Atlantic sediments may have been derived, at least at some localities, from fluvial transport (Kolbe, 1955 ; Melia, 1976 ; Mikkelsen, 1984 ; Gasse, 1989). Freshwater diatoms live in rivers and their increased influx in marine sediments can reflect increases in diatom productivity in inland water and/or changes in the river discharge. In that case, freshwater diatom peaks in Atlantic cores may indicate humid episodes.

The interpretation of freshwater diatoms in marine sediments raises the problems of identifying their source areas, the transporting agent, relative accumulation rates, and preservation. It is thus important to connect available data on freshwater diatoms in marine sediments with environment and diatom records from Africa, as shown below (after Gasse et al., 1989).

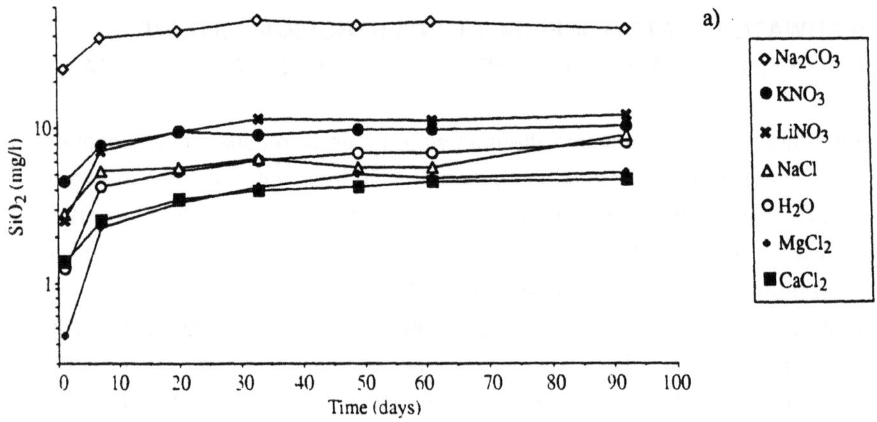

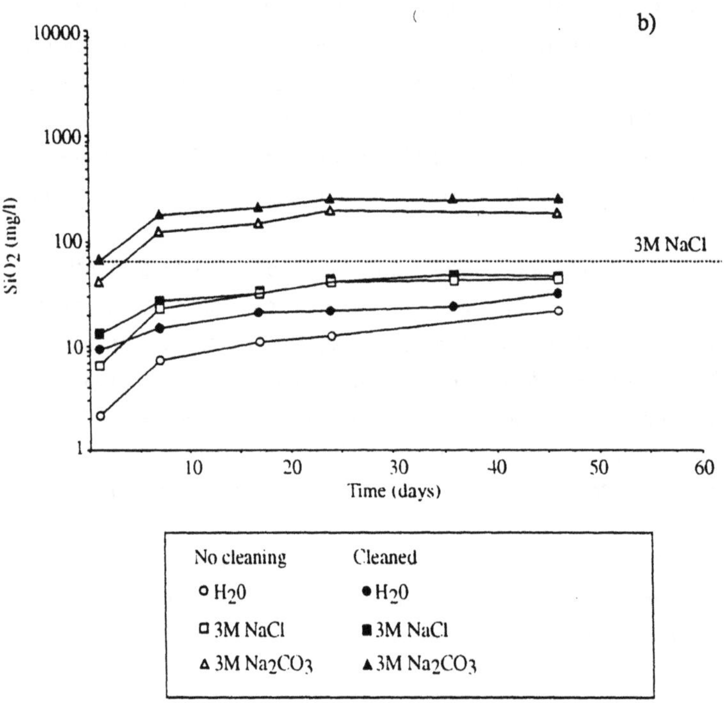

Figure 13

a) Changes in dissolved diatom silica concentration in various salt solutions againts time. Note log scale in ordinate. The experiment was performed on fossil diatoms from Lake Geneva. After Barker et al., 1993.

b) Silica concentation againts time for dissolution experiment based on living diatoms from Lake Geneva : a) uncleaned material, b) enzymatically cleaned (protease) material. After Barker et al., 1993.

5.1. Present-day sources of continental diatoms to the sea

Modern rivers

Diatom communities and rivers from west Africa can be roughly classified into three statistically derived groups on the basis of abundant taxa (Figs. 14, 15). Group 1 is found in small rivers with shallow acidic waters and very low mineral content. Group 2 is common in small rivers or littoral zones of large rivers with circumneutral waters. Group 3 is dominated by genus *Aulacoseira (A. granulata and varieties, A. agassizii, A., ambigua)* sometimes associated with *Stephanodiscus hantzschii* and *Cyclotella pseudostelligera.* It characterizes the plankton of large rivers. Genus *Aulacoseira* represents 66 to 80 % of the diatom population in the lower Congo and up to 98.5 % in the Niger river samples.

Diatom productivity in large rivers is high. In the low Congo, the water contains between 4 and 35 X 103 valves ml^{-1}. On the basis of the diatom contents in surface and bottom waters (mean value ≈ 10^4 valves ml^{-1}) and on the annual water discharge of the river (≈13 X 10^{11} m3 y^{-1}), the estimated freshwater influx to the sea is 13 X 10^{21} valves yr^{-1} (≈ 8.5 X 10 21 valves yr $^{-1}$ of *Aulacoseira).*

Lacustrine diatom-bearing sediments deflated in the Sahara and the Sahel

Late Quaternary diatom-bearing sediments (mainly Holocene) occur in a multitude of depressions through the Sahara and the Sahel. These sediments are submitted to intense deflation.

Great diversity in palaeoassemblages occur through space and time, due to the effects of local hydrological factors superimposed on those of major climatic fluctuations. In a given section, successive communities reflect changes in water depth and water chemistry during individual lacustrine cycle. Despite this complexity, palaeocommunities are statistically classified in two major groups (Fig. 16). Group I represents a continuum of assemblages which reflect changing water depth, and increasing alkalinity in water bodies of the carbonate-bicarbonate type.

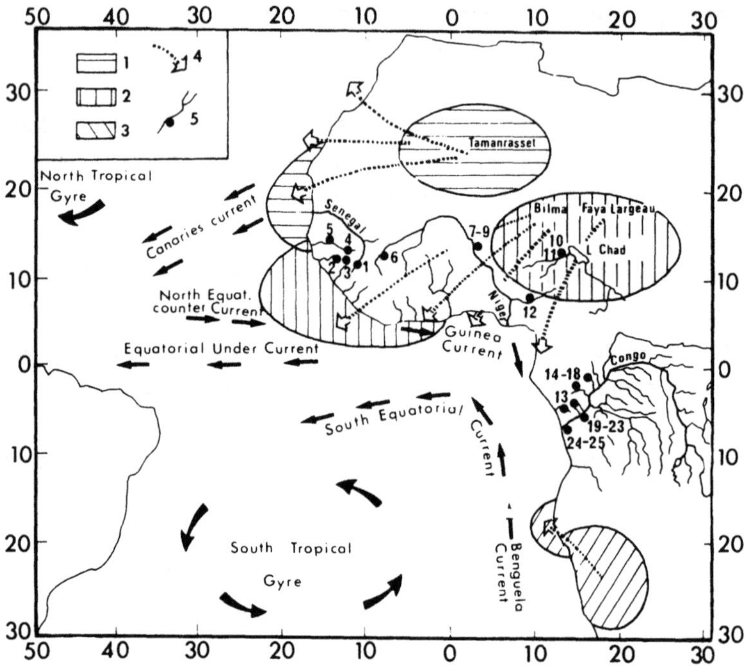

Figure 14. Present-day sources of freshwater diatoms in the Atlantic Ocean. Study area shown with zones having high occurrence of atmospheric haze, and approximate souce areas with dust trajectories: 1 = Northern Summer, 2 = Northern Winter, 3 = Southern Spring, 4 = dust trajectories. Ocean currents. 5 = Rivers with location of reference diatom samples. After Gasse et al., 1989.

As today, it predominates along the southern edge of the Sahara and in the Sahel. It is rich in *Fragilaria* spp. , *Aulacoseira* spp., *Stephanodiscus neoastraea* or *S. minutulus.* The species *Cyclotella ocellata* is abundant during the transgression of the Holocene lakes, and in Late Pleistocene sediments. Group II characterizes water of the chloride-sulfate type, and commonly saline. It is widespread in central and northern Sahara in depressions supplied by the Saharan aquifers. The most abundant and characteristic taxa are *Cyclotella caspia, Mastogloia* spp., *Amphora* spp. Genera *Stephanodiscus* and *Aulacoseira* are extremely rare or totally absent. The scarcity of *Aulacoseira* in oceanic core-top sediments north to 12°N (Pokras and Mix, 1985) may thus reflect the scarcity of this genus in the corresponding deflated sediments.

Despite the abundance of *Aulacoseira* spp. in both Sahelian lacustrine sediments and large rivers, each of these two possible sources of continental diatoms to the sea are characterized by some genera or species extremely rare or absent in the other one.

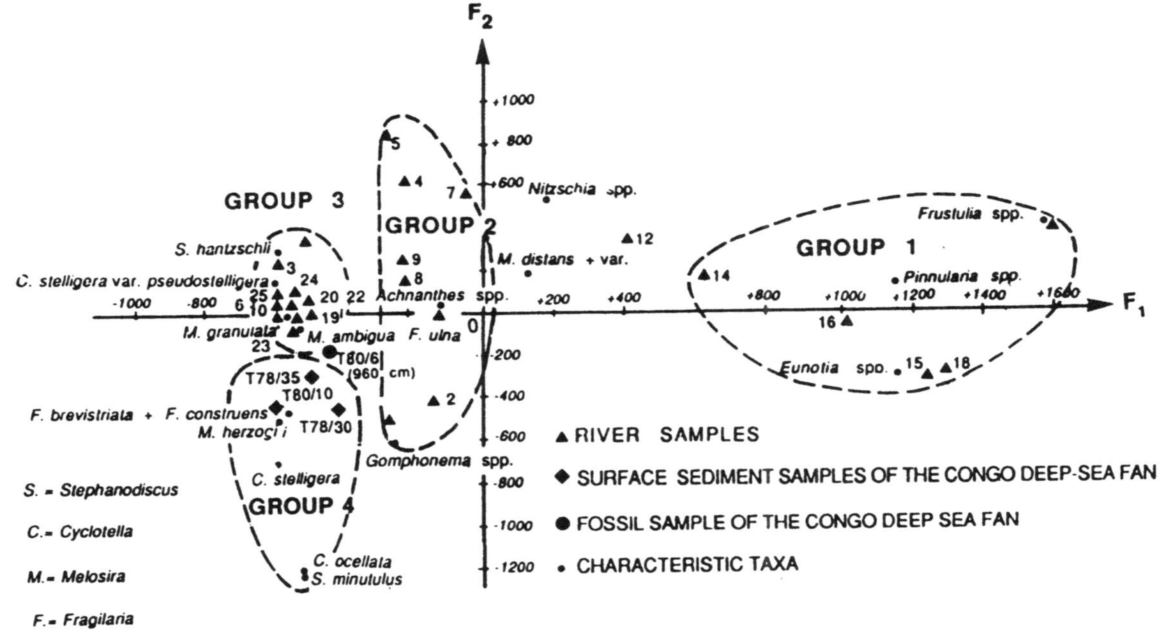

Figure 15. Factor analysis of correspondence based on diatom species percentages in river samples from West Africa (samples 1-25 located on Fig. 14), and in surfce sediment samples of Congo deep-sea fan (samples T78/30, T78/35, T80/10). The initial matrix (28 samples x 115 taxa) has been reduced by grouping species of similar ecology, belonging to a given genus (e.g., *Pinnularia* spp). Reduced matrix : 28 samples x 33 taxa. Projection on the factorial plane F1F2 (cumulated variance : 58.971 %) of the sample points (1-25) and of the most characteristic taxa (having a contribution greater than the mean). Note that Group 4 is greatly influenced by the contributions of *Cyclotella ocellata* and *Stephanodiscus minutulus*. The fossil sample T80/06 (depth : 940 cm) has been introduced as a supplementary (passive) element for comparison with modern assemblages. After Gasse et al., 1989.

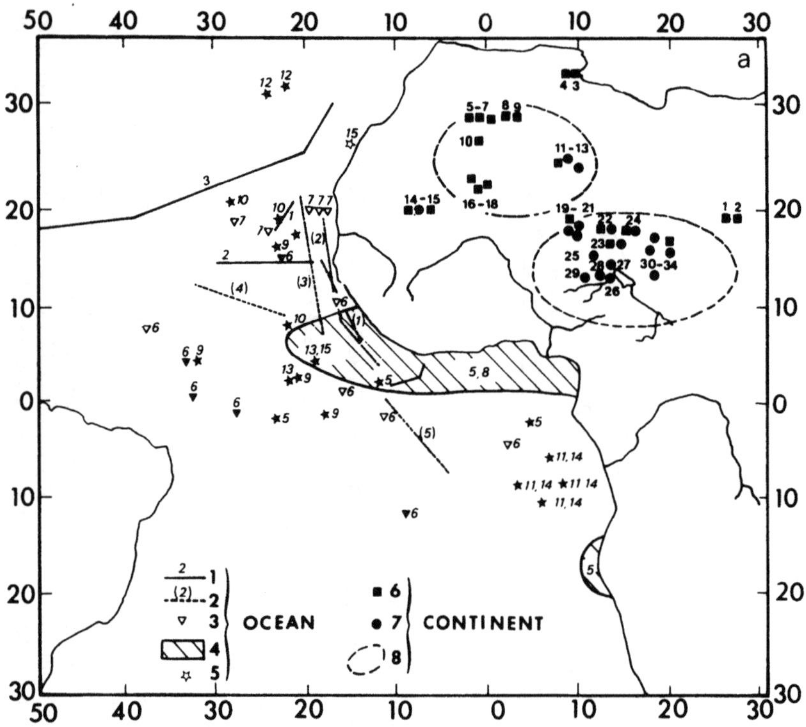

Figure 16. The freshwater diatom pattern in the Atlantic Ocean and the Late Quaternary continental diatom-bearing sediments. Atlantic : 1 = dust (literature data), 2 = dust (after Gasse et al., 1989), 3 = surface samples, 4 = areas with generally high abundance, 5 = core material. Filled symbols, dominated by *Aulacoseira granulata.* 6 = diatom assemblages of group I (of chloride and/or sulfate water) dominant, 7 = diatom assemblages of group II (of carbonate-bicarbonate water) dominant, 8 = major deflation areas. After Gasse et al., 1989.

5.2. Modern freshwater diatom influx in the intertropical Atlantic

Aulacoseira granulata is reported most commonly in dust samples collected over the Atlantic. However, detailed observations show that aeolian dust may contain diversified diatom species and that *Aulacoseira* is not always dominant. It is usually accompanied by lacustrine forms such as *Cyclotella ocellata, C. stelligera,* or *Stephanodiscus minutulus.*

In the Atlantic surface samples, an investigation by Maynard (1976) shows that *Cyclotella ocellata, Aulacoseira ambigua, A. granulata* and *Stephanodiscus "astraea "* are the predominant freshwater taxa. *Cyclotella ocellata* dominates the nearshore samples, while *A. granulata* dominates the more oceanic samples. The maximum abundance lies off Dakar with 1250 X 10^5 freshwater diatoms per gram carbonate-free sediment, and is dominated by *C. ocellata*. Both Melia (1984) and Pokras and Mix (1985) found high frequencies of freshwater diatoms between 0° and 10°N, especially in the Gulf of Guinea toward the Niger delta, but did not report on the species distribution. Maynard (1976) and Pokras and Mix (1985) relate the high abundance of freshwater diatoms in surface sediments to aeolian transport and particularly to the winter plume.

In the surface sediments of the Congo deep sea-fan, freshwater diatom transport is mainly due to turbidity currents and plume waters of the Congo river. The distribution of the freshwater diatoms clearly differs from that of the phytoliths which are windblown from soils. The surface assemblage (Fig. 15, Group 4) are close to those of the Congo river, but with higher percentages of *Cyclotella stelligera* and of *Fragilaria*, and lower frequencies of *Aulacoseira* spp. They also differ by the occurrence of *S. minutulus* and *C. ocellata*, taxa common in lakes but not recorded in the Congo river, and of phytoliths. This suggests that wind is also a present-day transporting agent.

5.3 Freshwater diatoms in the sediment cores of the Congo deep-sea fan

During the last 40,000 yr, freshwater diatom peaks observed in phytolith-free levels at ≈3000, 11,000, 35,000-40,000 yr BP can be related to humid, warm climatic phases in the lower Congo basin (Fig. 17) . The older freshwater peak occurring around 128,000 yr is also assumed to reflect a humid phase in equatorial zones, synchronous with the last interglacial period. High percentages of freshwater diatoms usually correlate with poor preservation of marine diatom valves, as found with surface sediments. The relatively low concentration of marine diatoms may be enhanced by their selective dissolution, or by a lower marine productivity during interglacial periods. A freshwater flora with perfectly preserved frustules is observed in a turbidite at about 300,000 yr. The flora exhibits clear analogies with the taphocenose of the surface sediments, but the absence of forms only found in lakes favours the hypothesis of a river origin. By contrast, the composition of the freshwater assemblages and the abundance of phytoliths in stage 2, 6 and 8 in coreT80/6 suggest important aeolian transport during these periods.

This example shows that a detailed identification of the allochtonous material allows to distinguish between the two possible sources of freshwater diatoms in marine sediments.

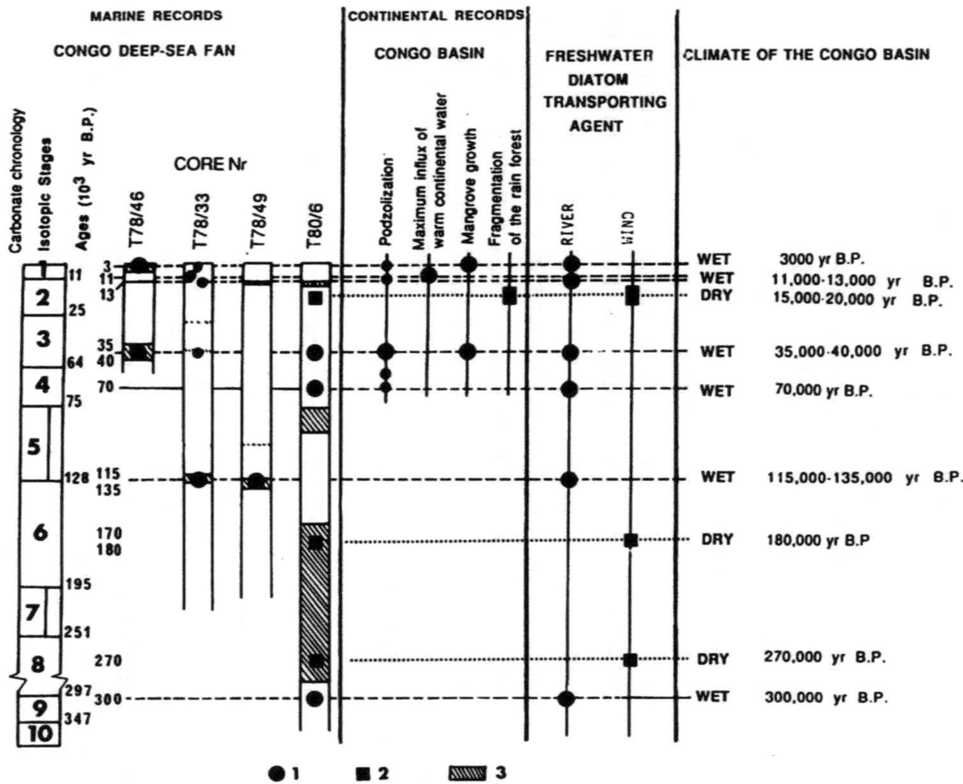

Figure 17. Tentative climatic interpretation of the freshwater diatom peaks in Quaternary cores of the Congo deep-sea fan in light of the climatic continental events recorded on the Congo basin. After Gasse et al., 1989.

6. CONCLUSIONS

This paper illustrates some applications of continental diatom studies for palaeoenvironmental, palaeoclimatic reconstruction. At large geographical scales, freshwater diatoms transported to the ocean inform on environmental conditions over the continents. A detailed analysis of the allochtonous material in marine sediments, connected with a knowledge of past and modern diatom communities on the continents, allow the identification of the source areas and of the transporting agent. This is especially important for nearshore material since freshwater diatoms in marine cores can reflect either enhanced winds and aridity, or reinforced river discharge associated to humid phases.

The examples selected here also demonstrate the utility of the diatom techniques in reconstructing past changes in continental hydrology. We shows that, in favourable cases, diatoms provide quantitative informations on changes in lake water depth and water concentration. It is obvious at a first glance that these hydrological variables depends on climatic conditions. The interpretation of past hydrology in terms of palaeoclimate is possible. This approach requires, however, a thorough knowledge of the regional hydrological and hydrogeological conditions, and of the hydrological and chemical processes which may damp or alter the climatic signal. If diatom-based proxy data is to reach its full potential, the causes of changes in water depth, water salinity and taphonomic modification need to be thoroughly understood through hydrological and chemical studies.

REFERENCES

Barker, P., Fontes, J.Ch., Gasse, F. & Druart, J.C., 1994 - Experimental dissolution of diatom silica in concentrated salt solutions and implications for palaeoenvironmental reconstruction. Limnol. Oceanogr., 39(1):99-110.

Barker, P., Roberts, C.N., Lamb, H.F., Van der Kaars, S. & Ben Kaddour, A., 1993 - Interpretation of Holocene lake-level change based on diatom assemblages from Lake Sidi Ali, Middle Atlas, Morocco. VIth Int. Palaeolimnology Symp., Canberra, April 1993.

Bergonzini, L. & Gasse, F.,. - Palaeohydrologic and palaeoclimatic interpretation of level fluctuations of Lake Tanganyika since the Last Glacial Maximum. Submitted to Quat. Res.

Birks, H.J.B., Line, J.M., Juggins, S., Stevenson, A.C. & Ter Braak, C.J.F., 1990 - Diatoms and pH reconstruction. Phil. Trans. Roy. Soc. Lond., B 327, 263-278.

Bonnefille, R., Chalié, F., Guiot, J. & Vincens, A., 1992 - Quantitative estimates of full glacial temperatures in equatorial Africa from palynological data. Climate Dynamics, 6:251-257.

Bultot, F., 1993 - Evaporation from a tropical lake : comparison of theory with direct measurements-comment. Journal of Hydrology, 143:513-519.

Chalié, F., 1992 - Paléoclimatologie quantitative en Afrique orientale depuis 40 000 ans. Thèse de l'Université de Droit, d'Economie et des Sciences Nat. d'Aix-Marseille, 212 pp.

Charles, D.F., Battarbee, R.W., Renberg, I., Van Dam, H. & Smol, J.P., 1989 - Palaeoecological analysis of lake acidification trends in North Africa and Europe using diatoms and chrysophytes. Acid Precipitations. Soils Aquatic Processes & Lake Acidifications, 4:207-276.

Charles, D.F. & Smol, J.P., 1990 - The Pirla II Project : Regional assessment of lake acidifications. Verh. Internat. Verein. Limnol., 24:474-480.

CLIMAP Project Members, 1976 - The surface of the ice-age. Earth Sci., 191:1131-1137.

Degens, E.T. & Hecky, R.E., 1973 - Palaeoclimatic reconstitution of Late Pleistocene and Holocene based on biogenic sediments from the Black Sea and a tropical African lake. Colloques Internationaux du CNRS, 219, 12.

Fritz, S.C., 1990 - Twentieth-century salinity and water-level fluctuations in Devils Lake, North Dakota : tests of a diatom-based transfert function. Limnol. Oceanogr., 35(8):1771-1781.

Fritz, S.C., Juggings, S., Battarbee, R.W. & Engström, D.R., 1991 - Reconstruction of past changes in salinity and climate using a diatom-based transfer function. Nature, 352:706-708.

Gasse, F., Talling, J.F. & Kilham, P., 1983 - Diatom assemblages in East Africa : classification, distribution and ecology. Rev. Hydrobiol. Trop., 16, 1:85-90.

Gasse, F. & Tekaia, F., 1983 - Transfer fuctions for estimating paleocological conditions (pH) from East African diatoms. Hydrobiologia, 103:85-90.

Gasse, F., 1986a - East African Diatoms. Taxonomy, ecological distribution. Bibliotheca Diatomologica, Bd. 11, 202 pp. J. Cramer, Stuttgart.

Gasse, F., 1986b - East African diatoms and water pH. In "Smol, J.P. et al. Eds., Diatoms and Lake Acidity. Developments in Hydrobiology", Junk Publ., Dordrecht, 29:149-168.

Gasse, F., 1987 - Diatoms for reconstructing palaeoenvironments and palaeohydrology in tropical semi-arid zones. Examples of some lakes from Niger since 12,000 B.P. Hydrobiologia, 154:127-163.

Gasse, F., Lédée, V., Massault, M. & Fontes, J.Ch., 1989 - Water-level fluctuations of Lake Tanganyika in phase with oceanic changes during the last glaciation and deglaciation. Nature, 342:57-59.

Gasse, F., Stabell, B., Fourtanier, E. & Iperen Van, Y., 1989 - Freshwater diatom influx in intertropical Atlantic : relationships with continental records from Africa. Quat. Res., 32:229-243.

Gasse, F. & Fontes, J.Ch., 1992 - Climatic changes in Northwest Africa during the last deglaciation. In "The Last deglaciation : absolute and radiocarbon chronologies", E. Bard & W.S. Wallace Eds., Eds. Nato ASI Series, Springer-Verlag Berlin, vol. 12, 295-325.

Griffiths, J.F., 1972 - Worl survey of climatology. Vol. 10, Climates of Africa, Ed. J.F. Griffiths Elsevier Pub. Cie, Amsterdam, 604 pp.

Haberyan, K.A. & Hecky, R.E., 1987 - The Late Pleistocene and Holocene stratigraphy and paleolimnology of Lakes Kivu and Tanganyika. Palaeogreogr., Palaeoclimatol., Palaeoecol., 61:169-197.

Hastenrath, S.L. & Kutzbach, J.E., 1983 - Paleoclimatic estimates from water and energy budgets of East African Lakes. Quaternary Research, 19:141-153.

Kolbe, R.W., 1955 - Sediment cores from the North Atlantic Ocean. Reports of Swedish Deep-Sea Expedition, 1947-1948, 7:151-184.

Kutzbach, J.E., 1980 - Estimates of past climate at Palaeolake Chad, North Africa, based on a hydrological and energy-balance model. Quat. Res., 14:210-223.

Kutzbach, J.E. & Street-Perrott, F.A., 1985 - Milankovitch forcing of fluctuations in the level of tropical lakes from 18 to 0 kyr B.P. Nature, London, 317:130-134.
Maynard, N.C., 1976 - Relationship between diatoms in surface sediments of the Atlantic Ocean and the biological and physical oceanography of overlying waters. Paleobiology, 2:99-121.
Melia, M.B., 1984 - The distribution and relationship between palynomorphs in aerosols and deep-sea sediments off the coast of Northwest Africa. Marine Geology, 58:345-371.
Mikkelsen, N., 1980 - Experimental dissolution of Pliocene diatoms. Nova Hedwigia, 33:893-907.
Parmenter, C. & Folger, D.W., 1984 - Eolian biogenic detritus in deep sea sediments : a possible index of equatorial Ice-Age aridity. Science, 185:695-698.
Pichon, J.J., Labracherie, M., Labeyrie, L. & Duprat, J., 1987 - A transfer function to estimate past sea surface conditions from core top distribution of marine diatom assemblages in Atlantic and W. Indian sectors of Southern Ocean. Palaeogeogr., Palaeoclimat., Palaeoecol., 61:79-95.
Pichon, J.J., Bareille, G., Labracherie, M., Labeyrie, L., Baudrimont, A. & Turon, J.L., 1992 - Quantification of the biogenic silica dissolution in Southern Ocean sediments. Quaternary Research, 37:115-139.
Pichon, J.J., Labeyrie, L., Bareille, G., Labracherie, M., Duprat, J. & Jouzel, J., 1992 - Surface water temperature changes in the high latitudes of the Southern hemisphere over the Last Glacial-Interglacial cycle. Palaeooceanography, 7:289-312.
Pokras, E.M., 1987 - Diatom record of Late Quaternary climatic changes in the eastern equatorial Atlantic and tropical Africa. Paleooceanography, 2(3):273-286.
Prentice, I.C., Guiot, J. & Harrison, S.P., 1992 - Mediterranean vegetation, lake levels and palaeoclimate at the Last Glacial Maximum. Nature, 360:658-660.
Prospero, J.M., 1981 - Arid regions as souces for mineral aerosols in the marine environments. In "Desert Dust : origin, characteristics, and effects of Man" (T.L. Péwé, Ed.), Geol. Soc. Amer. Spec. Paper, 186:71-86.
Sarthein, M., Erlenkeuser, H. & Zahn, R., 1982 - Termination I : The response of continental climate in the subtropics as recorded in deep-sea sediments. Bull. Inst. Géol. du Bassin d'Aquitaine, 31:393-407.
Shemesh, A., Burkle, C.L. & Froelich, P.N., 1989 - Dissolution and preservation of Antarctic diatoms and the effects on the sediments thanatocoenoses. Quat. Res., 31:288-308.
Stabell, B., 1986 - Variations of diatom flux in the eastern equatorial Atlantic during the last 400,000 years ("Meteor" cores 13519 and 13521). Mar. Geol., 72:305-323.
Stoffers, P. & Hecky, R.E., 1978 - Late Pleistocene-Holocene evolution of the Tanganyika Basin. Spec. Pub. Int. Ass. Sediments, 2:43-55.
Street, F.A. & Grove, A.T., 1976 - Environmental and climate implications of Late Quaternary lake-level fluctuations in Africa. Nature, 261:385-390.
Street-Perrott, F.A. & Roberts, N., 1983 - Fluctuations in closed-basin as an indicator of past atmospheric circulation pattern. In "Variations in the Global Water Budget" (Street-Perrott, F.A., Beran, M. & Ratcliffe, R.A.S., Eds.), Reidel, Dordrecht, 331-345.
Van Campo, E. & Gasse, F., 1993 - Pollen and diatom-inferred climatic and hydrological changes in Sumxi Co Basin (Western Tibet) from 13,000 yr B.P. Quat. Res., 39:300-313.

ANDEAN SNOWLINE EVIDENCE FOR COOLER SUBTROPICS AT THE LAST GLACIAL MAXIMUM

Geoffrey O. Seltzer
Byrd Polar Research Center
The Ohio State University
108 Scott Hall
1090 Carmack Road
Columbus, Ohio 43210
USA

INTRODUCTION

A common perception exists that glacier snowlines were depressed about 1000 m in American alpine regions during the Last Glacial Maximum (dated in marine sediments between 23 and 14 kyr BP by the $\delta^{18}O$ proxy for ice volume) as a result of a temperature reduction equivalent to a lapse rate of 5°- 6°C/km (e.g. Broecker and Denton, 1989). This contradicts reconstructions of sea-surface temperatures that are <2°C cooler in low-latitude regions (CLIMAP, 1981; Webster and Streten, 1978; Rind and Peteet, 1985; Rind, 1990). In this work I show that snowlines were not uniformily depressed 1000 m in the Andean subtropics and the revised estimates of temperature reduction are closer to the CLIMAP findings. However, glaciation in the tropical-subtropical latitudes of the Americas probably culminated twice during the late Pleistocene, more than 20 kyr BP and just prior to deglaciation about 14 kyr BP, making a direct comparison of the snowline evidence to the CLIMAP findings potentially misleading.

The evidence for temperature change in low-latitude regions during the Last Glacial Maximum is equivocal. Webster and Streten (1978) were the first to point out that there is a discrepancy between the CLIMAP reconstructions of sea-surface temperature and estimates of temperature reduction based on snowline and treeline depression in tropical Australasia. This resulted in questioning of the CLIMAP results (Rind and Peteet, 1985), which were subsequently confirmed, in part, by Prell (1985) and Broecker (1986). Recent pollen studies indicate that temperature reduction in tropical East Africa did not differ greatly from sea-surface values (Bonnefille et al., 1990; Van Campo et al., 1990), while for tropical Central and South America the differences are still large (Bush et al., 1990; Piperno et al., 1990; van der Hammen, 1988; Hooghiemstra et al., 1992).

The discrepancy is not easily resolved by invoking steeper lapse rates to explain cooler continental temperatures at the Last Glacial Maximum. Even in dry regions of the tropics the variation of temperature with elevation is close to the moist adiabatic lapse rate (e.g. Webster and Streten, 1978; Rind and Peteet, 1985). Thus it remains uncertain which proxies contain the

NATO ASI Series, Vol. I 22
Long-Term Climatic Variations
Edited by J.-C. Duplessy and M.-T. Spyridakis

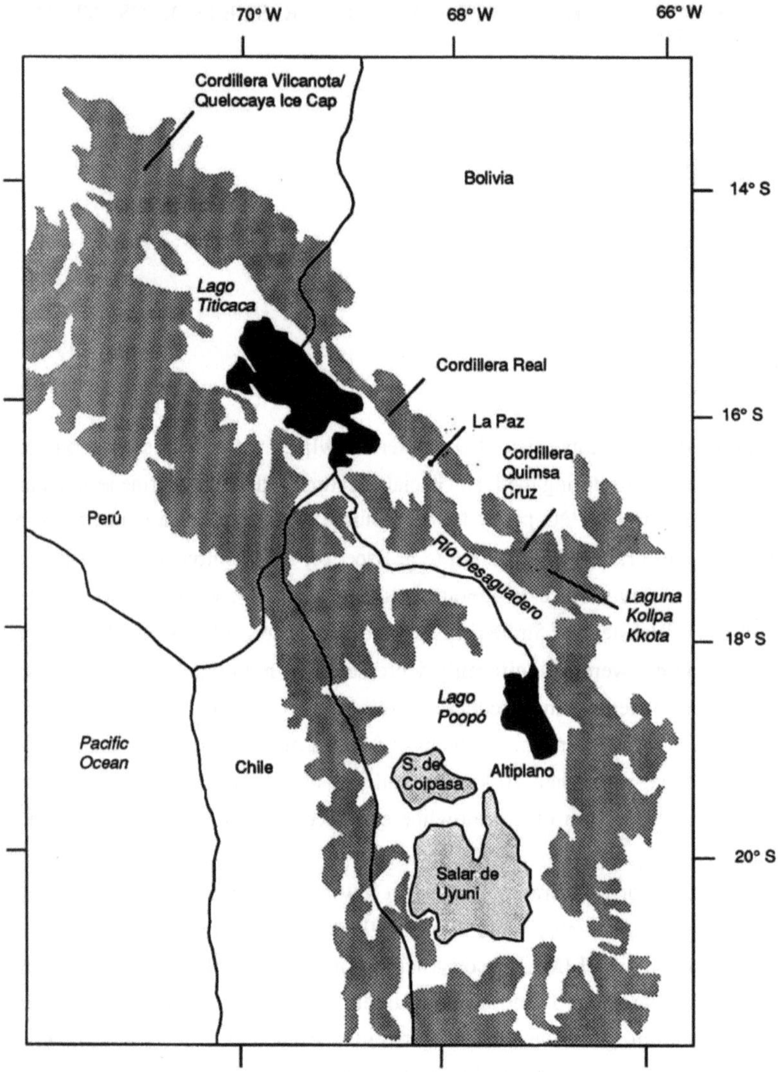

Figure 1. Physiographic map of the central Andes. Shaded area is > 4000 m in elevation.

error or whether a mechanism exists by which a small change in sea-surface temperatures could be consistent with significantly cooler temperatures over adjacent continents.

This unresolved problem stems from a lack of understanding of the climatic response of low-latitude regions to full-glacial conditions. It may also signify that we do not know how to interpret accurately the paleoclimatic proxies in terms of temperature change (e.g. Kutzbach and

Ruddiman, in press). However, sea-surface temperatures are an important boundary condition in general circulation models of paleoclimate whose output is highly dependent on the prescribed initial conditions (e.g. Rind and Peteet, 1985; Crowley, 1988). If the simulations are to be valid it is therefore important to resolve this discrepancy.

SNOWLINE DEPRESSION IN THE SUBTROPICAL ANDES

New evidence from Laguna Kollpa Kkota (17°26'S, 67°88'W, 4400 m) helps to constrain the magnitude of snowline depression during the late Pleistocene in the subtropical Andes. The lake is located on the western slope of the Eastern Cordillera in Bolivia (Fig. 1) and formed after deglaciation as terminal moraines dammed the local drainage. The glacier that occupied the valley was <1 km^2 in area and overrode bedrock primarily composed of Paleozoic shales. Radiocarbon dates from cores 7 m long obtained in the lake indicate that postglacial sedimentation began at least 20 kyr BP (Table 1). These dates also indicate that the watershed has not been glaciated over the last 20 kyr, as there are no moraines between the lake and the headwall of the valley. Thus the glacial geomorphology of the site can be used to estimate the maximum extent of snowline depression for the last 20 kyr.

Table 1. Radiocarbon dates from Laguna Kollpa Kkota, Bolivia.

Depth (cm)	Material	Date (yr B.P.)[a]	Laboratory Number
Core C			
205-206	Macrofossil	12,950 ± 120	CAMS-2389
	Macrofossil	12,610 ± 80	CAMS-2390
	Humic Acid Extract	12,660 ± 90	CAMS-2391
241-242	Macrofossil	13,130 ± 170	CAMS-2403
	Humic Acid Extract	12,900 ± 90	CAMS-2404
353-358	Bulk Organic Matter	13,960 ± 250[b]	BETA-50444
382-392	Bulk Organic Matter	13,290 ± 270[b]	BETA-50445
435-447	Bulk Organic Matter	18,130 ± 280[b]	BETA-51436
487-488.5	Humic Acid Extract	17,670 ± 120	CAMS-2392
	Humin Extract	19,480 ± 150	CAMS-2395
	Humin Extract	20,140 ± 160	CAMS-2433
488.5-490	Humic Acid Extract	17,580 ± 170	CAMS-2393
	Humic Acid Extract	19,440 ± 670	CAMS-2394
Core A			
404-411	Bulk Organic Matter	17,690 ± 780[b]	BETA-48225

[a] Accelerator-mass spectrometer dates except those of bulk organic matter.
[b] ^{13}C corrected date.

The moraines that bound Laguna Kollpa Kkota are older than those dated to a phase of late-Pleistocene glaciation in the Cordillera Real, Bolivia (Fig. 1). These moraines have a minimum age estimate of 10,460 ± 140 yr BP derived from basal lake sediments (Seltzer, 1992) and a maximum age estimate of 16,610 ± 130 yr BP provided by buried peat (Gouze et al., 1986). It is assumed that the late-Pleistocene moraines in the Cordillera Real are correlative with the culmination of glaciation in the Cordillera Vilcanota in southern Perú, dated to about 14 kyr BP (Mercer, 1984; Seltzer, 1990, 1992). Basal dates from lake sediments and peatlands throughout Perú and Bolivia suggest deglaciation from this late-Pleistocene glacial maximum began at least 12 kyr BP (Seltzer, 1990, 1992).

Estimates of snowline depression also support the hypothesis that the moraines at Laguna Kollpa Kkota predate those dated to about 14 kyr BP in the Cordillera Real. Based on the toe-to-headwall ratio method (Meierding, 1982) the snowline was at 4460 m when the Laguna Kollpa Kkota valley was glaciated and the snowline was depressed about 600 m from present (Fig. 2). The estimated snowline depressions at 14-12 kyr BP in the Cordillera Real and the Cordillera Quimsa Cruz are 300 ± 100 m and 465 ± 70 m, respectively (Seltzer, 1992; Müller, 1985). Thus it appears that glaciation at 14 kyr BP was less extensive then the earlier phase dated to >20 kyr BP. Snowline was never depressed lower than the crest of the headwall of the Laguna Kollpa Kkota valley (4560 m), or about 500 m below the modern snowline, over the last 20 kyr.

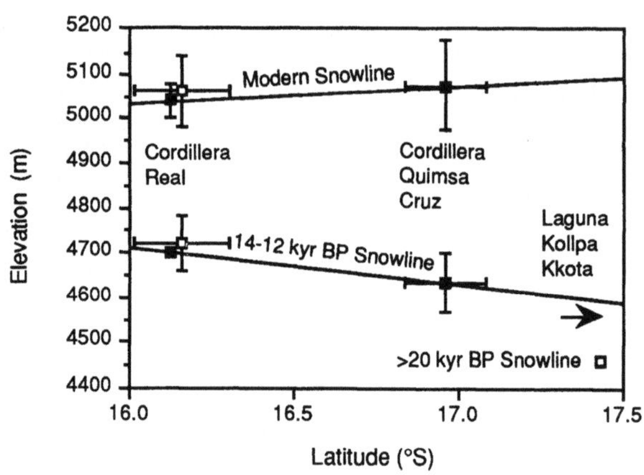

Figure 2. Modern and late-Pleistocene snowlines for the western slope of the Eastern Cordillera, Bolivia. The horizontal and vertical bars represent the latitudinal range of the sites and the variation in estimates of snowline, respectively. Snowline estimates are based on the accumulation area ratio (solid squares, AAR = 0.77) and the toe-to-headwall ratio (open squares, THAR = 0.37) methods. The arrow indicates the elevation of the crest of the headwall above Laguna Kollpa Kkota.

Snowline reconstructions for the eastern side of the Eastern Cordillera in Bolivia are more problematic because steep terrain and high precipitation cause erosion of glacial features. Also, there are no radiocarbon dates to provide age estimates of the moraines. However, glaciers clearly reached 3300 m, according to remnants of moraines and the transition from broad to narrow valleys, with a maximum snowline depression of about 1000 m based on the toe-to-headwall ratio method (Müller, 1985; Seltzer, 1992). If the glacial limit on the eastern side of the cordillera was reached at the same time as late-Pleistocene glaciation culminated on the western side (>20 kyr BP) then a snowline gradient existed across the divide that was steeper than today's. A similar relation has been observed in the Eastern Cordillera of the Peruvian Andes (Wright, 1983; Seltzer, 1987; Rodbell, 1992).

CLIMATIC INTERPRETATION

The role of precipitation in glaciation is important to consider in order to estimate accurately the temperature change from snowline depression. In the Bolivian Andes there are two lines of evidence that indicate how precipitation may have affected glaciation. The steeper east-west gradient in snowline during the late Pleistocene suggests glaciation was, in part, caused by an increase in precipitation. The moisture source for the region is the easterlies. A greater increase in precipitation on the eastern slope compared to the western slope of the Eastern Cordillera because of the orographic effect, and proportionally more snow compared to rain because of lower temperatures, could explain the steeper snowline gradient. Also, at least two large lakes covered much of the southern Altiplano in the late Pleistocene (Servant and Fontes, 1978), where today there are only *salars* (salt flats) (Fig. 1). In considering the possible climatic changes that could produce these lakes Hastenrath and Kutzbach (1985) used a simple energy balance model and Kesseler (1988) reasoned from the modern hydrologic balance of the region that precipitation would have to increase about 30 %. Given the proximity of the glacial valleys to these lakes and the modern aridity and low temperatures in the cordillera, it is difficult to reason that the increase in precipitation that produced the lakes did not simultaneously affect glaciation (e.g. Hastenrath, 1971). The dating of the high lake stands and the glacial sequence suggests they may have been time synchronous (Seltzer, 1992). Thus, late-Pleistocene glaciation in this area was probably associated with both a decrease in temperature and an increase in precipitation.

If we assume that snowline depression was solely a result of a temperature change without taking into account changes in precipitation, humidity, or net radiation (e.g. Porter et al., 1983; Seltzer, 1992), then a maximum snowline depression of 1000 m for the last 20 kyr on the eastern slope of the Eastern Cordillera of Bolivia, multiplied by a lapse rate of 6.5°C/km, suggests that temperature was reduced about 6.5°C. However, if the postulated increase of precipitation associated with glaciation was less significant to the west because of the orographic effect, then an estimate of temperature reduction of <3.9°C based on multiplying the lapse rate by the maximum snowline depression of 600 m on the western slope of the Eastern Cordillera would be closer to the true value of temperature reduction. Additionally, if snowlines were depressed by 300 ± 100 m, as on the western slope of the Cordillera Real about 14 kyr BP, and if it is assumed that this glacial advance was associated with the increase in precipitation that led to the formation of the last large lake on the Altiplano, then the temperature depression associated with the latest-Pleistocene glaciation would be 3.5°±1.6°C (Seltzer, 1992). These conclusions indicate that temperatures based on snowline depression were never reduced by more than a few degrees during the last 20 kyr in the Bolivian Andes.

Alternatively, temperatures during the Last Glacial Maximum may have been much colder, but the environment was drier, resulting in less extensive glaciation compared to 14 kyr BP and prior to 20 kyr BP (e.g. van der Hammen et al., 1981). Evidence exists throughout the tropical-subtropical latitudes of the Americas to suggest that the glacial maximum in this region predated that of the mid-latitudes of the Northern Hemisphere (about 18 kyr BP), and that a less extensive glaciation culminated just prior to deglaciation (Mercer and Palacios, 1977; van der Hammen et al., 1981; Wright, 1983; Heine, 1984; Mercer, 1984; White, 1986; Helmens, 1988; Seltzer, 1992; Rodbell, 1993). There are few moraines dated to 18 kyr BP, and the ones that are (e.g. Helmens, 1988) do not represent the glacial maximum. Therefore it is critical to evaluate the moisture conditions at 18 kyr BP to determine if a cold and dry climate could have existed that is not represented by the snowline proxy.

Comparison of temperature reconstructions based on snowline depression and low-elevation proxies also depends on the environmental lapse rate for the time period in question. A change in the lapse rate by only a few tenths of a degree becomes significant when extrapolated over 5 km of altitude. However, if glaciation were associated with an increase in precipitation in the Bolivian Andes, then lapse rates would be expected to have been higher, which is opposite to what is needed to explain the apparent discrepancy between temperature estimates at low and high altitudes.

CONCLUSIONS

Three factors need to be carefully considered when determining if there is a discrepancy between estimates of temperature reduction based on snowline depression and those of CLIMAP for sea-surface conditions in low latitudes. (1) The timing of glacial maxima, as it appears that low-latitude glaciation may actually be out of phase with the maximum ice volume registered by the $\delta^{18}O$ stratigraphy of marine carbonates, and a direct comparison with the CLIMAP findings may be misleading. (2) The amount of snowline depression needs to be determined by explicit means and the existence of snowline gradients across a region must be evaluated. (3) Climatic interpretation of snowline depression should at least consider the role of precipitation in glaciation to accurately estimate temperature change.

The glacial moraine proxy for paleotemperatures in the subtropical Andes does not appear to be in great conflict with the CLIMAP estimates of sea-surface temperature, considering the uncertainties in both estimates, which are at least 1°-2°C. However, the dating of the glacial moraine sequence needs to be improved because the snowline proxy may not be synchronous with the CLIMAP data. Also, the moisture conditions for the last 20,000 years in the region need to be further evaluated to determine if a dry but cold climate could have existed during the Last Glacial Maximum that is not reflected by the glacial-geologic data.

REFERENCES

Bonnefille, R., Roeland, J. C., and Guiot, J. (1990). Temperature and rainfall estimates for the past 40,000 years in equatorial Africa. Nature, 346, 347-349.

Broecker, W. S. (1986). Oxygen isotope constraints on surface ocean temperatures. Quaternary Research, 26, 121-134.

Broecker, W. S., and Denton, G. H. (1989). The role of ocean-atmosphere reorganizations in glacial cycles. Geochimica et Cosmochimica Acta, 53(10), 2465-2501.

Bush, M. B., Colinvaux, P. A., Wiemann, M. C., Piperno, D. R., and Liu, K. (1990). Late Pleistocene temperature depression and vegetation change in Ecuadorian Amazonia. Quaternary Research, 34(3), 330-345.

Crowley, T. J. (1988). Paleoclimate Modelling. In M. E. Schlesinger (Ed.), Physically-Based Modelling and Simulation of Climate and Climatic Change-Part II (pp. 883-949). Kluwer Academic Publishers.

CLIMAP (1981). Seasonal reconstructions of the Earth's surface at the last glacial maximum. Geological Society of America Map and Chart Series, MC-36: Geological Society of America.

Gouze, P., Argollo, J., Saliège., J. -F., and Servant, M. (1986). Interprétation paléoclimatique des oscillations des glaciers au cours des 20 derniers millénaires dans les régions tropicales: exemple des Andes boliviennes. Comptes Rendus de l'Académie des Sciences Paris, Série II, 303, 219-223.

van der Hammen, T. (1988). South America. In B. Huntley, & T. I. Webb (Eds.), Vegetation History (pp. 313-337). Kluwer Academic Publishers.

van der Hammen, T., Barelds, T., De Jong, H., and De Veer, A. A. (1981). Glacial sequence and environmental history in the Sierra Nevada Cocuy (Colombia). Palaeogeography, Palaeoclimatology, Palaeoecology, 32, 247-340.

Hastenrath, S. (1971). On the Pleistocene snow-line depression in the arid regions of the South American Andes. Journal of Glaciology, 10, 255-267.

Hastenrath, S., and Kutzbach, J. (1985). Late Pleistocene climate and water budget of the South American Altiplano. Quaternary Research, 24, 249-256.

Heine, K. (1984). Comment on "Plesitocene Glaciation of Volcano Ajusco, Central Mexico, and Comparison with the Standard Mexican Glacial Sequence" by Sidney E. White and Salvatore Valastro, Jr. Quaternary Research, 22(2), 242-246.

Helmens, K. F. (1988). Late Pleistocene glacial sequence in the area of the high plain of Bogotá (eastern cordillera, Colombia). Palaeogeography, Palaeoclimatology, Palaeoecology, 67, 263-283.

Hooghiemstra, H., Cleef, A. M., Noldus, G. W., and Kappelle, M. (1992). Upper Quaternary

vegetation dynamics and palaeoclimatology of the La Chonta bog area (Cordillera de Talamanca, Costa Rica). Journal of Quaternary Science, 7(3), 205-226.
Kessler, A. (1984). The paleohydrology of the late Pleistocene Lake Tauca on the southern Altiplano (Bolivia) and recent climatic fluctuations. In J. C. Vogel (Ed.), Late Cainozoic Palaeoclimates of the Southern Hemisphere (pp. 115-122). Rotterdam: A.A. Balkema.
Kutzbach, J. E., and Ruddiman, W. F. (in press). Model description, external forcing, and surface boundary conditions. In H. E. J. Wright, J. E. Kutzbach, T. I. Webb, W. F. Ruddiman, F. A. Street-Perrott, & P. J. Bartlien (Eds.), Global Climates Since The Last Glacial Maximum Minneapolis: University of Minnesota Press.
Meierding, T. C. (1982). Late Pleistocene glacial equilibrium-line altitudes in the Colorado Front Range: a comparison of methods. Quaternary Research, 18, 289-310.
Mercer, J. H. (1984). Late Cainozoic glacial variation in South America south of the Equator. In J. C. Vogel (Ed.), Late Cainozoic Paleoclimates of the Southern Hemisphere (pp. 45-58). Rotterdam: A.A. Balkema.
Mercer, J. H., and Palacios, M. O. (1977). Radiocarbon dating of the last glaciation in Peru. Geology, 5, 600-604.
Müller, R. (1985). Zur Gletschergeschichte in der Cordillera Quimsa Cruz (La Paz, Bolivia). Ph. D. thesis, Zurich.
Piperno, D. R., Bush, M. B., and Colinvaux, P. A. (1990). Paleoenvironments and human occupation in late-Glacial Panama. Quaternary Research, 33(1), 108-116.
Porter, S. C., Pierce, K. L., and Hamilton, T. D. (1983). Late Wisconsin mountain glaciation in the western United States. In H. E. Wright Jr., and S. C. Porter (Eds.), Late Quaternary environments of the United States (pp. 71-111). Minneapolis: University of Minnesota Press.
Prell, W. L. (1985). The stability of low-latitude sea-surface temperatures: An evaluation of the CLIMAP reconstruction with emphasis on the positive SST anomalies. (Contract No. DE-ACO2-83ER60167). U.S. Department of Energy.
Rind, D., and Peteet, D. (1985). Terrestrial conditions at the last glacial maximum and CLIMAP sea-surface temperature estimates: Are they consistent? Quaternary Research, 24, 1-22.
Rind, D. (1990). Puzzles from the tropics. Nature, 346, 317-318.
Rodbell, D. T. (1992). Late Pleistocene equilibrium-line altitude reconstructions in the northern Peruvian Andes. Boreas, 21, 43-52.
Rodbell, D. T. (1993). Subdivision of late Pleistocene moraines in the Cordillera Blanca, Perú based on rock weathering features, soils and radiocarbon dates. Quaternary Research, 39, 133-143.
Seltzer, G. O. (1987). Glacial history and climatic change in the central Peruvian Andes. Unpublished M.S. Thesis, University of Minnesota.
Seltzer, G. O. (1990). Recent glacial history and paleoclimate of the Peruvian-Bolivian Andes. Quaternary Science Reviews, 9(2/3), 137-152.
Seltzer, G. O. (1992). Late Quaternary glaciation of the Cordillera Real, Bolivia. Journal of Quaternary Science, 7, 87-98.
Servant, M., and Fontes, J.-C. (1978). Les lacs quaternaires des hauts plateaux des Andes boliviennes: premieres interpretations paleoclimatiques. Cahiers de l'ORSTOM, Série Géologique, 10, 9-23.
Van Campo, E., Duplessy, J. C., Prell, W. L., Barratt, N., and Sabatier, R. (1990). Comparison of terrestrial and marine temperature estimates for the past 135 kyr off southeast Asia: a test for GCM simulations of palaeoclimate. Nature, 348, 209-212.
Webster, P. N., and Streten, N. (1978). Late Quaternary ice age climates of tropical Australasia, interpretation and reconstruction. Quaternary Research, 10, 279-309.
White, S. E. (1986). Quaternary glacial stratigraphy and chronology of Mexico. Quaternary Science Reviews, 5(1-4), 201-205.
Wright, H. E., Jr. (1983). Late-Pleistocene glaciation and climate around the Junín Plain, central Peruvian Andes. Geografiska Annaler, 65A, 35-43.

EFFECTS OF SHORT AND LONG TERM CLIMATIC CHANGES ON PERMAFROST - SEDIMENTOLOGICAL DATA

Peter Worsley
Postgraduate Research Institute for Sedimentology
University of Reading
Box 227
Reading RG6 2AB
United Kingdom

1. Introduction

Permafrost or perennially frozen ground, is defined as 'the thermal condition in soil or rock of having temperatures below 0°C persist over at least two consecutive winters and intervening summer' (Brown and Kupsch 1974). It currently underlies approximately one fifth of the earth's terrestrial surface. During the cold stages of the Quaternary, permafrost extended equatorwards and developed in areas which currently enjoy relatively temperate climates. Consequently it is a phenomenon which has intrinsic interest for palaeoclimatologists, yet in comparison with the prominence given to glaciological reconstructions in the generation of G.C.M.s it has received relatively little attention. This assumes, of course, that its former presence and hence extent, can be identified via the evidence provided by the geological record. It has to be conceded that this is often a more difficult task in the field than the recognition of former glaciation limits and patterns.

We shall here consider the characteristics of modern permafrost and some of the vital processes associated with it. From this basis, the sedimentary structures generated by permafrost decay and growth will be discussed, concentrating on how the geologist can detect fluctuations in its extent. The field data is registered in both contemporary permafrost outcrops and in the stratigraphy of glacial stage successions in areas beyond the outer limits of permafrost and currently regarded as interglacial.

2. Modern permafrost.

Climate is fundamental to the formation of permafrost although its precise distribution is influenced by various terrain factors. The most diagnostic parameter is mean annual air temperature (m.a.a.t) for once the mean falls below 0°C the potential for permafrost creation exists. This was illustrated by Alexander and

NATO ASI Series, Vol. I 22
Long-Term Climatic Variations
Edited by J.-C. Duplessy and M.-T. Spyridakis

Worsley (1974) when they found permafrost occupying the core of a small 'Little Ice Age' (i.e. likely 18th century) end moraine in an environment with a present day m.a.a.t. of -1°C.

At any point on the earth's surface the annual temperature cycle is defined by the maximum and minimum temperatures attained within a given year (see Fig.1 for the form of the temperature profiles). Progressively with depth these annual fluctuations are attenuated until at approximately 12 m below the ground surface the immediate year to year values remain uniform. Deeper the overall temperature gradient will be positive due to the geothermal heat flux. The upwards projection of this gradient intercepts the ground surface close to the m.a.a.t.

Examination of Fig.1 will reveal three contrasting thermal regimes, two representing severe (A) and less severe (B) climates capable of sustaining permafrost and a non-permafrost one (C), experiencing only seasonal freezing restricted to the near surface. Comparison of A and B shows how the more extreme m.a.a.t. of A supports a shallower surfical zone of annual freezing and thawing i.e. the active layer, and a thicker zone of permafrost per se. Conversely B has a deeper active layer over a thinner permafrost layer. Geographically there is a continuous transition between the three states and conceptual visualisation of this expresses how the character of permafrost changes with latitude and with climatic change at a given site.

A schematic cross section to show the generalised form of permafrost in the northern hemisphere is shown in Fig.2. This reveals two contrasting situations for permafrost occurrence, mountains and lowlands, the first being subordinate in importance to the latter. Moving north from the equator the initial presence of permafrost is found within the high summits of mountains where the m.a.a.t. falls below 0°C. With increasing latitude the incidence of permafrost in mountains generally increases as the elevation at which the critical m.a.a.t. is achieved is lowered. The second situation is by far the most significant, for here permafrost is found below terrain without any appreciable relief. Once lowland permafrost is established increasing latitude is accompanied by two trends. The first is a progressive thickening and second, a concordant shallowing of the active layer. This is, of course, simply an expression of increasing climatic severity. As is apparent from the cross section, the main mass of permafrost is like a wedge with the steepest edge at the northern land mass limit just beneath the shores of the of the Arctic Ocean. The rapid disappearance of the permafrost beneath the ocean s due to the warming effect of the oceanic water. Any sub-seabed permafrost is

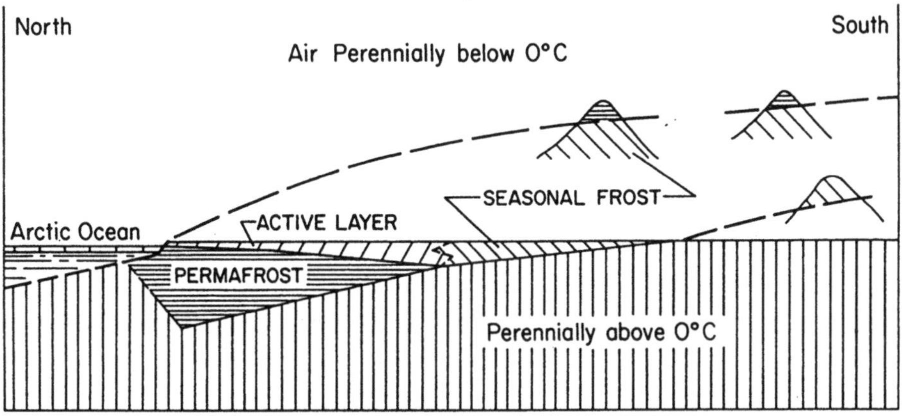

Fig.1. A schematic north-south cross section through the northern hemisphere to show the overall form of permafrost distribution. Two major kinds of occurrence, mountain and lowland are shown along with variations in the depth of the active layer.

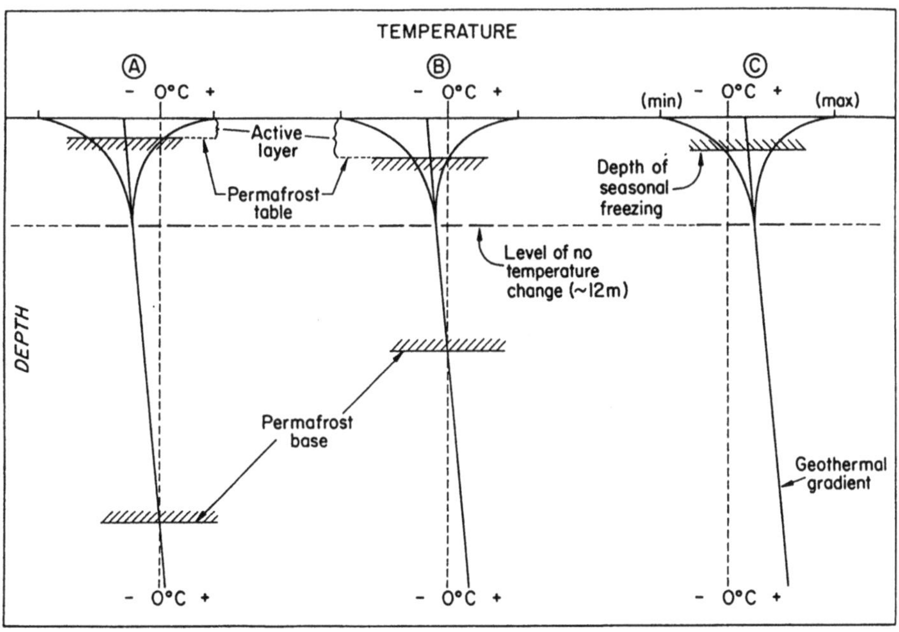

Fig.2. Ground temperature profiles for 3 locations - A,B and C along an idealised northern hemispheric north-south profile, with A and B having permafrost in the subsurface. Note the position of the O°C isotherm and how the position of this with respect to the geothermal gradient controls the depth of permafrost.

restricted to shallow coastal waters. During times of maximum ice sheet growth and allied global sea level low stands these would have been sub-aerially exposed to permafrost forming climatic regimes.

3. Critical physical processes allied to permafrost.

A wide range of geomorphological features are either unique to or enhanced by permafrost. Many, such as mass wasting landforms are rather ambiguous in their precise palaeoclimatic significance. Others such as pingos are certainly restricted to permafrost but their genesis and controlling parameters are still subject to research. The only widespread and reasonably well comprehended diagnostic process is that of thermal contraction which in turn generates ice wedges and sand wedges. The latter are seemingly rare and will not be considered further.

Ice wedges are primarily the result of annually occurring thermal contraction events which rupture the permafrost and thereby create linear cracks whilst the ground is being seasonally cooled. When the temperature increases expansion closes them up. Within the zone of annual ground temperature fluctuations (to about 12 m), rapid cooling can cause the tensile strength of the ground to be exceeded. The response to this is rupture within the permafrost, the cracks so generated extend upwards into the then frozen-up active layer and eventually to the ground surface. These cracks are obviously three dimensional and like many mediums which have been subject to rapid cooling, the cracks define polygonal patterns in plan form.

Open cracks at the ground surface during winter intercept any moving material such as wind blown snow or organic matter. In addition, hoar frost might condense directly onto the crack walls or snow meltwater might drain into the crack during the spring thaw. Any increment of mass will lead to stress concentrations in the vicinity of the crack during the phase of thermal expansion during the summer months. Annual repeated operation of this cyclic thermal expansion and contraction process at the same location leads to the cumulative build-up of the material which falls into or condenses onto the crack walls penetrating the permafrost. These bodies of ice and minor organic matter have wedge shaped cross profiles which typically might be 1-2 m wide at the top tapering to a point at 2-4 m. In plan the linear elements of the polygons are normally in the 10-20 m range.

It is important to appreciate that the ice wedge bodies do not extend to the ground surface, their tops usually correspond with the base of the active layer at the time of observation since the annual thaw within the active layer obliterates any ice vein which may form during contraction events.

4. Short term climatic changes.

A deepening active layer has a terminally destructive effect on ice wedges since they are progressively truncated at successively lower levels below the ground surface. This process is portrayed in Fig.3a. Thus in sections through permafrost it is sometimes possible to identify earlier phases of climatic amelioration where ice wedge bodies are well below the contemporary active layer depth. Where found, the tops correspond to what are known as 'thaw unconformities'. This does assume that the ground surface has been stable and no sedimentation has occurred i.e. the wedges are epigenetic in origin.

The converse is possible and a climatic deterioration might lead to a previously deep active layer becoming shallower with the permafrost table aggrading within the ground. In such circumstances reactivation of a pre-existing truncated ice wedge occurs, with new wedge growth related to a higher stratigraphic level than the thaw unconformity. This is shown in Fig 3b where in this example the aggradation of the permafrost has been a two-step process.

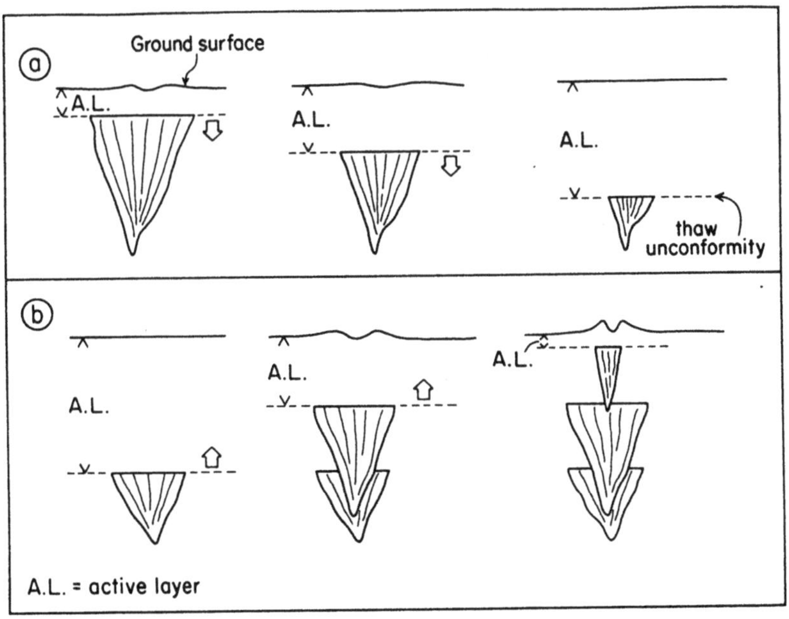

Fig 3. In 3a, the truncating effect of a deepening active layer on an ice wedge is shown. The depth of the thaw unconformity is directly related to active layer depth. In 3b the converse applies and at each step in the reduction of active layer depth, renewed growth of the ice wedge occurs to produce a form which might be identified as syngenetic whereas it is epigenetic since no sedimentation has occurred at the ground surface.

Some indication of the time involved in these changes might be gleaned from a consideration of the rates of ice wedge growth. There is general agreement that the maximum crack width imposes an upper limit to the ice wedge mass increase through the annual increment of material within it. Thus crude upper estimates of wedge growth can be derived knowing that the average crack width was 35 mm per year at an experimental site in Canada (Mackay 1986). Earlier studies suggested that cracking was a regular annual event although later observations have indicated that non-cracking years might be common. Hence a wedge with a width of one metre normal to the crack trend probably has a minimum age of some 300 years and is more likely to be twice that value. It must be emphasised that the data base on wedge cracking frequency is meagre.

Where sedimentation is active, the ground surface will aggrade and the permafrost table will extend upwards into the recently deposited materials (recall the climatic dependence of permafrost). The upward extension of ice wedges in unison with sediment accretion may lead to syngenetic ice wedges. While theoretically this may be possible, there is a lack of convincing descriptions of so-called syngenetic ice wedges and the difficulty of discriminating them from multistage or 'towered' ice wedges due to the thinning of the active layer (Fig.3b). A further problem concerns their preservation potential. It is difficult to envisage how the host 'overhangs' resulting from the stacked nature of both the syngenetic and epigenetic forms (see Fig.3b) might survive a casting event after permafrost decay. The resultant sedimentary structures are likely to be indistinguishable from the 'normal' types.

The above discussed ice wedge forms are not commonly exposed in permafrost environments, indeed permafrost per se is seen only where erosion is rapid. This is because the active layer normally protects the permafrost from the atmosphere and it must be removed if the permafrost table is to be exposed. The most favourable places to observe permafrost and associated ice wedge stratigraphy are either open coasts with no sea or beach ice or where rivers may be vigorously under-cutting their banks. The very presence of ice wedges is normally indicative of other excess ground ice contents (segregation ice) and the melting of all forms of ground ice promotes surface mass wasting and hence burial of the permafrost table. Thus the opportunities to study ice wedge stratigraphy in natural exposures are unfortunately limited.

5. Effects of long term climatic change.

Clearly a sustained period of climatic amelioration e.g. from a stadial to interstadial or interglacial, might lead to the total elimination of permafrost in a region and the attendant melting of all contained ground ice. Within a sequence of sediments the disappearance of contained ice bodies through melting must necessarily lead to the infilling or casting of the space formerly occupied by the ice.This mechanism must accompany the deepening of the active layer discussed previously and the upper parts of the ice wedges will become replaced by the sediments of the active layer. Since movements within the entire active layer tend to affect the whole unit and because of the prevailing high water contents during the thaw episode, casting of the ice wedge tops as such is not normally apparent.

A little-appreciated fact is that casting of ice wedges can occur independently of climatic change without the intervention of true active layer thickening. This may occur in two ways. First, where snow meltwater run off is concentrated along the surface troughs which invariably directly overlie the subsurface ice wedges. Erosion on the stream beds may exhume the underlying ice wedge tops and when this happens rapid thermal erosion of the wedges can lead to direct replacement by fluvially deposited sediment and slumped material from the immediately adjacent active layer. Second, active river channel migration can induce casting of ice wedges within the floodplain or adjacent low first terraces. This arises since river channels of a depth greater than 2-3 m are frequently accompanied by an unfrozen zone or talik within the sediments beneath the water flow. As the channel shifts the talik also migrates into sediments affected by permafrost.Thaw is induced by the high heat capacity of the river water and naturally any ice wedge bodies encountered by the extending talik will melt and casting might result. In fluvial sequences clearly truncated ice wedge casts are sometimes encountered and it is tempting to ascribe these to this mechanism. Accordingly, scattered intraformational ice wedge casts require critical evaluation before they are used as climatic change indicators.

Sedimentologically, this localised casting mechanism simulates ice wedge casting during climatic amelioration. Widespread regional permafrost degradation may be accompanied by drainage channels establishing themselves throughout the polygon networks. This is particularly so in coarse sediments such as gravels for two reasons. First, the gravels usually do not have major segregation ice development to promote excess pore water pressures during melting and second, the gravel once the interstitial ice has melted retains a sufficient strength for the walls between the melted ice wedges to maintain their slopes whilst the infill aggrades. If the host sediment had a finer grain size then both the ground ice contents would be greater and during thaw consolidation the material would be more likely to slump into the space formerly occupied by the ice wedges. The result would be that the casts would less faithfully reflect the form of the former ice wedge ice.

The field identification of true ice wedge casts is not a straight forward task, a problem which has been appreciated for over three decades (Johanson 1959). Despite this, non-periglacial specialists are apt to be over optimistic in the identification of ice wedge casts because a number of processes are capable of leading to the creation of structures which are wedge shaped in cross section. It is essential that the nature of the wedge's third dimension is established and that this has at least lateral continuity into the plane of the exposure. This latter expedient would eliminate many alleged ice wedge casts by simply demonstrating that in plan form the wedges are circular in nature and might be best described as 'carrotiforms' (see Younger 1992 for an example of a solutional feature misidentified as an ice wedge cast). Ideally a number of potential ice wedge casts need to be identified at a given site since their association with a polygonal ground plan dictates that they will not occur singly.

6. Repeated long term climate fluctuations.

To establish a record of long term permafrost history recourse is necessary to conventional stratigraphy. Ice wedge casts may be identified from sequences which are separated by interglacials or within the succession belonging to a single cold stage e.g. the Last Glacial of north west England (Worsley 1987). Assemblages of ice wedge casts at different stratigraphic horizons are able to demonstrate the recurrent formation and decay of regional permafrost. Sometimes glaciogenic sediments intervene between two permafrost episodes.

On occasion a horizon of ice wedge casts are encountered where they are penetrated by another set of ice wedge casts originating at a higher stratigraphic level in Pleistocene fluvial sediments.These clearly show two phases of permafrost with a hiatus represented by a paraunconformity corresponding to a period when it appears to have dissipated. The apparent paradox is the location of the second phase of ice wedge growth in precisely the same place as the earlier one, even though they are separated in time by a major climatic fluctuation and a phase of fluvial sedimentation sufficient to emplace at least 3 m of materials.Therefore the factors which might encourage the geographical coincidence of thermal contraction cracking despite the time gap and substantial climatic change require evaluation.

It may be speculated that the apparent `memory effect' is likely to be related to the discontinuity within the earlier sediments where the ice wedge casts are located. It will be recalled that the thermal contraction is initiated within the permafrost and under a tensile stress regime location of the initial ruptures are likely to reflect the zones of weakness in the sedimentary stack. As a result, the second generation of ice wedges grow in relation to cracking influenced by the buried ice wedge casts and as they deepen with time they will eventually grow into the earlier formed casts (see Fig.7.4 in Worsley 1987). Such final forms of later ice wedge casts penetrating earlier formed equivalents may be described as 'two cycle ice wedge casts'. These have been referred to by the rather cumbersome term 'stratigraphically superimposed ice wedge casts' by Seddon and Holyoak (1985). In any event they amply demonstrate the recurrence of permafrost in areas which are currently temperate. Careful compilation of stratigraphic relationships can yield valuable insights into cryohistory.

REFERENCES

Alexander MJ, Worsley P (1974) Stratigraphy of a Neoglacial end moraine in Norway. Boreas 2:117-142.

Brown RJE, Kupsch WO(1974) Permafrost terminology. Canada, National Research Council, Associate Committee

Geotechnical Research, Tech. Memo. 111: 62p.

Johnsson G (1959) True and false ice wedges in southern Sweden. Geogr. Annlr. 41:15-33.

Mackay JR (1986) The first 7 years (1978-1985) of ice wedge growth, Illisarvik experimental drained lake site, western Arctic coast. Can.J.Earth Sci. 23:1782-1795.

Seddon MB, Holyoak DT (1985) Evidence of sustained regional permafrost during deposition of fossiliferous Late Pleistocene river sediments at Stanton Harcourt (Oxfordshire England). Proc. Geol. Ass. 96:53-71.

Worsley P (1987) Permafrost stratigraphy in Britain - a first approximation. In Boardman, J. (ed) Periglacial processes and landforms in Britain and Ireland. Cambridge, Cambridge University Press 89-99.

Younger PL (1992) Quaternary geology and hydrogeology: the value of an interdisciplinary approach. Geoscientist 2:24-27.

University of Reading PRIS contribution no. 333.

BOX JENKINS MULTIVARIATE MODELLING AND CO-INTEGRATION: TWO STATISTICAL METHODS WITH POTENTIAL USEFULNESS IN CLIMATIC STUDIES

Alexandros E Milionis
Hellenic Air Force
Tactical Air Force Headquarters (ATA)
Meteorological Centre
Larisa, Thessaly,
Greece

1. Introduction

Climatic data are mainly in the form of time series. In an exploratory analysis of such data we often try to verify or discover causal relationships between two, or among a set of time series. We are also interested to know whether or not the level around which the series fluctuates is constant over time since this may have important physical implications; for the series of surface temperature, for instance, a change in level in the recent decades may indicate the existance of the greenhouse effect.

Over the last decade or so, the statistical methodology regarding these topics has been vastly developed, and its applications, mainly in econometrics, have been so far very successful. To a large extend, the transfer of this new knowledge to climatic data analysis has not yet been achieved, or at least not properly emphasised. There is little doubt that this knowledge can be of great assistance to the researcher of this field, particularly in such sensitive and troublesome areas as causality and spurious regressions.

Generally speaking, when a speaker is trying to explain an extensive field to an audience that is partly (presumably) new, he/she faces the problem of what are the appropriate topics to be mentioned and in what level of detail, given the inevitable limitation due to time. This talk will be an introductory, tutorial type one, and should

NATO ASI Series, Vol. I 22
Long-Term Climatic Variations
Edited by J.-C. Duplessy and M.-T. Spyridakis

by no means be considered as a complete review of the field. I will only try to present some methods which have been selected with the criterion of direct applicability to climatic data analysis. Several references will be quoted for further study.

2. Time series univariate modelling

A time series is a set of observations generated sequentially in time. If future values of a time series can be exactly determined by some mathematical function, for example such as

$$X_t = X_0 + Ut$$

which describes the distance X_t in a linear motion with constant speed U and initial position X_0 as a function of time t, the series is called deterministic. If future values can be described only in terms of a probability distribution, the series is called statistical. A stochastic process is a statistical phenomenon that evolves in time according to probabilistic laws. Hence, a statistical time series can be considered as a realisation (sample) of a stochastic process. If the stochastic process which generates a time series remains invariant through time, the series is called stationary. A less restrictive condition is to require the means, the variances, and the covariances of the stochastic process to be constant through time. This defines the so-called wide sense or weak stationarity. As Box and Jenkins (1976) point out there are an unlimited number of ways in which a process can be non-stationary. If a non-stationary process can become stationary after differencing d times ($d > 1$) the process is called homogeneous non-stationary or integrated of order d, denoted I(d). The so-called non-homogeneous or explosive non-stationarity (Box and Jenkins, 1976) will not be considered in this talk.

According to the Box-Jenkins "school" of time series analysis any homogeneous non-stationary time series Y_t , which has first been differenced d times to obtain stationarity, can be expressed as a function of past terms and past random uncorrelated innovations (shocks). The general form of such a model is (Box and Jenkins, 1976):

$$(1-B)^d Y_t = Z_t = c+\varphi_1 Z_{t-1}+\varphi_2 Z_{t-2}+ \ldots.+\varphi_p Z_{t-p}$$
$$- (\theta_1\varepsilon_{t-1}-\theta_2\varepsilon_{t-2}- \ldots.-\theta_q\varepsilon_{t-q})+ \varepsilon_t$$

where ε_t the current random shock, B the so-called backward shift operator, defined by the equation

$$B^k Y_t = Y_{t-k}$$

(1-B) is the difference operator as

$$(1-B)Y_t = Y_t - BY_t = Y_t - Y_{t-1}$$

$\varphi_1, \varphi_2, \ldots.,\varphi_p$ the so-called autoregressive coefficients, and $\theta_1, \theta_2,\ldots.\theta_q$ the so-called moving average coefficients.

According to this expression the current value of the series depends on the values of p previous terms of the same series, the values of q previous random schocks, and the current random shock. This model is called ARIMA(p,d,q), where ARIMA stands for Autoregressive, Integrated, Moving Average. If seasonal differencing of order D and seasonal components of seasonality S are also included, the model becomes: ARIMA(p,d,q)(P,D,Q)s where P and Q are analogous to p and q respectively.

Box-Jenkins univariate models are empirical models, created from the data, and the researcher should be aware of the following requirements:

a) The time series should be sufficiently long (at least 50 and preferably 100 successive observations should be used, according to the recommendation of Box and Jenkins (1976).

b) A certain iterative model-building strategy proposed by Box and Jenkins (1976) must always be followed. This process consists of (i) the identification, where a tentative model is identified from the character of the patterns of the autocorrelation and partial autocorrelation functions; denoted as ACF and PACF respectively, (ii) the estimation, which comprises estimation and significance testing of the coefficients of the model (iii) the diagnosis, where the whiteness of the residuals (i.e. the hypothesis that the residuals are uncorrelated random shocks) is examined, and (iv) the metadiagnosis, where the model is compared with other rival models. Example analyses in climatology can be found, for example, in Milionis, 1991, and Milionis and

Davies, 1992. Apart from the model building methodology described above, a number of alternative model selection criteria have been proposed (see Shibata, 1985 for a review). The logic of all these criteria is the same; to select the model that minimises the value of a parameter which is a function of the error variance and the number of the estimated parameters. This is done more or less automatically, so that the researcher has little or no control on the identification of a model. Since it is often the case that this methodology leads to overmodelling (i.e. order(s) of autoregressive and/or moving average polynomial(s) higher than necessary) and/or luck of physical meaning for some of the components of the selected model, I personally believe that the traditional Box-Jenkins model building strategy is preferable.

Once the univariate ARIMA model has been determined, it can be used, amongst others, as a first step in a methodology which examines causality between two or more time series. This methodology is described in the next section.

3. Testing for causality by using multivariate ARIMA modelling

After having developed the methodology to create univariate ARIMA models, any causal relationships between time series can be examined by using Box-Jenkins multivariate modelling. I will first give a formal definition for causality and then describe how to test for it using this methodology. Let us denote by $\sigma^2(Y_t)$ the variance of the predictive error of the Y_t series, when the calculation of it is based on the past values of Yt, $\sigma^2(Y_t,X_t)$ the same parameter but calculated using past values of both Y_t and X_t, and $\sigma^2(Y_t ,X'_t)$ the same parameter calculated using past values of Y_t and past and present values of X_t. Interchanging X_t and Y_t within the parentheses the meaning of the so-created parameters is in complete correspondence with the parameters defined above.

Then:

X_t (Granger) causes Y_t when

$$\sigma^2(Y_t,X_t) < \sigma^2(Y_t) \text{ and } \sigma^2(X_t,Y_t) =\sigma^2(X_t)$$

X_t instantaneously (Granger) causes Y_t when

$$\sigma^2(Y_t,X_t') < \sigma^2(Y_t,X_t)$$

In other words, Xt causes Y_t if it can explain variations in Yt after self-generated variations in Yt have been partialled-out. These ideas on causality have first been suggested by Granger (1969). Besides the cases mentioned here, there may be other more complex cases of causality, including feedbacks. For a more comprehensive and technical analysis see, for example, Pierce and Haugh (1977).

It is important to emphasise, at this point, that the time series nature of the data, i.e. the interdependence in a time series, must necessarily be taken into account when the relationships among time series are examined. High correlation coefficients by no means guarantee causal relationships. Common trends in two series may imply a causal relationship, but they may also be due to either spurious causes, or due to the internal but independent variations of the variables themselves. The very common regression models based on ordinary least squares (OLS) are inadequate to distinguish between these possibilities.

Box-Jenkins multivariate (transfer function) modelling on the other hand is a very good way to test for causality in time series and avoid spurious correlations because of its ability to control for past variations that are common in all series. This can be achieved in the following way (Box and Jenkins, 1976; Liu, 1988):

1. If the series Yt and Xt are not stationary they must be differenced until stationarity is obtained. Hence the series $W_t = (1\text{-}B)^d X_t$ and $Z_t = (1\text{-}B)^k Y_t$ will be considered as stationary hereafter.

2. Find the ARIMA model for Wt:

$$Wt = \varphi^{-1}(B)\Theta(B)\varepsilon_t$$

where

$$\varphi(B) = 1 - \varphi_1 B - \varphi_2 B^2 - \ldots\ldots. - \varphi_p B^p$$

and

$$\Theta(B) = 1 - \theta_1 B - \theta_2 B - \ldots\ldots\ldots - \theta_q B^q$$

the so-called autoregressive and moving average polynomials respectively.

3. Use the inverse model to filter Wt:

$$\varepsilon_t = \varphi(B)\theta^{-1}(B)Z_t$$

4 Use the same filter as in (3) to "prewhiten" Zt:

$$a_t = \varphi(B)\theta^{-1}(B)Z_t$$

The purpose of following steps (3) and (4) is to remove from the dependent variable Z_t that pattern of autocorrelation that is also common in the independent variable W_t.

5. Find the cross correlation function (CCF) between ε_t and a_t. The CCF for two realisations of stochastic processes at lag K can be estimated by the formula:

$$CCF(K) = S_{\varepsilon a}(K) / S_\varepsilon S_a$$

with

$$S_{\varepsilon a}(K) = \sum_{t=1}^{n-K} \{(\varepsilon_t - \bar{\varepsilon})(a_{t-k} - \bar{a})\}/n, \quad k=1,2,3,..,$$

$$S_a = \sum_{t=1}^{n}(a_t - \bar{a})/n$$

and

$$S_\varepsilon = \sum_{t=1}^{n}(\varepsilon_t - \bar{\varepsilon})/n$$

where n is the number of observations, and $\bar{\varepsilon}$ and $\bar{a}$ the means of the series ε_t and a_t respectively. The CCF, unlike the autocorrelation function is asymmetric; that is CCF(-K)=CCF(K). Therefore, the character of the CCF gives information about the direction of the relationship between the two series, i.e. it indicates the causality. In addition, as Box and Jenkins have shown, the values of the CCF at different lags are directly proportional to the impulse weights v_i , i = 0,1,2,... which are defined by the equation

$$Y_t = v_0 X_t + v_1 X_{t-1} + \ldots + v_k X_{t-k} + N_t$$

where Nt the noise component. Hence the character of the CCF provides also information about the strength of the relationship between two time series and is therefore a valuable tool. For these reasons it is strongly recommended that the CCF is considered when causality and transfer functions between time series of climatic data are examined.

An example analysis from the field of atmospheric sciences is given in Figure 1, which shows the plot of the CCF between surface smoke concentrations and surface wind speed at midday in Norwich, U.K. (from Milionis, 1991). Both series have been appropriately transformed before the CCF was calculated. As is seen in Figure 1 there is a significant negative correlation of magnitude 0.343 at lag 1. This indicates that at midday wind speed of the previous day has a strong causal effect on the average smoke concentration of the present day.

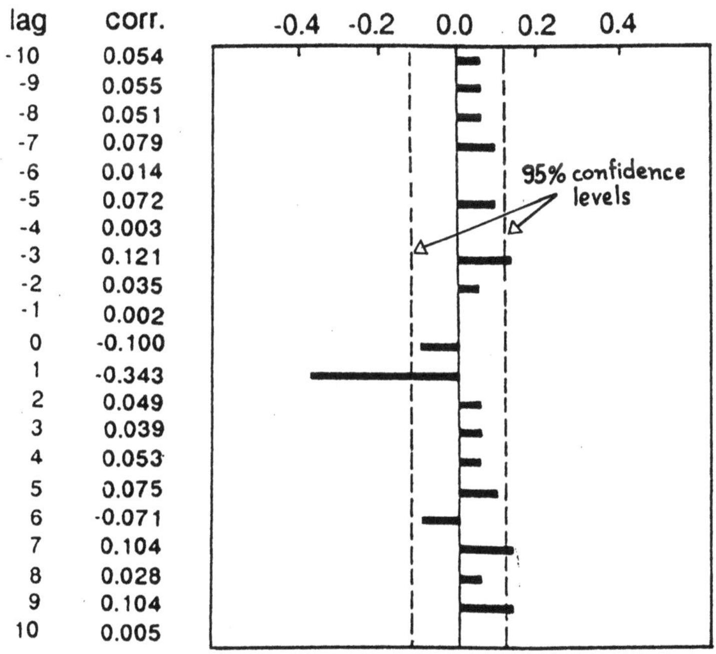

Figure 1. Plot of the CCF between the (appropriately transformed) series of surface wind speed and surface smoke concentration.

In addition to the Box-Jenkins approach, there is a second method for assessing Granger causality, mainly used by econometricians (see, for example, Pindyck and Rubinfeld, 1991). According to this method again we first difference both series, if necessary, to obtain stationarity and then we use autoregressive representation to express each of them as functions of previous lags of both series according to the equations:

$$Y_t = \sum_{i=1}^{m} a_i Y_{t-i} + \sum_{i=1}^{m} \beta_i X_{t-i} + \varepsilon_t$$

$$X_t = \sum_{i=1}^{n} c_i X_{t-i} + \sum_{i=1}^{n} d_i Y_{t-i} + e_t$$

It is possible by using the F-test to examine the hypothesis that a set of coefficients in the above equations are jointly significantly different from zero. Hence, if in the above equations $\hat{a}_i$, i= 1,2,.....,m are jointly significantly different from zero, while d_i, i = 1,2,...,n are not significantly different from zero, we conclude that Xt causes Yt.

Although this second method for assessing causality, also known as direct Granger method, is widely used and relatively easy to implement, I prefer the Box-Jenkins approach because the CCF provides a guideline for the identification of the relationship between the two series, while in the direct Granger method for the identification of the maximum lag of the past terms and the identification of the relationship in general, we go somehow blindly. For a discussion of other problems see, for example, Sims (1977). Of course there is no reason why parts of both methods cannot be combined together (further details in Milionis, 1991).

4. Elements of co-integration theory

In our previous discussion on the identification of an ARIMA model it was implied, although not directly mentioned, that whether or not a time series is non-stationary is indicated by the character of the series ACF and PACF. Of course the first piece of quantitative evidence regarding non-stationarity is given by the plot of the raw time series itself. However, sole reliance on the character of the ACF and PACF to determine whether or not a series needs differencing can lead to problems. This is because any observed time series is of finite size and the character of the ACF and PACF of the observed series may be very different than the theoretically expected ACF and PACF for a non-stationary process. Hence, there is often ambiguity regarding the order of differencing. This ambiguity generally increases as the size of the served series decreases. Dickey and Fuller have suggested another more formal test to determine the order of differencing, which I will describe below. For simplicity we will assume that for all cases non-stationarity will be of first order hereafter.

Let us consider the simple AR(1) model:

$$Y_t = aY_{t-1} + \varepsilon_t \qquad (4.1)$$

with ε_t white noise. For Y_t to be stationary in equation (4.1) it is necessary that $|a| < 1$. For practical purposes it is much easier to test whether or not a coefficient is different than zero so subtracting Y_{t-1} from both sides of equation (4.1) we have:

$$(1\text{-}B)Y_t = \beta Y_{t-1} + \varepsilon_t$$

with β = a-1. The conventional t-test for β, however, is valid only when $\beta<0$, and therefore it is inappropriate to test for non-stationarity. Modified tables for critical values of the t-statistic for the case of non-stationarity have been suggested by Dickey and Fuller and the corresponding test is called Dickey and Fuller (DF) or in its improved version Augmented Dickey and Fuller (ADF) test (Fuller, 1976; Dickey and Fuller, 1979).

It is well known that in many cases time series of climatological data are non-stationary. What is perhaps not well known to some researchers on this field is that regression models which use levels of non-stationary series can lead to very misleading conclusions. As Granger and Newbold (1974) have first shown, in such circumstances the usual significance testing of the regression coefficients can be seriously biased towards rejection of the null hypothesis of no relationship. Theoretical explanation of these findings was first given by Phillips (1986). There is one case, however, in which the usual least squares regression theory of stationary processes is still valid for non-stationary processes. This is when there exists a linear combination of these non-stationary processes which is stationary. For example, for two I(1) series Y_t and X_t if c and b in the equation

$$Y_t = c + b\,X_t + \varepsilon_t \qquad (4.2)$$

are such that ε_t is I(0) then Y_t - c - bX_t is also I(0). In such a case (Y,1,X) are the components of the so-called co-integrating set and (1, -c, -b) the components of the so-called co-integrating vector.

The concept of co-integration is of great importance in modern statistical theory and a lot of research has been devoted on that field over the last decade. It is worthy to examine it a little further in this talk. Let us consider equation (4.2) again. We have assumed that X_t, Y_t are both I(1) and can vary extensively on their own, but their linear combination Y_t - c - bX_t will be I(0) and will fluctuate around a certain level. Let us assume for U_t that

$$U_t = \rho U_{t-1} + \varepsilon_t$$

with $|\rho| < 1$ because U_t is I(0). The relationship between X and Y will be in equilibrium because a shock imparted to the system by U through ε will only have a temporary influence. Indeed after n periods its effect will be measured by ρ^n and since $|\rho|<1$ this effect dies away. So co-integration analysis looks for stationarity in deviations from equilibrium. If X_t and Y_t are co-integrated, there must be (Granger causality in at least one direction, as one variable can help forecast the other (Granger, 1991). Although in this talk we refer to I(1) series, co-integration theory applies in general to series of any order of integration and number (Engle and Granger, 1987).

To test for co-integration we first run a least squares regression among the I(1) series and then test whether or not the series of the residuals is stationary. The DF and ADF tests cannot be used in that case as the corresponding tables are valid for observed time series while the residuals are estimated. Appropriate tests have been suggested by Engle and Granger (1987) and Mackinnon (1991).

Before closing I would like to mention some practical points that can be used as a guideline for the researcher who intends to build a model using time series.

a) All series in a model must be of the same order of integration. The variance of an I(1) series is much greater than that of an I(0) series is much greater than that of an I(0) series, hence a combination of an I(1) and I(0) series will always be I(1). Therefore, in such a case the I(1) series must be differenced before it is used in the model.

b) If the time series to be used in a model are all I(1) then we first run a regression using OLS. If the residuals are I(1) then the series are not co-integrated, all of them should be differenced, and their first differences instead of their levels should be used in the model. No information is being lost by differencing because in this case no steady state equilibrium exists. c) If, again, all series are I(1) and the residuals from the OLS regression are I(0) the OLS estimates for the regression coefficients are the best ones. The estimates of their standard errors, however, are biased. Engle and Granger (1987) suggested a two-stage estimation in which in the second stage, the so-called error correction mechanism equation, all terms are I(0) and so no inferential difficulties appear.

REFERENCES

Box GEP, Jenkins GM (1976) Time series analysis: forecasting and control, revised edition, San Francisco, Holden-Day.

Dickey DA, Fuller WA (1979) Distribution of the Estimates for Autoregressive Time Series with a Unit Root, J. Amer. Statist. Assoc., 74: 427-431.

Engle RF, Granger CWJ (1987) Co-integration and Error Correction : Representation, Estimation and Testing, Econometrica 55 : 251-276.

Fuller WA (1976) Introduction to Statistical Time Series, New York, John Wiley.

Granger CWJ (1969) Investigating Causal Relations by Econometric Models and Cross-Spectral Methods, Econometrica : 37, 424-438.

Granger CWJ (1991) Developments in the Study of Co-integrated Economic Variables, 65-80, in Long Run Economic Relationships, Readings in Co-integration (eds Engle, R. F and Granger, C. W. J.), Oxford University Press.

Granger CWJ, Newbold P (1974) Spurious Regressions in Econometrics, J. Econometrics, 2 : 111-120.

Liu LM (1988) Box Jenkins Time Series Analysis, in BMDP statistical software manual, Vol. I, University of California Press.

Mackinnon, JG (1991) Critical Values for Co-integration Tests, In Long Run Economic Relationships, Readings in Co-integration (eds Engle, R. F. and Granger, C. W. J.), 267-276, Oxford University Press.

Milionis A E (1991) Climatology of Temperature inversions over Eastern England; Some Implications for Ground lLevel Air Pollution, Ph D thesis, University of East Anglia, Norwich.

Milionis AE, Davies TD (1992) Box-Jenkins Univariate Modelling for Climatological Time Series Analysis: an Application to the Monthly Activity of Temperature Inversions, Intern. J. Climat. (submitted).

Phillips PCB (1986) Understanding Spurious Regressions in Econometrics, J. Econometrics : 33, 311-340.

Pierce DA, Haugh L D (1977) Causality in Temporal Systems. Characterisations and a Survey, J. Econometrics, 5 : 265-293.

Pindyck, Rubinfeld DL (1991) Econometric Models and Economic Forecasts, McGraw-Hill, New-York.

Shibata R (1985) Various Model Selection Techniques in Time Series Analysis, in Handbook of Statistics, Volume 5 (eds Hannah, E. J., Krishnaiah, P, R. and Rao, M. M.), 179-187, North-Holland, Amsterdam.

Sims CR (1977) Exogeneity and Causal Ordering in Macroeconomic Models, in New Methods of Business Cycle Research: Proceedings from a Conference (ed. Sims C. R.), 23-44, Federal Reserve Bank of Minneapolis, Minneapolis.

III - UNDERSTANDING CLIMATIC CHANGES

MEASURING THE GAIN OF THE CLIMATE SYSTEM'S RESPONSE TO MILANKOVITCH FORCING IN THE PRECESSION AND OBLIQUITY BANDS

John Imbrie
Department of Geological Sciences
Brown University
Providence, RI 02912, U.S.A.

Using the amplitude of the Milankovitch radiation cycles for June at 65°N as a measure of the forcing, and the amplitude of oceanic $\delta^{18}O$ cycles as a measure of the system's response, we show that the system's gain in the precession and obliquity bands has been surprisingly strong and constant over the past 600,000 years.

1. INTRODUCTION

Many studies of deep-sea cores have demonstrated (1) that climatic cycles with periods near 23,000, 41,000, and 100,000 years dominate the climatic narrative of the past million years, and (2) that these climatic cycles are causally related to orbitally-driven variations in the annual cycle of insolation (see Imbrie et al., 1992 and references therein). This circumstance provides an important opportunity to discover and quantitatively model some of the key physical mechanisms that operate on these time scales to change the Earth's climate. But given the size and complexity of the global climate system, the inadequacy of our knowledge of the system's dynamics, and the incomplete nature of the records available for study, the scientific problem remains daunting.

A major source of difficulty in attacking this problem is that climatic variations, and therefore the processes which drive them, operate not only on the Milankovitch time scale, say from 20 kyr to 100 kyr (where 1 kyr = 1,000 years), but on time scales ranging from months to millions of years. One way of simplifying the descriptive and interpretive aspects of the ice-age problem, therefore, is to focus attention on the three main cycles by using standard techniques of spectral and filter analysis. We adopt such an approach in this paper, using essentially the same numerical techniques that have helped to identify and

NATO ASI Series, Vol. I 22
Long-Term Climatic Variations
Edited by J.-C. Duplessy and M.-T. Spyridakis

evaluate physical mechanisms that govern the semiannual, annual, quasi-biennial, and El Niño cycles (Rasmusson and Carpenter, 1982).

Specifically, we examine both the Milankovitch forcing and the system's response in two frequency bands where changes in the incoming radiation are large: the ~23-kyr precession band and the ~41-kyr obliquity band. Using the oceanic $\delta^{18}O$ record as a measure of the system's ice-volume response, and the intensity of incoming radiation in June at 65°N to represent the Milankovitch forcing, we ask two questions. First, is the system linear in radiation in each of these bands—i.e., is the governing differential equation linear? Finding that the answer to this question is "to first order, yes," we then measure the system's gain in each of these two bands and see how constant these gains have been over the past 600 kyr.

2. DEFINING THE FORCING AND RESPONSE

Variations in the Earth's orbit drive changes in the Earth's seasonal cycle of incoming radiation that affect every latitude band and season differently (Berger 1978). Thus the full Milankovitch forcing function is a complex varying forcing field of the form Q_{lst}, where Q is the incoming radiation at the top of the atmosphere in Wm^{-2}, l is the latitude, s is the season, and t is time. To simplify our analysis, we select a particular radiation curve that is widely believed to represent a dominant element in the Milankovitch forcing, namely, the radiation received in mid-June at 65°N (Imbrie et al., 1992). As shown in Figure 1, the amplitude of variations in this curve is much larger than the amplitude of the December radiation curve at the same latitude. Thus the Milankovitch forcing function here is essentially the modulation in the amplitude (i.e., the range) of the seasonal radiation cycle. Today, this range is 474 Wm^{-2}. At one recent extreme of the forcing function, at 128 ka, the range was 548 Wm^{-2}; at another extreme, 116 ka, the range was only 432 Wm^{-2} (ka = 1,000 years ago).

As shown in Figure 2, the dominant frequency components of this insolation curve have periods near 23 kyr (in the precession band), and near 41 kyr (in the obliquity band). These insolation variations are very much larger than those in the 100-kyr eccentricity band—where the climatic response, represented in this figure by oceanic $\delta^{18}O$, is as large as the response in the precession and obliquity bands combined. For this reason it is clear that the system's response in the 100-kyr band must involve strong nonlinearities. Setting the 100-kyr problem and the identification of its nonlinear mechanisms aside, we now focus on the system's forcing and response in the 23-kyr and 41-kyr bands, with the aim of determining how linear the system actually is at these frequencies.

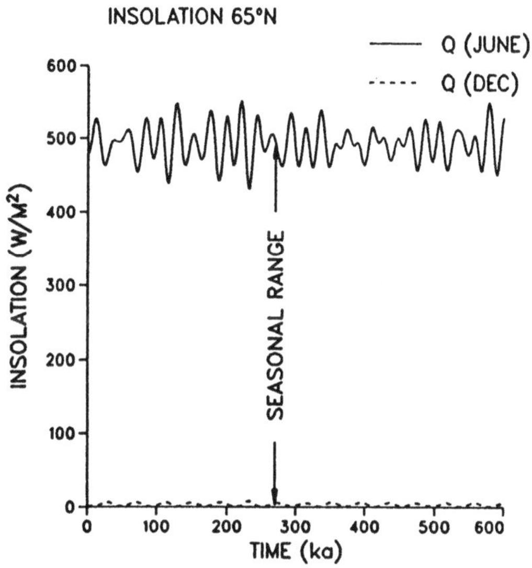

Fig. 1. Milankovitch forcing in June at 65°N (Berger and Loutre, 1991). The seasonal range of the annual cycle of insolation today is 474 Wm^{-2}. Over time, this range changes substantially, owing to modulations in the intensity of the incoming radiation, primarily in summer. At 128 ka, e.g., the range was 548 Wm^{-2}; at 116 ka, 432 Wm^{-2}.

3. RESULTS

3.1. Linearity of responses in the 23-kyr and 41-kyr bands.

The amplitude of the 23-kyr radiation cycle varies markedly over time (Figure 2) because it is exactly proportional to variations in eccentricity. Formally, this relationship is given by $P = e \sin \omega$, where P is an index of the state of the precession cycle, e the time-varying eccentricity, and ω the longitude of perihelion measured from the moving vernal point (Berger, 1978).

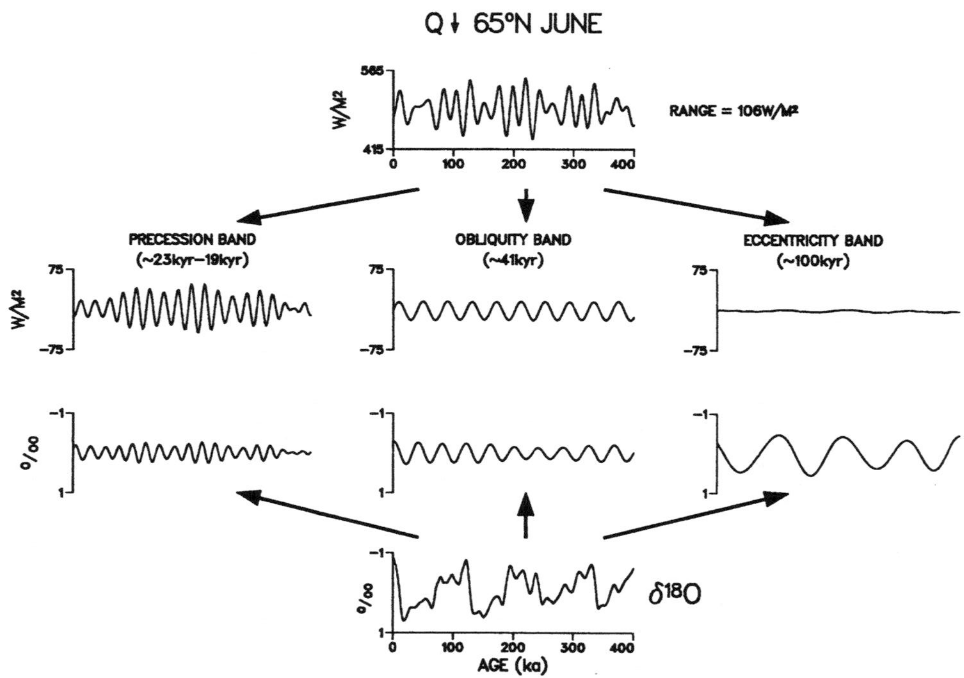

Fig. 2. Comparison of the Milankovitch forcing for June at 65°N, and the climatic response as measured by oceanic $\delta^{18}O$, in the precession (~23 kyr), obliquity (~41 kyr), and 100-kyr eccentricity bands. Using band-pass filters, the radiation and climate time series are partitioned into their dominant periodic components. Note that the amplitude of the radiation signal in the 100-kyr band (2 Wm^{-2}) is an order of magnitude smaller than in the other two bands (Berger, 1978).

Figure 3a (derived from the eccentricity curve) shows how the amplitude of the 23-kyr radiation cycle varies as a function of time. How closely do variations in the amplitude of the 23-kyr climate response, as observed in the oceanic $\delta^{18}O$ record and plotted as a dashed line in Figure 3b, follow this astronomical forcing function? The two curves are close to being exactly proportional—an observation that is at once compelling proof of the Milankovitch origin of this climatic cycle, and clear evidence (over this band of frequencies) that the climate system is essentially linear in radiation.

As shown in Figure 3c and d, the system under study is also essentially linear in the 41-kyr band. But here the linearity is not quite as strong as in the 23-kyr band.

3.2. Measuring the gain.

In studying physical systems that operate on an input to yield an output, the steady-state response to a unit change in the input is defined as the gain (Jenkins and Watts, 1968, page 12). In many engineering applications it is usually easier and more meaningful to measure gain on an operating system by cross-spectral analysis of input and output data rather than to conduct experimental runs with a step change in the input. In this statistical approach, the steady-state gain at a given frequency can be calculated as the ratio a_R/a_F, where a_R is the amplitude of the oscillating response and a_F is the amplitude of the forcing (Figure 4, top panel).

Using the amplitude of the radiation signal at June 65°N as a measure of the forcing in a given band, and the amplitude of the corresponding $\delta^{18}O$ signal as a measure of the system's response (envelopes in Figure 3), we now obtain smoothed estimates of the system's gain and plot them as a function of time in Figure 4 (lower panel). The values estimated in this way have remained fairly constant, the gain at 41 kyr being generally twice that at 23 kyr. This is equivalent to saying that the system has been approximately linear and stationary at these frequencies over the past half-million years. When examined in detail, however, the long-term trend in these curves suggests that the system has slowly evolved so as to become more sensitive to the obliquity forcing.

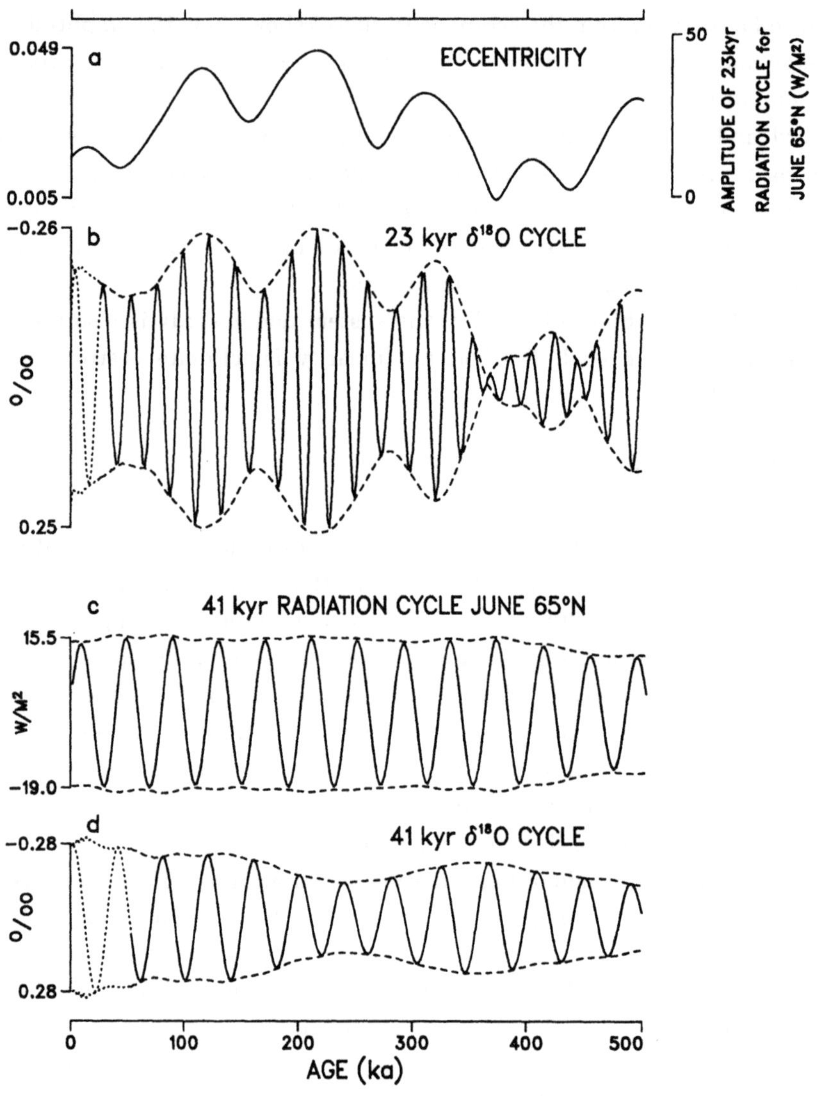

Fig. 3. Evidence that the climate system is linear in radiation over both the 23-kyr (a, b) and 41-kyr bands (c, d). (a) Right-hand scale shows how eccentricity modulates the amplitude of the 23-kyr radiation cycle in Figure 2. (b) An enlarged view of the 23-kyr $\delta^{18}O$ cycle in Figure 2. The strong, nearly linear relationship between the amplitude of this cycle (dashed lines) and the amplitude of the 23-kyr radiation cycle (which is proportional to eccentricity) shows that the glaciation process is essentially linear in this band. (c, d) Enlarged views of the 41-kyr radiation and $\delta^{18}O$ cycles in Figure 2. The approximately linear relationship between the envelopes of these cycles (dashed lines) shows that the process of glaciation is also essentially linear in this band. Envelopes are obtained from Hilbert transforms of the filtered time series (Meskó, 1984).

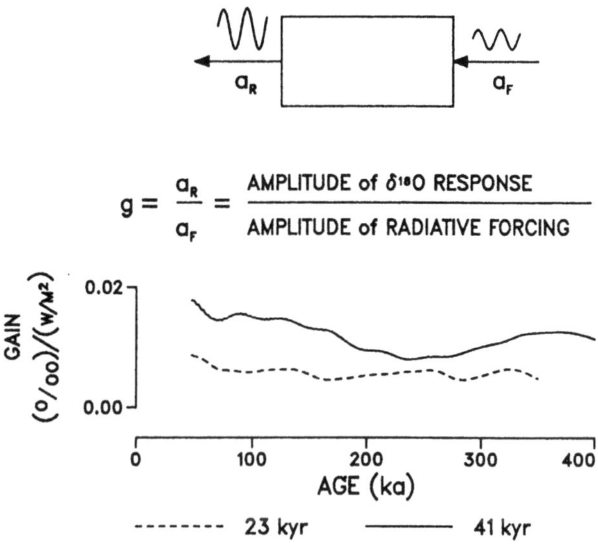

Fig. 4. Gain, g, as a function of time in the precession and obliquity bands. Using the amplitude of the radiation signal at June 65°N in a given band as a measure of the forcing, and the amplitude of the corresponding $\delta^{18}O$ signal as a measure of the system's response, smoothed estimates of the system's gain are obtained from the envelopes in Figure 3. Values have remained fairly constant, the gain at 41 kyr being generally twice that at 23 kyr.

4. CONCLUSIONS

Over the past half-million years, the climate system's response to the Milankovitch radiation forcing has been nearly linear and surprisingly constant in the precession and obliquity bands. However, when plots of the system's gain are examined in detail, the system is seen to have been slowly evolving over time so as to become more sensitive to the obliquity forcing.

ACKNOWLEDGMENTS. The work reported here was supported by National Science Foundation grants to Brown University, including SPECMAP grant ATM8812589 (Imbrie and Prell) ; ATM8706394 (Imbrie) ; ATM8516140 (Imbrie and Shackleton, Dec. 1985 to Dec. 1990); and a continuation of ATM8516140 (Jan. 1991 to —).

REFERENCES

Berger A (1978) Long-term variations of caloric insolation resulting from the Earth's orbital elements. Quat Res 9: 139-167

Berger A, Loutre MF (1991) Insolation values for the climate of the last 10 million years. Quat Sci Rev 10:297-317

Imbrie J, Boyle E, Clemens S, Duffy A, Howard W, Kukla G, Kutzbach J, Martinson D, McIntyre A, Mix A, Molfino B, Morley J, Peterson L, Pisias N, Prell W, Raymo M, Shackleton N, Toggweiler J (1992) On the structure and origin of major glaciation cycles, 1, Linear responses to Milankovitch forcing. Paleoceanography 7:701-738

Jenkins GM, Watts DG (1968) Spectral Analysis and Its Applications. Holden-Day, San Francisco

Meskó A (1984) Digital Filtering Applications in Geophysical Exploration for Oil. John Wiley, New York

Rasmusson EM, Carpenter TH (1982) Variations in tropical sea surface temperature and surface wind fields associated with the Southern Oscillation/El Nino. Mon Weather Rev 110:354-384

THE LAST TWO GLACIAL-INTERGLACIAL CYCLES SIMULATED BY THE LLN MODEL

A. Berger, C. Tricot[1], H. Gallée, Th. Fichefet and M.F. Loutre
Université Catholique de Louvain
Institut d'Astronomie et de Géophysique G. Lemaître
2 Chemin du Cyclotron
B-1348 Louvain-la-Neuve, Belgium

Abstract. A 2-dimensional model which links the atmosphere, the mixed layer of the ocean, the sea ice, the continents, the ice sheets and their underlying bedrock has been used to test the Milankovitch theory over the last two glacial-interglacial cycles. A series of sensitivity analyses have allowed us to better understand the internal mechanisms which drive the simulated climate system and in particular the feedbacks related to surface albedo and water vapour.

It was found that orbital variations alone can induce, in such a system, feedbacks sufficient to generate the low frequency part of the climatic variations over the last 122 kyr. These simulated variations at the astronomical time scale are broadly in agreement with reconstructions of ice-sheet volume and of sea level independently obtained from geological data. Imperfections in the simulated climate were the insufficient southward extent of the ice sheets and the too small hemispheric cooling at the last glacial maximum. These deficiencies were partly remedied in a further experiment by using the time-dependent atmospheric CO_2 concentration given by the Vostok ice core in addition to the astronomical forcing. In this transient simulation, 70 % of the northern hemisphere ice volume is related to the astronomical forcing and the related changes in the albedo, the remaining 30 % being due to the CO_2 changes. Analysis of the processes involved shows that variations of ablation are more important for the ice-sheet response than are variations of snow precipitation. A key mechanism in the deglaciation after the last glacial maximum appears to be the "ageing" of snow which significantly decreases its albedo. The other factors which play an important role are ice-sheet altitude, insolation, taiga cover, ice-albedo feedback, ice-sheet configuration ("continentality" and "desert" effect), isostatic rebound, CO_2 changes and temperature-water vapour feedback.

[1]present address: Institut Royal Météorologique, 3 av. Circulaire, B-1180 Bruxelles, Belgium.

NATO ASI Series, Vol. I 22
Long-Term Climatic Variations
Edited by J.-C. Duplessy and M.-T. Spyridakis

Numerical experiments have also been carried out with a 1-D radiative-convective model in order to quantify the influence of the CO_2 changes and of the water vapour feedback on the climate evolution of the northern hemisphere over the last 122 kyr. Results of these experiments indicate that 67 % of the simulated cooling at the last glacial maximum can be attributed to the astronomical forcing and the subsequent surface albedo increase, the remaining 33 % being associated with the reduced CO_2 concentration. Moreover, the water vapour feedback explains 40 % of the simulated cooling in all the experiments done.

The transient response of the climate system to both the astronomical and CO_2 forcing was also simulated by the LLN (Louvain-la-Neuve) 2.5-D model over the two last glacial-interglacial cycles. It is particularly significant that spectral analysis of the simulated northern hemisphere global ice volume variations reproduces correctly the relative intensity of the peaks at the orbital frequencies. Except for variations with time scales shorter than 5 kyr, the simulated long term variations of total ice volume are comparable to that reconstructed from deep sea cores. For example, the model simulates glacial maxima of similar amplitudes at 134 kyr BP and 15 kyr BP, followed by abrupt deglaciations. The complete deglaciation of the three main northern hemisphere ice sheets, which is simulated around 122 kyr BP, is in partial disagreement with reconstructions indicating that the Greenland ice sheet survived during the Eemian interglacial. The continental ice volume variations during the last 122 kyr of the 200 kyr simulation are, however, not significantly affected by this shortcoming.

1 Introduction to modeling paleoclimates

A hierarchy of climate models of increasing accuracy, dimensionality, and versability are being developed and applied (Hecht, 1985). Records of past climates can allow a broad testing of the capabilities of the climate models over a range of known changes in orbitally induced radiation. These tests may reveal weaknesses of the models; they will also suggest further process studies which will help improve the integrated understanding of the climate system.

Climate models can be divided into two very broad categories: (1) general circulation models (GCMs), and (2) statistical-dynamical models, which include energy balance models (EBMs) and 2-D zonally averaged dynamical models (Kutzbach, 1985). Both GCMs and statistical-dynamical models have been used for the modeling of palaeoclimates. GCMs have been used for simulating geographic features of palaeoclimates and for including, in the most explicit form, processes such as precipitation that depend on details of the atmospheric flow. However, because of the large computational costs that are involved, GCMs have only been used to provide "snapshot" views of the climate in equilibrium with some boundary conditions. These conditions include the seasonal and latitudinal changes of solar radiation at the top of the atmosphere (resulting from modifications in the Earth's orbit about the Sun), as well as variations in sea-surface temperatures, major ice sheets, atmospheric composition and turbidity, and locations of continents for simulations distant enough in the past (Prell and Kutzbach, 1987; Rind et al., 1989). On the other hand, statistical-dynamical models have a great potential for long-period climate simulations. As their dimensionality is reduced, they consume less computer time, thus allowing them to take into account all the different parts of the climate system in an interactive and time-dependent way. Therefore, these models can be used to simulate the transient response of the climate system to any "external" forcing or internal perturbation. For example, some of them have been coupled to slow responding portions of the climate system (e.g., ice sheets) in order to investigate the temporal evolution of climate on all time scales.

2 Simulations of temporal evolution of climate

Since the publication of the historic Hays et al. (1976) paper, a number of modeling efforts have attempted to explain the relation between astronomical forcing and climate change. Most of these modeling studies have focused on the origin of the 100,000-year cycle in ice volume. These studies have been motivated by the recognition that the amount of insolation perturbation at 100 kyr is not enough to cause a climate change of ice-age magnitude. Specific explanations for the large 100-kyr variance generally fall into two classes. In the first case, orbital variations are essential to ice volume fluctuations, but nonlinear interactions in the air-sea-ice system modify the signal. In the second case, ice volume fluctuations result from inherently nonlinear interactions in the air-sea-ice system, with orbital variations serving only to phase-lock the variations at the preferred time scales.

About the models where the ice volume flcutuations is primarily driven by orbital forcing, Imbrie et al. (1984) showed that at the 23-kyr and 41-kyr periods, ice volume responds linearly to orbital forcing, but at 100 kyr the effect is nonlinear. In particular, the 100-kyr power can be generated by transmission of 19-kyr and 23-kyr frequencies through a nonlinear system (Wigley, 1976), producing substantial power in both harmonics and 100-kyr subharmonics. Specific details of this nonlinear interaction vary considerably, but many mechanisms focus on the abrupt terminations of ice ages, which occur over an interval of about 10,000 years (Ruddiman and McIntyre, 1981; Hughes, 1987). Other approaches to the 100-kyr cycle focus on nonlinear interactions between accumulation/ablation, ice sheet flow, elastic lithosphere, and viscoelastic mantle (Oerlemans, 1982; Birchfield et al., 1982; Pollard, 1982; Hyde and Peltier, 1985).

In this paper we do not attempt to provide a comprehensive summary of all relevant modeling efforts in dynamical modeling but we feel it is interesting to document the model of Deblonde and Peltier (1991a,b) which differs rather substantially from the LLN model (Berger et al., 1990; Gallée et al., 1991, 1992) in the mix of ingredients that are proposed to account for the same phenomenon. This model is a global 1-level seasonal energy balance model which differentiates land from sea in a geographically realistic way and in which multiple ice sheets evolve in specific geographic sub-domains in response to the applied insolation anomalies. The model incorporates the influence of glacial isostatic adjustment, of elevation-desert feedback, and a realistic surface topography. The hydrological cycle is driven by a precipitation field that matches the present day precipitation within each of the cryospheric sub-domains and the ice sheet mass balances are computed as monthly averages on the basis of a geographically specific annual cycle that is diagnostically determined from the ice sheets. When it is initialized with ice free (except Greenland) northern hemisphere conditions beginning 124 kyr BP, this model successfully predicts the observed locations and the maximum thickness achieved by both the North American and the Northwest European ice complexes. With the same minimal set of interacting physical ingredients as were employed in the simpler model of Deblonde and Peltier (1990), however, the termination that commences near 18 kyr BP is not predicted. Therefore, without additional feedback loops, the model is unable to explain fully the deglaciation process.

A feedback that might achieve the desired effect in this regard is that associated with atmospheric dust. Ice cores and deep-sea sedimentary cores show that late glacial episodes are characterized by dry and dusty conditions for many thousands of years prior to major deglaciation events. When the albedo of the ablation zones in the Deblonde-Peltier model is modulated in a way implied by the dustiness records, complete deglaciations are then predicted to occur. In the LLN model, an impact of "ice-ageing" is invoked to effect a similar modulation of albedo, and this, too, effects the desired collapse of the ice sheet in response to the Milankovitch anomalies. For distinguishing which, if either, of these mechanisms are correct, it would appear that the only acceptable course is to describe more of the physics explicitly.

An entirely different approach to modeling the low frequency climate variability involves the hypothesis that glacial-interglacial fluctuations are a consequence of nonlinear "internal" interactions in a highly complex system (Nicolis, 1984; Saltzman, 1985).

Internally driven glacial-interglacial fluctuations can also be modeled more explicitly. For example, Ghil and colleagues (Ghil and Le Treut, 1981; Le Treut et al., 1988) modeled internally generated glacial-interglacial fluctuations which, when orbitally forced, produce characteristic Milankovitch periods, including harmonics (e.g., 10 kyr) and subharmonics (e.g., 100 kyr). In another example, Saltzman and Sutera (1987) incorporated a small number of variables (CO_2, surface and deep-water temperatures, and ice volume) in a model and were able to reproduce with some success both the 100-kyr cycle and the transition from dominant 41-kyr ice volume fluctuations prior to 900 kyr BP to dominant 100-kyr ice volume fluctuations after 900 kyr BP. The latter transition may also reflect the existence of "multiple equilibrium" states, whereby slowly changing boundary conditions can cause an abrupt transition in the climate state (e.g., North and Crowley, 1985).

Finally, Broecker et al. (1985, 1988) have postulated that "internal" glacial-interglacial fluctuations would not result from ice sheet nonlinearities but from instabilities in the ocean-atmosphere system (Broecker et al., 1985, 1988).

3 Climate system response to the astronomical forcing

It was suggested earlier (Berger, 1976 and 1979) that the time-evolution of the latitudinal distribution of the seasonal pattern of insolation is the key factor driving the behavior of the climate system where the complex interactions between its different parts amplify this orbital perturbation. That dynamical behaviour of the seasonal cycle suggests that time-dependent coupled climate models might be able to test whether or not the astronomical forcing can drive the long-term climatic variations. Such a time-dependent climate model including the atmosphere, the hydrosphere, the cryosphere and the lithosphere and their complex interactions must therefore be forced only by the astronomical variations of monthly insolations for each latitude, the so-called boundary conditions used in equilibrium atmospheric general circulation model experiments (ice sheets size and area, sea surface temperature, albedo ...) being all generated by the climate model itself.

Such a climate model which links the northern hemisphere atmosphere, ocean mixed layer, sea-ice and continents has been validated for the present-day climate (Gallée et al., 1991). It is a latitude-altitude model. In each latitudinal belt, the surface is divided into at most five oceanic or continental surface types, each of which interacts separately with the subsurface and the atmosphere. The oceanic surface types are ice-free ocean and sea-ice cover, while the continental surface types are the snow-covered and snow-free land and the Greenland ice sheet. The atmospheric dynamics is represented by a zonally averaged quasi-geostrophic model, which includes a new parameterization of the meridional transport of quasi-geostrophic potential vorticity and a parameterization of the Hadley sensible heat transport. The atmosphere interacts with the other components of the climate system through vertical fluxes of momentum, heat and water vapor. The model explicitely incorporates detailed radiative transfer, surface energy balances, and snow and sea-ice budgets. The vertical profile of the upper ocean temperature is computed by a mixed layer model which takes into account the meridional convergence of heat. Sea ice is represented by a thermodynamic model including leads and a new parameterization of lateral accretion. Simulation of the present climate shows that the model is able to reproduce the main characteristics of the general circulation (Gallée et al., 1991). The seasonal cycles of oceanic mixed layer, sea ice, and snow cover are also well reproduced. Sensitivity experiments show the importance of the meridional sensible heat transport by the Hadley circulation in the tropics, the seasonal cycle of the oceanic mixed-layer depth and sea ice formation in latitude bands where the average water temperature is above the freezing point.

4 The last glacial-interglacial cycle

This climate LLN model was then asynchronously coupled to a model of the three main northern hemisphere ice sheets and their underlying bedrock in order to assess the influence of several factors (processes and feedbacks), including snow surface albedo over the ice sheets, upon ice-age simulations using astronomically-derived insolation and CO_2 data from the Vostok ice core (Berger et al., 1990; Gallée et al., 1992).

The experiments that we have done so far, consider the northern hemisphere only. Before a model for the entire globe will be available, we assume that the variations of ice volume in the northern hemisphere dominate those of the southern hemisphere, and that the variations in the southern hemisphere ocean, sea ice and ice sheet will not affect significantly the climatic variations in the northern hemisphere at the time scale considered. In this version of the model used for paleoclimate reconstruction, there is no explicit deep ocean yet, although we recognize that deep water temperature has changed over the last glacial-interglacial cycle (Labeyrie et al., 1987) and that the change in heat stored by the ocean can be important. This ocean model is now developed and will be coupled to the present model (Hovine and Fichefet, 1993).

The model was first run with ice-sheet feedback by forcing it only with the astronomical insolation over the past 122 kyr (Berger, 1978) keeping the CO_2 concentration at the Eem level ($\sim$ 270 ppmv). Large variations of ice volume are simulated between 122 and 55 kyr BP, with a rapid latitudinal extension of the North American and Eurasian ice sheets starting at 120 kyr BP. The model simulates a maximum ice volume of 40 10^6 km^3 at roughly 20 kyr BP and a total deglaciation as well, this simulation of both the glaciation and deglaciation being a crucial test of the efficiency of the model. The simulated evolution of the three northern ice sheets is generally in phase with geological reconstructions. The major discrepancy between the simulation and observations lies in the temperature variations and in the ice-sheet extent, a discrepancy linked to an overestimation of the CO_2 concentration which was assigned an interglacial value at the last glacial maximum. A final set of experiments therefore addresses the effect of CO_2 on the climate of the last glacial maximum. This experiment made by forcing the model with both insolation and CO_2 variations over the last 122 kyr (Barnola et al., 1987) significantly improved the temperature results and the ice volume reconstructions obtained in the insolation-only experiment (Gallée et al., 1992).

The simulated total ice volume deviation from present day value (assumed to be 30.5 10^6 km^3, that is, 27.9 10^6 km^3 for Antarctica and 2.6 10^6 km^3 for Greenland; Hughes et al., 1981) is displayed in Figure 1 as a function of time compared to the variation of the global seawater oxygen isotopic ratio given by Labeyrie et al. (1987) and Duplessy et al. (1988). The latter values are scaled and plotted such that their maximum corresponds to the deviation from present-day of the last glacial maximum total ice volume (taken as 48.6 10^6 km^3, reconstructed by Marsiat and Berger, 1990). The simulated deviations are

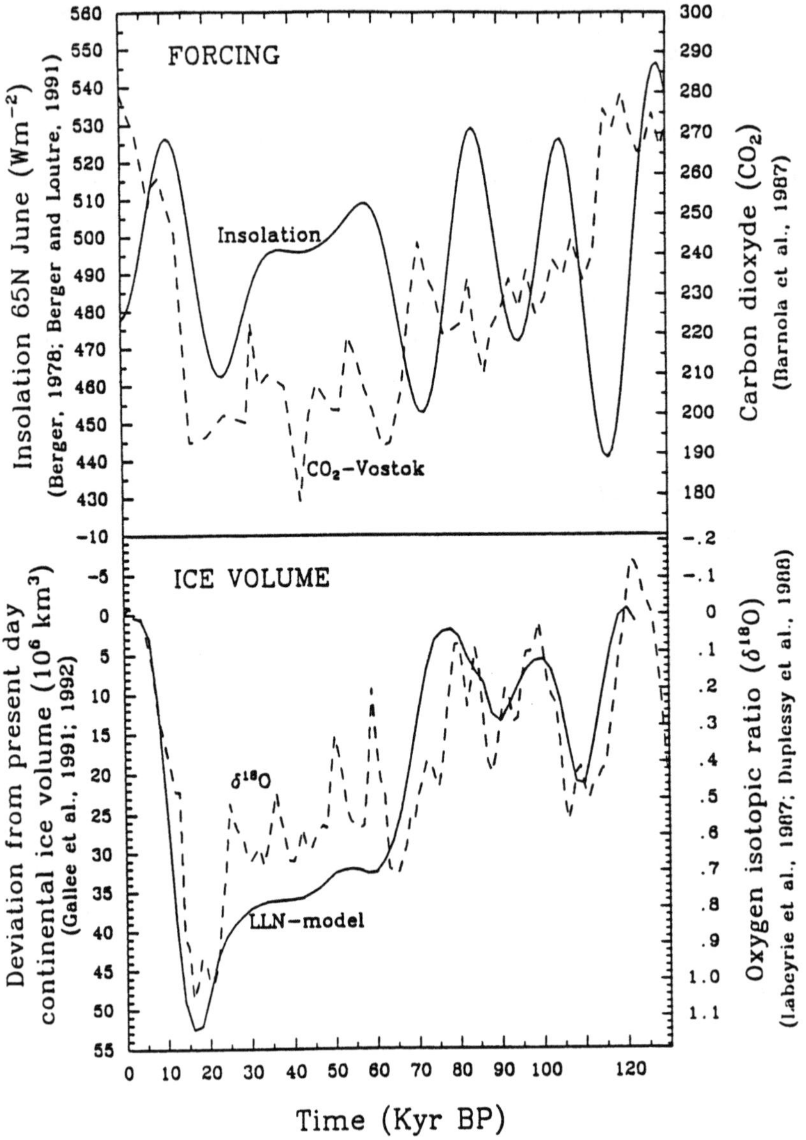

Figure 1. Simulated total ice volume deviation from present-day value assumed to be 30.5 10^6 km^3 (lower panel, full line, Gallée et al., 1992). This is compared to the variation of the global sea water oxygen isotopic ratio (lower panel, dashed line) given by Labeyrie et al. (1987)). These values are scaled and plotted such that their maximum corresponds to the deviation from present-day of the last glacial maximum total ice volume taken as 48.6 10^6 km^3 (Marsiat and Berger, 1990). The upper panel provides the long term variations of the June insolation at 65°N (full line, Berger, 1978) and of the Vostok CO_2 concentration (dashed line, Barnola et al., 1987).

obtained by adding the northern hemisphere ice volume changes simulated by the model for each of the individual ice sheets (Eurasia, Greenland, North America) and the changes in Antarctica reconstructed as follows. We assumed (1) that the Antarctic ice volume at 122 kyr BP and from 6 kyr BP to the present was the same as now (J. Oerlemans and D. Raynaud, personal communication, 1987); (2) that 18 kyr ago, it was 9.8 10^6 km^3 larger than today (Hughes et al., 1981); and (3) that intermediate values are given by linear interpolation. (Assumption (2) seems to be in agreement with the recent results obtained by Huybrechts (1990) from a three-dimensional ice sheet model; variation of his Antarctic simulated ice volume between 16 kyr BP and to-day, which depends however on sea level variation rather than on precipitation variations, amounts to 7.5 10^6 km^6).

The main characteristic of Figure 1 is the similarity between the reconstructed and the simulated global ice volumes, except for variations with time scales shorter than 5 kyr and for an overestimation of the calculated ice volume during isotopic stage 3 when compared to Labeyrie et al. (1987). Large ice volume oscillations are found between 122 kyr BP and 55 kyr BP, while the deglaciation is abrupt both in reconstruction and in simulation. The ice volumes for each ice sheet simulated by the LLN-model are also in good agreement with independent reconstructions made from proxy data by Mangerud (1991) and Boulton et al. (1985). At 19 kyr BP, the simulated temperature of the Northern Hemisphere is 3.4°C colder than the present day simulated value, and the simulated total ice volume deviation is 53 10^6 km^3.

5 The feedback mechanisms

One important feature of this LLN model is that it simulates both the glaciation and deglaciation of the last two climatic cycles. This is due to a number of feedbacks introduced in the model. Although these represent only a subset of all the feedbacks acting in the climate system, it is useful to analyze their relative contributions.

5.1 Entering the glaciation

At 60°N, insolation starts to decrease at 133, 127 and 121 kyr BP for March, June and September respectively. Summer insolation therefore peaked around 123 kyr BP at high northern latitudes. The minimum is reached 11 to 12 kyr later. For June, insolation decreases from 545 Wm^{-2} to 440 Wm^{-2}, a decrease of almost 20 %. These latitudes and months are characterized by a strong precession signal whereas in December obliquity dominates the spectrum at this latitude (insolation maxima and minima are alternatively 147, 129, 115 and 88 kyr BP).

Each time this type of decrease in insolation occurs, positive feedbacks amplify the response of the climate system to such changes in the external forcing, an amplification which leads to ice sheets accumulating not only up to the time of minimum insolation but continuing to accumulate until a few thousand years later (4-5 kyr) due to the inertia of the slow-response part of the system. During these latter periods, negative feedbacks progressively slow the build-up of the ice sheets until to they begin to retreat and melt following the ice maximum.

When insolation decreases, surface temperature decreases which delays the melting of snow fields in high latitudes. At the same time, taiga is replaced by tundra which increases the albedo of the vegetated surface covered by snow. Both the snow fields and tundra are therefore leading to an increase of the surface albedo creating a positive feedback. This is reinforced by the subsequent decrease of water vapour content of the atmosphere that results in a decrease in the downward infrared radiation at the surface. But this cooling is also responsible for decreases of the upward infrared, of the latent heat and of the sensible heat fluxes at the surface which feedback negatively on surface temperature.

During this initiation phase, which lasts roughly 4 kyr, the zonal mean temperature cooling at the surface leads to a cooling in the atmosphere which, with the insolation decrease, impacts the ice-surface energy budget. The ice-surface temperature does not reach the melting point, even in summer. This means that we have a net accumulation at the surface, ablation being suppressed. As a consequence, the ice sheets are growing which depresses the lithosphere below. With the altitude of the ice sheets increasing, the temperature at their surface decreases (a positive feedback) which progressively decreases the snowfall (a negative feedback) and finally stops the growth of the ice sheet. At the same time, the ice sheets extend to the south. Because of the continentality effect, snowfall at the top of the ice sheets in the interior of the continents decreases (a negative feedback), slowing down the southwards motion of the southern front of the ice sheets. Finally, the maximum volume of ice is reached, 110 kyr BP, 4 kyr later than the minimum of insolation.

5.2 The deglaciation process

As insolation had already started to increase 4 kyr before the ice maximum, this increase starts to be important at 110 kyr BP, particularly at the southern edge of the ice sheets which begin to melt. At the same time, the decreased snow fall triggers the "snow ageing" process which is more efficient in regions of the ice sheets where the temperature is above -10°C. This significantly decreases the albedo of both non-melting and melting snow cover, mainly to the south. In the meantime, snow accumulation has remained positive over the northern part of the ice sheets creating a N-S flux of ice within the ice sheets. At the southern edge, melting is faster than ice flow and isostatic rebound is not fast enough to level up the ice sheet which could prevent future net ablation above. This is not the case north of the ice sheets where isostatic rebound maintains the ice sheets high allowing the North to South flux of ice to continue to be efficient. As a consequence, the ice sheet volumes decrease, the heights of the ice sheets decrease, temperature at their surfaces increases which increases the global ablation. This leads to a decrease of the zonal surface albedo and a further replacement of tundra by taiga which feedbacks positively on the albedo decrease, directly and by reducing the albedo of the continental surfaces covered by snow. This leads finally to a minimum of ice volume reached about 5 kyr later than the insolation maximum.

6 The CO_2 - water vapour - albedo feedback mechanisms at the LGM

The comparison of the results obtained from the transient experiments using the Milankovitch and the Milankovitch + CO_2 forcings shows that, at the Last Glacial Maximum, the long term CO_2 changes are responsible for roughly 50 % of the temperature change and 30 % of the ice volume change. In an attempt to understand the role of the temperature-albedo feedback and the importance of the greenhouse gases (water vapour and CO_2), several sensitivity experiments have been made (Gallée et al., 1992; Tricot, 1992 and Berger et al., 1992).

In particular, the physical processes responsible for the cooling simulated at the LGM under the forcing of both insolation and CO_2 were investigated with a 1-D radiative convective (RCM) climate model (Tricot, 1992). The main reason for using a RCM is the number of experiments to be done to isolate the individual contribution of the main processes involved, the potentiality of a RCM for reaching such an objective (Ramanathan and Coakley, 1978) and the fact that it requires much less computational power than the 2.5-D model.

6.1 The Radiative-Convective model

Radiative-convective models (RCMs) are one-dimensional (horizontally-averaged) vertical column models that are used to study the globally averaged annual mean temperature (T) profile. They can yield a first estimate of the likely change in this globally averaged temperature profile due to a change of the internal or external radiative forcing. A comprehensive review of the use of RCMs for climate modelling is given by Ramanathan and Coakley (1978).

The RCMs are based on the thermodynamic equation

$$\frac{dT}{dt} = \frac{g}{C_p}\frac{dF}{dp} \tag{1}$$

where F is the sum of the net fluxes of shortwave (solar) and longwave (terrestrial) radiation, g is the acceleration due to gravity, C_p is the specific heat at constant pressure and p is the pressure.

The model equilibrium is achieved using a forward time marching numerical scheme (Manabe and Strickler, 1964). The iteration technique starts from an assumed initial global mean vertical temperature distribution. The pressure grid extends from the ground to the 0.5 hPa level and contains 19 gridpoints. A new temperature profile is calculated from Eq.(1) as a function of the net radiative heating/cooling computed in the 18 layers.

In the stratosphere, the temperature is in radiative equilibrium and hence determined only by a local balance between solar heating and terrestrial cooling. The solar heating depends mostly on ozone concentration and the longwave cooling is largely controlled by carbon dioxide and is primarily dependent on local temperature.

In the troposphere, a non radiative heat transfert (heat convection) must also be taken into account. A "convective adjustment scheme" is performed in the RCM at each iteration. If within any layer the computed lapse rate is greater than a prescribed critical lapse rate, the lapse rate is adjusted back to the critical lapse rate. This process is done such that the layer undergoing convective adjustment conserves energy. The surface temperature is calculated from the net radiative flux at the surface and the total convective heat flux dissipated in the troposphere during the convective adjustment procedure. In the standard version of the RCM, if the model's humidity (given by the mixing ratio, r) is allowed to vary, the prescribed vertical distribution for the relative humidity, HR, proposed by Manabe and Wetherald (1967) is used with a fixed relative humidity at the surface ($HR_s = 0.8$):

$$HR = HR_s \, \frac{\frac{p}{p_s} - 0.02}{0.98} \quad \text{and therefore,} \quad r = 0.622 \, \frac{HR \, e_s(T)}{p - HR \, e_s(T)}$$

The radiative schemes are discussed in detail in Tricot (1992) and their main characteristics are given here in Annex A and B.

As a first validation of the RCM, the quality of the solar and terrestrial radiative schemes (discussed in detail in Tricot, 1992) has been tested by comparison with the results of line-by-line radiative models used in the framework of the Intercomparison of Radiation Codes in Climate Models (ICRCCM, see Ellingson and Fouquart, 1991). For a large range of surface and atmospheric conditions - e.g., surface albedo, solar zenith angle and gaseous concentrations -, the radiative parameterizations agree very well with the most sophisticated radiative models available today. We refer to Tricot (1992) for a detailed discussion of these comparisons.

As a further validation of the RCM, a mean global temperature profile was calculated for normal climate conditions. The computation was performed assuming a surface albedo of 0.1, a critical lapse rate of -6.5 K km^{-1}, a standard ozone profile and a cloud filling in half the layer located between 700 and 600 hPa, with an optical thickness of 6.5. The water vapour profile is internally computed assuming a relative humidity of 80 % at the surface. This parameterization of the water vapour profile, based on a prescribed vertical distribution for the relative humidity, allows a non linear positive feedback between the temperature and the water vapour amount in the troposphere. In Figure 2, the RCM results for the present climate conditions are compared with the temperature profile of the US standard atmosphere (McClatchey et al., 1972). The two profiles show a general good

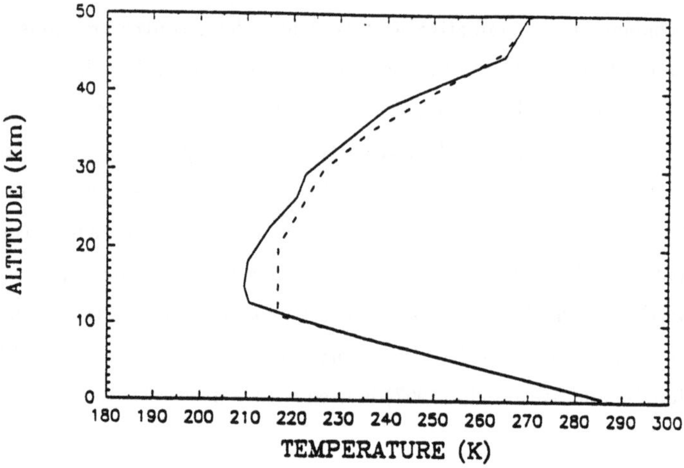

Figure 2. Vertical profiles of temperature from 0 to 50 km for present-day. The full line represent the simulation by the radiative-convective model of Tricot (1992); the dashed line is from the U.S. standard atmosphere model by McClatchey et al. (1972). The RCM results are obtained with 10 layers in the troposphere and 8 layers in the stratosphere, a surface albedo of 0.1, a solar constant of 1368 Wm^{-2}, a cloud layer between 700 and 600 hPa with our optical thickness of 6.5 and a cloudiness of 0.5, 330 ppmv for the CO_2 concentration, 1750 ppbv for the CH_4 concentration, 280 ppbv for the N_2O concentration, a water vapour profile given by a fixed relative humidity and a standard ozone profile.

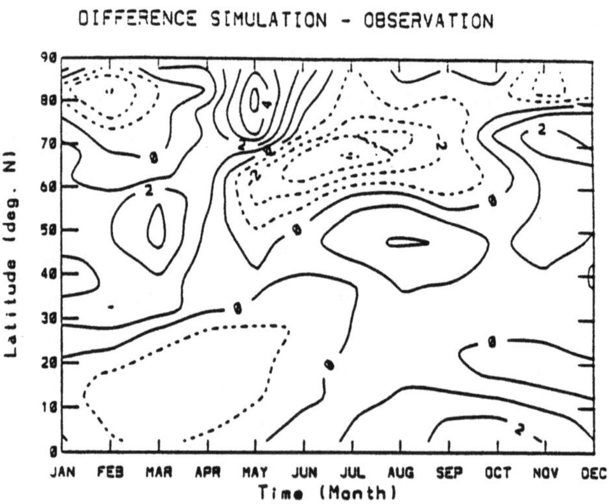

Figure 3. Seasonal change of the deviation from present-day observations (Schneider and Thompson, 1979) of the zonal mean air temperature at the surface of the Earth calculated by the LLN1 version of the LLN 2.5-D climate model (Tricot, 1992).

agreement, except for the simulated lower stratosphere which is too cold. But, it is known that the US standard atmosphere overestimates by some 10 K the actual hemispherically averaged temperature in that atmospheric region (Ramanathan and Dickinson, 1979).

From these validation tests, we conclude that the RCM satisfactorily simulates the present global climate. Furthermore, the direct radiative forcing induced by change in the values of the model parameters seems to be correctly estimated.

Sensitivity experiments undertaken with the RCM are discussed in Tricot (1992). In particular, when the CO_2 concentration is doubled (from 330 ppmv to 660 ppmv), the instantaneous perturbations of the net radiative budget (before the action of feedbacks) are estimated to be 3.72 Wm^{-2} at the tropopause and 1 Wm^{-2} at the surface. At equilibrium, the temperature increases by 1.8 K at the surface and in the troposphere and decreases in the stratosphere above 20 km. The tropospheric warming is 10 % larger if the absorption of solar radiation by CO_2, mainly in the stratosphere, is not taken into account. The sensitivity of the RCM to CO_2 change is very similar to that calculated in other sensitivity studies using RCMs with the same basic assumptions (e.g., Ramanathan, 1981). Without allowing the water vapour feedback to operate, the warming is reduced to 1.2 K.

Even if RCMs are highly simplified models of the climate system, their comparative simplicity permits an easier and a more complete understanding of the feedbacks (e.g., surface albedo and water vapour feedbacks) than the more comprehensive, and therefore more complex, 2-D and 3-D (GCMs) climate models.

6.2 The modified version of the 2.5-D LLN model of the climate system

As the purpose here is to quantify the relative contribution of each of the main climatic feedbacks, some improvements in the LLN model were first made by Tricot (1992) in such a way that the sensitivity of this modified version, designated as LLN1, to changes in trace gas concentrations will be closer to the sensitivity of the RCM itself. These improvements are summarized in Table 1 and concern mainly the radiative part of the model, involving new infrared and solar radiation schemes described in Annex A and B, and an explicit computation of the stratospheric temperature. Climatological data about

Table 1: Modification of the LLN model

Processes	Gallée et al. (1991) LLN	Tricot (1992) LLN1
Infrared radiation	Morcrette (1984)	a simplified version of the statistical model narrow-band where only six spectral band were considered (see Table B.1), and the transmissivity functions given by Shibata and Aoki, 1989
Solar radiation	Tricot and Berger (1988)	the parameterization of the Rayleigh and aerosol scattering are from Fouquart (1986) and the multiple scattering by the cloud water droplets from Joseph et al. (1976).
Tropopause	Calculated from T_s from Sellers (1983)	fixed from observations (Oort, 1983)
Stratospheric temperature	equal to the temperature at the tropopause	given by an energy budget calculated from radiative fluxes and a dynamical dating fixed for each ϕ and season (Oort, 1983)
Clouds	- equivalent average cloud as a function of ϕ from London (1957) for annual mean and from Berlyand et al. (1980) for the seasonal variations. - fixed height and optical thickness as a function of ϕ from Chou et al. (1981)	- equivalent cloud for each ϕ and season, above the ocean and the continents (Warren et al., 1986; 1988) - height from Chou et al. (1981) - optical thickness = 5.3
Aerosols	WMO (1986)	Tanré et al. (1984) Harvey (1988)
Oceanic transport	Sellers (1973) with a constant diffusion coefficient	Harvey (1988) with a diffusion coefficient function of ϕ.

cloud and aerosols also have been revised. LLN1 has been validated for the present-day climate. The differences from observations are more or less similar to those described in Gallée et al. (1991) and obtained from the LLN original model (see for example Figure 3 which provides the deviations from observations of the seasonal pattern of the simulated zonal mean temperature). The sensitivity of LLN1 to a doubling of CO_2 concentration ($\Delta T = 1.97K$) is similar to the sensitivity of the original version of the LLN model ($\Delta T = 2K$). Indeed, the weaker radiative perturbation in LLN (see Gallée et al., 1992) is compensated by allowing the tropopause height to be a function of the surface temperature. That induces a positive feedback not present in LLN1 where the tropopause level is fixed.

The cooling at the LGM simulated by LLN1 using the surface boundary conditions calculated by the LLN transient simulation, the seasonal cycle of insolation at 18 kyr BP and a CO_2 concentration of 194 ppmv amounts to 4.5 K. This cooling is associated with an increase of the planetary albedo from 31 to 32.6, an increase of the surface albedo from 16.2 to 19.3 and a decrease in the annually - averaged vertically - integrated water vapour concentration from 2.44 to 1.99 g cm^{-2}.

At the LGM, in addition to the insolation and CO_2 concentration variations, the hemispherically-averaged changes in the surface albedo and in the vertical distribution of water vapour are available as part of the LLN1 2.5-D model diagnostics. These changes were inserted one by one or in combination into the RCM. This model was therefore able to compute the change in the hemispheric mean surface temperature induced by the prescribed perturbations. The usual convective adjustment procedure was used, but with the mean temperature profile predicted by the 2.5-D model taken to be the critical lapse rate in the troposphere.

In the radiative-convective model, the water vapour profile is here calculated as for LLN1 through the Smith (1966) formulation of the humidity at level p:

$$q(p) = q_s \left(\frac{p}{p_s}\right)^x$$

- q_s is the specific humidity at the surface calculated from T_s and the relative humidity at the surface.

- p_s is the surface pressure.

- x is a prescribed parameter which is a function of time and latitude, fixed in LLN1 from the climatological humidity values at the surface and at 700 hPa given by Oort (1983). In the RCM, averaged values taken from LLN1 simulations for the surface relative humidity and x are kept fixed.

The optical thickness of the cloud situated between 700 and 600 hPa was fixed to 5.6 and the cloudiness to 0.6, values which are very similar to the hemispherically - and annually averaged - values of LLN1.

6.3 CO_2, insolation and water vapour

Insolation, CO_2 and surface albedo obtained from LLN1 at the LGM were then used to form the radiative convective model in order to investigate the importance of the water vapour feedback in particular.

Experiment 1 (Table 2) with RCM is related to the present-day climate and leads to a mean temperature of 288.32 K, very close to the 288.36 K obtained from LLN1, the surface albedo being prescribed to its LLN1 value.

In Experiment 2, only the surface albedo is changed to its LGM LLN1 value, the CO_2 and water vapour concentration being kept constant to their present-day value. In such a case, if we accept the Milankovitch hypothesis according to which the insolation variations modify the snow and ice covers in the Northern Hemisphere high latitudes, changing only α_s in the RCM is equivalent to looking for the "natural" cryospheric response of the climatic system to the insolation variations. The simulated cooling of 1.8 K is important and corresponds to a direct radiative forcing of -5.6 Wm^{-2} (i.e. the net radiative budget at the tropopause has decreased - less trapping - by 5.6 Wm^{-2}).

If we allow the water vapour feedback to amplify this perturbation (experiment 3), the cooling amounts to 3 K which corresponds to a gain factor of 1.66, R being given by (see Appendix C):

$$\Delta T_s = R\,\Delta T_{s_0}$$

where ΔT_{s_0} is the response of the RCM without any water vapour feedback. The smaller amount of water vapour in this cooler atmosphere leads indeed to a weakening of the

Table 2. Sensitivity experiments of a RCM to CO_2, water vapour and surface albedo (Tricot, 1992).

EXPERIMENTS WITH RCM	T_s K	ΔQ Wm^{-2}	α_{pl} (%)	$< H_2O >$ (g/cm^2)	ΔT_{so} K	ΔT_s K	R
1 • PRESENT-DAY CLIMATE $CO_2 = 330$ ppmv $\alpha_s = 16.24$ H_2O computed by RCM	288.32		32.85	2.29			
2 • $CO_2 = 330$ ppmv $\alpha_s = 19.25$ H_2O fixed as in 1 •	286.52	-5.62	34.43	2.29	-1.8	-1.8	1.
3 • $CO_2 = 330$ ppmv $\alpha_s = 19.25$ H_2O computed by RCM	285.33	-5.62	34.66	1.88	-1.8	-2.99	1.66
4 • $CO_2 = 194$ ppmv $\alpha_s = 16.24$ H_2O fixed as in 1 •	287.38	-2.91	32.85	2.29	-0.94	-0.94	1.
5 • $CO_2 = 194$ ppmv $\alpha_s = 16.24$ H_2O computed by RCM	286.75	-2.91	32.97	2.07	-0.94	-1.57	1.67
6 • $CO_2 = 194$ ppmv $\alpha_s = 19.25$ H_2O fixed as in 1 •	285.58	-8.53	34.42	2.29	-2.74	-2.74	1.
7 • $CO_2 = 194$ ppmv $\alpha_s = 19.25$ H_2O computed by RCM	283.81	-8.53	34.77	1.70	-2.74	-4.51	1.65

T_s is the annually - and hemispherically - averaged temperature at the surface

ΔQ is the initial radiative perturbation at the tropopause (W m^{-2}) after a possible modification of the CO_2 concentration and of the surface albedo α_s (from their values in exp. 1 •)

α_{pl} is the planetary albedo

$< H_2O >$ is the water vapour concentration in g cm^{-2}

ΔT_{so} is the response of RCM without H_2O feedback

f_{H_2O} and R are respectively the feedback factor for H_2O and the gain factor with

$$\Delta T_s = \frac{1}{1-f} \Delta T_{so} = R \, \Delta T_{so}$$

greenhouse effect. It combines with the larger planetary albedo induced by the increase of the surface albedo to produce a total cooling 66 % larger than without the water vapour feedback. As a consequence, the water vapour induces a positive feedback mainly through its impact on the greenhouse effect and, in a much smaller amount, through an increase in the planetary albedo related to a smaller absorption of the solar radiation in the atmosphere by the water vapour.

If we now decrease the CO_2 concentration from "present-day" value (330 ppmv) to the LGM value (194 ppmv), the climatic response is a cooling of 0.94°C or 1.57°C respectively without (experiment 4) and with (experiment 5) the water vapour feedback. This corresponds to a gain factor of 1.67 very similar to the value obtained in experiment 3 for the albedo increase. This might be a coincidence and it must be tested in more sophisticated models where geographical contrasts do exist. The change of the surface albedo at the LGM (experiment 3) is indeed largely related to the change of the surface cover in middle and high latitudes, whereas the CO_2 concentration change (experiment 5) is a worldwide phenomenon.

The last 2 experiments combine the changes in the CO_2 concentration and in the surface albedo. When the water vapour feedback is switched off, the cumulative effect of these forcings leads to a total cooling which is the sum of the individual coolings: -2.74 = -1.8 - 0.94. This is expected through a linearization of the problem in the absence of the main non-linearity related to water vapour feedback. Allowing this water-vapour feedback to amplify the direct initial perturbation leads to a total cooling of 4.51 K (experiment 7), very close to the cooling of 4.46 K calculated by the LLN1 model for the LGM boundary conditions. Even in this case, this total response is more or less the sum of the individual responses (experiments 3 and 5): -4.51 $\simeq$ -2.99 - 1.57 because the perturbations are small in amplitude compared to the fundamental state (a few degrees against 288 K). Table 2 thus illustrates the linear theory of feedbacks when the external forcings are due to albedo and CO_2 changes, the only feedback being related in our study to the water vapour concentration change.

In terms of the feedback parameter, λ, used in the theory of the greenhouse effect intensification, we have

$$\Delta T_{s_{equilibrium}} = \frac{\Delta Q}{\lambda} = \frac{G_0 \Delta Q}{1-f} = \frac{\Delta T_{s_0}}{1-f} = R\,\Delta T_{s_0}$$

The feedback factor, $f = 1 - \frac{1}{R}$, for the water vapour is estimated from experiments 3,

5 and 7 to be roughly 0.4. λ in experiments 2, 4 and 6 is equal to 3.1 and in experiments 3, 5 and 7 it amounts to roughly 1.9; this leads to a positive feedback of the water vapour equal to 1.2, a value close to the value (1.4) obtained in the greenhouse theory (e.g., Berger and Tricot, 1992). It happens therefore that the water vapour feedback intensity is about independent of the specific character of the prescribed changes.

All these sensitivity tests done with a RCM using the value of the parameters simulated at the last glacial maximum by the LLN1 2.5-D model lead to the following conclusions for the northern hemisphere:

1. The cooling using only the astronomical forcing - but allowing for the water vapour feedback - amounts to 3°C, 67 % of the 4.5°C cooling resulting from both the astronomical and CO_2 forcings. It must be stressed that what we call astronomical forcing includes both the change in insolation and the consequent changes in surface and planetary albedo.

2. If the CO_2 concentration is kept fixed to its interglacial level (330 ppmv), the astronomical-albedo forcing without the water vapour feedback (WVF) explains 60 % of the 3°C cooling, the water vapour feedback being therefore responsible for 40 % of the total cooling in this Milankovitch alone experiment.

3. In the astronomical + CO_2 experiment, the direct effect of these forcings (i.e. without the WVF) explains 60 % (2.7°C) of the 4.5°C cooling. As in 2., the WVF is therefore responsible for 40 % of the cooling. If we discriminate between the astronomical-albedo forcing (AA) and the CO_2 contribution (CO_2), we can see that without the WVF, AA explains 40 of the 60 % of the total cooling. The contributions to the WVF (40 % of the total cooling) are 27 % for AA and 13 % for CO_2.

4. In this astronomical + CO_2 cooling, it may also be interesting to analyse the proportion which is due to water vapor for each sub-experiment. Considering only the astronomical-albedo forcing, the WVF is responsible for 27 of the 67 % of the global cooling, the remaining 40 % is the direct effect of AA. In the CO_2 alone experiment, the WVF explains 13 of the 33 % of the global cooling, the direct effect of CO_2 accounting for the remaining 20 %.

In summary, the general conclusions are:

1. If we discriminate between the astronomical-albedo forcing and the CO_2 forcing, AA is responsible for 67 % and CO_2 for 33 % of the cooling, whatever experiment is considered:

without WVF	:	-2.7 = -1.8 - 0.9
with WVF	:	-4.5 = -3.0 - 1.5
due to WVF	:	-1.8 = - 1.2 - 0.6

2. If we discriminate between the direct effect of the astronomical-albedo and/or CO_2 forcings and the water vapour feedback contribution, AA and/or CO_2 accounts for 60 % and the WVF for 40 % of the cooling, whatever experiment is considered:

without CO_2	(- 3.0 = - 1.8 - 1.2)
with CO_2	(- 4.5 = - 2.7 - 1.8)
due to CO_2 alone	(- 1.5 = - 0.9 - 0.6)

7 The two last glacial-interglacial cycles experiment

An obvious experiment (called the 200 kyr experiment) to be done with the LLN 2.5-D model was to force it by astronomically-driven insolations and by prescribed CO_2 concentrations over the last 200 kyr (Gallée et al., 1993), in order to confirm the ability of the model to sustain glacial-interglacial cycles. On the other hand, possible shortcomings of the 200 kyr simulation could help us to determine missing or misrepresentated processes in this model.

The Vostok ice core does not provide a reconstruction of CO_2 concentration variations before 150 kyr BP, so that the model was forced by CO_2 concentrations adapted from Shackleton et al. (1992). The Shackleton et al. (1992) data have been modified by taking the Barnola et al. (1987) data between 19 kyr BP and present time and by adding to each value before 19 kyr BP the mean difference of -22ppmv (221ppmv -243ppmv), calculated over the last glaciation (between 120 and 20 kyr BP), between the Barnola et al. (1987) and Shackleton et al. (1992) data.

The simulation starts assuming that the ice sheets are melted at 200 kyr BP, because this time corresponds to an interglacial in marine $\delta^{18}O$ records. Two glacial maxima comparable in amplitude are simulated, at 134 and 15 kyr BP. The similar amplitude of these glacial maxima is also found in the marine $\delta^{18}O$ records (e.g. Imbrie et al., 1984, see Fig. 4). Other important maxima in the continental ice volume occur in the model output at 180, 109 and 59 kyr BP. The phase lags between the insolation minima at 65°N and ice volume maxima are respectively 6, 5, 6, 11 and 8 kyr for the ice volume maxima at 180, 134, 109, 59 and 15 kyr BP.

A further comparison of the results of the 200 kyr experiment with the SPECMAP record is made in Figure 4. The variations correlate well and a spectral analysis of both time series gives spectral peaks of similar amplitude for the 100, 41, 23 and 19 kyr periods. The correct relative intensity of the spectral peaks for the simulated ice volume is probably the most important result of this 200 kyr experiment as it demonstrates the ability of the LLN model to reproduce the 100 kyr cycle. Moreover, the LLN model simulates a deglaciation centred on 130 kyr BP and a stable interglacial with high sea-level lasting from about 127 kyr BP to 115 kyr BP. This duration agrees well with geological evidence for the last interglacial (Müller, 1989), whereas a major weakness both of the SPECMAP record and of previous model simulations (e.g. Imbrie and Imbrie, 1980; Berger et al., 1981) is precisely their inability to depict the several thousand-year long stable interglacials.

It is also interesting to see that the simulated annual mean northern hemisphere temperature decreases significantly (by 3°C) between 200 and 180 kyr BP and between 80 and 60 kyr BP. Important increases (of 4°C) occur at the beginning of each interglacial, namely between 134 and 125 kyr BP and between 15 kyr BP and present time. These important variations in the simulated annual mean northern hemisphere temperature occur when both insolation and CO_2 concentration variations lead to important variations in the continental ice volume.

Differences between the variations of the simulated northern hemisphere ice volume for the 122 kyr (see sections 4 and 5 of this paper) and the 200 kyr experiments are the most important between 122 and 80 kyr BP, with a larger ice volume simulated in the 122 kyr experiment. This larger ice volume is in closer agreement with reconstructions obtained from deep sea cores. Moreover in the 200 kyr experiment, a complete deglaciation of the Greenland ice sheet is simulated during the Eemian interglacial. This is in contradiction with recent analysis of Greenland ice cores (Souchez et al., 1990; Reeh et al., 1991), which indicates that the Summit region of the Central Greenland ice sheet seems to have survived the warm climate of the Eemian interglacial. This is also in opposition to the hypothesis

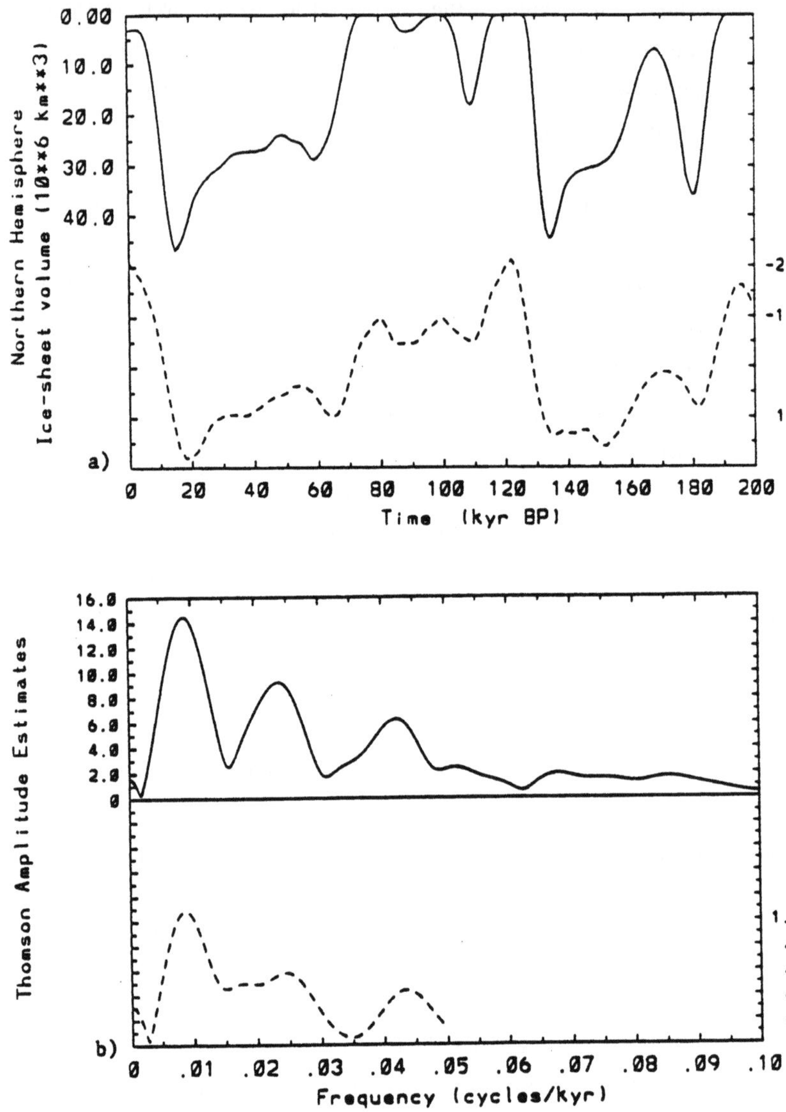

Figure 4. (a) Variations over the last 200 kyr of (i) the simulated ice volume of the northern hemisphere (solid curve) (Gallée et al., 1993), and (ii) $\delta^{18}O$ variation as reconstructed in the SPECMAP time series (short-dashed curve) (Imbrie et al., 1984; Martinson et al., 1987). The forcing used in the simulation is the insolation variation at the top of the atmosphere and the CO_2 variation adapted from Shackleton et al. (1992). (b) Spectral amplitude in the Thomson multi-taper harmonic analysis (Thomson, 1990) of (i) the simulated ice volume of the northern hemisphere for the 200 kyr experiment (solid curve), and (ii) $\delta^{18}O$ variation over the last 200 kyr as reconstructed in the SPECMAP time series (short-dashed curve).

done for the initiation of the LLN model at 122 kyr BP in the 122 kyr experiment where a volume equal to 2/3 of the present Greenland ice sheet was assumed (Gallée et al., 1992), a value in agreement with a recent simulation done by Létréguilly et al. (1991).

The differences in the ice volume simulated after the Eemian interglacial occur because the climate simulated during the Eem is warmer if the Greenland ice sheet is melted at this time. A larger negative insolation anomaly is then necessary for the ice-sheet initiation. Over the North American ice sheet in the 65–70°N latitude band, the ice ablation decrease amounts to 1.03 mie/yr between 122 and 120 kyr BP in the 122 kyr experiment, while it amounts to 0.54 mie/yr in the 200 kyr experiment (mie = meter of ice equivalent). The snow precipitation increase amounts to 0.04 mie/yr between 122 and 120 kyr BP in the 122 kyr experiment, while it amounts to 0.08 mie/yr in the 200 kyr experiment. In this area, ice-sheet initiation occurs at 120 kyr BP in the 122 kyr experiment, while in the 200 kyr experiment it occurs only at 119 kyr BP, when snow precipitation and ice ablation amount to 0.43 mie/yr and 0.39 mie/yr respectively.

Finally, an important difference between the atmospheric CO_2 forcing in the 200 kyr and 122 kyr experiments has been found just before and during isotopic stage 5e. There is a substantial phase shift between the Barnola et al. (1987) and Shackleton et al. (1992) reconstructions at this time. In particular, the CO_2 increase (before this interglacial) in the Barnola reconstruction is earlier than Shackleton's reconstruction by 8 kyr. More precisely, the maximum in CO_2 concentration occurs at 126 kyr BP in the Shackleton et al. (1992) data, adopted for the 200 kyr experiment, but occurs at 134 kyr BP in the Barnola et al. (1987) data. In order to assess this impact, a sensitivity experiment has been done by forcing the model with the Barnola et al. (1987) CO_2 data and starting at 150 kyr BP with the ice sheet configuration as it is simulated in the 200 kyr experiment at this time. This experiment is referred to as the 150 kyr experiment.

It is remarkable that the CO_2 lag does not significantly affect the time of the ice volume maximum: this maximum occurs at 134 kyr BP in the 200 kyr experiment and at 136 kyr BP in the 150 kyr experiment (for comparison, the age given in the SPECMAP chronology, Martinson et al., 1987, is 135 kyr BP). This suggests that in our model, the timing of the simulated ice volume variations is more tightly locked to the timing of insolation variations than to the timing of CO_2 variations.

At the penultimate glacial maximum, the CO_2 lag impacts more significantly upon the ice volume than its phase. The penultimate ice volume maximum in the 150 kyr experiment is lower than that of the 200 kyr experiment by 6 $10^6 km^3$. The 200 kyr experiment gives two glacial maxima of comparable amplitude, a feature which is in agreement with the SPECMAP reconstruction. In the 150 kyr experiment the continental ice volume at the last glacial maximum is very similar to that of the 200 kyr experiment, and thus of larger amplitude than that of 136 kyr BP. Consequently, the difference in the simulated penultimate glacial maximum ice volumes favours the CO_2 chronology proposed by Shackleton et al. (1992) and used in the 200 kyr experiment, although there is no doubt that as regards actual CO_2 concentration the direct measurements of Barnola et al. (1987) are to be preferred.

8 Conclusions

Simulation of the advances and retreats of the ice sheets over the last two glacial-interglacial cycles indicates that the orbital parameter variations are able to trigger feedbacks in the LLN model which then induce significant climatic variations at that time scale. We would argue that (i) because the LLN model reproduces correctly the power spectrum of the reconstructed climatic variations over the last two glacial-interglacial cycles; (ii) because it gives a more realistic depiction of the Eemian interglacial (isotopic stage 5e) and of its length than previous models and (iii) because its sensitivity to the CO_2 forcing and to the water vapour feedback is similar to that of the more complex GCMs, it is its chronology rather than the SPECMAP chronology that should be the target for testing new radiometric dating methods.

Sensitivity tests to external forcing, boundary conditions and types of processes considered with their parameterizations, lead to the following conclusions in order of importance:

- the behavior of the latitudinal and seasonal distributions of insolation is determinant; no climatic changes are generated if the insolation is kept constant;

- the snow albedo parameterization over the snow and ice fields, in particular the snow-ageing process (i.e. snow albedo as a function of the snowfall frequency), plays an important role in shaping the 100 kyr-cycle through its influence in the ablation process;

- variations in the ice ablation has a more significant impact on the ice-sheet response than variations in snow precipitation. The ice ablation is lagging behind the summer insolation forcing, being amplified by the desert-altitude effect and the snow ageing;
- the variation of the area covered by the continental tundra amplifies the response of the climate model to the forcing.

Considering all the physical processes included in this model, many parameters had to be used. All of them were taken from the literature with their range of uncertainty. Among the possible physical values of these parameters, we have selected those which allow a good representation of the present-day climate, and lead to a reasonable simulation of the long term climatic variations over the last glacial-interglacial cycle. Further measurements to narrow the range of plausible values for each parameter would confirm whether or not the value we have selected is close to the most probable.

An extended simulation covering the last 200 kyr has shown that these selected values remain valid for reproducing the last two glacial-interglacial cycles. It must also be stressed that the model results are quite robust against the physical processes included.

It must be pointed out that the physical processes included in the LLN model are probably not the only ones to play a significant role in the climate system and this simulation must be regarded first of all as an exercise to learn how the coupling of the different components of the climate system works at the astronomical time scale, and to test parameterizations which have appeared in the literature for each individual subsystem. Given that the model has many parameters connected to the numerous processes that are parameterized within it, the fact that it fits more or less one set of data (the present-day climate, for example) is not a definitive test of its physical validity. The most significant result is, however, that among the large number of possible combinations of parameter values allowing a good representation of the present-day climate, only a very limited set allows at the same time reasonable palaeoclimatic reconstruction at the astronomical time scale. It seems unlikely therefore that, by chance alone, such a non-linear numerical model can simulate the present-day latitudinal and seasonal cycles of many climate variables (Gallée et al., 1991) and, simultaneously, slight changes over the last few centuries (Smits et al., 1993), changes as large as those projected for the next few decades (Fichefet et al., 1989), and the gross features of the low frequency part of the last two glacial-interglacial cycles (Gallée et al., 1992, 1993).

The non-linearities in the climate system can however introduce spurious responses which can be masked sometimes by compensating errors made in the different parameterizations. Moreover, the potential model errors might be compensated by errors in the geological and/or CO_2 chronologies. An improvement of the $\delta^{18}O$ time scale, in particular, will therefore help in determining possible shortcomings of the model and consequently in refining our knowledge of the various processes which can be inferred from model studies.

In view of the relatively good simulations obtained by the LLN model, there are many important implications that need to be explored. First, were the results fortuitous, or rather, were most of the important processes captured in two dimensions by neglecting such factors as changes in deep water formation ? Second, in view of the critical importance of snow - albedo - temperature - vegetation feedbacks on the ice-sheet initiation and deglaciation processes, can the parameterizations of these feedbacks be improved ? Third, looking to the sensitivity of all the models used to test the astronomical theory, how are we to distinguish which, if either, of the mechanisms are correct ? It would appear that the only acceptable way is to describe more of the physics, chemistry and biology explicitly.

9 Acknowledgements

H. Gallée is supported by the Belgian Scientific Research Programme on Antarctica of the Prime Minister's Science Policy Office, Th. Fichefet is supported by the National Fund for Scientific Research (Belgium), and M.F. Loutre is supported by Contract CEA BC-4561 of the Commissariat Français à l'Energie Atomique. This research was sponsored partly by the Climate Programme of the Commission of the European Communities under Grants EV4C-0052-B (GDF) and EPOC-0004 (EDB). Computations with the palaeoclimate model were done using an IBM RS/6000-550 workstation under a study contract between IBM Belgium and the Catholic University of Louvain in Louvain-la-Neuve (1992-1994).

10 References

Aoki T., 1978a. Semi-direct band models for the transmittance calculation. *J. Met. Soc. Japan*, 56(2), pp. 112-119.

Aoki T., 1978b. Semi-direct random band model for exponential-tailed S^{-1} intensity distribution. *J. Met. Soc. Japan*, 56(5), pp. 508-515.

Barnola J.-M., Raynaud Y., Korotkevich Y.S., and Cl. Lorius, 1987. Vostok ice core provides 160,000-yr record of atmospheric CO_2. *Nature*, pp. 408–414.

Berger A., 1976. Long-term variations of daily and monthly insolation during the last Ice Age. *EOS*, 57(4), pp. 254.

Berger A., 1978. Long-term variations of daily insolation and Quaternary climatic changes. *Journal of Atmospheric Science*, 35(12), pp. 2362-2367.

Berger A., 1979. Insolation signatures of Quaternary climatic changes. *Il Nuovo Cimento*, 2C(1), pp. 63-87.

Berger A., Guiot J., Kukla G., and P. Pestiaux, 1981. Long term variations of monthly insolation as related to climate change. *Geologischen Rundschau*, Bd. 70(2), pp. 748–758.

Berger A., Gallée H., Fichefet Th., Marsiat I., and C. Tricot, 1990. Testing the astronomical theory with a coupled climate-ice sheet model. In: *Geochemical Variability in the Oceans, Ice and Sediments. Palaeogr., Palaeoclimatol., Palaeoecol.*, 89(1/2), L.D. Labeyrie and C. Jeandel (eds), Global and Planetary Change Section, 3(1/2), pp. 125-141.

Berger A. and M.F. Loutre, 1991. Insolation values for the climate of the last 10 million years. *Quaternary Science Reviews*, 10 n°4, pp. 297-317.

Berger A., and C. Tricot, 1992. The greenhouse effect. *Surveys in Geophysics*, 13, pp. 523-549.

Berger A., Gallée H., and C. Tricot, 1992. Glaciation and deglaciation mechanisms in a coupled 2-D climate - ice sheet model. *J. of Glaciology*, (in press).

Berlyand T.G., Strokina L.A. and L.Y.E. Greshnikova, 1980. Zonal cloud distribution on the Earth. *Meteorol. Gidrol.*, 3, pp. 15-23.

Birchfield G.E., Weertmann J., and A.T. Lunde, 1982. A model study of high-latitude topography in the climatic response to orbital insulation anomalies. *J. Atmos. Sci.*, 39, pp. 71-87.

Boulton G.S., Smith G.D., Jones A.S., and J. Newsome, 1985. Glacial geology and glaciology of the last mid-latitude ice sheets. *J. Geol. Soc. London*, 142, pp. 447-474.

Broecker W.S., Peteet D.M., and D. Rind, 1985. Does the ocean-atmosphere system have more than one stable mode of operation ? *Nature*, 315, pp. 21-26.

Broecker W.S., Andree M., Wolfi W., Oeschger H., Bonani C., Kennett J., and D. Peteet, 1988. The chronology of the last deglaciation: implications to the cause of the Younger Dryas event. *Paleoceanog.*, 3(1), pp. 1-19.

Chou M.D., 1986. Atmospheric solar heating rate in the water vapor bands. *J. Climate Appl. Meteor.*, 25, pp. 1532-1542.

Chou M.D., 1990. Parameterizations for the absorption of solar radiation by O_2 and CO_2 with application to climate studies. *J. Climate*, 3, pp. 209-217.

Chou S.H., Curran R.J., and G. Ohring, 1981. The effects of surface evaporation parameterization on climate sensitivity to solar constant variations. *J. Atmos. Sci.*, 38, pp. 931-938.

Clough S.A., Kneizys F.X., Davies R., Gamache R., and R.H. Tipping, 1980. Theoretical line shape for H_2O vapor: application to the continuum. In: *Atmospheric Water Vapor*, A. Deepak, T.D. Wilkerson and L.H. Ruhnke (eds), pp. 25-46, Academic Press, NY.

Deblonde G., and W.R. Peltier, 1990. A model of late Pleistocene ice sheet growth with realistic geography and simplified cryodynamics and geodynamics. *Climate Dynamics*, 5 n°, pp. 103-110.

Deblonde G., and W.R. Peltier, 1991a. Simulations of continental ice sheet growth over the last glacial-interglacial cycle: experiments with a one level seasonal energy balance model including realistic geography. *Journal of Geophysical Research*, 96, pp. 9189-9215.

Deblonde G., and W.R. Peltier, 1991b. A one-dimensional model of continental ice volume fluctuations through the Pleistocene: implications for the origin of the Mid-Pleistocene climate transition. *Journal of Climate*, 4(3), pp. 318-344.

Duplessy J.Cl., Labeyrie L., and P.L. Blanc, 1988. Norwegian sea deep water variations over the last climatic cycle: Paleo-oceanographical implications. In: *Long and Short Term Variability of Climate*, pp. 83-116, Earth Sci. Ser., edited by H. Wanner and U. Siegenthaler, Springer-Verlag, New York.

Ellingson R.G., and Y. Fouquart, 1991. The intercomparison of radiation codes in climate models: an overview. *J. Geophys. Res.*, 96(D5), pp. 8925-8927.

Fichefet Th., Tricot C., Berger A., Gallée H., and I. Marsiat, 1989. Climate studies with a coupled atmosphere-upper ocean-ice sheets model. *Philosophical Transactions of the Royal Society of London*, A329, pp. 249-261.

Fouquart Y., 1986. Radiative transfer in climate modeling.In: *Proceedings of the NATO-ASI on Physically-based modeling and simulation of climate and climatic change*, M. Schlesinger (ed.), pp. 223-283, Erice, 11-23 May 1986, D. Reidel Publ. Company, Dordrecht, Holland.

Fouquart Y., and B. Bonnel, 1980. Computations of solar heating of the Earth's atmosphere: a new parameterization. *Beitr. Phys. Atmosph.*, 53, pp. 35-62.

Gallée H., van Ypersele J.P., Fichefet Th., Tricot C., and A. Berger, 1991. Simulation of the last glacial cycle by a coupled sectorially averaged climate - ice-sheet model. I. The Climate Model. *J. Geophys. Res.*, 96, pp. 13,139-13,161.

Gallée H., van Ypersele J.P., Fichefet Th., Marsiat I., Tricot C., and A. Berger, 1992. Simulation of the last glacial cycle by a coupled, sectorially averaged climate - ice-sheet model. II. Response to insolation and CO_2 variation. *J. Geophys. Res.*, 97 n° D14, pp. 15,713-15,740.

Gallée H., Berger A., and N.J. Shackleton, 1993. Simulation of the climate of the last 200 kyr with the LLN 2D-model. In: *Ice in the Climate System*, R. Peltier (ed.), proceedings of the NATO Advanced Research Workshop, Aussois, France, September 7-11, 1992. (in press).

Ghil M., and H. Le Treut, 1981. A climate model with cryodynamics and geodynamics. *J. Geophys. Res.*, 86, pp. 5262-5270.

Godson W.L., 1953. The evaluation of infra-red fradiative fluxes due to atmospheric water vapour. *Quart. J.R. Met. Soc.*, 79, pp. 367-379.

Harvey L.D.D., 1988. On the role of high latitude ice, snow, and vegetation feedbacks in the climatic response to external forcing change. *Clim. Change*, 13, pp. 191-224.

Hays J.D., Imbrie J., and N.J. Shackleton, 1976. Variations in the earth's orbit: pacemaker of the ice ages. *Sciences*, 194, pp. 1121-1132.

Hecht A., 1985. Paleoclimatology. A retrospective of the past 20 years. In: *Paleoclimate Analysis and Modelling*, A. Hecht (ed.), pp. 1-26, Wiley, New York.

Hovine S. and Th. Fichefet, 1993. A model study of the glacial oceanic circulation. In: *Long Term Climatic Variations - Data Modelling*, R. Peltier (ed.), proceedings of the NATO Advanced Study Institute, Siena (Italy), September 27 - October 11, 1993. (in press).

Hughes T.J., 1987. Ice dynamics and deglaciation models when ice sheets collapsed. In: *North America and Adjacent Oceans During the Last Deglaciation*, W.F. Ruddiman and H.E. Wright Jr. (eds), pp. 183-220, The Geology of North America, v.K-3 Geol. Soc. Am., Boulder, CO.

Hughes T.J., Denton G.H., Anderson B.G., Schilling D.H., Fasthook J.L., and C.S. Lingle, 1981. The last great ice sheets: a global view. In: *The Last Great Ice Sheets*, G.H. Denton and T.J. Hughes (eds), pp. 275-317, Wiley Interscience Publ., USA.

Huybrechts Ph., 1990. The Antarctic ice sheet during the last glacial-interglacial cycle: a 3-D model experiment. *Ann. of Glaciol.*, 14, pp. 115-119.

Hyde W.T., and W.R. Peltier, 1985. Sensitivity experiments with a model of the Ice Age: the response to harmonic forcing. *Journal of Atmospheric Sciences*, 42 n° 20, pp. 2170-2188.

Imbrie J., and J.Z. Imbrie, 1980. Modeling the climatic response to orbital variations. *Science*, 207, 943–953.

Imbrie J., Hays J.D., Martinson D.G., McIntyre A., Mix A.C., Morley J.J., Pisias N.G., Prell W.L., and N.J. Shackleton, 1984. The orbital theory of Pleistocene climate: support from a revised chronology of the marine $\delta^{18}O$ record. In: *Milankovitch and Climate*, A. Berger et al. (eds), pp. 269-305, D. Reidel, Dordrecht, Holland.

Johnson R.B., and .E.E. Branstetter, 1974. Integration of Planck's equation by the Laguerre-Gauss quadrature method. *J. Opt. Soc. Am.*, 64(11), pp. 1445-1449.

Joseph J.H., Wiscombe W.J., and J.A. Weinman, 1976. The delta-Eddington approximation for radiative flux transfer. *J. Atmos. Sci.*, 33, pp. 2452-2459.

Kutzbach J.E., 1985. Modelling of paleoclimates. *Adv. Geophys.*, 25A, pp. 159-196.

Labeyrie L.D., Duplessy J.Cl., and P.L. Blanc, 1987. Variations in mode of formation and temperature of oceanic deep waters over the past 125,000 years. *Nature*, 327, pp. 477-482.

Letréguilly A., Reeh N., and P. Huybrechts, 1991. The Greenland ice sheet through the last glacial-interglacial cycle. *Palaeogeogr., Palaeoclimatol., Palaeoecol., (Global Planet.Change Sect.)*, 90, pp. 385–394.

Le Treut H., Portes J., Jouzel J., and M. Ghil, 1988. Isotopic modeling of climatic oscillations: implications for a comparative study of marine and ice core records. *J. Geophys. Res.*, 93, pp. 9365-9383.

London J., 1957. A study of the atmospheric heat balance. *Final Report, contract AF*, 19(122)-165, Dept. Meteor. Oceanogr., New York University, 99pp.

Manabe S., and R.F. Strickler, 1964. Thermal equilibrium of the atmosphere with a convective adjustment. *J. Atmos. Sci.*, 21, pp. 361-385.

Manabe S., and R.T. Wetherald, 1967. Thermal equilibrium with a given distribution of relative humidity. *J. Atmos. Sci.*, 24, pp. 241-259.

Mangerud J., 1991. The last glacial history of Scandinavia between the last interglacial and the last glacial maximum. In: *Klimageschichtliche Probleme der Letzten 130,000 Jahre*, B. Frenzel (ed.), pp. 307-300, Akademie der Wissenschaften und der Literatur, Mainz, G. Fisher Verlag, Stuttgart, New York.

Marsiat I., and A. Berger, 1990. On the relationship between ice volume and sea level over the last glacial cycle. *Climate Dynamics*, 4, pp. 81-84.

Martinson D.G., Pisias N.G., Hays J.D., Imbrie J., Moore T.C., and N.J. Shackleton, 1987. Age dating and the orbital theory of the ice ages: development of a high-resolution 0 to 300,000-year stratigraphy. *Quat.Res.*, 27, pp. 1–27.

McClatchey R.A., Fenn R.W., Selby J.E.A., Volz F.E., and J.S. Garing, 1972. Optical properties of the atmospheres. *Rep. AFCRL - 72 - 0497.* Environmental Research Paper 411, Air Force Cambridge Res. Lab., Mass., 3rd ed., 108p.

Morcrette J.J., 1984. Sur la paramétrisation du rayonnement dans les modèles de la circulation générale de l'atmosphère. *Thèse de Doctorat d'Etat*, Université des Sciences et des Techniques de Lille, 630, 373pp.

Müller H., 1989. Palynologische Untersuchung eemzeitlicher Ablagerungen einer 15 km westlich Sylt niedergebrachten Kernbohrung. *Probleme der Küstenforschung im südlichen Nordseegebiet*, 17, pp. 119–124.

Nicolis C., 1984. Self-oscilations, external forcings, and climate predictability. In: *Milankovitch and Climate*, A. Berger, J. Imbrie, J. Hays, G. Kukla and B. Saltzman (eds), pp. 637-652, D. Reidel, Dordrecht, Holland.

North G.R., and T.J. Crowley, 1985. Application of a seasonal climate model to Cenozoic glaciation. *J. Geol. Soc. (London)*, 142, pp. 475-482.

Oerlemans J., 1982. Glacial cycles and ice-sheet modelling. *Clim. Change*, 4, pp. 353-374.

Oort A.H., 1983. Global atmospheric circulation statistics 1958-1973. *NOAA professional paper 14*, 180pp.

Pollard D., 1982. A simple ice sheet model yields realistic 100 kyr glacial cycles. *Nature*, 296, pp. 334-338.

Prell W.L. and J.E. Kutzbach, 1987. Monsoon variability over the past 150,000 years. *J. Geophys. Res.*, 92, pp. 8411-8425.

Ramanathan V., 1981. The role of ocean-atmosphere interactions in the CO_2 climate problem. *J. Atmos. Sci.*, 38, pp. 918-930.

Ramanathan V., and J.A.Jr. Coakley, 1978. Climate modeling through radiative convective models. *Rev. Geophys. Space Phys.*, 16, pp. 465-489.

Ramanathan V., and R.E. Dickinson, 1979. The role of stratospheric ozone in the zonal and seasonal radiative energy balance of the Earth troposphere system. *J. Atmos. Sci.*, 36, pp. 1084-1104.

Reeh N., Oerter H., Letréguilly A., Miller H., and H.W. Hubberten, 1991. A new detailled ice-age oxygen-18 record from the ice-sheet margin in central West Greenland, *Palaeogeogr., Palaeoclimatol., Palaeoecol., (Global Planet.Change Sect.)*, 90, pp. 373–383.

Rind D., Peteet D. and G. Kukla, 1989. Can Milankovitch orbital variations initiate the growth of ice sheets in a general circulation model ? *J. of Geophys. Res.*, 94, D10, pp. 12,851-12,871.

Roberts R.E., Selby J.E.A., and L.M. Biberman, 1976. Infrared continuum absorption by atmospheric water vapour in the 8-12 μm window. *Appl. Opt.*, 15(19), pp. 2085-2090.

Ruddiman W.F., and A. McIntyre, 1981. Oceanic mechanisms for amplification of the 23,00-year ice-volume cycle. *Science*, 212, pp. 617-627.

Saltzman B., 1985. Paleoclimate modeling. In: *Paleoclimate Analysis and Modeling*, A.D. Hecht (ed.), pp. 341-396, Wiley-Interscience, New York.

Saltzman B., and A. Sutera, 1987. The mid-Quaternary climatic transition as the free response of a three-variable dynamical model. *J. Atmos. Sci.*, 44, pp. 236-241.

Schlesinger M., 1989. Model projections of the climatic changes induced by increased atmospheric CO_2. In: *Climate and Geo-Sciences*, A. Berger, S. Schneider and J.Cl. Duplessy (eds), pp. 375-416, NATO ASI, Series C vol. 285, Kluwer Academic Publishers, Dordrecht, Holland.

Schneider S.H., and S.L. Thompson, 1979. Ice ages and orbital variations: some simple theory and modelling. *Quat. Res.*, 12, pp. 188-203.

Sellers W.D., 1973. A new global climate model. *J. Appl. Meteorol.*, 12, pp. 241-253.

Sellers W.D., 1983. A quasi-three-dimensional climate model. *J. Clim. Appl. Meteor.*, 22, pp. 1557-1574.

Shackleton N.J., Le J., Mix A., and M.A. Hall, 1992., Carbon isotope records from Pacific surface waters and atmospheric carbon dioxide. *Quat.Sci.Rev.*, 11, pp. 387–400.

Shibata K., and T. Aoki, 1989. An infrared radiative scheme for the numerical models of weather and climate. *J. Geophys. Res.*, 94(12), pp. 14,923-14,943.

Smith W.L., 1966. Note on the relationship between total precipitable water and surface dew point. *J. Appl. Meteor.*, 5, pp. 726-727.

Smits I., Fichefet Th., Tricot C., and J.P. van Ypersele, 1993. A model study of the time evolution of climate at the secular time scale. *Atmosfera*, (in press).

Souchez R., Lemmens M., Lorrain R., Tison J.-L., Jouzel J., and D. Sugden, 1990. Influence of hydroxil-bearing minerals on the composition of ice from the basal zone of an ice sheet. *Nature*, 345, pp. 244–246.

Tanré D., Geleyn J.F., and J. Slingo, 1984. First results of the introduction of an advanced aerosol-radiation interaction in the ECMWF low resolution global model. In: *Aerosols and their climatic effects*, H.E. Gerber and A. Deepak (eds), pp. 133-177, A. Deepak Publ., Hampton, USA. ¡item[] Thomson D.J., 1990. Quadratic-inverse spectrum estimates: applications to palaeoclimatology. *Phil. Trans. R. Soc. London*, A, 332, pp. 539-597.

Tricot C., 1992. Sur la contribution des gaz à effet de serre dans les changements climatiques à long terme. *Thèse de doctorat*, Faculté des Sciences, Université Catholique de Louvain, Louvain-la-Neuve. Unpublished manuscript.

Tricot C., and A. Berger, 1988. Sensitivity of present-day climate to astronomical forcing. In: *Long and Short Term Variability of Climate*, pp. 132-152.

Warren S.G., Hahn C.J., London J., Chervin R.M., and R.J. Jenne, 1986. Global distribution of total cloud cover and cloud type amounts over land. *DOE/ER/60085-41, NCAR Technical Notes, NCAR/TN-273/STR*, Boulder, CO, 29 pp. + 200 maps.

Warren S.G., Hahn C.J., London J., Chervin R.M., and R.J. Jenne, 1988. Global distribution of total cloud cover and cloud type amounts over the ocean. *DOE/ER-0406, NCAR Technical Notes, NCAR/TN-317+STR*, Boulder, CO, 42 pp. + 170 maps.

Washington W.M., and D.L. Williamson, 1977. A description of the NCAR GCM's in General Circulation Models of the atmosphere. In: *Methods in Computational Physics*, J. Chang (ed.), 17, pp. 111-172, Academic Press.

Wigley T.M.L., 1976. Spectral analysis and the astronomical theory of climatic change. *Nature*, 264, pp. 629-631.

World Meteorological Organization, 1986. A preliminary cloudless standard atmosphere for radiation computation. *Rep. WCP-112, WMO*, World Climate Res. Programme, 53pp.

ANNEX A: The Shortwave Radiation Scheme

The solar radiation computations in the RCM and in the LLN1 climate model are based on the SUNRAY scheme developped by Fouquart and Bonnel (1980). The following processes are taken into account: absorption by H_2O, CO_2, O_2 and O_3, Rayleigh scattering, absorption and scattering by cloud and aerosol layers, and reflection by the surface. One single spectral interval (0.25 - 4 μm) is considered to compute the solar radiative fluxes. The parameterizations given by Fouquart (1986) is used for the spectrally averaged optical thickness for Rayleigh scattering and for the spectrally averaged ozone transmittance calculated for each layer as a function of pressure, depth and O_3 amount. The spectrally averaged transmittances for H_2O, CO_2 and O_2 are taken from Chou (1986, 1990), with allowance for overlapping of H_2O and CO_2 absorption bands.

The computation of the solar fluxes and heating rates in the atmospheric layers involves three distinct steps.

1. The first one consists in calculating the transmittance and reflectance for each homogeneous layer. These quantities are obtained with the delta-Eddington approximation for the cloud layer, whereas first order expansions of the analytical solutions of the two-streams formulation (e.g., Fouquart, 1986) are used for clear-sky layers to take into account the attenuation by the air molecules and aerosols. For each layer, transmittance and reflectance are given as a function of the optical thickness, the single scattering albedo, the asymmetry factor and the effective solar zenith angle. To account for the decrease of the direct solar beam and the corresponding increase of the diffuse part of the downward radiation by the upper scattering layers, an effective solar zenith angle is calculated at each level as a function of the total cloud cover above that level. In the case of a clear sky, downward radiation in each layer is treated as a collimated radiation with the same zenith angle as the incident radiation at the top of the atmosphere. For a cloudy sky, the same assumption is made for the upper layer, while the downward radiation into and below the cloud is assumed to be diffuse, the same being assumed for the upward radiation in both cloudy and clear conditions. For diffuse radiation, the mean path length is classically taken to be 1.66 times larger than that of the vertical path. Differences between the reflectances and transmittances for radiation incident from above and below each

layer are therefore taken into account by using different incidence direction of the solar radiation.

2. The second step is to combine the layers, using the ascending method presented by Fouquart and Bonnel (1980) and extended by Fouquart (1986). Starting from the layer near the surface and accounting for the combined reflectance of the underlying layers, the reflectance R_{j+1}^{+} at level $j+1$ and the global transmittance T_j^- of layer j are given by (assuming an upward numbering of the layers and levels):

$$R_{j+1}^{+} = R_j + R^- T_j T_j^* / (1 - R_j^* R^-) \quad j = 1, m \qquad (A.1)$$

$$T_j^- = T_j / (1 - R_j^* R^-) \quad j = 1, m \qquad (A.2)$$

where R_j and T_j are the reflectance and transmittance of the isolated layer j for the radiation incident from above, R_j^* and T_j^* are the reflectance and transmittance of the isolated layer j for the radiation incident from below and R^- is the combined reflectivity of the underlying layers. For the layer near the surface, R^- is equal to the surface albedo.

The denominator of Eqs (A.1) and (A.2) accounts for the multiple reflections between the layers. This term is especially important when the reflection between layers is large which will typically be the case when a thick cloud overlies a bright surface (e.g., a snow or sea-ice covered surface).

3. Third, the downward fluxes, $F_{sj}^{\downarrow}$, and upward fluxes, $F_{sj}^{\uparrow}$, at level j are computed using the method presented by Fouquart and Bonnel (1980) and treating properly the multiple reflections between the layers (assuming here a downward numbering of the levels):

$$F_{sj}^{\downarrow} = F_{s1}^{\downarrow} \prod_{k=1}^{j-1} T_k^- \quad j = 2,\ m+1 \qquad (A.3)$$

$$F_{sj}^{\uparrow} = F_{sj}^{\downarrow} R_j^+ \quad j = 1,\ m+1 \qquad (A.4)$$

where $F_{s1}^{\downarrow}$ is the downward solar radiation incident at the top of the atmosphere.

Scattering is the main complicating factor in the solar computations. Since scattering and molecular absorption occur simultaneously, the exact amount of absorber along the path length is unknown and the band models are not usable to compute transmission. The photon path distribution method is available to calculate the probability $p(u)$ that a photon contributing to the radiative flux F_c in the conservative case (without absorption) has encountered an amount of absorber between u and $u + du$. With this distribution, the radiative flux F_ν at frequency ν is related to that of the conservative case by:

$$F_\nu \;=\; F_c \int_0^\infty p(u)\; exp(-k_\nu u) du \qquad (A.5)$$

and the spectrally averaged radiative fluxes over a spectral range $\Delta\nu$ can be calculated with the help of a band model transmission $t_{\Delta\nu}(u)$:

$$F_{\Delta\nu} \;=\; \frac{1}{\Delta\nu} \int_{\Delta\nu} F_\nu\, d\nu \;=\; F_c \int_0^\infty p(u)\; t_{\Delta\nu}(u) du \qquad (A.6)$$

The problem reduces to that of finding the distribution function $p(u)$. The method consists first in solving the scattering problem in the conservative case (i.e., to compute F_c) and, second, in solving the scattering problem in an "a priori" manner, that is for a set of arbitrarly fixed absorption coefficients k, thus giving a set of simultated radiative fluxes $F(k)$; an inverse Laplace transform is then performed on (A.5) to give $p(u)$.

The main advantage of the method is that, except for purely clear atmosphere, i.e. without molecular scattering or aerosols, the distribution $p(u)$ is smooth enough for (A.5) and (A.6) to give accurate results even if $p(u)$ itself is not known very accurately. Indeed the distribution function $p(u)$ needs not to be calculated explicitly (Fouquart and Bonnel, 1980). Since absorption in the water vapour bands is mainly strong, the saturation of the absorption bands must be taken into account. Fouquart and Bonnel determined an effective amount of absorber u_e, given by

$$u_e \;=\; ln\; (F(k)/F_c)/k \qquad (A.7)$$

In eq.(A.7), k is chosen so as to approximate the spectrally averaged transmission in clear atmosphere, that is

$$exp\left(-k\,\frac{u}{\mu_0}\right) \;=\; t_{\Delta\nu}\;(u/\mu_0)$$

where u is the total amount of absorber in a vertical column and $\mu_0 \;=\; \cos Z_0$, with Z_0 the solar zenith angle. This particular choice of the absorption coefficient allows to account for the saturation effect in the "a priori" scattering calculation. Finally, the knowledge of the vertical distribution of the effective amount of absorber, u_e, allows to compute the actual fluxes at each level, considering the scattering/absorption interactions.

ANNEX B: The Longwave Radiation Scheme

The longwave radiation scheme used in the RCM and in the LLN1 climate model follows a wide-band formulation of the radiative transfer equation which was designed for use in GCM by Shibata and Aoki (1989).

A balance between flux accuracy and computer burden is achieved by dividing the longwave spectrum into six wide wavenumber intervals (Table B.1). In a given wide spectral region i, the upward and downward fluxes are given at a level of height z, respectively, by:

$$F_i^{\uparrow}(z) = \pi\, B_i(z) - \int_0^z \pi\, \frac{dB_i}{dz'}(z')\, t_{dB_i}(z,z')\, dz' \qquad (B.1)$$

and

$$\begin{aligned} F_i^{\downarrow}(z) &= -\pi\, B_i(Z) t_{B_i}(Z,z) + \pi\, B_i(z) \\ &+ \int_z^Z \pi\, \frac{dB_i}{dz'}(z')\, t_{dB_i}(z',z)\, dz' \end{aligned} \qquad (B.2)$$

where B_i is the Planck function integrated on the wide spectral interval i, $t_{B_i}(z,z')$ and $t_{dB_i}(z,z')$ are two normalized spectrally averaged diffuse transmissivity functions for flux through a slab between levels z and z', Z is the height of the top of the model atmosphere.

These equations are derived from the radiative transfer equation assuming that the downward flux at the top of the model atmosphere is zero and that the air close to the surface is at the same temperature as the surface itself. The vertical integrations in Eqs (B.1) and (B.2) are solved numerically by a trapezoidal rule except for the adjacent layers in the case of large absorption, where the transmissivity function is assumed to be closely approximated by the two-parameter random model in the strong absorption regime (Shibata and Aoki, 1989). The Planck fluxes and their derivatives relative to the temperature in a wide spectral interval are evaluated in Eqs (B.1) and (B.2) with a Gauss-Laguerre quadrature (Johnson and Branstetter, 1974).

Table B.1. The characteristics of the six spectral intervals used in the wide-band formulation of the longwave radiative transfer.

Spectral interval	Limits cm^{-1}	Absorbers
1	0-500	H_2O (lines + continuum)
2	500-800	H_2O (lines + continuum) CO_2 O_3 N_2O
3	800-1150	H_2O (lines + continuum) CO_2
4	960-1090	O_3
5	1150-1400	H_2O (lines + continuum) CH_4 N_2O
6	1400-3000	H_2O (lines + continuum)

The line absorptions by H_2O, CO_2, CH_4, N_2O and O_3 are separately treated for each gas. The absorption by the H_2O continuum is parameterized following the results of Clough et al. (1980) and its dependency on temperature in the three wide bands between 500 and 1400 cm^{-1} is taken into account according to the formulation of Roberts et al. (1976). The total fluxes at each level are obtained by summing the partial fluxes from each spectral region. The effects of overlapping between gaseous absorbers is treated by making the usual assumption that the transmissivity functions of gases may be multiplied together to give the overlapped values.

In a given wide spectral region i, the Planck function cannot be considered independent of the frequency and two normalized spectrally averaged transmissivity functions are then defined for each absorber. The first one corresponds to the weighted average of the transmission function by the Planck function:

$$t_{B_i}(p,T,u) = \frac{\int_{\Delta\nu_i} B_\nu(T)\, t_\nu(p,T,u)\, d\nu}{\int_{\Delta\nu_i} B_\nu(T)\, d\nu} \qquad (B.3)$$

and is used to calculate the first term in Eq. (B.2) dealing with the top of the atmosphere. The second band transmissivity corresponds to the weighted average of the transmission function by the derivative of the Planck function

$$t_{dB_i}(p,T,u) = \frac{\int_{\Delta\nu_i} \frac{dB_\nu(T)}{dT}\, t_\nu(p,T,u)\, d\nu}{\int_{\Delta\nu_i} \frac{dB_\nu(T)}{dT}\, d\nu} \qquad (B.4)$$

and is used to calculate the integrals in Eq.(B.1) and (B.2). In Eq. (B.3) and (B.4), p, T, and u are, respectively, the pressure, the temperature and the absorber amount; and $\Delta\nu_i$ is the width of the wide spectral interval i.

For each absorber, Tricot (1992) has presented fast parameterizations of the spectrally averaged transmissivity functions given by Eqs (B.3) and (B.4) based on results obtained from a detailed semi-direct random band model (Aoki, 1978a,b) including 600 spectral intervals with widths of 5 cm^{-1}. The narrow random model uses the statistical band parameterization of Malkmus to compute the average transmission on a small spectral interval and takes into account the wing contributions of the lines centered inside and outside the spectral interval.

In the model atmosphere, the normalized transmissivity functions must be calculated over inhomogeneous path lengths with varying temperature and pressure. The atmospheric inhomogeneity is treated following the general method of Godson (1953), based on the fact that the absorber amounts are additive in a homogeneous path. Shibata and Aoki (1989) detailed the numerical procedure used also in the present longwave scheme.

The modification of the clear-sky fluxes due to the cloud layer is simply introduced following the method discussed by Washington and Williamson (1977) and the cloud cover is supposed, for simplicity, to behave as a blackbody, as is the surface of the Earth. The first step is to calculate the fluxes corresponding to a clear-sky atmosphere between the different levels along the vertical. The fluxes are then reevaluated assuming an overcast cloud layer of unity emissivity. Finally, the actual fluxes are derived from a linear combination of the fluxes obtained for the clear and overcast atmospheres.

ANNEX C: Feedback Parameter and Gain Factor

Let N be the net energy flux at a given level, which may be, for example, the surface or the tropopause, and the surface temperature, T_s, the only independent variable. N is a function of T_s and of external, E, and internal, I, parameters. E is a vector of quantities whose changes can lead to a change in climate, but which are independent of climate. I is a vector of quantities that can change as the climate changes and feed back to modify the climatic change (Schlesinger, 1989; Berger and Tricot, 1992).

E includes, for example, the solar constant, the volcanic dust and for the present purpose the CO_2 concentration (although it may change as a result of climatic change).

I contains all the variables of the climate system, except T_s. If T_s is the only dependent variable in a model, as it is in energy balance models, $I = I(T_s)$.

A small change in the energy flux, ΔN, can therefore be expressed as:

$$\Delta N = \sum_i \frac{\partial N}{\partial E_i} \Delta E_i + \left(\frac{\partial N}{\partial T_s} + \sum_j \frac{\partial N}{\partial I_j} \frac{dI_j}{dT_s} \right) \Delta T_s \qquad (C.1)$$

or

$$\Delta N = \Delta Q - \lambda \Delta T_s = \Delta Q - (\lambda_0 \; F) \; \Delta T_s \qquad (C.2)$$

which defines the direct radiative forcing ΔQ, the feedback parameter λ and the feedback factor $\frac{F}{\lambda_0}$ by direct identification with (C.1).

After the equilibrium is reached in response to the external forcing ΔQ, $\Delta N = 0$ and we have

$$\Delta T_s = \frac{\Delta Q}{\lambda_0 - F} = \frac{\Delta Q}{\lambda}$$

ΔQ is the change in N due to a change in the external quantifies,

$-\lambda_0 \Delta T_s$ is the change in N due to the change of T_s alone,

$F \Delta T_s$ is the change in N due to the change in the internal quantities through their dependance on T_s,

$G_0 = \frac{1}{\lambda_0}$ is the gain (output/input) of the climate system when the feedback factor $f = G_0 F = 0$. Thus G_0 is the zero-feedback gain and $(\Delta T_s)_0 = G_0 \Delta Q$ is the temperature change without the feedbacks.

G and ΔT_s are the gain and temperature change with the feedbacks

$$\Delta T_s = \frac{\Delta T_{s0}}{1-f}$$

For a 0-D planetary energy balance model,

$$N = (1-\alpha_p)\,\frac{S}{4} - \epsilon\,\sigma\,T_s^4$$

where S is the solar constant, ϵ the effective emissivity of the Earth-atmosphere system and α_p the planetary albedo. For $\alpha_p = 0.3$, $\frac{S}{4} = 342$ Wm^{-2} and $T_s = 288$ K,

$$\lambda_0 = \frac{\partial N}{\partial T_s} = \frac{(1-\alpha)S}{T_s} \sim 3.33 \;\; Wm^{-2}\, K^{-1}$$

or $G_0 = 0.3$ K $(W\, m^{-2})^{-1}$.

The response of the climate system to the direct radiative forcing for a 2 x CO_2 when only the surface temperature is allowed to change is therefore:

$$\Delta T_{so} = \frac{\Delta Q}{\lambda_0} = 1.12\mathrm{K}$$

if ΔQ is supposed to be equal to 3.72 Wm^{-2} as it is calculated from our RCM.

In this standard version of the RCM-2 x CO_2 experiment the only feedback analysed is related to water vapour. The response of the RCM to a 2 x CO_2 forcing with and without the water vapour feedback being 1.76 K and 1.16 K respectively, the gain factor R is equal to 1.52:

$$R = \frac{\Delta T_s}{\Delta T_{s0}} = \frac{G}{G_0} = \frac{1}{1-f} = 1.52$$

and therefore the feedback factor: $f = 1 - \frac{1}{R} = 0.34$
and the gain: $G = \frac{\Delta T_s}{\Delta Q} = 0.46$.

f begin positive ($0 < f < 1$, $R > 1$), the feedback is positive. Actually, the sign of the response is always the same as the sign of the forcing, even when there is a negative feedback ($f < 0$, $R < 1$) regardless of its magnitude. Finally, from

$$\Delta T_s = \frac{\Delta Q}{\lambda} = \frac{G_0}{1-f}\,\Delta Q$$

$\lambda = 2.11$ Wm^{-2} K^{-1} and the water vapour feedback is estimated to decrease λ_0 by 1.2 Wm^{-2} K^{-1}. The value of all these parameters are very close to what is usually found from other models (Schlesinger, 1989).

PYSICS OF THE ICE AGE CYCLE

W. Richard Peltier
Department of Physics
60 St. George St.
University of Toronto
Toronto, Ontario
Canada M5S 1A7

During the last half of the Pleistocene epoch, beginning roughly 900,000 yrs ago, the variability evidenced in the climate system has been dominated by the 100 kyr ice age cycle first recognized by Broecker and Van Donk (1970). The sequence of events during this time has been markedly improved recently by employing the Milankovitch hypothesis of the origin of ice ages to refine our knowledge of their chronology! Various physical models of this long timescale climate oscillation have been proposed that may be distinguished either as "wet", in the sense that they rely in large part upon internal forcing due to variation in the strength of the thermohaline circulation (e.g. Broecker and Denton, 1989) or as "dry", in which case they posit a secondary role for the oceans. Following a brief review of developments in the area of ice age chronology, I will discuss recent and ongoing attempts to resolve the issue as to whether the ice age cycle is primarily a "wet" or a "dry" phenomenon.

Introduction

Since its first recognition by Broecker and Van Donk (1970) in $\delta^{18}0$ data from deep sea sedimentary cores, the issue of the explanation of the 100 kyr ice-age-cycle that has dominated climate system variability for the last 900 kyr of Pleistocene time has remained outstanding. Although the seminal paper by Hays et al. (1976) has often been misconstrued as establishing the validity of the astronomical theory of Milankovitch (e.g. 1941), in fact nothing could be further from the truth. What this analysis did establish was that the spectral signature of the summertime seasonal orbital insolation anomaly was indeed impressed indelibly upon deep sea

NATO ASI Series, Vol. I 22
Long-Term Climatic Variations
Edited by J.-C. Duplessy and M.-T. Spyridakis

$\delta^{18}0$ records. These records, which are understood to be proxy for continental ice volume (Shackleton, 1967), reveal statistically significant concentrations of spectral power at the period of 41 kyr that is contained within the summertime seasonal insolation anomaly due to the change in orbital obliquity, and at 19 kyr, 21 kyr and 23 kyr due to the contribution to insolation change by eccentricity modulation of the precessional cycle. However, Milankovitch's theory posited a direct, which is to say linear, link between variations in summertime insolation and variations in continental ice volume. If this theory were correct then all significant variance in deep sea $\delta^{18}0$ records would be found at these orbitally significant periods. In fact more than 60% of the variance in typical deep sea $\delta^{18}0$ records over the last 700 kyr of earth history is associated with the 100 kyr ice age cycle mentioned above and there is no significant power at this period in the insolation anomaly time series that Milankovitch believed to be responsible for inducing the ice ages (e.g. Hyde and Peltier 1987). Understanding the ice age cycle therefore clearly obliges us to identify the nonlinear interaction(s) within the climate system that are responsible for transforming the applied orbital insolation forcing (which has no significant 100 kyr component) into the observed ice volume response whose spectrum is dominated by the low frequency 100 kyr oscillation.

In order to provide focus for the discussion to follow, I show on Figure 1a the $\delta^{18}0$ time series from the highest resolution deep sea sedimentary record that is presently available. This record, from ODP Site 677 in the Panama Basin off the west coast of South America, has been extensively discussed from the perspective of chronology by Shackleton et al. (1990). Also shown beneath the top plate on this Figure are a number of different statistical characterizations (discussed in the figure caption) of the change of the statistical character of the record with time from about 2 Myr ago (the beginning of the record) to the present. Clearly evident, even by visual inspection, is the fact that the last 900 kyr of $\delta^{18}0$ (ice volume) variability has been dominated by the 100 kyr ice-age-cycle in which ice volume rises slowly to a maximum over a period of about 90 kyr and then "terminates" in an episode of rapid decrease that occurs on a time scale of about 10 kyr. Prior to about 900 kyr ago, however, the 100 kyr cycle is essentially absent from the record. A time dependent spectrum of the original time series (Figure 2), often referred to as a digital sonogram, demonstrates that in the earliest Myr of time all significant variance in the $\delta^{18}0$ time series is found in the spectral lines expected on the basis of the astronomical theory whereas, following 900 kyr BP (before present), the time series is dominated

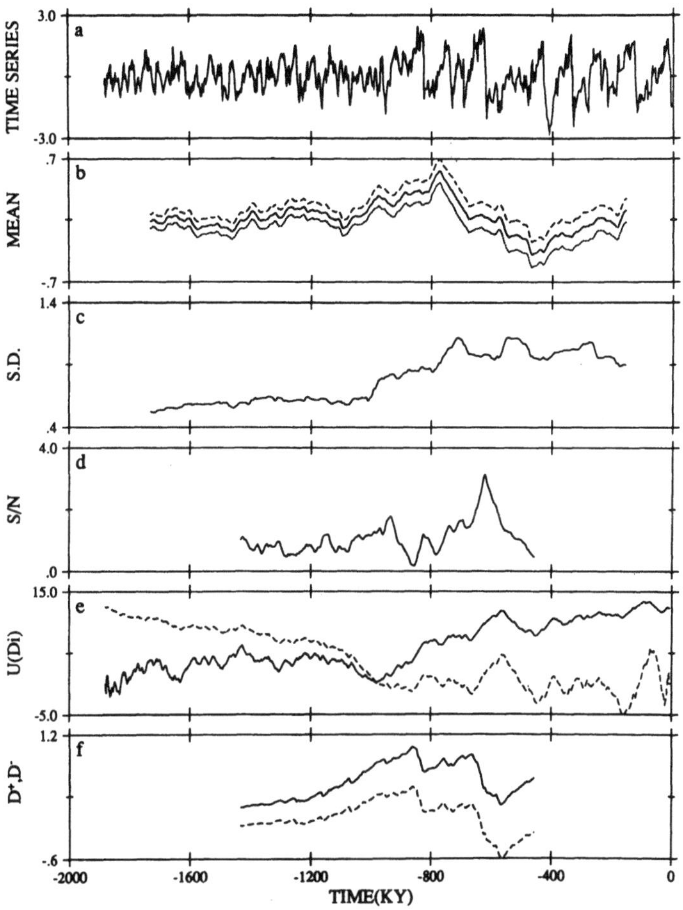

Figure 1. Statistical analysis of ODP site 677 $\delta^{18}0$ record with the depth age curve corresponding to model II discussed below. (a) Time series. (b) Running mean of the time series shown in (a) with segment length of 300 kyr. (c) Changing standard deviation of the time series shown in (a) computed for a segment length of 300 kyr. (d) Signal over noise ratio averaged over a range of segment lengths of 300 to 450 kyr. (The confidence level is 99%). (e) Mann-Kendall rank statistic, U(Di), for forward (solid) and retrograde (dashed) time series. (f) Test for jump in variance, D^+ (solid line) and D^- (dashed line), averaged over the same segment lengths as in (d).

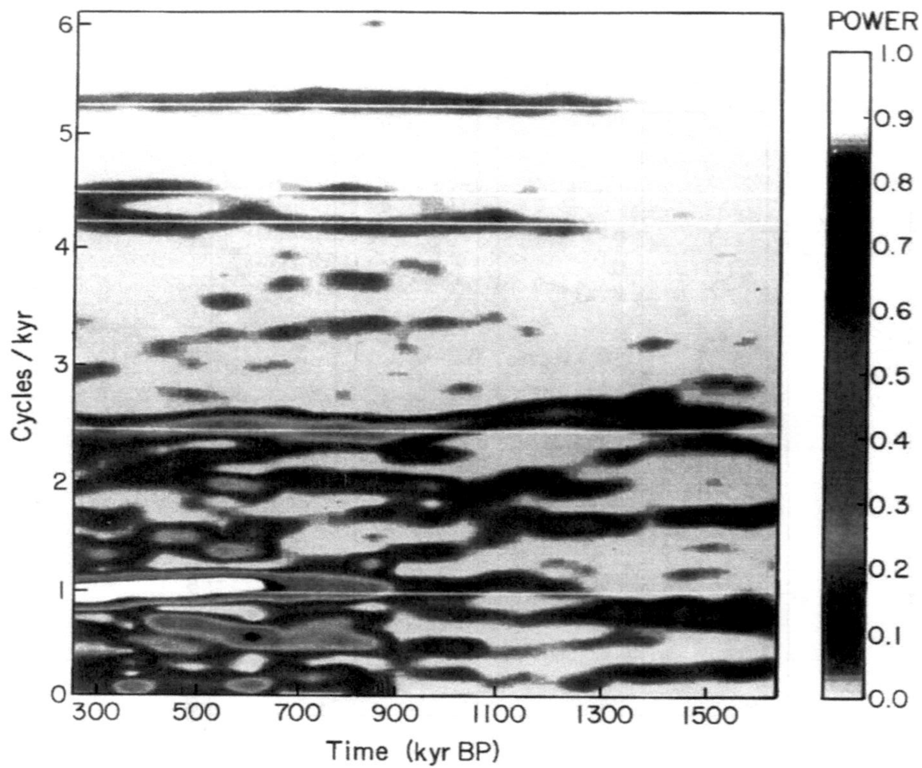

Figure 2. Digital sonogram of the ODP 677 time series shown in Figure 1. This time dependent power spectrum has been computed using a running time window 500 kyr in length. The horizontal (white) reference lines are at the orbitally significant periods of 19, 21, 23, 41 and 100 kyr. Note that the strong 100 kyr component of climate system variability is not evident until about 900 kyr BP.

by the "unexplained" 100 kyr ice-age-cycle. In the remainder of this paper I will review recent theoretical attempts to explain the ice age cycle, following a brief review of the issue as to how an appropriately accurate chronology may be established for deep sea sedimentary $\delta^{18}0$ data such as those shown in Figure 1.

Ice-Age Chronology

In the previously referred to seminal analysis of Hays et al. (1976) which first conclusively demonstrated, through spectral analysis of appropriate $\delta^{18}0$ time series from deep sea cores, that the periodicities predicted by the Milankovitch theory were indeed evident in these ice volume proxy records, the time scale for each record was fixed by appeal to magneto-stratigraphy. What one did was to find the depth in the core at which the induced magnetization in the sedimentary grains reversed direction. This depth, which is quite easily observed in cores taken from sufficiently high geographic latitude, was taken to correspond to an age of 730,000 ± 10,000 yrs on the basis of the previously well established chronology of magnetic field reversals produced on the basis of age and magnetic measurements in land based sequences of volcanic rocks (e.g. Cox et al. 1963). Assuming a constant sedimentation rate, unique to each core, one could thereby construct a linear map from depth to time and thereby convert the $\delta^{18}0$ vs. depth data into a $\delta^{18}0$ vs. time series such as the one shown on Figure 1a. This practise essentially reached is apogee with publication of the so called SPECMAP timescale by Imbrie et al. (1984). This timescale was believed to be accurate for ages slightly beyond the age of the Brunhes-Matuyama boundary at 730,000 ± 10,000 yrs and was based upon a stacked sequence of records. In the later analysis of individual sedimentary cores sufficiently long to reach further back in time, (e.g. DSDP 607, Raymo et al. 1990) this timescale was simply transferred to the cores by the correlation of appropriate control points and thereafter extended to greater depth.

With the availability of the very high resolution $\delta^{18}0$ data set from the ODP 677 site (Shackleton and Hall, 1989), for which no magnetostratographic control was available, an alternate method for establishing an appropriate timescale suggested itself to the author. This was essentially an extension of the method employed to refine the youngest part of the SPECMAP scale by Martinson et al. (1987) who adjusted the timescale in a relatively short sedimentary record so as to achieve maximum/shallowest time domain coherence between an appropriate orbital insolation time series and the $\delta^{18}0$ signal in the deep sea core by bandpassing both the insolation record and the $\delta^{18}0$ record through a filter with a 41 kyr passband and then adjusting the timescale in the deep sea core so as to achieve maximum time domain coherence between the two filtered time series. Since the $\delta^{18}0$ record from the ODP 677 core was of sufficiently high resolution that the triplet of lines in the above described eccentricity - precession band should also be well resolved, a somewhat more sophisticated version of the Martinson approach became a viable objective method of determining an age model for this core.

This new method, results of which were first reported by the author at a meeting on the Milankovitch theory that was held at the Lamont-Doherty Observatory in 1988, consisted of iteratively refining the timescale through the following sequence of steps.

1. A first guess timescale was assumed to correspond to the SPECMAP scale in the shallowest portion of the record and was extended to the deepest level by assuming constant sedimentation rate.
2. The resulting time series was bandpassed through a filter with a central period of 41 kyr. This bandpass filter is shown in the power spectrum diagram of Figure 3 in terms of its spectral response characteristics.
3. The summertime seasonal insolation time series for an assumed control latitude, say 55°N, was next bandpass filtered through the same 41 kyr filter described in 2.
4. The timescale employed in step 2 was then adjusted so as to achieve best time domain coherence between the 41 kyr signals in the filtered $\delta^{18}0$ and insolation time series in the top most part of the record.
5. Using the updated timescale from step 4 on which to represent the $\delta^{18}0$ data, the ODP 677 record was next bandpassed through a filter, also shown on Figure 3 in terms of its spectral response characteristics, with a central period of 21 kyr.
6. The same caloric summertime seasonal insolation signal employed in 3 was next bandpassed through the same 21 kyr filter applied to the $\delta^{18}0$ record in step 5.
7. The revised timescale developed by obliquity tuning in step 4 was then further adjusted over the same top-most-part of the record so as to achieve an equally high quality time domain fit to the eccentricity-precession signal as that achieved in step 4 at the obliquity period.
8. The iterative sequence of steps 4-7 was then repeated, moving ever further down the core in each successive repetition of the sequence.

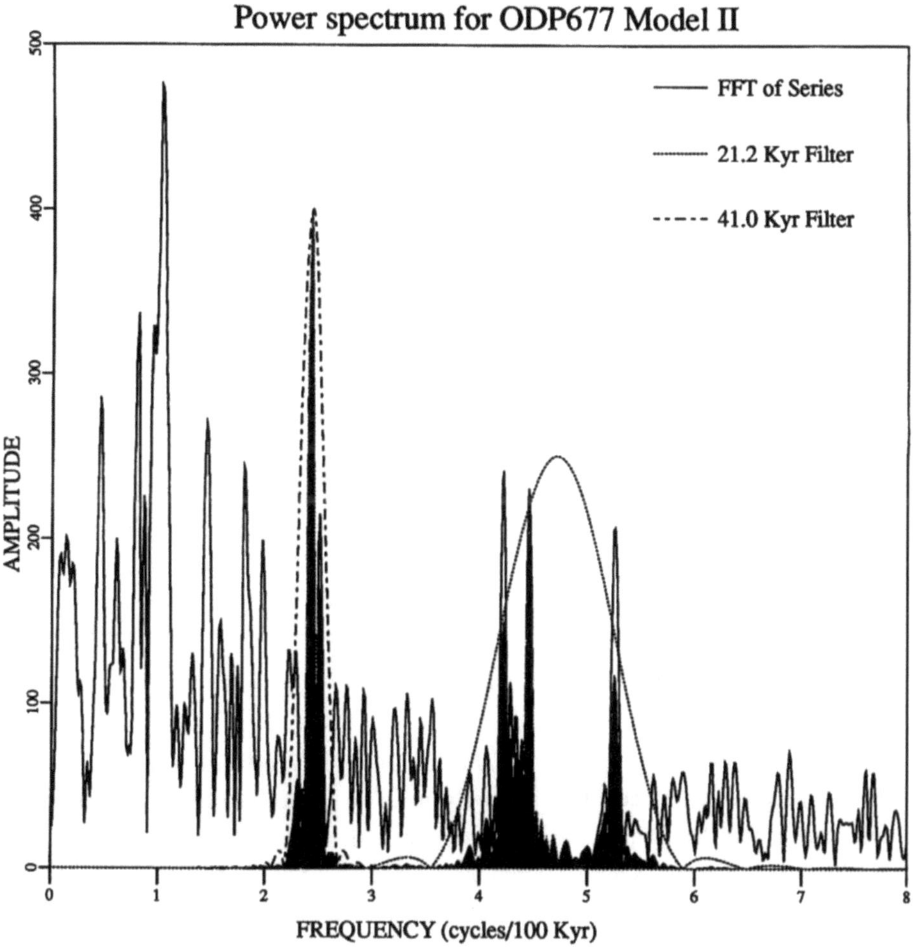

Figure 3. Power spectrum of the entire ODP 677 record using the chronostratigraphy corresponding to age model II on Figure 4. The dash-dotted and dotted curves represent the spectral transfer functions of the bandpass filters used to tune the chronostratigraphy in the obliquity and eccentricity-precession bands respectively. The solid black regions represent the spectrum of ODP 677 seen through these filters.

Figure 4 compares four different timescales for the ODP 677 record. The first, labelled I, is the timescale obtained by tuning the record solely to the 41 kyr obliquity period while that labelled II, the preferred timescale, is obtained by tuning both to obliquity and to the eccentricity-

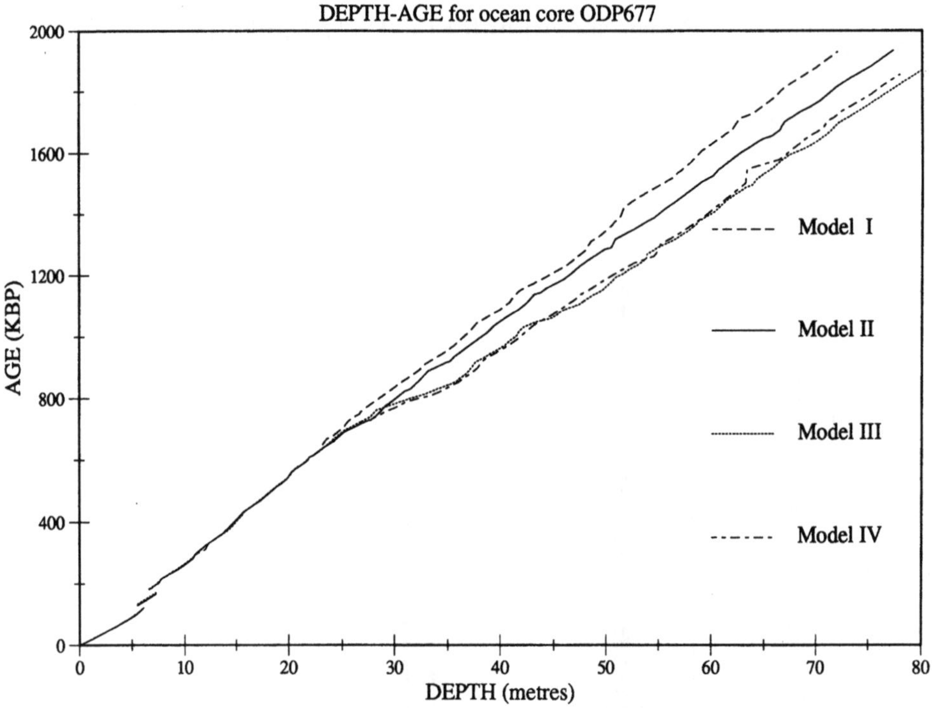

Figure 4. Four age models for ODP 677. Model I is the result obtained by tuning only in the obliquity band. Model II is iteratively tuned to produce optimum time domain coherence in both the obliquity and eccentricity-precession bands. Model III was obtained by Shackleton (personal communication) by transferring the chronostratigraphy from the DSDP 607 core to ODP 677. Model IV is a variant on model III.

precession signal. This timescale is characterized by an almost constant sedimentation rate. In comparison, the timescale labelled III is the timescale delivered by transferring the timescale from DSDP core 607 to the ODP 607 core by correlating "like" features. This timescale and its near relative labelled IV is clearly characterized by a marked change in sedimentation rate near 700 kyr B.P. The "quality" of the constant sedimentation rate timescale labelled II may be assessed by inspection of the time domain coherence between the obliquity and eccentricity precession signals in the $\delta^{18}O$ time series and in the orbital insolation time series. These comparisons are shown in Figures 5 and 6 respectively. Although it cannot be seen as

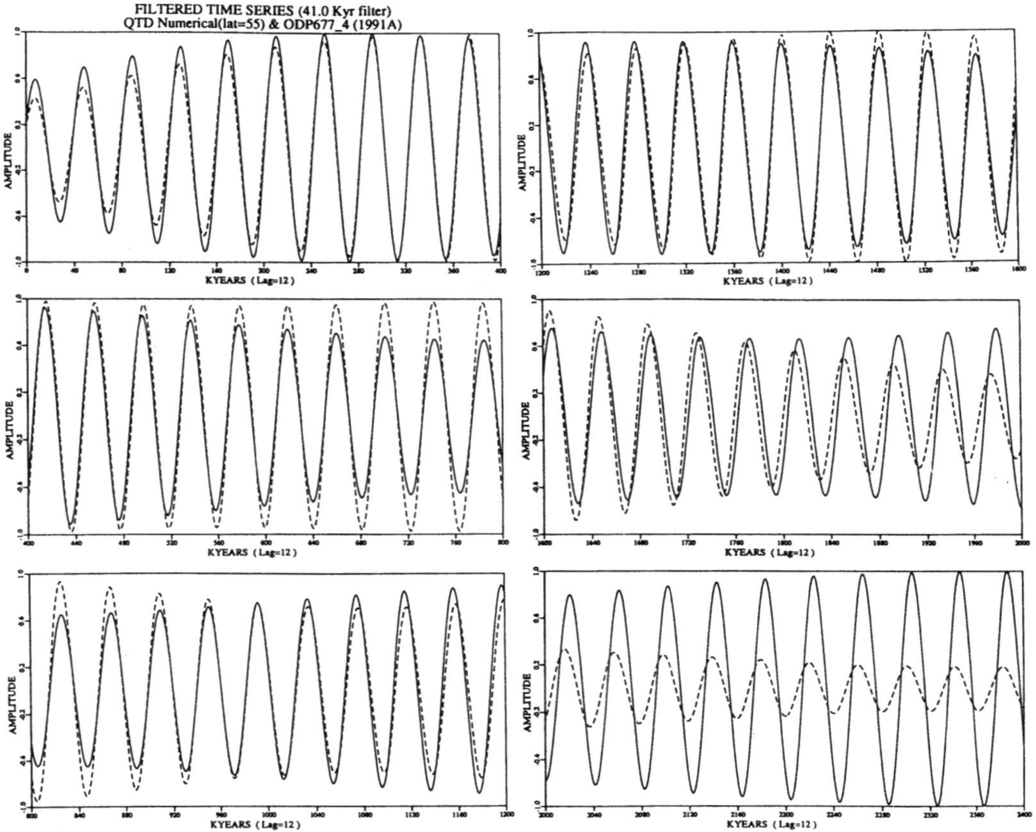

Figure 5. The 55°N insolation signal according to the Quinn et al. (1991) astronomical reconstruction bandpassed through the 41 kyr filter (solid) compared to the ODP 677 record on age model II (dashed) bandpassed through the same filter.

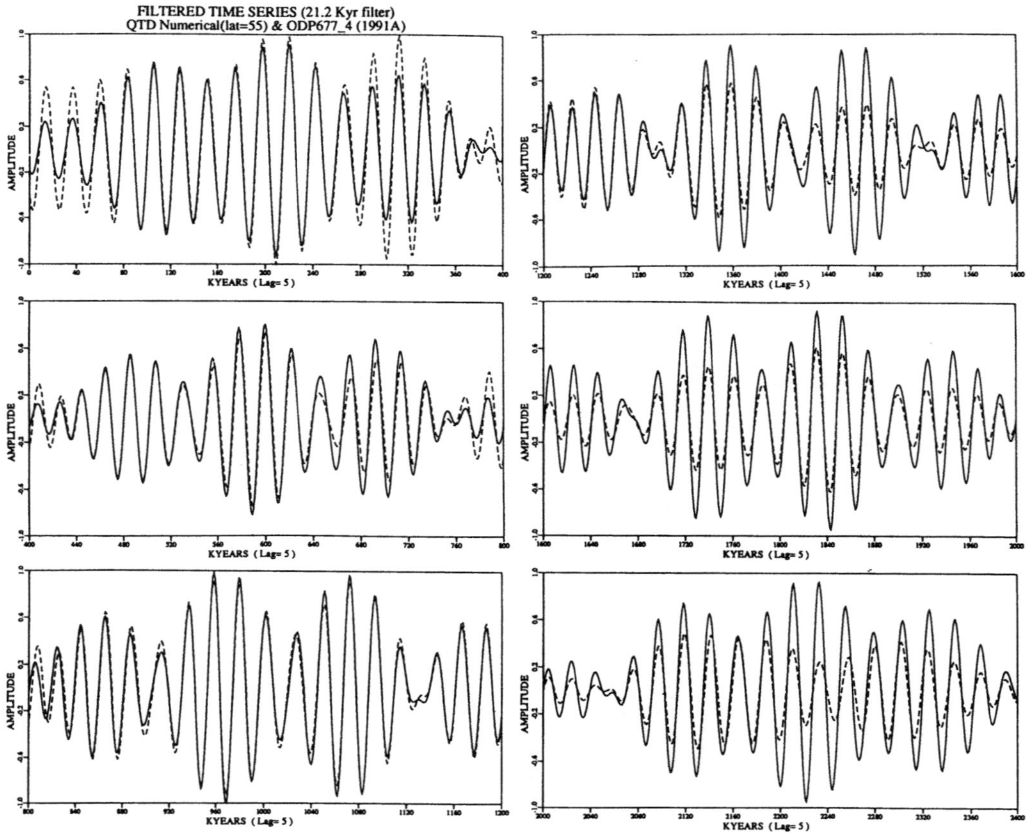

Figure 6. Same as for Figure 5 but comparing the astronomical signal with ODP 677 in the eccentricity-precession band.

remarkable that it is possible to achieve the nice time domain coherence between insolation and $\delta^{18}O$ at the period of the 41 kyr "singlet" (Figure 5) it is in fact remarkable that the characteristic beating in time due to the interference between the three component lines in the eccentricity - precession "multiplet" can be so accurately recovered in the $\delta^{18}O$ data by an appropriate choice of timescale. Inspection of Figure 4 furthermore shows that the timescale that delivers this extraordinary time domain coherence between $\delta^{18}O$ and insolation is one which deviates minimally from a constant sedimentation rate timescale. Comparing this timescale with that transferred from the DSDP 607 (Atlantic) core of Ruddiman and Raymo (1988) it is clear that the latter implies the existence of a marked change in sedimentation rate between 700 and 800 kyr BP.

If one accepts the new almost constant sedimentation rate timescale labelled II on Figure 4, however, then one is obliged to face the objection that this would imply that the previously determined radiometric age for the Brunhes-Matuyama reversal is too young by about 7%, i.e. somewhat more than an obliquity cycle. This was first pointed out in the analysis presented by Shackleton et al. (1990) based upon very much less detailed tuning of the ODP 677 timescale than discussed above. Although there was some initial reluctance in the community to accept this possibility, Baksi et al. (1993) have very recently demonstrated, by employing the very accurate $^{39}Ar/^{40}Ar$ step heating procedure, that in fact the previously accepted age of the Brunhes-Matuyama boundary was indeed too young by precisely the 7% suggested on the basis of the above described Milankovitch tuning procedure. This must be seen as an extraordinary demonstration of the power of the Milankovitch insight that orbitally induced changes of summer seasonal insolation could be responsible for causing substantial changes of continental ice volume. Even though the astronomical theory of Milankovitch cannot explain the strong variability of ice volume at the 100 kyr period that has dominated climate system history for the last 900 kyr, the orbital signal is nevertheless clearly evident in $\delta^{18}O$ records throughout the Pleistocene epoch. Theoretical efforts to explain how this signal is transformed to create the 100 kyr oscillation will be discussed in the following subsections.

A model of the 100 kyr cycle and the mid-Pleistocene transition

In attempting to understand the 100 kyr cycle it is clearly important to understand not only how a 100 kyr response might be engendered in the climate system by an insolation forcing that contains essentially no power at this period, but it is also equally important to seek to understand why the 100 kyr component of the response was essentially absent prior to 900 kyr ago even though the insolation signal in this earlier period differed insignificantly from that in more recent time. In a recent attempt to address these questions, DeBlonde and Peltier (1991) have further developed the model originally introduced by Peltier (1982) and employed in a sequence of initial experiments by Hyde and Peltier (1985, 1987). This model couples the three basic ingredients illustrated schematically in Figure 7, namely a circumpolar ice sheet whose thickness is a function of latitude only, the viscously delayed response of the earth to the loading of its surface by this ice sheet, and the orbital insolation forcing of Milankovitch that is assumed to enter the system via a proportional latitudinal motion of the so-called equilibrium line. The latter separates the region of accumulation from the region of ablation.

The mathematical realization of this model consists of the vertically integrated continuity equation for ice mass, in the from:

$$\frac{\partial H}{\partial t} = \frac{\lambda}{a\sin\theta}\frac{\partial}{\partial\theta}\left[\frac{\sin\theta\, H^5}{a}\left(\frac{\partial h}{\partial\theta}\right)^2\frac{\partial h}{\partial\theta}\right] + G(\theta,t) \tag{1}$$

in which $\lambda = C\,(\rho_i g)^3/5$ with C a constant that depends upon the microphysical properties of ice, ρ_i is ice density and g is the surface gravitational acceleration. The dependent variables in (1), H and h, are respectively ice thickness and the height of the surface of the ice sheet above mean sea level. H is therefore decomposed in the sum $H(\theta,t) = h(\theta,t) + h'(\theta,t)$ as on Figure 7 in which h' is the depression of the solid surface of the earth below its equilibrium level due to the weight of the ice load. The latter deflection is in part elastic and in part viscous and its evolution with time may be accurately calculated using the detailed theory of glacial isostatic adjustment developed in Peltier (1974, 1976, 1985). This theory delivers the evolution equation for the field $h'(\theta,t)$ required to complete (1) as

$$\frac{\partial h'(\theta,t)}{\partial t} = \frac{3\rho_i}{\rho_a}\left\{\sum_{\ell} P_\ell \int P'_\ell \sin\theta' \left[\sum_j -r_j^\ell s_j^\ell e^{-s_j^\ell t} D_j^\ell(\theta',t) + \sum_j r_j^\ell H(\theta',t)\right] d\theta' \right.$$
$$\left. + \sum_{\ell} P_\ell \int P'_\ell \sin\theta'\, q_\ell^E \frac{\partial H(\theta',t)}{\partial t}\, d\theta' \right\} \tag{2}$$

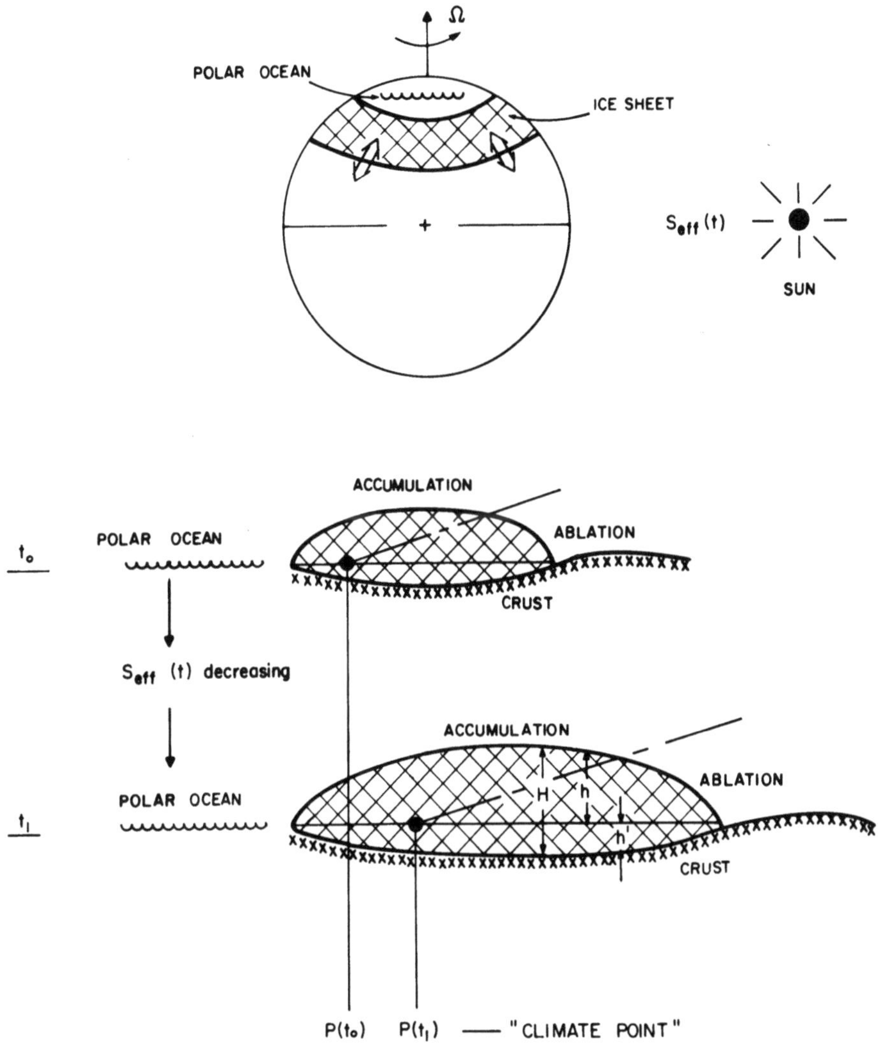

Figure 7. Schematic diagram indicating the physical components of the one dimensional nonlinear model of the ice-ages that has been designed to explain the appearance of the 100 kyr cycle in the ODP 677 record.

in which

$$D_j^\ell\,(\theta',t) = \int_o^t H(\theta',t')\, e^{s_j^\ell t'}\, dt'$$

so that

$$\frac{\partial D_j^\ell}{\partial t} = H(\theta',t)\, e^{s_j^\ell t} \tag{3}$$

In order to integrate the dynamical system embodied in equations (1)-(3) one first steps (3) forward in time to update the D_j^ℓ. In the second step (2) is integrated forward to a obtain a new field h′(θ,t). Finally the nonlinear diffusion equation (1) is integrated forward to update H(θ). In (2) the P_ℓ are conventional Legendre polynomials, ρ_a is the average density of the earth, the r_j^ℓ and s_j^ℓ are the amplitudes and inverse relaxation times of the discrete set of normal modes of viscous gravitational relaxation that are required to characterize the viscous part of the glacial isostatic adjustment process (Peltier 1974, 1976, 1985) and the q_ℓ are the elastic surface load Love numbers of degree ℓ that appear in this theory.

In (1) the mathematical form employed for the accumulation function G(θ,t) is simply (e.g. Oerlemans 1980, Pollard 1982):

$$G(x,t) = \begin{cases} a(h-E) - b\,(h-E)^2, & for\ (h-E) \le 1500m \\ 0.56m\ yr^{-1} & ,\ for\ (h-E) > 1500m \end{cases} \tag{4}$$

where E is the so-called equilibrium line that tilts upwards to the south as on the sketch in Figure 7. North of the point of intersection of this line with the surface of the ice sheet is the accumulation zone. The ablation zone is to the south. On the equilibrium line itself where h=E, G≡0. Following Milankovitch (1941) we assume that positive anomalies in caloric summer seasonal insolation ΔQ induce a rise in the equilibrium line and thus an increase in the scale of the ablation zone and thus a diminution of ice mass. The position of the equilibrium line is computed according to the representation (with x=sinθ):

$$E(x,t) = E_o - s\,(x-x_o) + k\Delta Q(t) \tag{5}$$

The anomaly ΔQ(t) should be computed using the results of Quinn et al. (1991) but I shall illustrate the response of the above described model using the 55°N forcing from Berger (1978). The constants a,b and s,k in (4) and (5) are listed in DeBlonde and Peltier (1991). In order to obtain an adequate simulation of the ODP 607 record shown on Figure 1 we are obliged to further augment the above structure with an additional ad hoc feedback loop, the possible origins of which I shall discuss below. This feedback loop amounts to an acceleration of the ablation process whenever the model embodied in the above equations delivers an ablation rate that exceeds some critical value. As in Pollard (1983) we define an instantaneous melting rate as:

$$\dot{M} = (-3.0 \, x \, 10^{-7} \, m^{-1}) \, \frac{dC}{dt} \tag{6}$$

in which C is the instantaneous cross-sectional area of the ice sheet in Figure 7. When $\dot{M}$ exceeds some critical value, 12 cm yr^{-1} is employed here ($\dot{M}$ being expressed in terms of an equivalent rate of sea level rise), then it is sharply increased to 20 cm yr^{-1}. This is equivalent to a raising of the equilibrium line by 500-1000 m and thus to a temperature rise of 3-6°C, assuming an atmospheric lapse rate of 6.5°C km^{-1} (see DeBlonde and Peltier 1991 for more detailed discussion). As we will see, the nature of the response that the model delivers to the applied insolation forcing depends critically upon whether or not this additional feedback loop is engaged or disengaged.

An example of the response of the model to the Milankovitch input for 55°N latitude is shown in Figure 8a along with the same statistical characterizations of the time series as were shown on Figure 1 in connection with the discussion the ODP 677 record. Visual inspection of this synthetic ice volume record demonstrates that it has the same qualitative features as the deep sea $\delta^{18}0$ record from the Panama basin. In order to achieve this high quality fit to the observations what has been done is to "switch-on" the above described additional feedback loop at 900 kyr BP with simultaneous changes in E_0 and k of 250 m and 11 m $(W \, m^{-2})^{-1}$. When this additional feedback is turned off as it is in the first half of the simulated record then the response to orbital forcing is just as predicted by Milankovitch with significant power only at the periods present in the incoming insolation. Once the feedback has been turned-on, however, the dominant response is at the 100 kyr period that is characteristic of the ice-age cycle. The issue then clearly arises as to what this ad hoc positive feedback in the model should be taken to represent. In the model of Pollard (1983) it was imagined to be due to calving into proglacial lakes during the

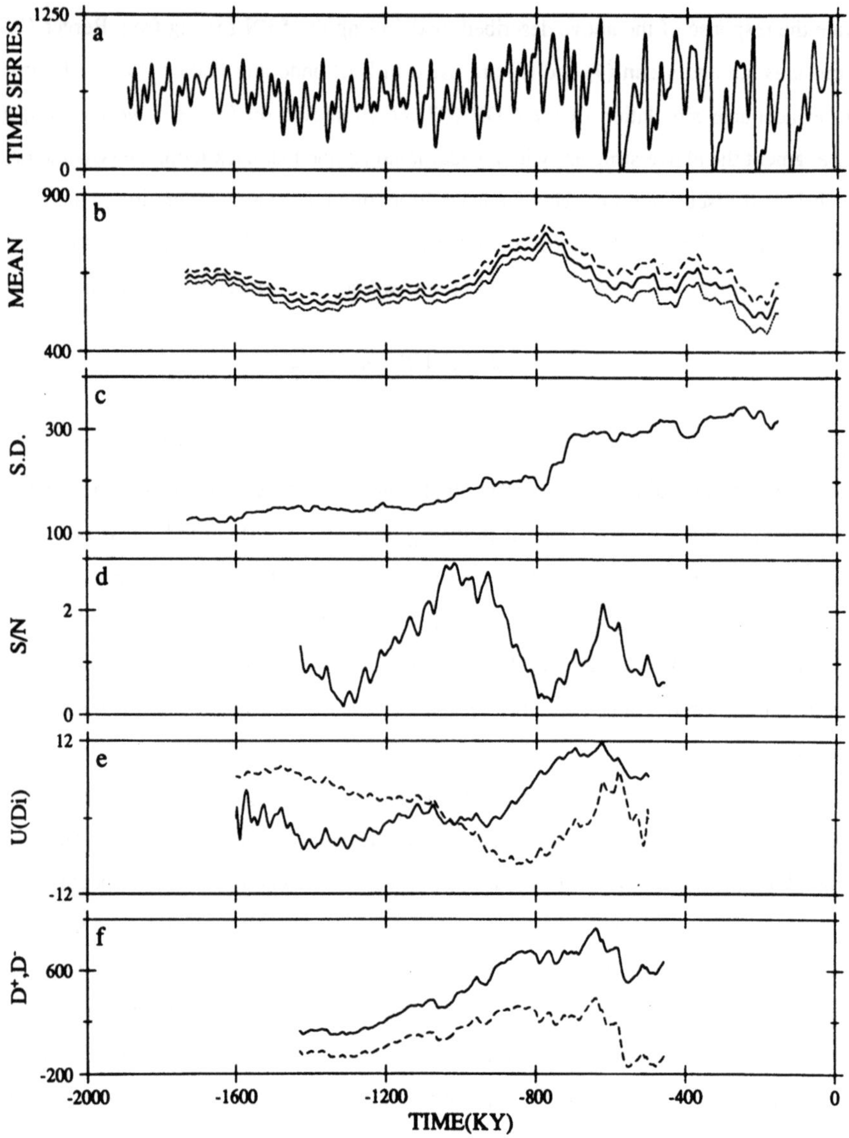

Figure 8. Same as for Figure 1 but the time series in plate (a) is a simulation of the ODP 677 record based upon the integration of the one dimensional non-linear model described in the text.

retreat of a large (e.g. Laurentide) ice sheet. The heat capacities of such lakes are so small, however, that it is hard to imagine how they could provide the forcing required. Other candidates are of course possible to imagine and of these it will suffice to mention two only, one "dry", the other "wet". In the category of "dry" feedbacks the most plausible seems to me to be that associated with terrigenous dust (e.g. De Angelis et al. 1987) since it is now well established both on the basis of ice core records and deep sea sedimentary cores that each of the strong 100 kyr pulses of ice sheet advance and retreat were characterized by extremely high dust levels in the full glacial phase of the cycle that preceded the termination. Peltier and Marshall (1993) have recently tested this hypothesis and found it to be tenable. This is clearly a "dry" mechanism. Of the possible "wet" mechanisms, I believe that the idea of a marine based instability of the Laurentide ice margin such as was undoubtedly responsible for the recently discovered Heinrich events (e.g. Heinrich 1988) is the most deserving of further attention.

One way in which one might hope to better understand which of these mechanisms (or some other) is actually responsible for the 100 kyr cycle is simply to make the above described model more physically realistic without adding any further physical processes. This is the idea to which I turn in the next subsection.

Multiple ice sheets in a global energy balance model: the origin of terminations

The one dimensional model described in the last subsection clearly has a number of unrealistic characteristics. Foremost amongst these is that, lacking realistic geography, it is unable to distinguish the west European ice complex from the North American complex. The ice thickness distributions within these two main continental ice sheets, as recently deduced on the basis of the inversion of postglacial relative sea level data by Tushingham and Peltier (1991), are shown on Figure 9. There is very little evidence to support the notion that there was significant ice in Eastern Siberia and this fact may on the surface appear to be unexpected. In proceeding, as I shall do in the present section, to add detail to the model of the last Section, my goal will be in part to explain this apparent paradox and to demonstrate how far we can in fact go in explaining the detailed thickness distribution for glacial maximum shown in Figure 9. An equally important goal, however, is to discover whether or not, with a slightly more complex model describing the interaction of the same physical processes previously considered, we might discover a more

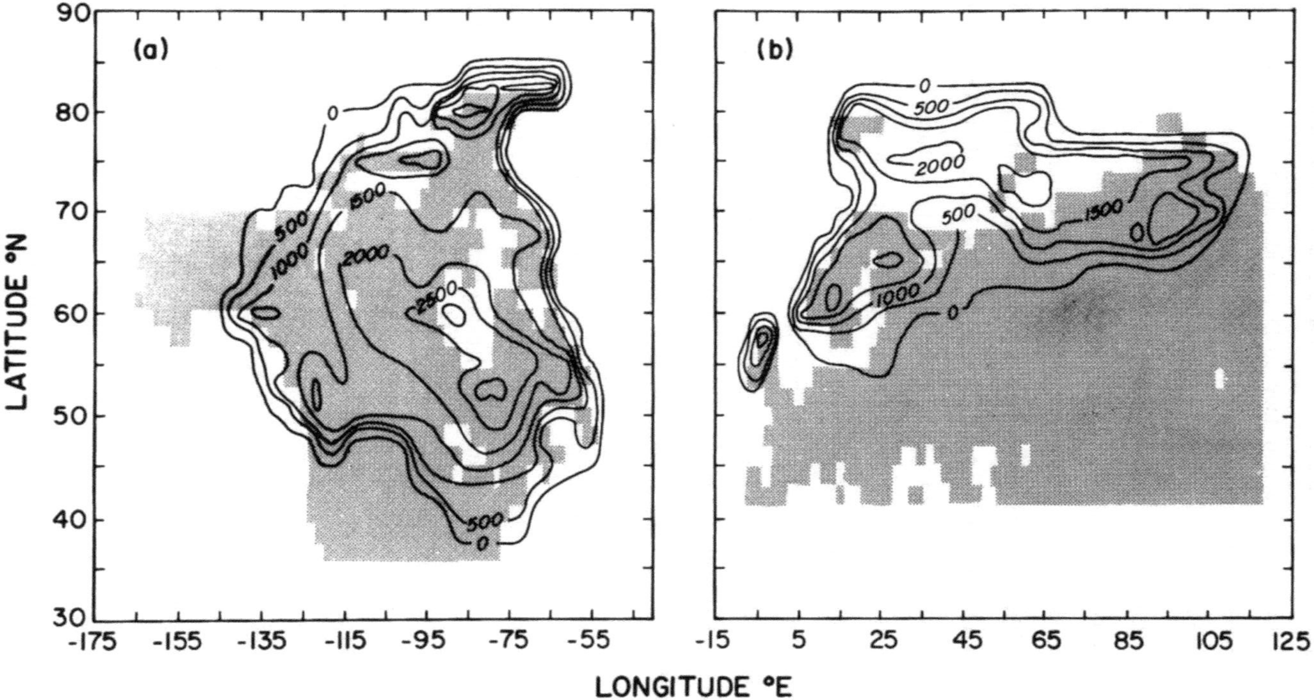

Figure 9. Thickness isopleths for the North American (a) and Fennoscandian (b) ice complexes at last glacial maximum according to the deconvolution of post glacial relative sea level records by Tushingham and Peltier (1991).

satisfactory explanation of the origins of the termination process.

In this more realistic model the same evolution equation for ice thickness is employed as previously and glacial isostatic adjustment is also explicitly described, though both of these processes are taken to be both longitude and latitude dependent rather than latitude dependent only. The main enhancement of the dynamical system consists of the explicit solution for the longitude and latitude dependence of the annual cycle of the sea level temperature field through the integration of the following energy balance equation:

$$C(\underline{r})\,\frac{\partial T(\underline{r},t)}{\partial t} - \underline{\nabla}_h \cdot \left[D(x)\,\underline{\nabla}_h T(\underline{r},t)\right] + A + BT(\underline{r},t) = \frac{Q}{4}\,a(\underline{r})\,S(x,t) \qquad (7)$$

in which $C(\underline{r})$ is the effective heat capacity (equal to C_ℓ over land, C_w over water and C_i over sea ice), D(x) is a diffusion coefficient intended to represent all forms of heat transport (baroclinic eddies etc.), A+BT represents the longwave emission to space with the empirical coefficients A and B determined by fitting to modern satellite data, Q = 1360 W/m^2 is the solar constant and $a(\underline{r})$ is the coalbedo (the fraction of the solar energy per unit area and time that is absorbed by the earth) and S(x,t) is the solar distribution function. With D expanded to degree 4 in Legendre polynomials the linear energy balance model embodied in (7) is essentially that of North et al. (1983). It is linear in the sense that $a(\underline{r})$ is not a function of T and so, for example, sea ice extent cannot be carried as a prognostic variable.

The ice sheet model to which this linear energy balance model is linked is essentially a two-dimensional version of (1) and takes the form:

$$\frac{\partial H}{\partial t} = \underline{\nabla}_h \cdot (D_i\,\underline{\nabla}_h h) + G(\theta,\phi,t) \qquad (8)$$

in which the highly nonlinear diffusion coefficient is just:

$$D_i = B_1\,H^{m+1}\,(\underline{\nabla}_h h \cdot \underline{\nabla}_h h)^{\frac{(m-1)}{2}}$$

with $B_1 = \lambda(\rho_i g)^m$ and the stress-exponent m ≈ 3 based upon experimental observations of ice rheology. The dependent variables H, h, and G are as previously. Now in two spatial dimensions the accurate model of glacial isostatic adjustment embodied in (3) is quite numerically demanding to integrate and I have therefore elected to replace it for the purpose of initial experiments with the simpler local adjustment model embodied in the equation:

$$\frac{\partial h'}{\partial t} = - \frac{(h' - h_o')}{\tau} + (\frac{\rho_i}{\rho_e \tau}) H \tag{9}$$

in which $h' = H-h$ is the depth of the bedrock measured positive below a given reference level, ρ_e is the density of the bedrock and h_o' is the bedrock depression which is assumed to be in isostatic balance in the absence of a load, essentially the present day topography relative to mean sea level. τ is the isostatic adjustment timescale.

Now the critical part of the above described model concerns the manner in which the ice sheet and energy balance components are coupled together and this is taken to occur through the mass balance function $G(\theta,\phi,t)$. This function is taken to consist of two parts, an accumulation function defined as

$$\begin{aligned} G_c\ (m/month) &= P(\theta,\phi)\, L_m\,, \quad T_c(\theta,\phi) \le 0 \\ &= 0 \qquad\qquad\qquad , \quad T_c(\theta,\phi) > 0 \end{aligned} \tag{10}$$

in which the T_c are now monthly average values of the temperature determined by solving (7) for the "instantaneous" annual cycle and corrected for height, $P(\theta,\phi)$ in m/yr is the present day annually averaged precipitation and L_m is the duration of a month. The accumulation function in (10) is further adjusted for the so-called elevation desert effect by correcting it to:

$$G_{cc} = G_c\ x\ (2^{-max(o,[h(km)-2])})\ , \tag{11}$$

following Budd and Smith (1981). To obtain the net mass balance at every surface point in the model we next compute an ablation function in the form:

$$G_g(m/month) = max(0, aT_c(^{\circ}C) + bR_m(W/m^2) + c) \cdot (10/\rho_i[km/m^2]) \tag{12}$$

In this expression the field $T_c\ (\theta,\phi)$ is the same as previously while R_m is the monthly mean insolation at the top of the atmosphere. The parameters a,b,c have been obtained by fitting present day ablation rates on glaciers (e.g. Pollard 1983).

With all parameters of the model now specified it can be directly integrated to simulate the evolution of the space-time distribution of ice from the last interglacial (Eemian) period to the present, with R_m (θ,t) being determined by the orbital variations. The model is in fact run in "a-synchronous" mode, beginning with the construction of an equilibrium annual cycle by solution of (7). The ice sheet and bedrock adjustment equations are next integrated forward for 1000 years of model time, during which the annual cycle is assumed to be stationary. A new equilibrium annual cycle is next calculated using the updated surface ice distribution to modify the albedo function and the new annual cycle is again held stationary while the ice sheet and bedrock equations are again integrated forward for 1000 years. The optimum time step separating recomputations of the annual cycle has been determined by numerical experimentation to be the longest compatible with a sufficiently accurate prediction of the history of ice sheet evolution.

A typical example of a prediction of the distribution of surface ice at LGM based upon this model (after DeBlonde and Peltier 1991b) is shown on Figure 10. Comparing this distribution with that previously inferred by Tushingham and Peltier on the basis of the inversion of relative sea level histories (Figure 9) it is clear both on the basis of the geographical location of the continental ice sheets and on the basis of their maximum thicknesses that the above described model rather accurately describes these aspects of the LGM ice cover. The reasons for the success of the model are also quite easily understood. Laurentian ice is prevented from invading the American sector of the North American continent by the latitudinal insolation gradient which drives ablation rates to extremely high levels south of about 45°N latitude. Similarly the Fennoscandian ice sheet is inhibited from expanding eastward by the extremely large amplitude of the annual cycle that obtains in the interior of the vast Eurasian continent. Summers there are simply too warm to permit perennial snow cover to develop. The remaining fractions of the perimeters of Laurentian and Fennoscandian ice are coincident with the continental shelves, beyond which water depth increases rapidly leading to extremely high calving rates and confinement of the ice sheets to their proximate continents.

Although the above described successes of the energy balance coupled ice sheet model described in this section must be seen as encouraging there are also aspects of the model integration, discussed more fully than is possible here in DeBlonde and Peltier (1991b), that are considerably less than satisfactory. In particular the present model suffers the same flaw as was

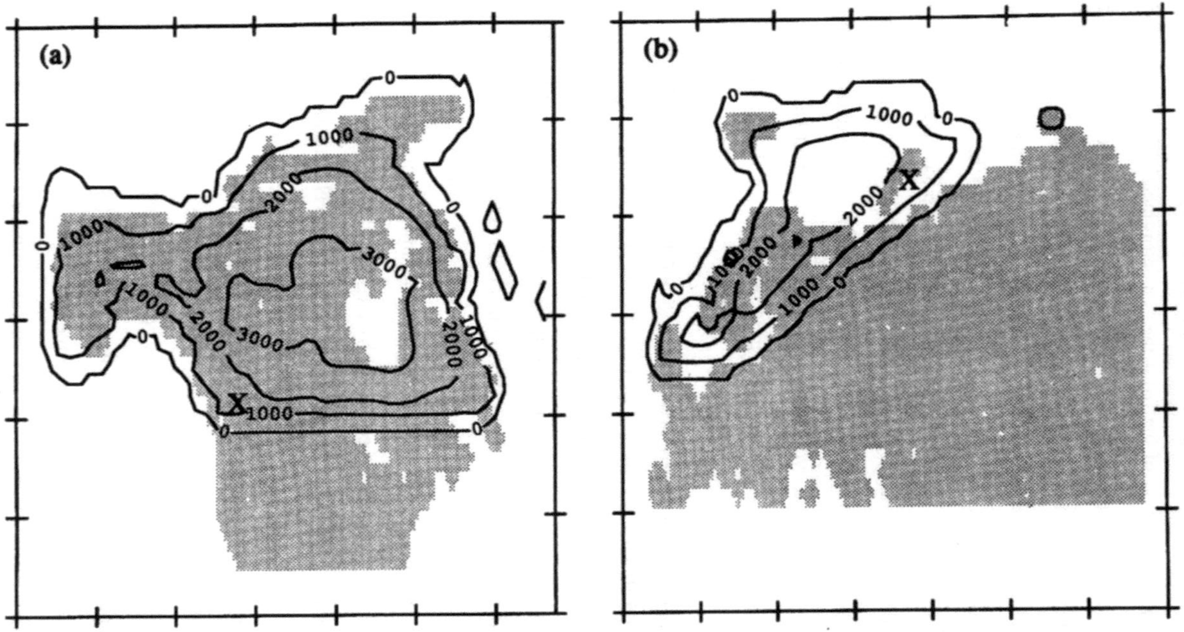

Figure 10. Ice thickness at glacial maximum for the Laurentian (a) and Fennoscandian (b) regions predicted by the ice sheet coupled energy balance model described in the text. These plates should be compared with the "observed" distributions shown on Figure 9.

revealed by the integrations of the one-dimensional analogous model discussed in the last Section. This concerns the inability of the model to properly explain the termination process. Although ice volume does diminish significantly following LGM in response to the orbital forcing alone, this diminution is insufficient to explain the observed Holocene distribution of land ice. Although the use of a nonlinear version of the energy balance model embodied in (7), as discussed in DeBlonde et al. (1992), can somewhat improve this critical aspect of the simulation, especially if representations are included in the model of the additional forcing associated with CO_2 variations and heat flux effects associated with changes in the strength of the thermohaline circulation (DeBlonde and Peltier 1993) the complete collapse of the Laurentide ice sheet still proves difficult to understand. Additional feedbacks, either "wet" or "dry", still appear to be required to understand the near complete deglaciations of Canada and Fennoscandia that occurred. What these additional required feedbacks might be is addressed in the following Section.

Conclusions

The results discussed in the previous sections of this paper illustrate a number of recent advances that have been achieved in our understanding of the 100 kyr ice age cycle that has dominated climate system variability for the last 900 kyr of earth history. The chronology of ice volume changes through the full Pleistocene epoch is now rather well determined by the advanced tuning strategy discussed above that has been so successfully applied to the ODP 677 core from the Panama Basin. This record shows that the 100 kyr oscillation "turned-on" smoothly in the time interval from 900 kyr BP to about 700 kyr BP. Prior to 900 kyr BP the variance in the ice volume record was contained almost entirely in the orbitally related spectral peaks corresponding to the periods 41 kyr, 23 kyr, 21 kyr and 19 kyr. Such variability is just as expected on the basis of the Milankovitch hypothesis with ice volume linearly related to the summertime (caloric) seasonal insolation anomaly. Subsequent to this mid-Pleistocene climate transition the 100 kyr cycle has dominated the records of $\delta^{18}0$ change. This cycle is not expected on the basis of the Milankovitch hypothesis which is therefore not in detailed accord with observations.

In order to understand the origins of the 100 kyr cycle one is obliged to introduce an appropriate nonlinearity into the physical model of climate system behaviour that governs the response on the very long timescales of the evolution of the planetary orbit. Analysis of a simple

1-D model that couples the nonlinear physics of the accumulation and flow of a continental ice sheet with the physics of glacial isostatic adjustment demonstrates that a large fraction of the variability in ice volume may be thereby explained but strong 100 kyr oscillations only appear when an additional feedback loop is added to the model such that meltback is enhanced when it exceeds some critical threshold. This model provides no means whereby the precise origins of the required feedback may be assessed, however, making the model less satisfactory than one might hope.

Increasing the precision with which the ice physics and accumulation processes are described, especially when the energy balance model to which the ice sheets are linked is made nonlinear, does improve the quality of the "terminations" delivered by the climate model somewhat though not sufficiently to allow us to claim that the model provides a good fit to the data. As discussed in the last section of this paper, even adding explicit descriptions of the "known" feedbacks associated with variations in the concentration of CO_2 and with the strength of the North Atlantic heat pump that is linked to the strength of the thermohaline circulation fail to explain, in the non-linear energy balance context, the near complete "termination" of the Laurentide ice sheet that marked the onset of Holocene time.

In my view there remain two excellent candidates for the remaining feedback needed to induce total deglaciation of the dominant Laurentian ice mass. The first of these, a "dry" mechanism, involves the feedback onto surface albedo that is expected to be caused by the marked increase of terrigenous dust load in the atmosphere that was apparently characteristic (DeAngelis et al., 1988) of the time of glacial maximum, not only of the last glacial maximum recorded in ice cores but also of previous glacial maxima as recorded in deep sea sedimentary cores. This increase in terrigenous dust load is always observed to precede the termination and could constitute the missing feedback as recently argued in the analyses described by Peltier and Marshall (1993). A possible wet mechanism that could be competitive, in my view, with this dry mechanism might consist of a marine instability of the Laurentide ice sheet, perhaps the same instability, now aided by insolation forcing, that is presumably responsible for the so-called Heinrich events (Heinrich, 1988) that have recently been identified in deep sea cores from the North Atlantic basin and which appear to record, in the quasi-cyclic appearance of ice rafted debri, occasions of intense instability of the eastern flank of the Laurentide ice sheet during the time of maximum ice cover. Which of these two equally plausible feedback loops is principally responsible for the disappearance of Laurentide ice is as yet unclear but prospects are rather good that further application of the model based strategy described here, for inferring what the observations can be taken to imply, can be expected to lead to a deepening of understanding.

References

Baksi, A.K., V. Hsu, M.O. McWilliams and E. Farrar, 1992, $^{40}Ar/^{39}Ar$ dating of the Brunhes-Matuyama geomagnetic field reversal. *Science*, **256**, 356-34.

Berger, A., 1978, Long term variations of calorific insolation from the Earth's orbital elements. *Quat. Res.*, **9**, 139-167.

Broecker, W.S. and J. Van Donk, 1970, Insolation changes, ice volumes and the $\delta^{18}0$ record in deep sea cores. *Rev. Geophys. Space Phys.*, **8**, 169-198.

Broecker, W.S. and G.H. Denton, 1989, The role of ocean-atmosphere reorganizations in glacial cycles. *Geochim. Cosmochim. Acta.*, **53**, 2465-2501.

Budd, W.F. and I.N. Smith, 1981, The growth and retreat of ice sheets in response to orbital radiation changes, sea level, Ice and Climate Changes, IAHS Publ. **131**, 369-409.

Cox, Allan, R.R. Doell and G.B. Dalrymple, 1963, Geomagnetic polarity epochs and Pleistocene geochronometry. *Nature*, **198**, 1049-1051.

De Angelis, M., N.I. Baskov and V.N. Petrov, 1987, Aerosol concentrations over the last climatic cycle (160 kyr) from an Antarctic ice core. *Nature*, **325**, 318-321.

DeBlonde, G. and W.R. Peltier, 1991a, A one dimensional model of continental ice volume fluctuations through the Pleistocene: Implications for the origin of the mid-Pleistocene climate transition. *J. Climate*, **4**, 318-344.

DeBlonde, G. and W.R. Peltier, 1991b, Simulations of continental ice sheet growth over the last glacial-interglacial cycle: experiments with a one-level seasonal energy balance model including realistic geography. *J. Geophys. Res.*, **96**, 9189-9215.

DeBlonde, G., W.R. Peltier and W.T. Hyde, 1992, Simulations of continental ice sheet growth over the last glacial-interglacial cycle: experiments with a one level seasonal energy balance model including seasonal ice-albedo feedback. *Global and Planetary Change*, **98**, 37-55.

DeBlonde, G. and W.R. Peltier, 1993, Late Pleistocene ice age scenarios based upon observational evidence. *J. Climate*, **6**, 709-727.

Hays, J.D., J. Imbrie and N.J. Shackleton, 1976, Variations in the Earth's orbit: pacemaker of the ice ages. *Science*, **194**, 1121-1132.

Heinrich, H., 1988, Origin and consequences of cyclic ice rafting in the Northeast Atlantic Ocean during the past 130,000 years. *Quat. Res.*, **29**, 142-152.

Hyde, W.T. and W.R. Peltier, 1985, Sensitivity experiments with a model of the ice-age cycle: the response to harmonic forcing. *J. Atmos. Sci.*, **42**, 2170-2188.

Hyde, W.T. and W.R. Peltier, 1987, Sensitivity experiments with a model of the ice-age cycle: the response to Milankovitch forcing. *J. Atmos. Sci.*, **44**, 1351-1374.

Imbrie, J., N.J. Shackleton, N.G. Pisias, J.J. Morley, W.L. Prell, D.G. Martinson, J.D. Hays, A. McIntyre and A.C. Mix, 1984, The orbital theory of Pleistocene climate: support from a revised chronology of the marine $\delta^{18}0$ record. In *Milankovitch and Climate*, A. Berger and coeds., 269-305.

Martinsen, D.G., N. Pisias, J.D. Hays, J. Imbrie, T.C. Moore and N.J. Shackleton, 1987, Age dating and orbital theory of the ice ages: development of a high resolution 0-300,000 - year chronostratigraphy. *Quat. Res.*, **27**, 1-30.

Milankovitch, M., 1941, Canon of insolation and the ice age problem, and glacial cycles, K. Serb. Acad. Beorg. Spec. Publ. 132.

North, G.R., J.R. Mengel and D.A. Short, 1983, Simple energy balance model resolving the seasons and the continents: Application to the astronomical theory of the ice ages. *J. Geophys. Res.*, **88**, 6576-6586.

Oerlemans, J., 1980, Model experiments on the 100,000 year glacial cycle. *Nature*, **287**, 430-432.

Peltier, W.R., 1974, The impulse response of a Maxwell earth. *Rev. Geophys. and Space Phys.*, **12**, 649-669.

Peltier, W.R., 1976, Glacial isostatic adjustment II. The inverse problem. *Geophys. J.R. astron. Soc.*, **46**, 669-706.

Peltier, W.R, 1982, Dynamics of the ice age earth. *Adv. Geophys.*, **24**, 1-146.

*Peltier, W.R., 1985, The LAGEOS constraint on deep mantle viscosity: results from anew normal mode method for the inversion of viscoelastic relaxation spectra. *J. Geophys. Res.*, **90**, 9411-9421.

*Peltier, W.R. and S. Marshall, 1993, Glacial terminations and terrigenous dust: experiments with a nonlinear ice sheet coupled energy balance model, J. Climate, submitted.

Pollard, D., 1982, A simple ice sheet model yields realistic 100 kyr glacial cycles. *Nature*, **296**, 334-338.

Pollard, D., 1983, A coupled climate - ice sheet model applied to the Quaternary ice ages. *J. Geophys. Res.*, **88**, 7705-7718.

Quinn, T.R., S. Tremaine and M. Duncan, 1991, A 3 million year integration of the Earth's Orbit. *Astronomical Journal*, **101**, 2287-2305.

Raymo, M.E., W.F. Ruddiman, N.J. Shackleton and D.W. Oppo, 1990, Evolution of global ice volume and Atlantic-Pacific $\delta^{13}C$ gradients over the last 2.5 M.Y. *Earth and Planet. Science Lett.*, **97**, 353-368.

Shackleton, N.J. 1967, Oxygen isotope analysis and Pleistocene temperatures readdressed. *Nature*, **215**, 15-17.

Shackleton, N.J. and M.A. Hall, 1989, Stable isotope history of the Pleistocene at ODP Site 677. K. Becker, H. Sakai et al., Eds., Proc. ODP Sci. Results, **3**, College Station, Tx.

Shackleton, N.J., A. Berger and W.R. Peltier, 1990, An alternative astronomical calibration of the lower Pleistocene timescale based upon ODP Site 677. *Trans. Roy. Soc. Edinburgh; Earth Sciences*, **81**, 251-261.

Tushingham, M. and W.R. Peltier, 1991, ICE-3G: A new global model of late Pleistocene deglaciation based upon geophysical predictions of post-glacial relative sea level change. *J. Geophys. Res.*, **96**, 4497-4523.

A MODEL STUDY OF THE GLACIAL OCEANIC CIRCULATION

S. Hovine and Th. Fichefet
Institut d'Astronomie et de Géophysique G. Lemaître
Université Catholique de Louvain
Chemin du Cyclotron, 2
B-1348 Louvain-la-Neuve
Belgium

ABSTRACT. A primitive equation model zonally averaged over the Atlantic, Indian and Pacific basins is used to investigate the oceanic general circulation of the last glacial maximum. In addition to wind forcing, the model circulation is driven by restoring the sea-surface temperature and salinity to prescribed values. Under present-day boundary conditions, the model reproduces the global conveyor belt. When glacial conditions for temperature and salinity are used to force the model, significant changes occur in the circulation patterns. Most of the deep Atlantic is filled with water originating from the model Southern Ocean. Deep-water production completely stops between 60 and 70°N in the Atlantic. However, the surface-water salinity around 50°N is sufficiently high to permit deep convection to a maximum depth of 2000 m, as a source of intermediate and deep waters. Open convection occuring between 30 and 40°N, contributes to the ventilation of these waters. In the North Pacific, the model simulates a weak intensification of the intermediate-water production. It is worth pointing that all these changes compare favourably with the existing geological reconstructions.

1. Introduction

At the last glacial maximum (LGM), 18,000 years ago, continental ice sheets extended southward to 41°N on the American side of the Atlantic and to 52°N on the European coast (CLIMAP, 1981). The polar front in the North Atlantic, which follows today a meridional trend just east of Greenland, followed a zonal trend along 43°N. This configuration is similar to that of the modern Southern Ocean, and it points to an equivalent barring of meridional heat transport at the ocean surface above 42°N (Ruddiman and McIntyre, 1977). The surface salinity distribution was also significantly different from the present one. The recent reconstruction made by Duplessy et al. (1991) indicates that the modern relatively saline surface waters of the high-latitude North Atlantic were replaced by low-salinity surface waters. The reconstruction also shows the presence of a sharp salinity gradient that coincides with the polar front. However, high-salinity sur-

NATO ASI Series, Vol. I 22
Long-Term Climatic Variations
Edited by J.-C. Duplessy and M.-T. Spyridakis

face waters were present near 30–40°W, north of the polar front. Labeyrie et al. (1992) suggest that the hydrographic front was sufficiently unstable to allow some of the saline subtropical waters to penetrate into the central northern Atlantic. Isotope measurements in deep-sea cores reveal that these modifications in the ocean surface characteristics were accompanied by marked changes in the oceanic general circulation (e.g., Boyle and Keigwin, 1987; Duplessy et al., 1988). The purpose of the present work is to investigate these circulation changes with a zonally averaged, three-basin ocean model.

2. Description of the model

To model the global ocean circulation, an idealized ocean model has been used consisting of three zonally averaged ocean basins, representing the Atlantic, Indian, and Pacific oceans. A brief summary of the model characteristics is presented here; further details can be found in Fichefet and Hovine (1993).

The Atlantic basin extends from 90°N to 80°S, the Indian from 25°N to 70°S, and the Pacific from 65°N to 80°S. These three basins are connected between 40 and 65°S by a circumpolar channel through which mass, heat, and salt are exchanged. Within each individual basin, the model formulation is based on the zonally averaged form of the primitive equations expressed in spherical coordinates. In the zonally averaged u-momentum equation, the east–west density gradient $\Delta\rho$ is parameterized following the parameterization proposed by Wright and Stocker (1991):

$$\Delta\rho \quad = \quad -\epsilon\Delta\Lambda\sin 2\phi\frac{\partial\rho}{\partial\phi}, \tag{1}$$

where $\Delta\Lambda$ is the angular width of the basin, ϕ is the latitude and ϵ is a closure parameter. Between 55 and 65°S, this closure parameter is taken equal to 0 in the upper 3000 m and to 0.15 at greater depths to account for the quasi-absence of continental barrier in this zone (Wright and Stocker, 1992). A zero value is also used at all depths between 85 and 90°N. Elsewhere, we take $\epsilon = 0.5$.

3. The control experiment

In order to assess the simulation capability of the model described above, a control run with present-day surface boundary conditions was performed. This run was started from an ocean at rest with uniform potential temperature $T = 2$°C and salinity $S = 34.7$ ppt. The surface wind stress (τ^λ,τ^ϕ) over each ocean basin was derived from the monthly mean data of Hellerman and Rosenstein (1983) and the sea-surface temperatures and salinities were relaxed on a time scale of 10 days to the monthly and seasonal mean surface data of Levitus (1982) zonally averaged over the basin considered. The model was integrated forward in time for 10,000 years. The solution at this stage will be referred to as the equilibrium solution. It is worth pointing that this steady state remains stable upon a switch from restoring to mixed boundary conditions.

Figure 1 displays the annual mean meridional overturning streamfunction simulated at equilibrium in the Atlantic and Pacific basins. Deep water is formed in the Atlantic between 60 and 70°N at a rate of about 22 Sv (1 Sv = 10^6 m^3 s^{-1}), in qualitative agreement with current estimates (e.g., Gordon, 1986). Only about 5 Sv of this water recirculates within the Atlantic, the remainder is exported into the circumpolar channel at depths exceeding 2500 m. Below 1500 m, deep water penetrates into the Indian basin at a rate of about 8 Sv, and into the Pacific basin at a rate of about 18 Sv. This water then upwells over a broad region in both basins and finally returns to the Atlantic via the circumpolar channel as intermediate and upper water masses. This makes the model thermohaline circulation relatively consistent with the conveyor-belt circulation popularized by Broecker (1987, 1991). Note, however, that there is no Antarctic Bottom Water cell at great depths in the Atlantic. It is possible that under more favourable conditions for density forcing in the immediate vicinity of the Antarctic continent, such a circulation would develop (e.g., England, 1992; Wright and Stocker, 1992). Superimposed to the thermohaline circulation are vigorous wind-driven cells in the upper thermocline. These Ekman overturnings are particularly intense in the Pacific. It should be noted that the clockwise cell located just north of the equator in this basin is mainly responsible for the formation of North Pacific intermediate water in the model.

Further evidence of the model performance for present-day boundary conditions is given in Fichefet and Hovine (1993). This paper demonstrates that the model shows acceptably good agreement with enough aspects of the behaviour of the real ocean to permit useful studies of the ocean in the geological past.

4. The LGM experiment

In order to investigate the nature of the glacial oceanic circulation, an LGM experiment was conducted with the model. As in the control case, the model was spun up from rest under restoring boundary conditions. The surface wind stresses were kept fixed to their present-day values owing to the lack of reliable informations on their behaviour during the last ice age. With regard to the relaxation temperatures and salinities, the standard values were modified daily by an amount $\Delta T^\star$ or $\Delta S^\star$. The daily values of $\Delta T^\star$ were deduced from the February and August sea-surface temperature changes between the LGM and the present time given by CLIMAP (1981) and zonally averaged over each ocean basin. For $\Delta S^\star$, we used throughout the year, north of 10°S in the Atlantic, the summer sea-sur-

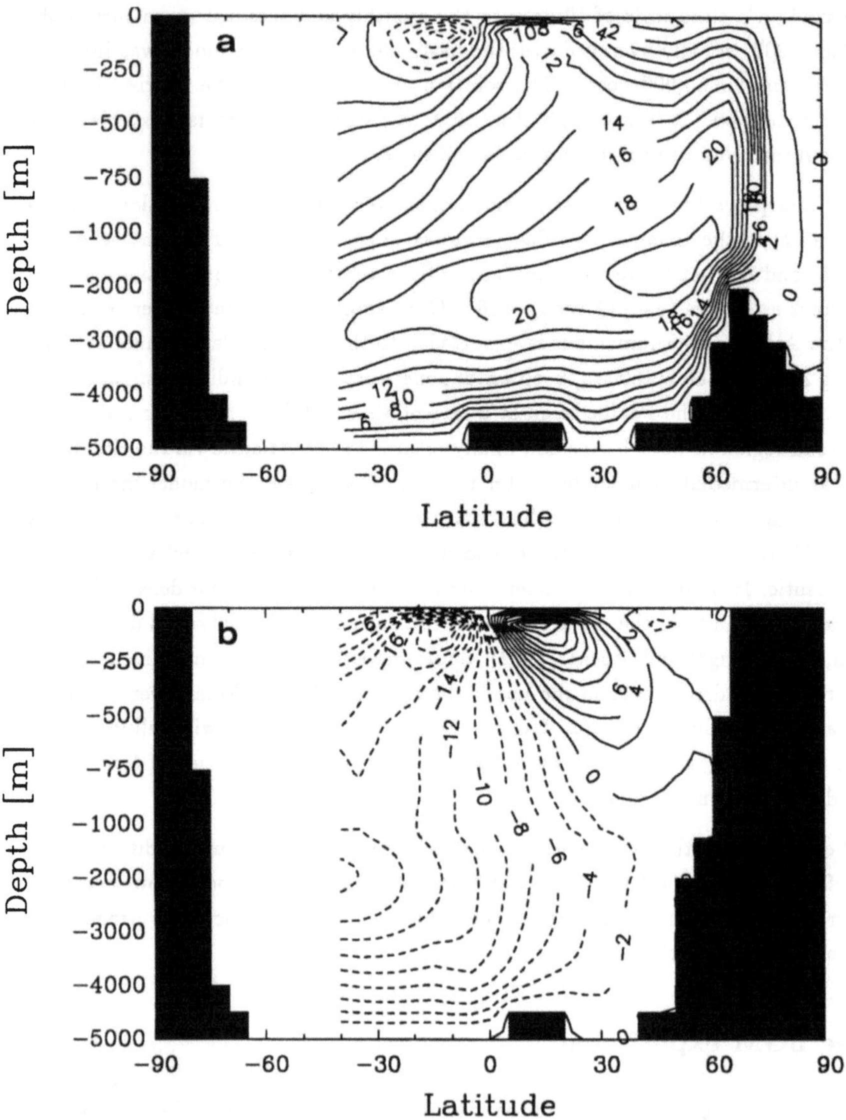

Figure 1. Contours of the annual mean meridional overturning streamfunction in the Atlantic (a) and Pacific (b) basins for the steady state obtained under present-day surface boundary conditions. Solid contours indicate clockwise circulation; units are Sv (1 Sv = 10^6 m^3 s^{-1}).

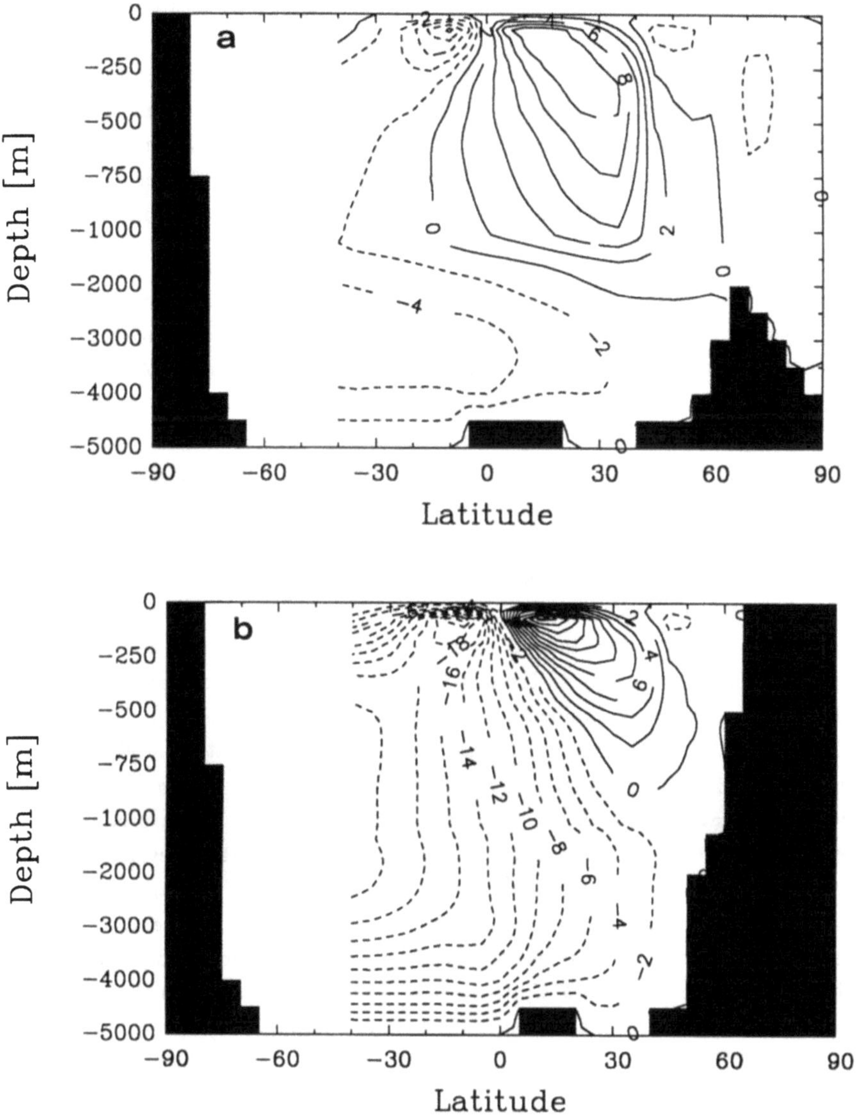

Figure 2. Contours of the annual mean meridional overturning streamfunction in the Atlantic (a) and Pacific (b) basins for the steady state obtained under glacial surface boundary conditions. Solid contours indicate clockwise circulation; units are Sv.

face salinity differences between the LGM and the present time produced by Duplessy et al. (1991) and zonally averaged. Elsewhere, a salinity anomaly of +1 ppt was imposed. This estimate, based on the fact that the mean glacial salinity was higher than the modern one by about 1 ppt (e.g., Duplessy et al., 1991), is consistent with a preliminary global reconstruction of the surface-water salinity distribution during the LGM made by J.-C. Duplessy (personnal communication, 1992). The model is integrated under this forcing for 10,000 years. Once again, the steady state obtained remains stable upon switching to mixed boundary conditions.

The annual mean meridional overturning streamfunction simulated at equilibrium is depicted in Figure 2 for the Atlantic and Pacific basins. A comparison with Figure 1 indicates that the largest changes in the circulation patterns occur in the Atlantic. The most prominent feature is that, below 2000 m, most of the Atlantic basin is filled with Antarctic Bottom Water, so that the relative contribution of water from the Southern Hemisphere as compared to that from the Northern Hemisphere is much greater than in the control experiment. Another important change is that the region located between 60 and 70°N stops forming deep water, which follows from the stratification due to the low surface-water salinity. Around 50°N, the surface-water salinity is sufficiently high to permit deep convection to a maximum depth of 2000 m, as a source of intermediate and deep waters. Open convection taking place between 30 and 40°N contributes to the ventilation of these waters up to the equator. All these features compare favourably with the geological evidence obtained from deep-sea sediment cores (e.g., Duplessy et al., 1980; Boyle and Keigwin, 1987; Duplessy et al., 1988; Curry et al., 1988). In the Pacific basin, the model circulation undergoes only small changes. One notes, however, a deep-water ventilation significantly higher than the modern one and a production of North Pacific intermediate water slightly enhanced. This last feature results from the surface cooling and salinity increase specified in the North Pacific, and agrees with the geological reconstruction of Duplessy et al. (1988). As far as the temperature field is concerned, Figure 3 shows that the use of glacial boundary conditions induces a cooling of the global deep ocean of about 2°C, which is in general agreement with the geological estimates (e.g., Chappell and Shackleton, 1986; Labeyrie et al., 1987).

5. Conclusions

A zonally averaged, three-basin ocean model has been used to study the World Ocean circulation during glacial times.

Under present-day boundary conditions, the model predicts a circulation and water-mass properties that are in general agreement with the observations. In particular, deep-water formation occurs in the Atlantic between 60 and 70°N and around Antarctica as in the reality, and a global conveyor belt of plausible strength is established.

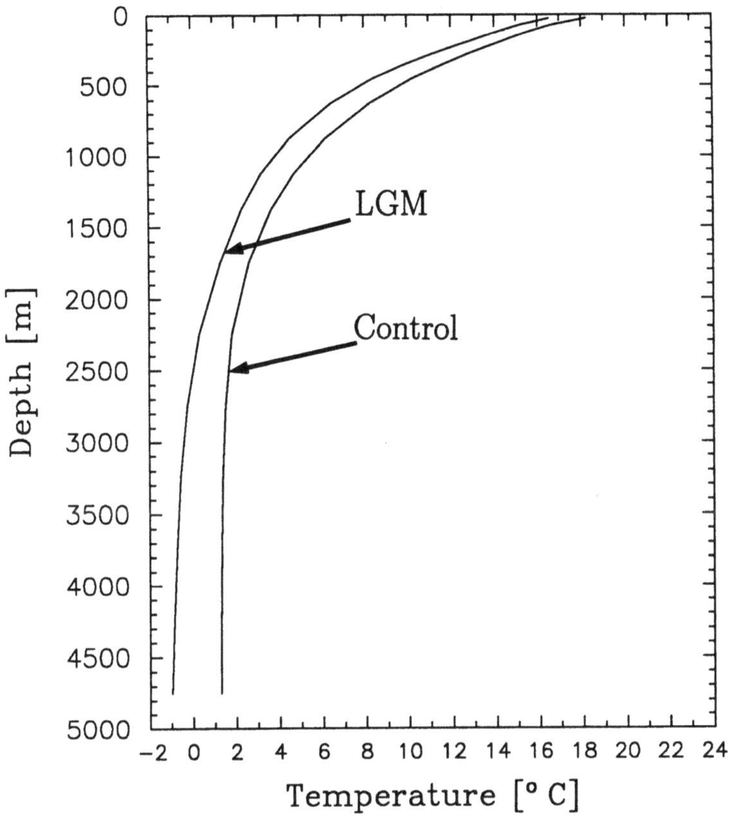

Figure 3. Annual mean potential temperature averaged over the three ocean basins from the control run and the LGM experiment.

When glacial conditions for temperature and salinity are used to force the model, the following features are simulated: 1) a pronounced intrusion of Antarctic Bottom Water below 2000 m in most of the Atlantic basin; 2) a cessation of deep-water formation between 60 and 70°N in the Atlantic; 3) a modest production of intermediate and deep waters (to a maximum depth of 2000 m) in the central northern Atlantic and a ventilation of these waters up to the equator; 4) a weak intensification of the formation of North Pacific intermediate water; and 5) a cooling of the global deep ocean of about 2°C. All these changes agree qualitatively with the geological evidence obtained from deep-sea sediment cores.

The results of the present study need, of course, to be confirmed by carrying out further experiments with a more reliable global reconstruction of the glacial sea-surface salinity distribution when it will become available. Additional experiments should also be conducted with surface wind stresses taken from an AGCM that has run under glacial boundary conditions to quantify the impact of the changes in this forcing component on the LGM oceanic circulation.

Acknowledgments

Valuable comments on this work were received from J.-C. Duplessy. Th. Fichefet is sponsored by the National Fund for Scientific Research (Belgium) and S. Hovine is partly supported by the Climate Programme of the Commission of the European Communities under Contract EPOC-0004 (EDB). These supports are gratefully acknowledged.

References

Boyle, E.A., and L.D. Keigwin, 1987: North Atlantic thermohaline circulation during the last 20,000 years linked to high latitude surface temperature. *Nature*, **330**, 35–40.

Broecker, W.S., 1987: Unpleasant surprise in the greenhouse? *Nature*, **328**, 123–127.

Broecker, W.S., 1991: The great ocean conveyor. *Oceanography*, **4**, 79–89.

Chappell,J., and N.J. Shackleton, 1986: Oxygen isotopes and sea level. *Nature*, **324**, 137–140.

CLIMAP Project Members, 1981: Seasonal reconstruction of the earth's surface at the last glacial maximum. *Geol. Soc. Am. Map. Chart Serv.*, **MC–36**.

Curry, W.B., J.-C. Duplessy, L.D. Labeyrie, and N.J. Shackleton, 1988: Changes in the distribution of $\delta^{13}C$ of deep water ΣCO_2 between the last glaciation and the Holocene. *Paleoceanography*, **3**, 317–341.

Duplessy, J.-C., J. Moyes, and C. Pujol, 1980: Deep water formation in the North Atlantic Ocean during the last ice age. *Nature*, **286**, 476–482.

Duplessy, J.-C., N.J. Shackleton, R.G. Fairbanks, L.D. Labeyrie, D. Oppo, and N. Kallel, 1988: Deepwater source variations during the last climatic cycle and their impact on the global deepwater circulation, *Paleoceanography*, **3**, 343–360.

Duplessy, J.-C., L.D. Labeyrie, A. Juillet-Leclerc, F. Maitre, J. Duprat, and M. Sarnthein, 1991: Surface salinity reconstruction of the North Atlantic Ocean during the last glacial maximum. *Oceanologica Acta*, **14**, 311–324.

England, M.H., 1992: On the formation of Antarctic Intermediate and Bottom Water in ocean general circulation models. *J. Phys. Oceanogr.*, **22**, 918–926.

Fichefet, Th and Hovine S., 1993: The glacial ocean: A study with a zonally averaged, three-basin ocean circulation model. In "Proceedings of the NATO Advanced Research Workshop on Ice in the Climate System", in press.

Gordon, A.L., 1986: Interocean exchange of thermocline water. *J. Geophys. Res.*, **91**, 5037–5046.

Hellerman, S., and M. Rosenstein, 1983: Normal monthly wind stress over the World Ocean with error estimates. *J. Phys. Oceanogr.*, **13**, 1093–1104.

Labeyrie, L.D., J.-C. Duplessy, and P.-L. Blanc, 1987: Variations in mode of formation and temperature of oceanic deep waters over the past 125,000 years. *Nature*, **327**, 477–482.

Labeyrie, L.D., J.-C. Duplessy, J. Duprat, A. Juillet-Leclerc, J. Moyes, E. Michel, N. Kallel, and N.J. Shackleton, 1992: Changes in the vertical structure of the North Atlantic Ocean between glacial and modern times. *Quat. Sc. Rev.*, **11**, 401–413.

Levitus, S., 1982: *Climatological Atlas of the World Ocean*, NOAA Prof. Paper 13, U.S. Dept. of Commerce, Washington, DC, 173 pp.

Ruddiman, W.F., and A. McIntyre, 1977: Late Quaternary surface ocean kinematics and climatic change in the high-latitude North Atlantic. *J. Geophys. Res.*, **82**, 3877–3887.

Wright, D.G., and T.F. Stocker, 1991: A zonally averaged ocean model for the thermohaline circulation. Part I: Model development and flow dynamics. *J. Phys. Oceanogr.*, **21**, 1713–1724.

Wright, D.G., and T.F. Stocker, 1992: Sensitivities of a zonally averaged global ocean circulation model. *J. Geophys. Res.*, **97**, 12,707–12,730.

OGCM-constraints to PM's

M. Lautenschlager[1] and E. Maier-Reimer[2]
[1] Deutsches Klimarechenzentrum GmbH
[2] Max-Planck-Institut für Meteorologie
Bundesstraße 55
D–20146 Hamburg
Germany

Abstract

We examined the response of our ocean models of physical circulation and geochemical tracer distributions on atmospheric glacial forcing. The CLIMAP [1981] sea surface temperature (SST) was taken directly as a boundary condition. The windstress and freshwater flux were derived from the ice age response of an atmospheric general circulation model (AGCM). Near the surface, the ocean response in temperature and circulation reflects primarily the imposed glacial forcing fields. The simulated deep ocean response, especially in the North Pacific, is in conflict with the observational evidence. The glacial changes in the physical ocean circulation appear to be qualitatively as derived from sediment cores, but, probably, highly overestimated. The misfit between model and data is established clearly in the distribution of $\delta^{13}C$, where a direct comparison with sediment core data is possible. The simulated AGCM freshwater flux, which is linked for example by the evaporation to the imposed SST field, is suspected to be the most probable reason for this conflict.

Introduction

Great efforts have been made during the last 15 years to reconstruct the surface climate of the last glacial maximum (LGM) from a very heterogeneous set of original data [CLIMAP, 1981]. As a result, we have now inter alia global maps of surface temperature for summer and winter and of ice coverage which can be taken to drive an AGCM. The glacial boundary conditions for the ocean general circulation model (OGCM), surface temperature, wind stress, and freshwater flux, are then available from the simulation of the atmospheric glacial response.

Substantial progress has been made in modeling the atmospheric response to LGM boundary conditions [e.g. Kutzbach and Guetter, 1986; Broccoli and Manabe, 1987; Rind, 1987; Lautenschlager and Herterich, 1990]. A rigorous intercomparison of these models is performed within PMIP (Paleoclimate Model Intercomparison Project)[Joussaume, this volume]. Much less has been done on the glacial ocean circulation. Aside from the somewhat inferential studies of Miller and Russell [1989] and Keffer et al. [1988], there has been no attempt to model the three-dimensional glacial response of the ocean and couple a global carbon model to a three-dimensional simulation of the ocean 18,000 years before present (18 ka).

NATO ASI Series, Vol. I 22
Long-Term Climatic Variations
Edited by J.-C. Duplessy and M.-T. Spyridakis

In this study (modification of Lautenschlager et al. [1992]), we simulate the three-dimensional circulation of the ice age ocean and the effect on the ocean carbon cycle. Our approach is to force an OGCM with the results from an AGCM, which in turn has been forced with LGM boundary conditions from CLIMAP [1981]. The model generated output – upwelling, $\delta^{13}C$, $\Delta^{14}C$ – are compared qualitativley to the deepsea record. Most of the comparisons involve data collected since original publication of the CLIMAP [1981] sea surface temperature (SST) field.

As will be seen in the results section, there are some significant differences between our model results and observations. Some readers may question the utility of reporting negative results. Although such objections might seem reasonable, we believe there are three justifications for our actions: 1) there are some areas where in fact there are agreements between models and observations; 2) it is useful to report the results as a benchmark for comparison of future runs with other models or different boundary conditions; 3) the most significant model-data discrepancy can be traced to a critical boundary condition, the CLIMAP [1981] Pacific SST field. Our results underline those of an earlier study [Rind and Peteet, 1985] as to the need for their reevaluation, but our results also extend the magnitude of the model-data discrepancies for 18ka not only to tropical land areas, but also to the oceanic thermohaline circulation and preservation of deepsea sediments.

Model chain

The Earth's climate is determined by the interaction of atmosphere, biosphere, ocean, and cryosphere. Simulations of future climate changes due to the increase of atmospheric carbon dioxide and other greenhouse gases are only reliable if the ability of the numerical models to simulate the present climate is shown as well as the ability of the models to simulate climatic states far away from the present one. Present-day available models are tuned to simulate present-day climate. Therefore, it is a priori not clear that these models sucessfully simulate climatic states different from the present. Up to now, coupled ocean-atmosphere models suffer from the fact that the separately developed models do not balance each other with respect to the errors in the fluxes across the air sea interface. Consequently, the climate of the coupled model exhibits a drift even in attempts to simulate the present climate. A systematic tuning of free parameters to remove such drifts would require so many experiments that even with todays computer facilities it is not feasible. It is convenient to introduce flux corrections (Sausen et al., 1988) in which mismatches of the fluxes of the present days climate are added to the fluxes in the coupled system. By this technique the submodels can be coupled to a rather stationary model of the present climate which allows, at least, the investigation of the effect of small perturbations. For the simulation of a completely different climate state, however, with potentially increased impact of nonlinearities, the flux correction is conceptually not applicable. It is therefore still critical to look at the behaviour of the subsystems. In the absence of a reliable real climate model we investigate the response of individually developed (and tested) models for different aspects of the climate system. The models are cobnsecutively run in the order given by the direction of forcing causality (Fig. 1). The CLIMAP reconstruction of SST is based on sedimentary information (note that the use of empirical transfer functions can be interpreted as a modeling approach, too). With this SST-field we drive our

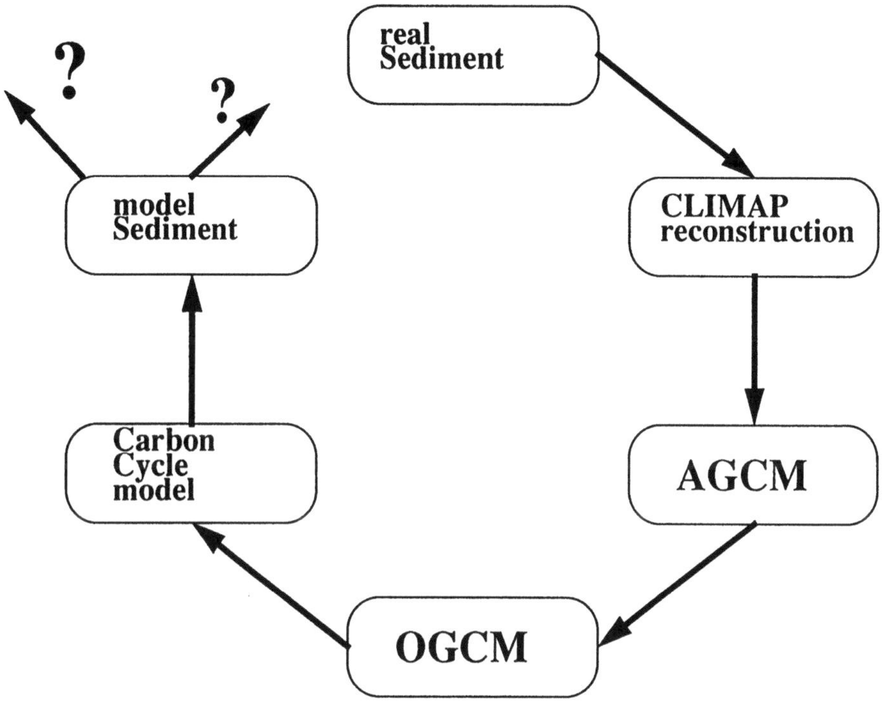

Figure 1: Flow chart of model succession

atmosphere model (as part of PMIP). The output of the atmosphere model drives the oceanic circulation model in which we examine the response of the carbon cycle. The sediment part of the carbon cycle model, finally, enables us to perform direct comparison of data with model results. Since each submodel has been proven to produce reasonable results, we should expect that the model sediment comes close to the real sediment. We show that this is not true.

AGCM experiment

The AGCM used is the T21 model developed at the European Centre for Medium-Range Weather Forecasts (ECMWF) in Reading [Louis, 1984] and modified for climate simulations at the Meteorological Institute of Hamburg University [Dümenil and Schlese, 1987]. The T21 AGCM is based on the primitive equations including radiation, a hydrological cycle and a three layer soil model, determining soil temperature and soil moisture. The model equations are solved in the spectral domain, truncated at wavenumber 21. This corresponds to a horizontal resolution of 5.6° in latitude and longitude. The vertical coordinate is resolved by 16 levels in a hybrid coordinate system: the terrain-following

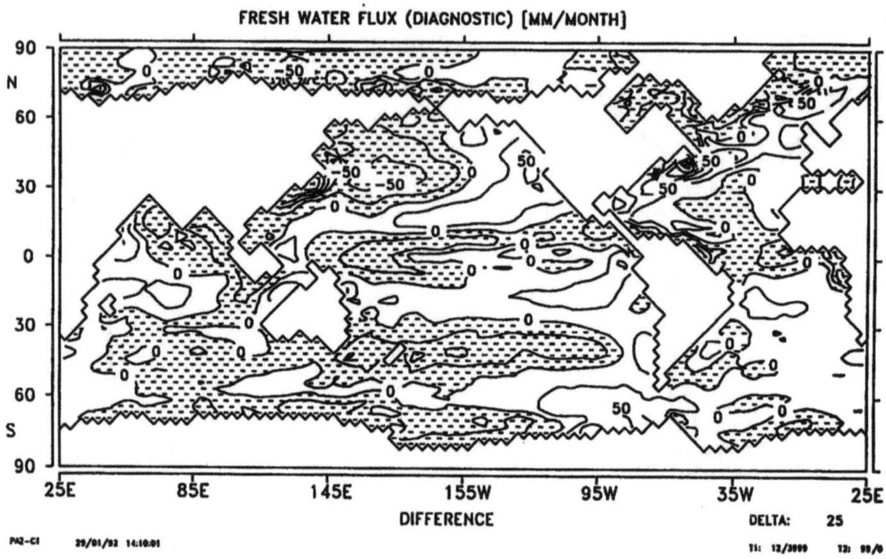

Figure 2: Annual mean fresh-water flux anomaly (LGM minus present-day) calculated from the T21 ice-age experiment. Contour intervall: 25 mm/month, negative anomalies are dashed.

sigma coordinate at low model levels is transformed to pressure coordinate at higher levels yielding surfaces of constant pressure in the stratosphere.

Two T21 experiments were performed over six annual cycles, a present-day control and the ice age response [Lautenschlager and Herterich, 1990]. For the LGM, the lower boundary conditions were obtained from CLIMAP [1981]. The last five model years were used to calculate the climate means and anomalies (ice age minus present). These five year mean anomalies were taken as driving forces for the ocean experiments, including computation of the annual mean freshwater flux anomaly (Fig. 2).

The freshwater flux into the ocean is defined as precipitation minus evaporation plus river run-off.

OGCM experiment

The OGCM used is the large scale geostrophic global model developed at the Max-Planck-Institut für Meteorologie [Maier-Reimer and Hasselmann, 1987]. The model physics is based on the conservation of salt, heat and momentum. Additionally, a thermodynamic sea ice model with simplified rheology is included. The prognostic equation for the vertical component of momentum is replaced by the hydrostatic approximation. The discretization in time is written in a rigorously implicit way (Euler-backward differences). The resulting set of linear equations is solved iteratively for the baroclinic modes and by elimination for the barotropic mode. The formulation almost completely suppresses gravity waves which would require a rather short time step in conventional circulation models with an explicit discretization in time. The model uses a time step of 30 days. The effective horizontal

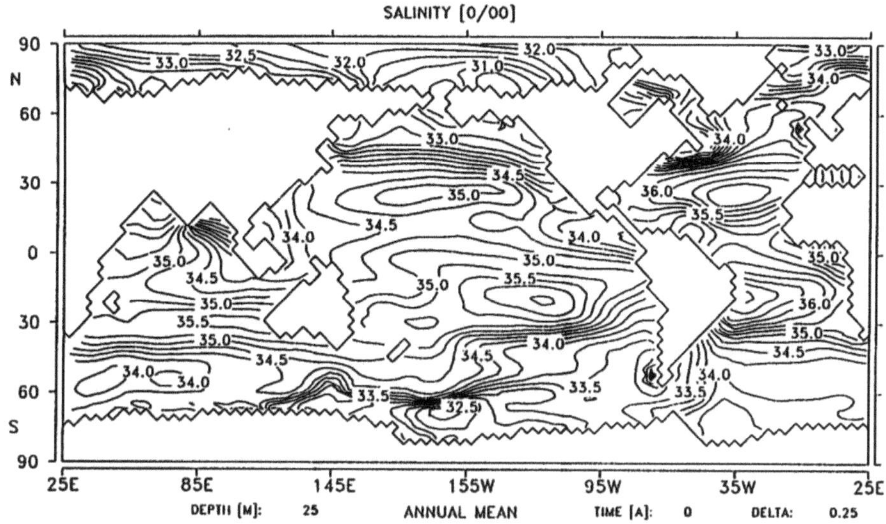

Figure 3: Annual average of surface salinity from modern freshwater fluxes. Contour intervall: 0.25 $^0/_{00}$.

model resolution is 3.5°and the vertical is resolved by 11 layers, six layers above 1000 m depth and five layers below. Details are given in Maier-Reimer et al. [1993].

The OGCM was integrated with present-day boundary conditions until equilibrium was achieved. Then, after 10,000 model years, the salinity boundary condition was changed into freshwater flux which had been diagnostically determined from the salinity boundary condition. Further model integration for 750 years did not show any changes in the present-day equilibrium ocean circulation. Fig.3 shows the resulting surface salinity which stably remains at the previously prescribed Levitus salinity. This ocean circulation was taken as present-day control experiment for comparison to the 18 ka experiments ("ice age runs"). Then the modern boundary conditions were changed to the ocean ice age forcing and the OGCM was integrated for 4000 years until a new equilibrium was achieved. For the SST the CLIMAP [1981] estimates are prescribed directly. The CLIMAP [1981] SST's were choosen as forcing in both models, the AGCM and the OGCM, for two reasons: 1) the two models were kept somewhat at the climate state inferred from the data; 2) it is impossible to integrate the T21-AGCM and the OGCM for some thousand years in the coupled mode. The AGCM anomalies of wind stress and freshwater flux were added to the present-day boundary fields to compile the ocean ice age forcing. All OGCM experiments as well as the AGCM runs were integrated including the annual cycle. An overall surface salinity increase due to glacial sea level reduction does not affect the surface salinity gradients and was neglected in these experiments.

A major point of note in our study is that we force the ocean with freshwater fluxes, computed from the AGCM precipitation–minus–evaporation results, rather than attempt any paleo–estimates of salinity. For the ocean there are up to now only crude approxi-

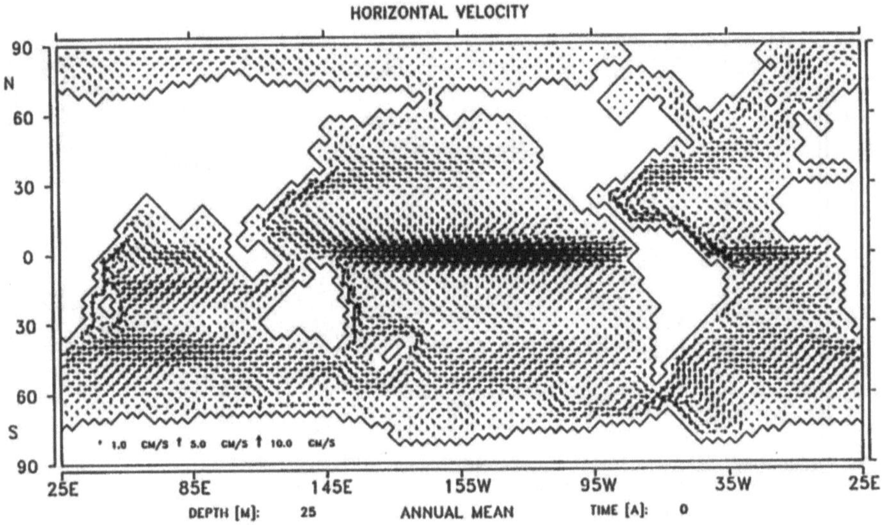

Figure 4: Annual average of surface horizontal velocity for modern boundary conditions.

mations of the contrast in surface salinity of the major oceans at the LGM as compared to the modern ocean. On the other hand, it is widely accepted that even small changes of salinity in the regions of deepwater formation affect crucially the deep circulation. For example, the pronounced difference between the modern Atlantic and Pacific is attributed exclusively to the very different freshwater budgets in the northern parts, characterized by the high precipitation in the northern Pacific [Warren, 1983] which is fed primarily by the evaporation excess in the Atlantic [Broecker et al., 1990]. The modern oceanic conveyor belt is driven primarily by the cooling of salty water in the northern Atlantic [Gordon, 1986]. Even for constant fluxes the Atlantic is believed to have the potential for at least two different modes of operation [Broecker et al., 1985]. This conjecture, which was anticipated from the analysis of an extremely simplified heat-salt oscillator [Stommel, 1962], has been strongly supported by the results of numerical experiments with OGCM's [Bryan, 1986; Manabe and Stouffer, 1988].

Some attempts have been made to derive salinity changes from the oxygen isotope relation in foraminiferal communities [Duplessy et al., 1991]. Unfortunately, this technique fails in cold water where the lack of fractionation during brine formation and the disappearance of calcite producers in cold water disturbes the well established oxygen-salinity correlation of temperate waters. It is for that reason impossible to force an OGCM directly with reconstructed global surface salinity. Even though we cannot utilize this option, we can examine the effects of salinity changes on the thermohaline circulation with our $\delta^{13}C$ and $\Delta^{14}C$ proxies. These will be discussed in the next section.

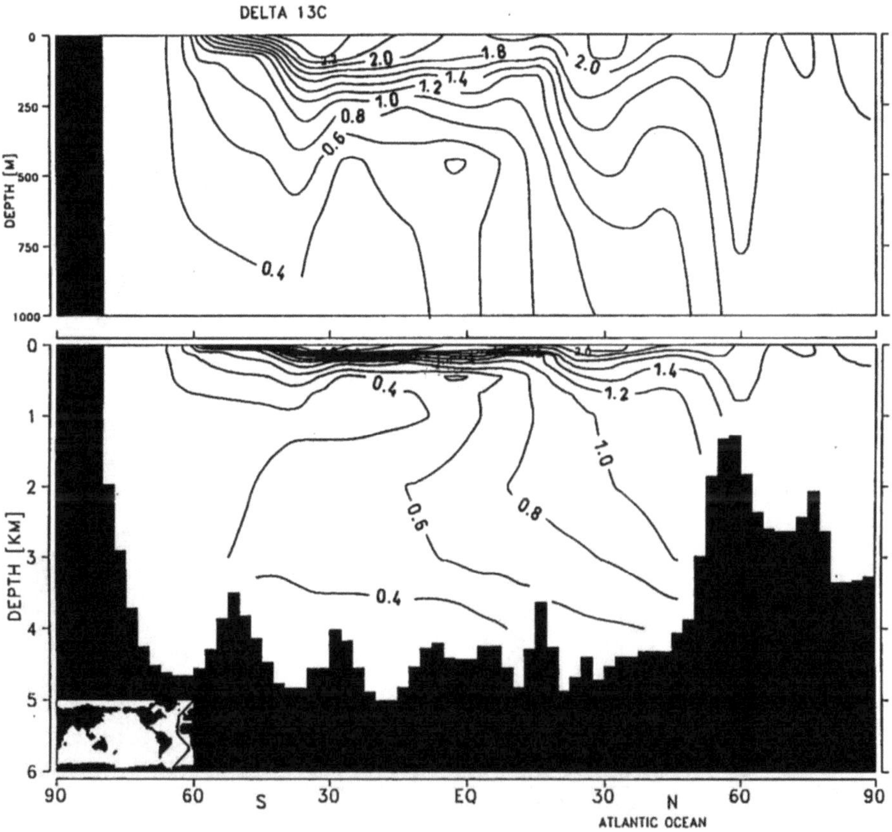

Figure 5: Distribution of $\delta^{13}C$ in the eastern Atlantic for modern boundary conditions. Contour intervall: $0.2^0/_{00}$

The carbon cycle model

The carbon cycle model computes the distribution of $\sum CO_2$ of ^{12}C, ^{13}C and ^{14}C as the result from nutrient limited new production in the given current field. Distinction is made between calcite production and the formation of soft tissue. Sinking particles of organic matter (POC) are remineralized provided there is enough oxygen. The calcite pool interacts with a bioturbated sediment layer. The variables of the model are alkalinity, phosphate as a limiting nutrient, oxygen, dissolved CO_2, POC, and calcite. In all carbon tracers the three isotopes are treated separately. In the atmosphere the CO_2 concentrations are supposed to be zonally mixed with a meridional diffusion of one hemisphere/year. In the gas exchange between ocean and atmosphere, the C^{13} is fractionated according

$$F^{13} = (1.02389 - 9.483/(T + 273.16))F^{12} \sum {}^{13}CO_2/\sum {}^{12}CO_2$$

where T is the local SST and F denotes the sea-to-air fluxes. For the implementation into soft tissue material we assume a constant fractionation of $-20^0/_{00}$. For radiocarbon

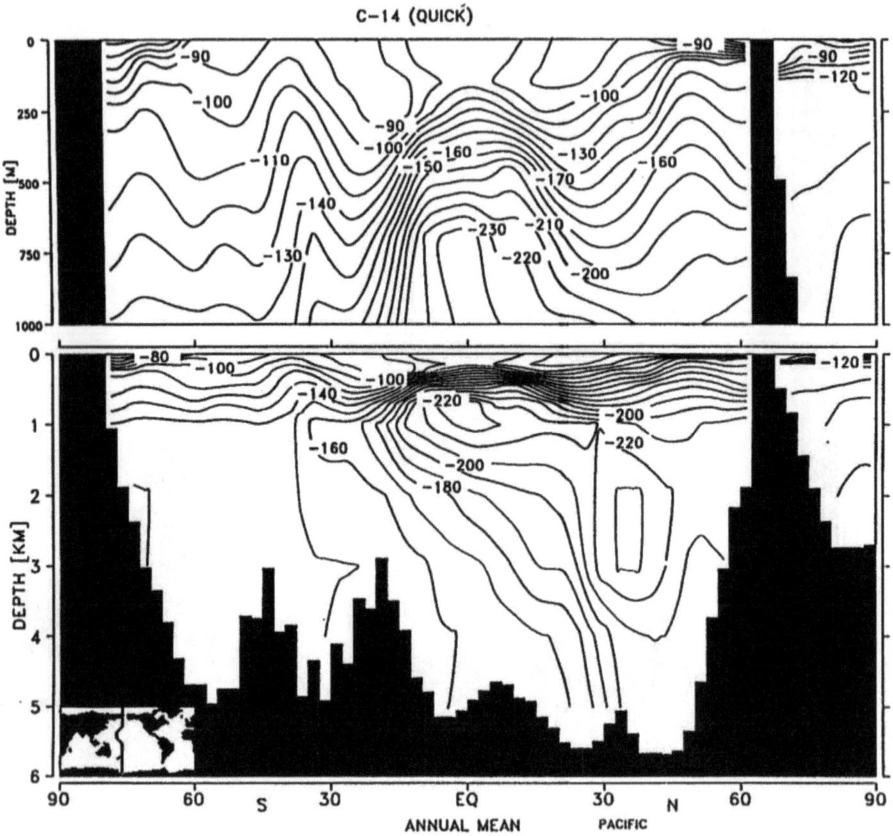

Figure 6: Distribution of $\Delta^{14}C$ in the western Pacific for modern boundary conditions. Contour intervall: $10^0/_{00}$. ^{14}C – age in years: $\Delta t \simeq -8033\ ln(1+\Delta^{14}C/1000)$

we assume the square of these fractionation factors. Details of the model are given in Bacastow and Maier-Reimer [1990] and in Maier-Reimer and Bacastow [1990]. The inventories of the tracers were tuned carefully to produce a stationary preindustrial CO_2 concentration of 275 ppm for the control run. The mixing ratio of the isotopes in dissolved carbon gives an information of organic carbon productivity and, thus, nutrient distribution and of deep ocean circulation. $\delta^{13}C$ reflects the biological side – it is almost perfectly inverse to phosphate [Maier-Reimer and Bacastow, 1990]. $\Delta^{14}C$ after correction by the fractionation effects represents an integral measure of the time elapsed since the water mass last contacted the atmosphere. By comparing planktonic with benthic foraminifera that are found in the same sediment layer, the radiocarbon age difference between surface and bottom gives an information on the turnover time of the ocean, with an uncertainty of approximately 200 years [Lal and Suess, 1983].

The model simulates the modern distribution of $\delta^{13}C$ on a section in the eastern Atlantic, (Fig. 5) comparing well with Kroopnick's [1985] evaluation of the GEOSECS $\delta^{13}C$ data. The model also simulates the present distribution of $\Delta^{14}C$. The section of the western Pacific, (Fig. 6) for instance compares well with the GEOSECS data [Ostlund and

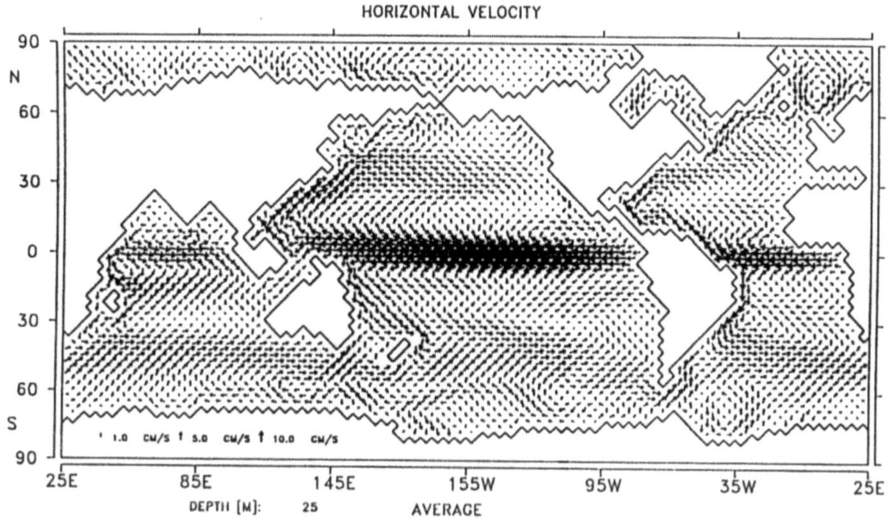

Figure 7: Annual average of surface horizontal velocity. OGCM response to ice-age boundary conditions.

Stuiver, 1980]. The deep Pacific has generally too high $\Delta^{14}C$ by approximately $20^0/_{00}$. We attribute this to a too strong mixing around Antarctica.

Results

Comparison of simulated surface currents for the control run (Fig. and for the ice age response experiment (Fig. 7), show that in the ice age experiment, the North Atlantic polar front migrated from its present position (65°N) to 45°N and the Gulf Stream crossed the Atlantic at 40°N. The Brazil current and the Atlantic equatorial current were intensified during LGM, and the surface currents in the North Pacific and in the Arctic ocean were modified in agreement with glacial topography. The Bering Strait was closed because of a sea level reduction in the range of 100 – 150 m. In the ice age response, the strength of the Kuroshio increased and the strength of the East Australian current decreased compared to the present. The simulated glacial Indian ocean currents showed a more zonal structure compared to the present. The Antarctic circumpolar current did not change noticeably north of the margins of permanent glacial sea ice coverage.

The glacial ocean response shows a slightly increased upwelling in the equatorial East Atlantic and a decreased upwelling in the equatorial East Pacific. This reflects the glacial trade wind response in the AGCM: stronger trades in the East Atlantic and weaker trades in the East Pacific. Upwelling also decreased slightly in the Arabian Sea and in Bay of Bengal, regions where the the glacial summer monsoon circulation was reduced in the AGCM response.

As discussed earlier, salinity greatly affects the thermohaline circulation. The differences

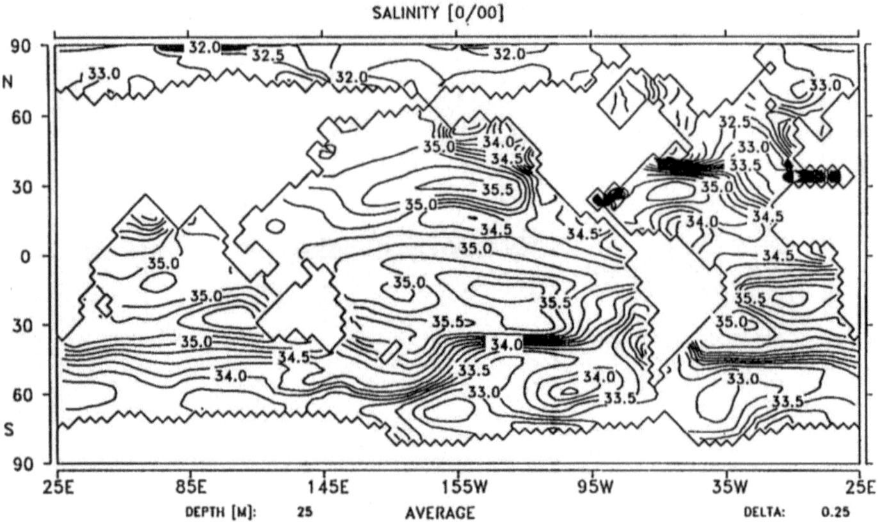

Figure 8: Annual average of surface salinity. OGCM response to ice-age boundary conditions. Contour intervall: 0.25 $^0/_{00}$.

in surface salinities for the control run and for the ice age response (Fig. 8) are directly related to the imposed freshwater anomalies. The larger glacial sea ice thickness around the North Pole decreased the total amount of salinity in the uppermost model layer. The modern salinity contrast between North Atlantic and North Pacific, with higher salinities in the North Atlantic, was reversed in the glacial ocean response. The North Pacific surface water contained around $2^0/_{00}$ more salt than the North Atlantic water (Fig. 10). The modern salinity contrast between the equatorial Atlantic and the equatorial Pacific of $0.5^0/_{00}$ was reduced in the 18 ka simulation. The glacial surface salinity seemed to be slightly increased in the Weddell Sea and slightly decreased in the Ross Sea compared to the control run. The zonally averaged meridional circulation for the Atlantic (Fig. and for the Pacific (Fig. 11, 12) shows that compared to the control experiment, the simulated glacial contributions to deep water production of the Atlantic and of the Pacific were reversed. In the North Atlantic, deepwater production completely stopped in the ice age simulation and in the North Pacific, thermohaline circulation started intensively. (This reversed pattern has a remarkable similarity to the first attempts to simulate the present climate with coupled OAGCM's [Bryan et al., 1975]). Compared to the present, the glacial bottom water was younger in the equatorial Pacific ($\sim$ 300 years, C^{14}-age) and older in the equatorial Atlantic ($\sim$ 300 years, C^{14}-age). The deep sea temperature was reduced by 2 – 3°C in the Atlantic, by 1 – 2°C in the Pacific and by 2°C in the Indian ocean.

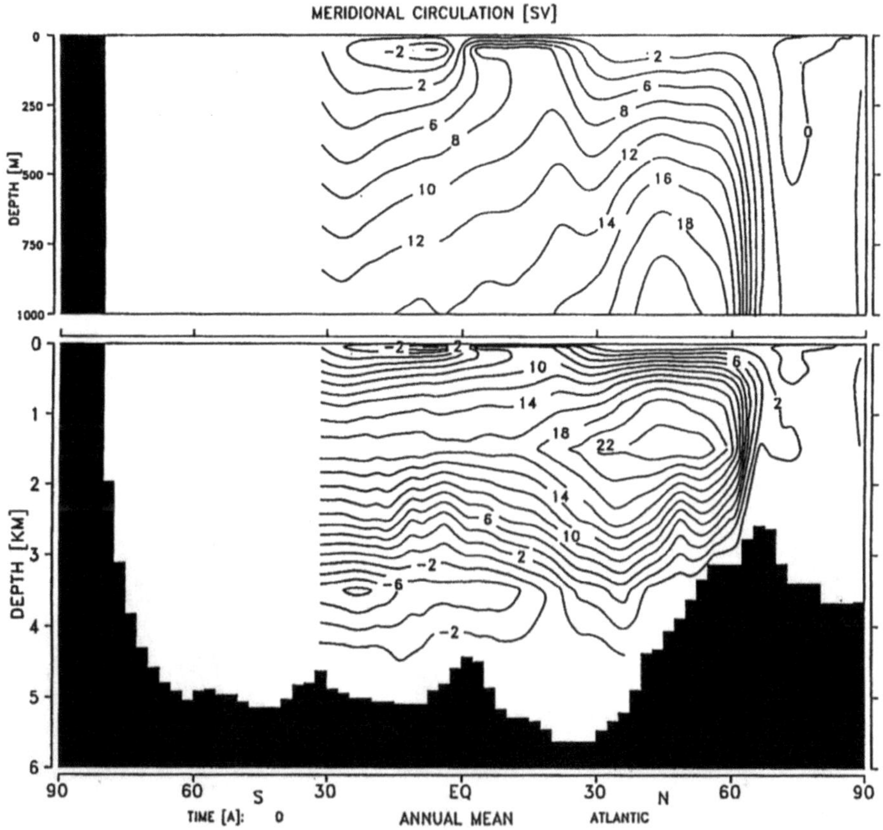

Figure 9: Annual mean of averaged Atlantic meridional circulation for modern boundary conditions. Contour intervall: 2 Sv

Discussion

Ocean model results can be compared with geological evidence obtained from deep sea sediment cores. Comparing the simulations with these evidence there is some (potential) agreement but also severe disagreement obtained.

Model–data agreements

The simulated temperature reduction of 1 – 3°C in the deep ocean fits into the suggested temperature decrease of 1 – 2°C [Chappell and Shackleton, 1986; Labeyrie et al., 1987].

The model agrees with the observations in the following areas. The North Atlantic polar front and Gulf Stream are displaced from their present positions to 45°N and 40°N, respectively [McIntyre et al., 1976]. Surface salinity in the subpolar North Atlantic may have been lower than present [McIntyre et al., 1976]. Upwelling increased in the equatorial Atlantic [Sarnthein et al., 1988].

The model indicates reduced glacial production of North Atlantic deep water, in agree-

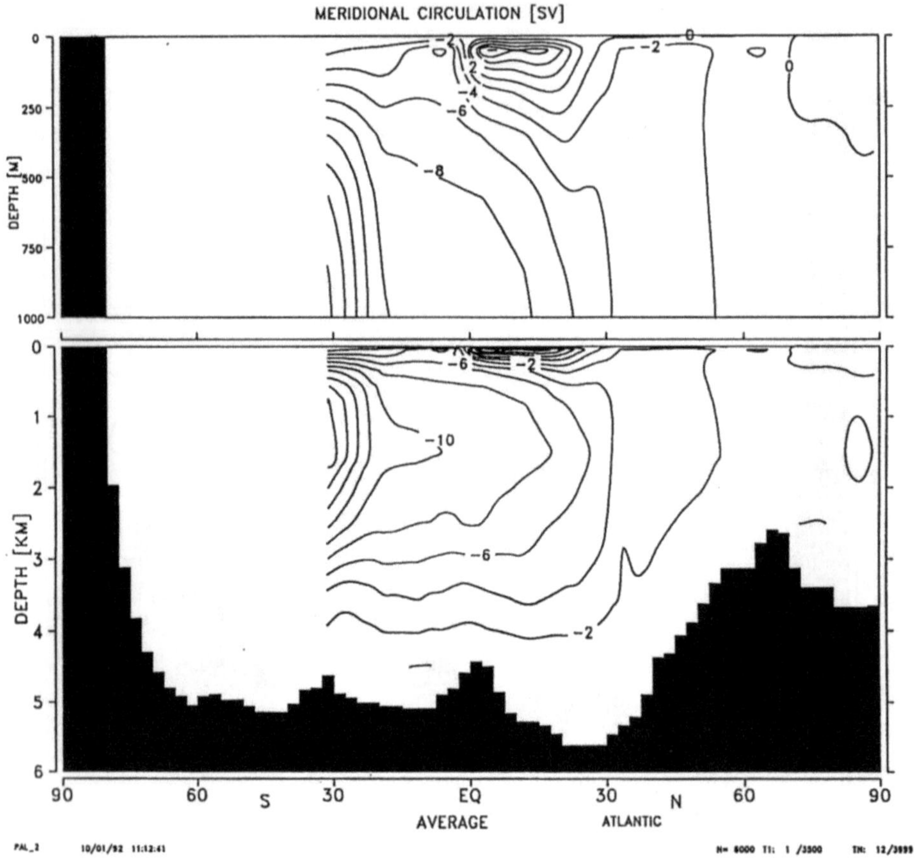

Figure 10: Annual mean of averaged Atlantic meridional circulation. OGCM response to ice-age boundary conditions. Contour intervall: 2 Sv .

ment with Cd/Ca and $\delta^{13}C$ which indicates a reduction of 30 – 50% while the intermediate water was stronger ventilated [Boyle and Keigwin, 1987; Crowley, 1983; Duplessy et al., 1988]. The thermohaline circulation stopped in the Norwegian Sea but seemed to be active in the northeast Atlantic, possibly on the European continental shelf [Duplessy et al., 1980]. The deepwater circulation was dominated by the northward flow of Antarctic bottom water [Curry et al., 1988; Duplessy and Shackleton, 1985]. Apart from an overestimated reduction of thermohaline circulation and the correlated underestimated ventilation of intermediate water the OGCM response fits into the outlined conception of the glacial Atlantic. The model calculated only minor changes in the strength of the Anarctic outflow between the present and the LGM.

Data indicate a glacial reduction of the Indian summer monsoon and of the upwelling in the Arabian Sea. The LGM surface salinity gradients increased in both, the Arabian Sea and the Bay of Bengal [Duplessy, 1982]. Both models, the AGCM and the OGCM, fit in these data. The AGCM simulated the glacial reduction of the summer monsoon circulation and the OGCM computed consistently the glacial upwelling reduction and the surface salinity increase. Mesoscale ocean studies for this area agree with the presented

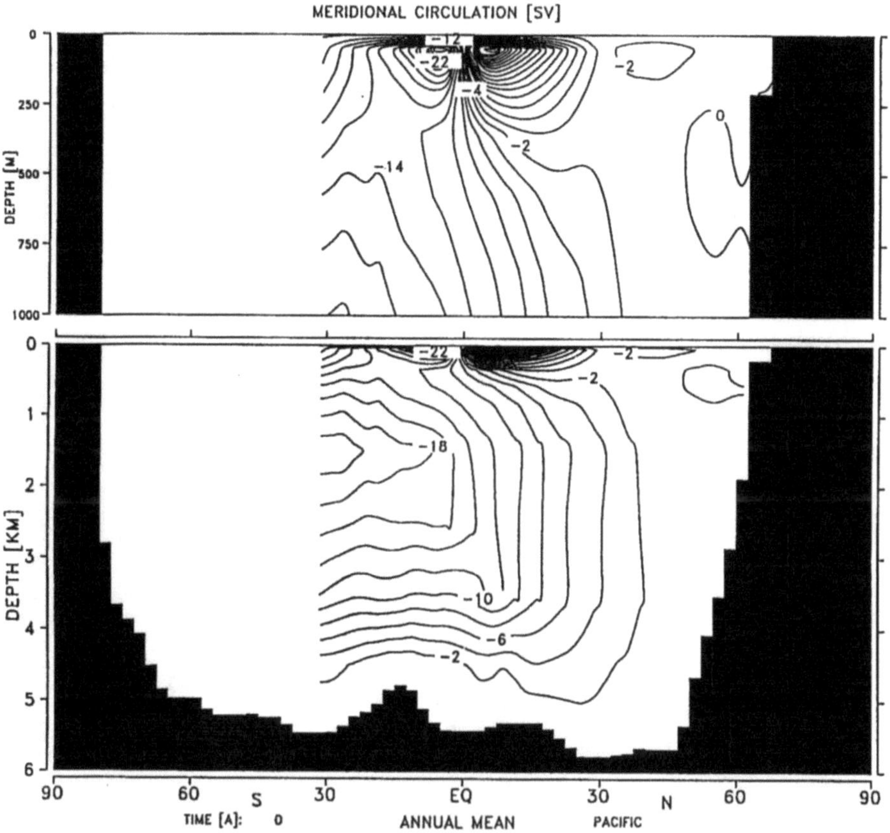

Figure 11: Annual mean of averaged Pacific meridional circulation for modern boundary conditions. Contour intervall: 2 Sv

OGCM results [Luther et al., 1990].

Apart from regional differences between model and data like the missing increased equatorial upwelling at 18 ka in the Pacific the glacial ocean response at the surface is comparable to the geological evidence. We emphasize, however, that this agreement is directly influenced by the prescribed boundary conditions for the AGCM step.

Model-data disagreements

The major descrepancies between data and model appear in the glacial deep ocean response. Carbon isotope and Cd/Ca data indicate that the intermediate waters in the North Pacific and North Atlantic should have been more strongly ventilated at the LGM and the North Atlantic thermohaline circulation should have been substantially reduced [Boyle and Keigwin; 1987; Duplessy et al., 1988]. The glacial source of global deep water should have been from Antarctica with comparable strength as observed today [Oppo and Fairbanks, 1987; Duplessy et al., 1988]. This sketch yields a globally reduced deep water overturn at 18 ka. The meridional stream function in the Atlantic of our glacial run (Fig.

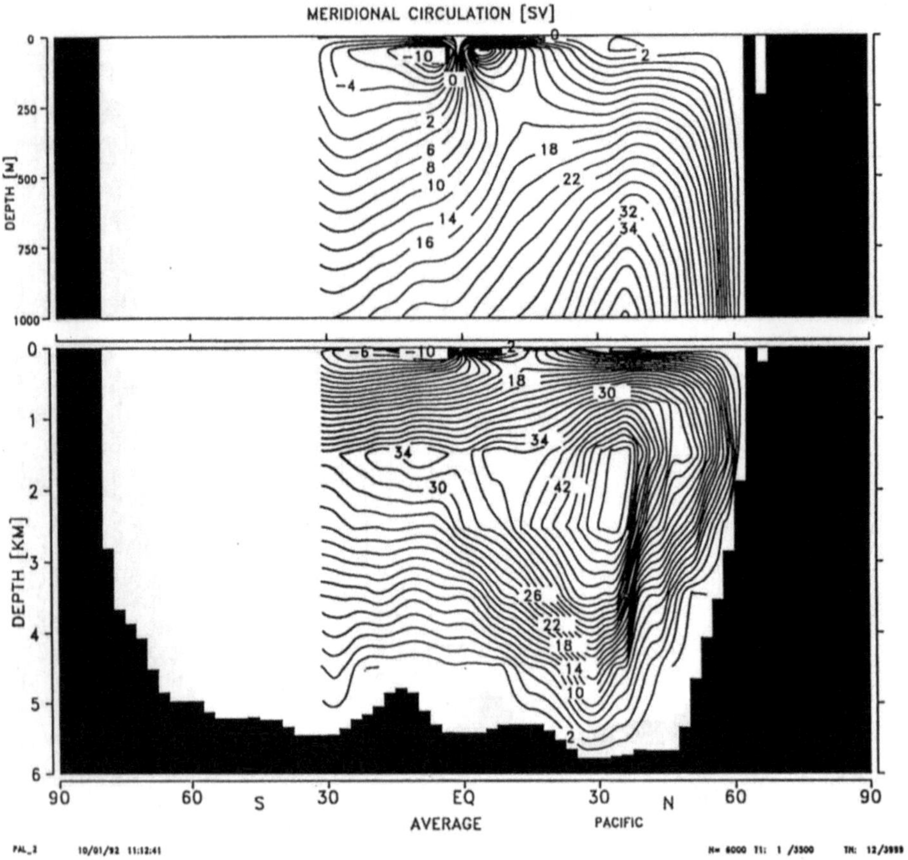

Figure 12: Annual mean of averaged Pacific meridional circulation. OGCM response to ice-age boundary conditions. Contour intervall: 2 Sv

10) seems to be roughly compatible with this outline while the meridional stream function in the Pacific (Fig. 12) shows incompatability.

A comparison of our carbon isotopes simulation based on the LGM response reveal discrepancies with data that are much more pronounced than the foregoing qualitative arguments. Whereas Duplessy et al. [1988] described a $\delta^{13}C$ distribution similar in structure as the modern one, but with a much shallower spreading of the North Atlantic deep water (rather an intermediate water than a deep water), our model produces a pronounced minimum of $\delta^{13}C$ at 500 m depth (Fig. 13).

In the northern Pacific, the situation is similarly inconsistent: the meridional stream function shows a pronounced overturning which would be compatible with the ideas of stronger production of intermediate water; the age difference between pelagic and benthic foraminifera [Broecker et al., 1988; Shackleton et al., 1988], however, is in clear contrast to our LGM response on radiocarbon distribution. The model, in fact, predicts (Fig. 14) younger glacial deep water in the Pacific than at present.

At this stage of model verification we have to state clearly that our model results are wrong.

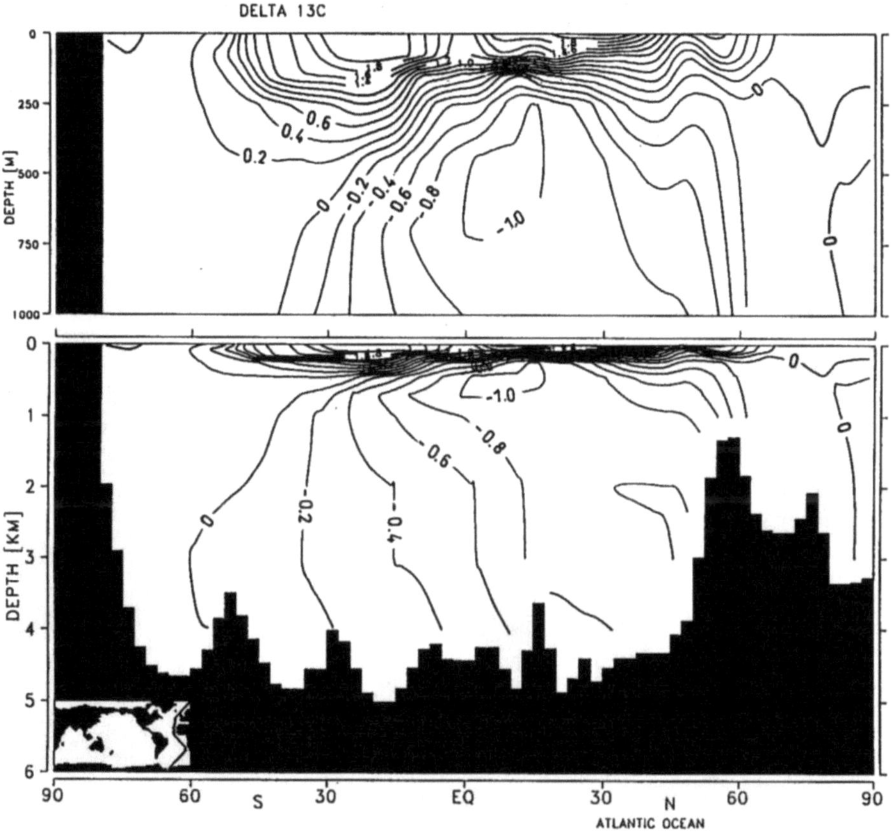

Figure 13: Distribution of $\delta^{13}C$ in the eastern Atlantic for the ice age boundary conditions Contour intervall: $0.2^0/_{00}$

Analysis of model-data disagreements

The freshwater flux taken from the glacial AGCM response were identified as the most probable reason for the "dissimulation". It would be very useful to get more information about glacial ocean surface salinitiy, especially in the North Pacific, to verify the regional T21 LGM response in the freshwater flux. The glacial Aleutian pressure system was simulated 20° farther eastwards than in the present-day control run. Therefore, the simulated freshwater anomaly was positive in the eastern North Pacific and negative in the western part where the glacial deepwater production was simulated by the OGCM. The glacial location of the Aleutian pressure system could not be obtained from geological data in Alaska [Kutzbach and Wright, 1985]. However, the T21 pattern of freshwater flux anomaly are also simulated by other AGCM's using 18 ka CLIMAP boundary conditions [Joussaume, 1989; Miller and Russell, 1990].

The boundary forcing of the AGCM in the North Pacific should be connected with the prescription of the large marine ice extent. A recent re-examination of the primary geological data in that region provided indication that the CLIMAP reconstruction of ice

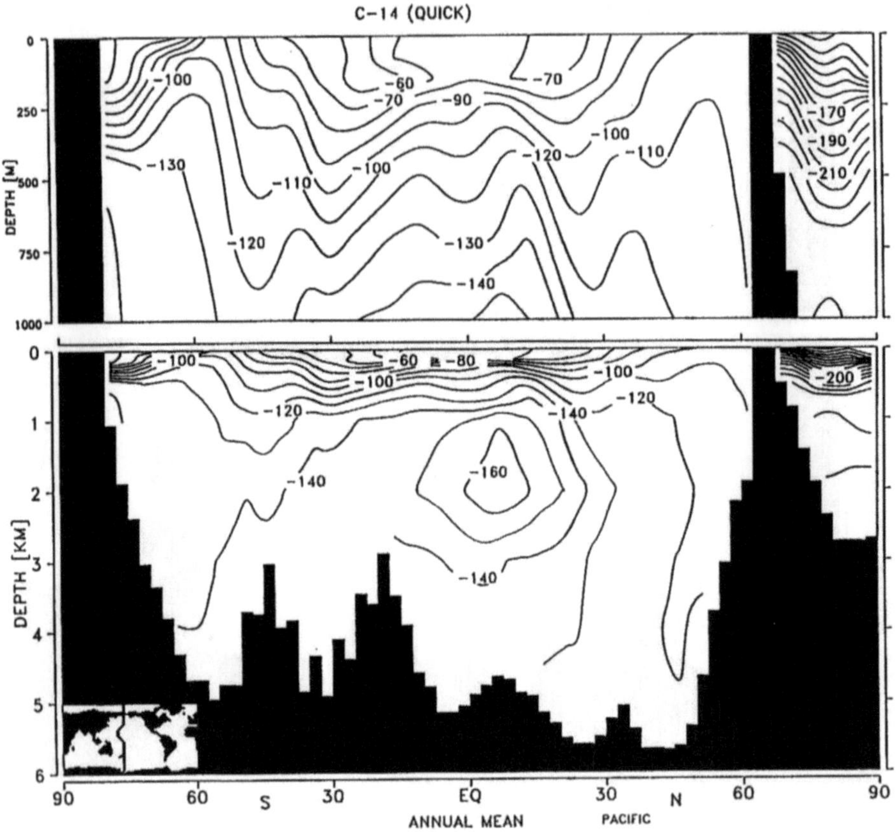

Figure 14: Distribution of $\Delta^{14}C$ in the western Pacific for ice age boundary conditions. Contour intervall: $10^0/_{00}$. ^{14}C – age in years: $\Delta t \simeq -8033\ ln(1 + \Delta^{14}C/1000)$

coverage may have been overestimated [Ruddiman, pers. comm.]. If so, we can expect that with a revised sea ice reconstruction our model results would exhibit a smaller departure from the modern ocean simulation than the LGM response discussed in this paper.

The CLIMAP SST field has also been challenged because of disagreements between AGCM results and observations over tropical land areas [Webster and Streeten, 1978; Rind and Peteet, 1985]. Although initial attempts at independently validating the CLIMAP SST's appear to have been supportive [Prell, 1985; Broecker, 1986], a more recent reexamination of the transfer functions in the Pacific does suggest that the SST field may need revision [Chen et al., 1991].

Thus, the critical dependency of climate model results to the CLIMAP SST's is not just based on atmospheric models and tropical land areas, where in some cases the observations themselves are sparse. Our results indicate that, when ocean models are considered, model-data discrepancy may also extend to the validation of results with deep-sea sediments. Although more work is clearly required on OGCM studies of the ice age circulation, our results suggest that a satisfactory reconcilation with the sediment record may require not only a reevaluation of CLIMAP SST reconstruction, but also of the glacial ice sheets.

Conclusions

The atmosphere is the most sensible part of the climate system, but it represents only one component. Carbon dioxide is now widely accepted to act not only as a forcing but also as a dynamic component of the climate system. For the analysis of past climate states, it represents the most important component since it is subject to direct measurements. We have demonstrated a strong caveat for any effort of climate modelling: For AGCM's, the precipitation fields belong to the most uncertain aspects of the results; for the Ocean, however, they represent a crucial contribution to the forcing of the thermohaline circulation.

Acknowledgements

We appreciated helpful discussions especially with T. J. Crowley, but also with W. S. Broecker, J. C. Duplessy, K. Hasselmann, K. Herterich, and M. Sarnthein. The computer graphics are carefully prepared for publication by H. Höck and M. Grunert. This Project was supported by the Bundesministerium für Forschung und Technologie.

References

Bacastow, R. and E. Maier-Reimer, 1990: Ocean - circulation model of the carbon cycle. Climate Dynamics **4**, 95-126.

Boyle, E. A. and L. D. Keigwin, 1987: North Atlantic thermohaline circulation during the last 20 000 years: Link to high latitude surface temperature. Nature **330**, 35-40.

Broccoli, A. J. and S. Manabe, 1987: The influence of continental ice, atmopsheric CO_2, and land albedo on the climate of the last glacial maximum. Climate Dynamics **1**, 87-99.

Broecker, W. S., T.-H. .Peng, J. Jouzel and G. Russel, 1990: The magnitude of global fresh-water transports of importance to ocean circulation. Climate Dynamics, **4**, 73-79.

Broecker, W. S., M. Andree, G. Bonani, W. Wolfli, H. Oeschger, M. Klas, A. Mix and W. Curry, 1988: Preliminary estimates for the radiocarbon age of deep water in the glacial ocean. Paleocean. **3**, 659-669.

Broecker, W. S., 1986: Oxygen isotope constraints on surface ocean temperatures. Quat. Res. **26**, 121-134.

Broecker, W. S., D. M. Peteet and D. Rind, 1985: Does the ocean-atmosphere system have more than one stable mode of operation? Nature **315**, 21-26.

Bryan, F., 1986: High-latitude salinity effects and interhemispheric thermohaline circulations. Nature **323**, 301-304.

Bryan, K., S. Manabe, and R. C. Pacanowski, 1975: A global ocean-atmosphere climate model Part II: The oceanic circulation. J. Phys. Oceanogr. **5**, 30-46

Chappell, J. and N. J. Shackleton, 1986: Oxygen isotopes and sea level. Nature **324**, 137-140.

Chen, M.-T., J. Imbrie and W. L. Prell, 1991: Estimating sea-surface temperatures from tropical Pacific planktonic foraminifers: Comparison of Imbrie-Kipp method and modern analog technique. EOS, Transactions, AGU, Spring Meeting 1991, Vol.72, No.17, p.157.

CLIMAP Project Members, 1981: Seasonal reconstruction of the earth's surface at the last glacial maximum. Geol. Soc. Amer. Map Chart. Ser., MC-36.

Crowley, T. J., 1983: Calcium-carbonate preservation patterns in the central North Atlantic during the last 150,000 years. Mar. Micropaleontol. **6**, 97-129.

Curry, W. B., J. C. Duplessy, L. D. Labeyrie and N. J. Shackleton, 1988: Changes in the distribution of δ^{13}C of deep water $\sum CO_2$ between the last glaciation and the Holocene. Paleocean. **3**, 317-341.

Dümenil, L. and U. Schlese, 1987: Description of the general circulation model. From: Climate simulations with the T21 - model in Hamburg, Ed.: G. Fischer, Meteor. Inst. Large Scale Atmosph. Model. Rep. No. 1.

Duplessy, J. C., L. Labeyrie, A. Jullet-Leclerc, F. Maitre, J. Duprat and M. Sarnthein, 1991: Surface salinity reconstruction of the North Atlantic Ocean during the last glacial maximum. Oceanologica Acta **14**, 311-324.

Duplessy, J. C., N. J. Shackleton, R. G. Fairbanks, L. Labeyrie, D. Oppo and N. Kallel, 1988: Deepwater source variations during the last climate cycle and their impact on on the global deepwater circulation. Paleocean. **3**, 343-360.

Duplessy, J. C. and N. J. Shackleton, 1985: Response of global deep-water circulation to Earth's climatic change 135 000 – 107 000 years ago. Nature **316**, 500-507.

Duplessy, J. C., 1982: Glacial to interglacial contrasts in the northern Indian Ocean. Nature **295**, 494-498.

Duplessy, J. C., J. Moyes and C. Pujol, 1980: Deep water formation in the North Atlantic Ocean during the last ice age. Nature **286**, 476-482.

Gordon, A. L., 1986: Interocean Exchange of Thermocline Water. J. Geoph. Res. **91,C4**, 5037-5046.

Joussaume, S., 1989: Simulations du climat du dernier maximum glaciire à l'aide d'un modèle de circulation générale de l'atmosphère incluant une modélisation du cycle des isotopes de l'eau et des poussières d'origine désertique. Thèse de Doctorat d'Etat, 507p., Univ. of Paris VI, France.

Keffer, T., D. G. Martinson and B. H. Corliss, 1988: The position of the Gulf Stream during Quaternary glaciations. Science **241**, 440-442.

Kroopnick, P. M., 1985: The distribution of ^{13}C of $\sum CO_2$ in the World Oceans. Deep Sea Research **32**, 57-84

Kutzbach, J. E. and P. J. Guetter, 1986: The influence of changing orbital parameters and surface boundary conditions on climate simulations for the past 18000 years. J. Atmos. Sci., **43**, 1726-1759.

Kutzbach, J. E. and H. E. Wright, 1985: Simulation of the climate 18 000 years BP: Results for the North American / North Atlantic / European sector and comparison with the geological record of North America. Quatern. Sci. Rev. **4**, 147-187.

Labeyrie, L. D., J. C. Duplessy and P. L. Blanc, 1987: Variations in mode of formation and temperature of oceanic deep waters over the the past 125 000 years. Nature **327**, 477-482.

Lal, D. and H. E. Suess, 1983: Some comments on the exchange of CO_2 across the air-sea interface. Journ. Geophys. Res. **88**, 3643-3646.

Lautenschlager, M., U. Mikolajewicz, E. Maier-Reimer and C. Heinze, 1992: Application of Ocean Models for the Interpretation of Atmospheric General Circulation Model Experiments on the Climate of the Last Glacial Maximum. Paleoceanography **7**, 769-782.

Lautenschlager, M. and K. Herterich, 1990: Atmospheric Response to Ice-Age Conditions – Climatology near the Earth's Surface. Journ. Geophys. Res. **95**, 22547-22557.

Louis, J.-F., 1984: ECMWF forecast model: physical parameterization. ECMWF Research Department, Research manual 3.

Luther, M. E., J. J. O'Brien and W. L. Prell, 1990: Variability in Upwelling Fields in the Northwestern Indian Ocean, 1. Model Experiments for the Past 18,000 Years. Paleoceanography **5**, 433-445.

Maier-Reimer, E., U. Mikolajewicz and K. Hasselmann, 1993: Mean Circulation of the Hamburg LSG OGCM and its Sensitivity to the Thermohaline Surface Forcing. J.Phys. Oceanogr. **23**, 731 - 757.

Maier-Reimer, E. and R. Bacastow, 1990: Modelling of Geochemical Tracers in the Ocean .In: M.Schlesinger (ed.) Climate-Ocean Interaction, Kluwer press, Dordrecht, pp.233-267.

Maier-Reimer, E. and K. Hasselmann, 1987: Transport and storage of CO_2 in the ocean - an inorganic ocean-circulation carbon cycle model. Climate Dynamics **2**, 63-90.

Manabe, S. and R. J. Stouffer, 1988: Two stable equilibria of a coupled ocean-atmosphere model. Journ. Climate **1**, 841-866.

McIntyre, A., N. G. Kipp, A. W. H. Bé, T. Crowley, T. Kellogg, J. V. Gardner, W. L. Prell and W. F. Ruddiman, 1976: Glacial North Atlantic 18 000 years ago: A CLIMAP reconstruction. Geol. Soc. Amer. Mem. **145**, 43-76.

Miller, J. R. and G. L. Russell, 1990: Oceanic freshwater transport during the last glacial maximum. Paleoceanography **5**, 397-407.

Miller, J. R. and G. L. Russell, 1989: Ocean heat transport during the last glaciation. Paleocean. **4**, 141-155.

Oppo, D. W. and R. G. Fairbanks, 1987: Variability in the deep and intermediate water circulation of the Atlantic during the last 25 000 years: Northern Hemisphere modulation of the Southern Ocean. Earth Planet. Sci. Lett. **86**, 1-15.

Östlund, H. G. and M. Stuiver, 1980: GEOSECS Pacific radiocarbon. Radiocarbon **22**, 25-33.

Prell, W. L., 1985: The stability of low-latitude sea-surface temperatures: An evaluation of the CLIMAP reconstruction with emphasis on the positive SST anomalies. U.S. Dept. Energy Cont. Rep., Cont. No. DE-ACO2-83ER60167.

Rind, D., 1987: Components of the Ice Age circulation. J. Geophys. Res. **92, D4**, 4241 - 4281.

Rind, D. and D. Peteet, 1985: Terrestrial conditions at the last glacial maximum and CLIMAP sea-surface temperature estimates: Are they consistent? Quat. Res. **24**, 1-22.

Sarnthein, M., K. Winn, J. C. Duplessy and M. R. Fontugne, 1988: Global variations of surface ocean productivity in low and mid latitudes: influence on CO_2 reservoirs of the deep ocean and atmosphere during the last 21 000 years. Paleocean. **3**, 361-399.

Sausen, R., K. Barthel and K. Hasselmann, 1988: Coupled ocean-atmosphere models with flux correction. Climate Dynamics **2**, 145-163.

Shackleton, N. J., J. C. Duplessy, M. Arnold, P. Maurice, M. A. Hall and J. Cartlidge, 1988: Radiocarbon age of the last glacial Pacific deep water. Nature **335**, 708-711.

Stommel, H., 1962: Thermohaline Convection with two stable regimes of flow. Tellus **13** 224-230

Warren, B. A. 1983: Why is no deep water formed in the North Pacific? J.Mar. Res. **41** 327 - 347

Webster, P. N. and N. Streeten, 1978: Late Quaternary ice age climates of tropical Australia, interpretation and reconstruction. Quat. Res. **10**, 279-309.

MID-LATITUDE DEPRESSIONS DURING THE LAST ICE-AGE

Paul J. Valdes and Nicholas M. J. Hall
Department of Meteorology,
University of Reading,
Reading RG6 2AU, U.K.

Abstract

A GCM is used to investigate the distribution, strength and characteristics of mid-latitude depressions during the last glacial maximum (LGM) and to understand their role in the formation and maintenance of continental ice sheets. The model results show that during the northern hemisphere winter, there is an increase in the storm track activity. This results in significant changes in rainfall, and accumulation on the edge of the ice sheet. However in summer, there is net melting at the ice sheet edges, but accumulation at high altitude.

1 Introduction

Many simple models of ice sheet growth (see for example Suarez and Held, 1979, Schneider and Thompson, 1979, Pollard, 1984, Peltier and Hyde, 1984 and more recently Gallee *et al*, 1991 and Satlzman and Verbitsky, 1992) are couched in terms of zonal or sectoral means, with crude representations of ice accumulation, typically only depending on the height of the ice sheet. This allows long integrations and facilitates simple diagnosis. However, such models are flawed to the extent that they represent crudely, if at all, changes in the atmospheric circulation and the supply of moisture. Factors such as this could be of pivotal importance to the maintenance or onset of a glaciated

NATO ASI Series, Vol. I 22
Long-Term Climatic Variations
Edited by J.-C. Duplessy and M.-T. Spyridakis

state. The large, continental ice sheets estimated by the CLIMAP project, for example, would have a profound impact on the circulation, and potentially, on the distribution of mid-latitude depressions carried therein. The simpler models cannot easily represent the effects of mid latitude depressions. To study these issues requires the use of a general circulation model (GCM) but these can only be used for 'snapshot' simulations (Rind, 1987, Kutzbach and Guetter, 1986, Manabe and Broccoli, 1985, Broccoli and Manabe, 1987, Lautenschlager and Herterich, 1990, Joussaume, 1992). Studies such as these have largely been concerned with the last glacial maximum and have generally used the CLIMAP boundary conditions (1981).

A fundamental but unresolved question is whether the growth of the leading edge of an ice sheet occurs at the expense of higher latitude ice. Do the rain or snow bearing depressions track across the ice or do they stay near its edge ? Recently Shinn and Barron (1989) modelled the climate sensitivity to ice sheet size and configuration. Some considerable uncertainty remains in both the extent and height of the northern hemisphere ice sheets and they considered two possible extremes. The results showed that in the maximum ice sheet case, the Atlantic storm track shifted north and extended further into the Arctic than for the minimum case. Thus this result would suggest that the ice sheets were not starved of precipitation. However the GCM had a relatively low resolution, R15, and was run for perpetual January mode only. The equivalent experiment for July is of interest as the presence of extensive ice sheets suggests possibly large temperature gradients, even in the northern hemisphere summer.

In this paper, we report on simulations of both northern hemisphere winter (February) and summer (August). We will particularly concentrate on the role of mid-latitude depressions in transporting moisture onto the northern hemisphere ice sheets. The model will be briefly described in section 2, and the main results for the two sets of simulations will be described in sections 3 and 4. The paper will conclude with a short discussion and description of planned future work.

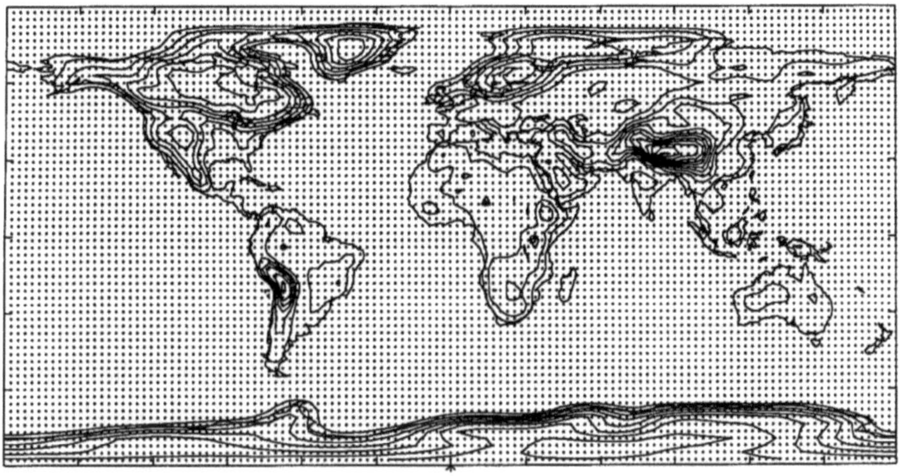

Figure 1: The orography and land sea mask taken from the CLIMAP data set as used in the LGM simulation. In addition, the grid points on which physical and non-linear processes are evaluated are shown.

2 Model Description

The GCM is a version of that used by the UK Universities Global Atmospheric Modelling Programme (UGAMP), and is based on the forecast model of the European Centre for Medium Range Weather Forecasting (ECMWF). The version we are using is nearly identical to that currently being used in the Atmospheric Model Intercomparison experiment (AMIP).

It is a spectral model, using a triangular truncation at total wavenumber 42. The physical parameterisations are evaluated on a longitude/latitude grid of 128 x 64 points. However it should be noted that the equivalent resolution is somewhat less than this grid indicates. The model has 19 levels in the vertical, 5 of which are within 150 mb of the surface (*i.e.* in the boundary layer). Figure 1 shows the land/sea mask used in the model, together with the model representation of the CLIMAP elevation for the LGM. It also shows the grid spacing.

The model includes a relatively standard set of parameterisations. The radiation scheme is that of Morcrette (1990), and includes a predictive cloud scheme based on relative humidity criteria (Slingo, 1987). A parameterization of the effect on the large scale flow of small scale atmospheric gravity waves generated by sub-grid scale orography is also included. This is referred to as gravity wave drag (Palmer *et al*, 1986). The vertical advection uses a total variance diminishing scheme (Thuburn, 1993). This scheme improves the representation of vertical moisture transports, but has the disadvantage of being relatively dissipative. An important modification included in this version of the model is that the convection scheme is based on the work of Betts and Miller (1986) and Slingo and Blackburn (1992). This scheme has resulted in a substantial improvement of the mean tropical circulation, compared to the results using the original Kuo convection scheme. It has also improved the variability of the model, particularly the temporal characteristics of tropical convection, and mid-latitude low frequency disturbances (such as blocking highs). The land surface parameterization prescribes surface albedo and roughness length, and uses a three layer diffusive soil model to calculate surface temperature and moisture. There is a no-flux boundary condition at the bottom of the soil model (approximately 6m thick). The model sea surface temperatures are prescribed and are based on CLIMAP reconstructions, both for present day and the LGM. The position of the sea ice limits are given but no temperatures over the sea ice itself. To calculate this, we use a simple sea ice model identical to that in Slingo *et al* (1985). This is a purely thermodynamic model in which heat is stored by and diffused across a slab of ice which is two metres thick.

So far we have performed four simulations, all of which use perpetual solar forcing (*i.e.* no annual cycle). The runs are:

(I) Perpetual February forcing, present day orbital parameters and boundary conditions for sea surface temperature, sea ice extent, surface albedo, surface elevation, (as given by CLIMAP). Carbon dioxide concentrations were set to 270 ppm.

(II) Perpetual February forcing and LGM (21000 years BP) orbital parameters and boundary conditions (as given by CLIMAP). Carbon dioxide concentrations were set to 190 ppm.

(III) and (IV) As in (I) and (II) except for perpetual August forcing with August boundary conditions.

It is worth noting that the CLIMAP reconstructions include those for present day, as well as the LGM. However these boundary conditions are significantly different to those normally used in the UGAMP model. The surface albedo differs by up to 0.1. In addition, the surface elevation is a mean value over a 2×2 grid. This treatment is different to the standard UGAMP model, which uses 'envelope' orography, in which the mean heights of the mountains are increased by a value proportional to the variance of the sub-grid scale orography to represent the blocking effects of high mountains.

Each experiment has been run for 432 days, and we will present average results for the last 360 days of the integrations.

3 Perpetual February Experiments

Figure 2 shows the model simulation of mean surface air temperature (at a height of 2m above surface) for the perpetual February simulations for (a) Present day and (b) LGM conditions. The 0^0C isotherm approximately shows the limit of the sea ice and is prescribed from the CLIMAP data set. Considering the simplicity of the sea ice model and the lack of a seasonal cycle, the present day surface temperatures are remarkably good. This is particularly true for the Arctic region. Observations of January surface temperatures (Schutz and Gates, 1971), show Arctic regions dropping to -25 to -35 C, very similar to the model simulations. Similarly, land temperatures over North America and Eurasia are close to observations. southern hemisphere temperatures are not quite so realistic. The model has Antarctic temperatures dropping as low as -50^0C, whereas observations suggest temperatures nearer -35^0C. However, this discrepancy occurs only in areas of high orography. In regions of low elevation, the model is closer to observations.

The simulation for the LGM shows enhanced temperature gradients in the northern hemisphere, particularly over the northern Atlantic. In these regions, the temperatures have dropped by up to 40^0C. This is comparable to

(a)

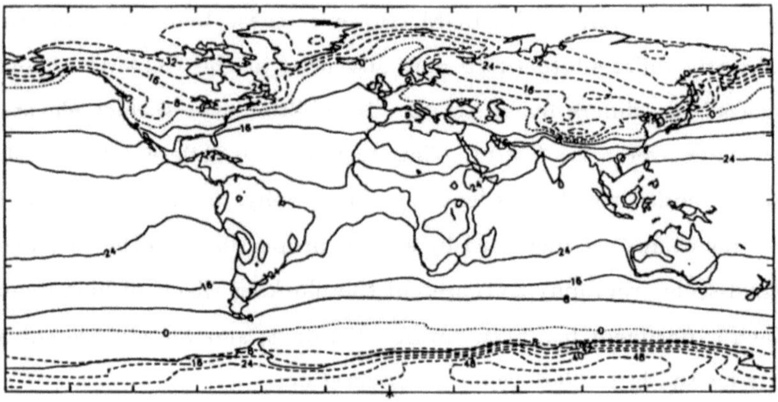

(b)

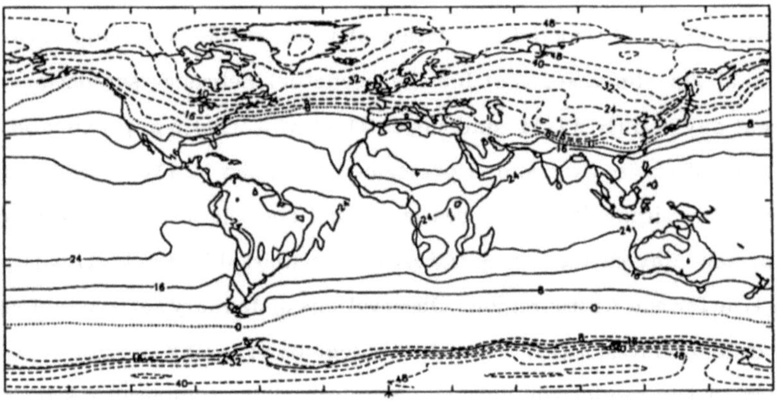

Figure 2: Mean surface temperature for the perpetual February model simulations using (a) present day boundary conditions, and (b) LGM boundary conditions. The contour interval is 8°C and negative values are dashed. The sea surface temperature and sea ice limit are specified from CLIMAP data.

previous studies (*e.g.* Joussaume, 1992). Temperatures decrease rapidly with distance poleward of the ice edge and the enhanced temperature gradient extends across the entire ocean. The gradient is also strong over North America. This appears, in part, to be a product of the high elevations of the Laurentide ice sheet which cools to below -56 C. Similar low temperatures are found over the Greenland and Scandinavian ice sheets.

In the southern hemisphere, CLIMAP data shows the ice edge to be considerably further north at the LGM than for the present day. However, the model simulations show a very flat temperature gradient between the ice edge and approximately 65-70^{0}S. This is caused by a mismatch between the ice model and the boundary conditions. The ice model does not allow the ice to melt. The surface is only allowed to warm to -2^{0}C, after which it is artificially constrained to remain constant. Thus the flat gradient is showing that the ice extent given by CLIMAP is too large for consistency with the model. If the simulated ice edge were free to move, then the model would considerably contract the area of sea ice in the high southern latitudes. The reasons for this discrepancy are difficult to ascertain. A possible candidate is that the perpetual solar forcing is strongly affecting the simulation. It is reasonable to expect that if a seasonal cycle were included, the mismatch may not be as extreme. However it should be stated that, using a similar argument, the northern hemisphere may warm. Another possible cause of the discrepancy is the lack of any dynamics within the sea ice model. A number of studies have suggested that, for present day conditions, sea ice dynamics is important in the Antarctic, and does indeed have the effect of extending the sea ice margin.

These changes are important for the mid-latitude depressions. The altered temperature gradients in the northern hemisphere result in modified activity in the storm track regions. Figure 3 shows the transient eddy kinetic energy per unit mass at 250 mb for (a) present day and (b) LGM conditions. This has been temporally filtered to include only the relatively fast moving mid-latitude depressions with time scales for growth and decay of the order of six days or less. using the filter described in Hoskins *et al*, 1989. The maxima over

(a)

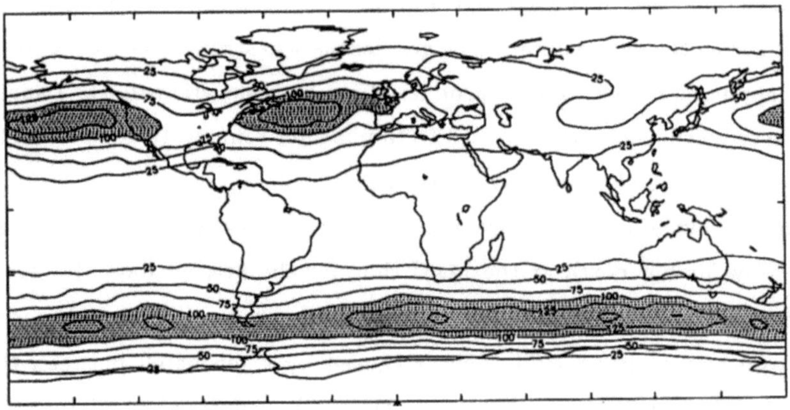

(b)

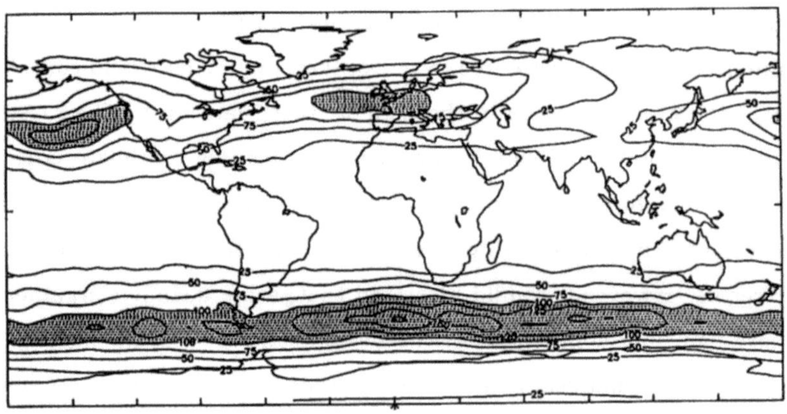

Figure 3: Mean high pass eddy kinetic energy per unit mass at 250 mb for the perpetual February model simulations for (a) present day and (b) LGM. The contour interval is 25 m^2s^{-2}. Values greater than 100 m^2s^{-2} are stippled.

the Pacific, Atlantic, and southern hemisphere oceans are commonly used as locators of the 'storm tracks'.

The present day simulation is in reasonable agreement with observations (Hoskins *et al.*, 1989). The peaks in the Pacific and Atlantic are similar to observations, but both storm tracks are further south than observed, by about 5^0. The eastern end of the Atlantic storm track decreases more rapidly than observed. This latter feature is a common problem in many climate models (Boer *et al*, 1992). The intensity of the southern hemisphere storm track is also in good agreement with observations with the peak in the southern Indian Ocean, but again the model storm tracks are 5^0 too far south.

The two northern hemisphere storm tracks show very significant changes in the LGM simulation. In the Pacific, the largest change occurs at the end of the storm track. The peak intensity is unaltered, but there is an equatorward shift of about 5^0. This is consistent with a similar equatorward shift in the ice edge, and hence in the maximum surface temperature gradient. Thus the Pacific storm track appears to be following the edge of the sea ice.

The Atlantic storm track is more complicated. For the LGM simulation, the region of maximum eddy kinetic energy is considerably more confined meridionally, but extends much further into Europe. Thus over the Atlantic, the eddy kinetic energy is considerably reduced, whereas over Europe there is a substantial increase. This increase extends well into the continent, as far as 90^0E. These changes are generally consistent with the changed surface temperature gradient. The temperature gradients at the edge of the sea ice result in mid-latitude depressions closely following the edge of the ice. The sharpness of the storm track is a result of the relative efficiency with which the depressions can transport heat, and the relatively narrow meridional band over which heat needs to be transferrred with such strong temperature gradients.

The changes in the storm tracks for the LGM are relatively modest in the southern hemisphere. The zonally averaged kinetic energy increases by approximately 10 percent. The maximum increase occurs between 60^0W and 90^0E and appears to be associated with increased advection of warm, tropical air from the Atlantic. James and Anderson (1983) have shown that, for present day climate, the position of the southern hemisphere storm track is strongly linked to the supply of moisture from the tropical Atlantic.

Thus changes in both the Pacific and Atlantic storm tracks are substantial, and closely associated with changes in temperature gradients. Diabatic processes (in particular release of latent heat) may also be playing a role.

The substantial modifications to the storm tracks give rise to changes in the precipitation distribution. As in most GCM's, two types of precipitation are parameterized, namely that associated with convection, and that associated with condensation in regions of large scale ascent. The former process is closely related to sea surface temperature, the latter with precipitation from mid-latitude depressions. Figure 4 shows the model precipitation for (a) Large scale ascent, present day; (b) Convection, present day; (c) Large scale ascent, LGM and (d) Convection, LGM. In the northern hemisphere storm track regions, the LGM simulation shows considerably reduced convective precipitation but enhanced large scale precipitation. Associated with the Atlantic storm track, there is effectively no convective rainfall over the sea ice, but a considerable increase in the large scale precipitation extending into southern Europe. For the Pacific storm track, there is greatly enhanced orographic rainfall over the northern Rockies and an increase of convective rainfall over the ocean. The changes are broadly consistent with the changes in eddy activity and surface temperature.

Finally we show the predicted net accumulation of snow in the last 270 days of the LGM integrations, Figure 5 shows this calculation for the northern hemisphere ice sheets. The units are metres of water equivalent (*i.e.* the amount of water in the form of ice). The GCM does not have a sophisticated treatment of snow and ice (for instance, it is assumed that snow remains uncompacted, even if it is very thick) and so the net accumulation should be treated with some considerable caution. In addition, the results will be very sensitive to the seasonal cycle. However hopefully it will give a robust indication of regions of accumulation and melting, particularly in regions where the changes are large.

(a)

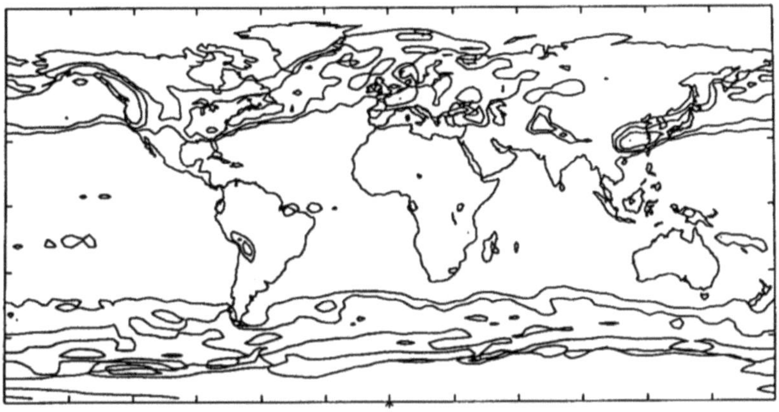

(b)

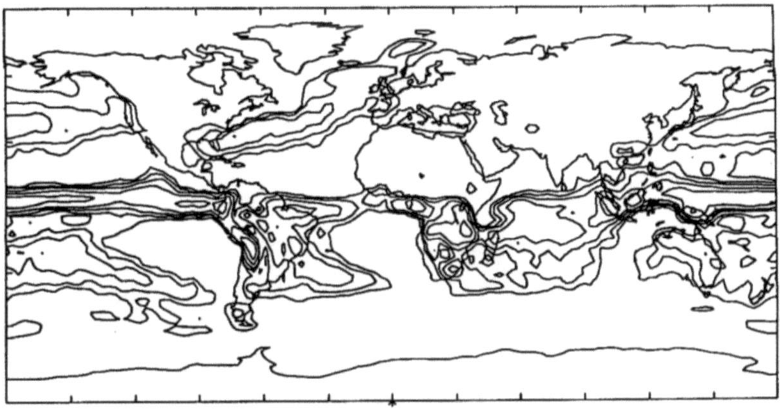

(c)

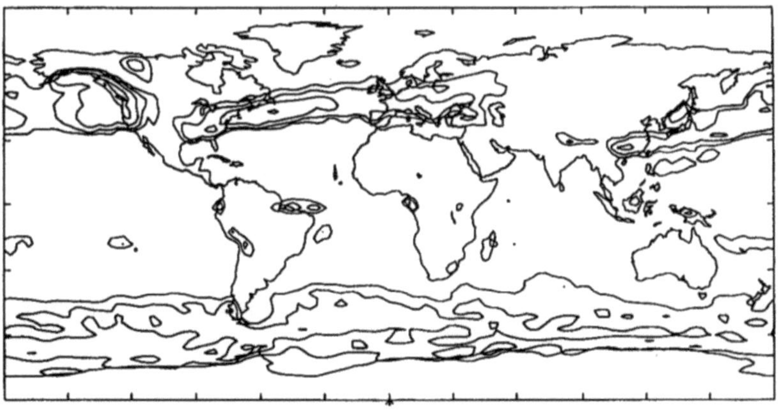

(d)

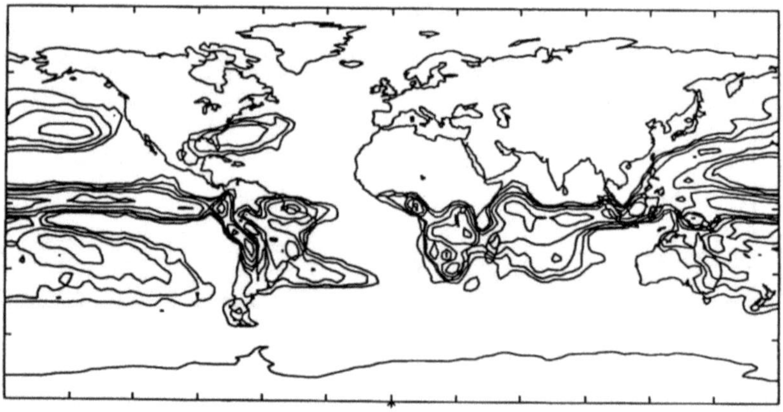

Figure 4: Simulated precipitation for perpetual February conditions, (a) Large scale processes, present day, (b) Convective processes, present day (c) Large scale processes, LGM, and (d) Convective processes, LGM. Contour intervals are successively doubling (i.e. 1, 2, 4, 8, 16 mm/day).

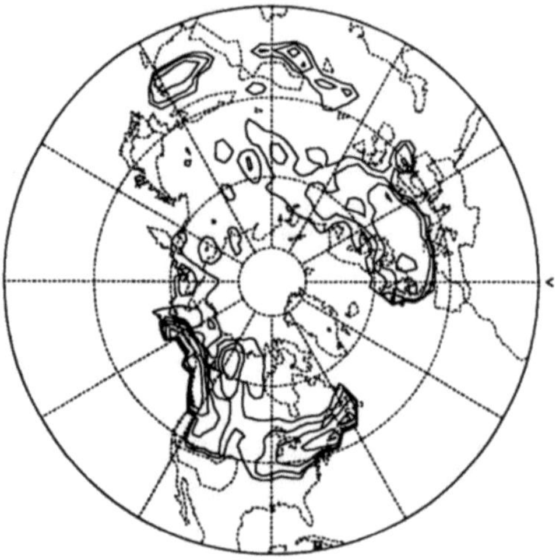

Figure 5: Net accumulation of snow (in metres of water equivalent) in the final 270 days of the integration for prepetual February conditions at the LGM. Contour intervals are successively doubling (i.e. 0.1, 0.2, 0.4, 0.8, 1.6 m). Negative contours are dashed.

For both ice sheets, there is accumulation virtually everywhere, both in the interiors of the glaciers and at the ice edges. However, the largest changes occur at the ice edge. There is substantial accumulation over the northern Rockies, over the eastern seaboard of the U.S.A. and over southern Europe. This accumulation is closely related to the changes in storm tracks and precipitation already discussed.

There is also net accumulation of snow over the Tibetan plateau and the edge of the Antarctic continent (not shown). The former result is of interest in that the existence of a Tibetan ice sheet is controversial (Kuhle, 1987). The GCM suggests that winter conditions are suitable for an ice sheet to form but clearly, from the model results to date, we have no indication as to whether this ice sheet would last through the year. Therefore as such, the model results are not surprising.

4 Perpetual August Experiments

We will present a very much briefer discussion of these simulations.

Figure 6 shows the simulated mean August surface temperatures for (a) present day, and (b) LGM. The present day results are in reasonable agreement with observations. The LGM results show very strong temperature gradients in the southern oceans, suggesting that this time the sea ice extent is consistent with the model simulation. In the northern hemisphere, there are significantly stronger temperature gradients over both the Pacific and Atlantic oceans. In addition, the peak temperature gradient at the eastern end of the Atlantic storm track is significantly further north. The strong temperature gradients over the land are the result of the ice sheet elevation.

The changes in temperature gradients again result in important changes in the storm tracks. Figure 7 shows the eddy kinetic energy for (a) present day and (b) LGM. The present day simulation is reasonable. The northern hemisphere storm tracks are very well simulated. The maximum of the southern

(a)

(b)

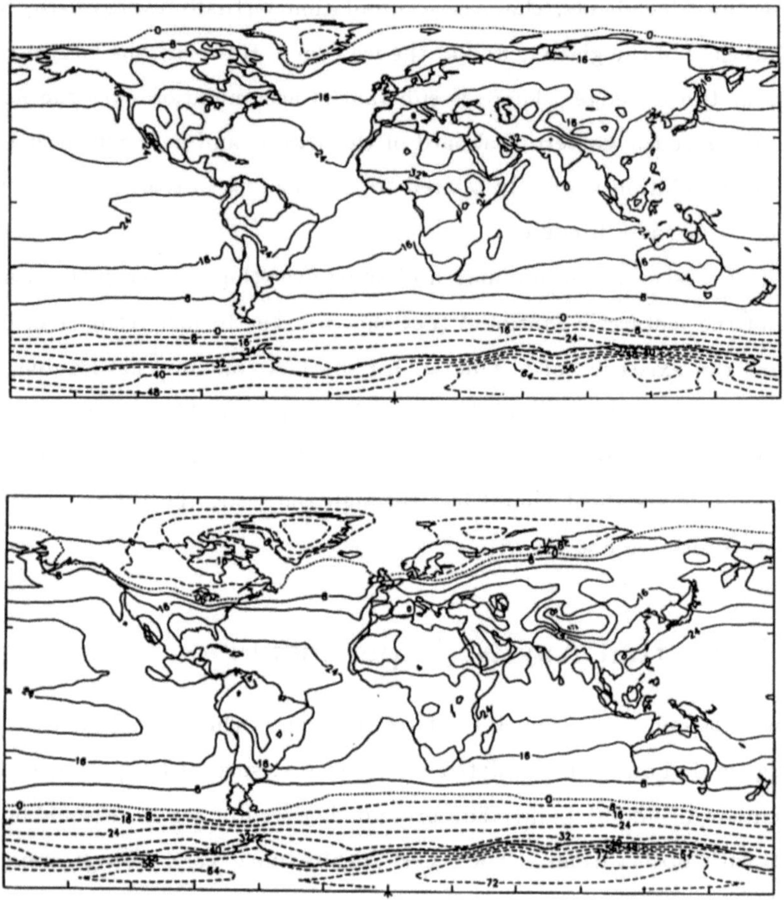

Figure 6: Mean surface temperature for the perpetual August model simulations for (a) present day and (b) LGM The contour interval is 8^0C and negative values are dashed. The sea surface temperature and sea ice limit are specified from CLIMAP data.

(a)

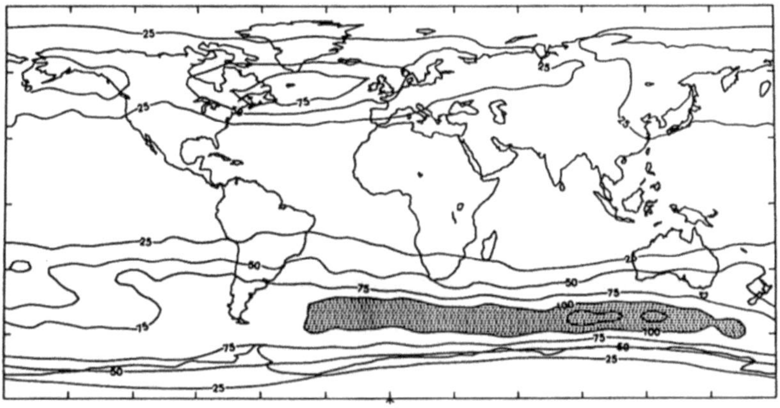

(b)

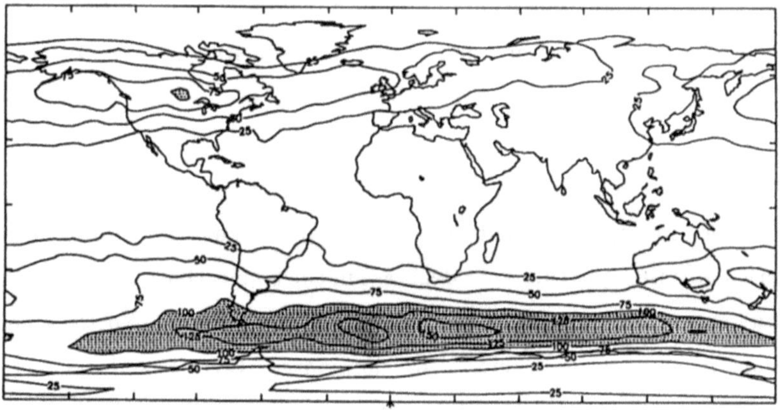

Figure 7: Mean high pass eddy kinetic energy per unit mass at 250 mb for the perpetual August model simulations for (a) present day and (b) LGM. The contour interval is 25 m^2s^{-2}*. Values greater than 100* m^2s^{-2} *are stippled.*

hemisphere storm track is well captured in both strength and position but, like the February simulations, is too far south by about 5^0. The model also does not successfully reproduce the splitting of the storm track seen over the Australian region. This split is strongly associated with an increased frequency of anticyclonic blocks over New Zealand, and this GCM, like many others, simulates anticyclonic blocks poorly. In the LGM simulation the enhanced temperature gradients in the southern hemisphere result in a substantial increase in the intensity of the storm tracks, by up to 50 percent. This increase covers a broad range of longitudes from 120^0W to 120^0E. The storm track is also confined to a narrower latitude band, resulting in significant decreases in eddy kinetic energy on the equatorward side. In the northern hemisphere, the enhanced Pacific temperature gradient results in an increase in the summer storm track by a substantial amount but a decrease in the eddy kinetic energy associated with the western end of the Atlantic storm track. In the eastern side of the Atlantic, there is a small increase in the strength of the storm tracks.

The changes in the storm tracks result in significant changes in the northern hemisphere precipitation patterns as shown in figure 8. Over both North America and Europe, there is a decrease in the convective rainfall but increases in the large scale precipitation. In north west and central North America the increase in large scale precipitation dominates. Over Scandinavia, the total precipitation is unaltered, with approximately equal but opposite changes in large scale and convective amounts. However the net snow accumulation (figure 9) shows that during the LGM, the precipitation in the Scandinavian region was falling as snow resulting in a significant accumulation of snow in the center of the ice sheet. Similar results also occur on the western side of the North American ice sheet. These regions in the February simulation also show accumulation. Thus it seems likely that these will be regions of annual accumulation. Taken at face value, the accumulation rates are consistent with the formation of a 2 km ice sheet on glacial/interglacial time scales. Further south, near the edges of the ice sheets, there are substantial regions of melting.

(a)

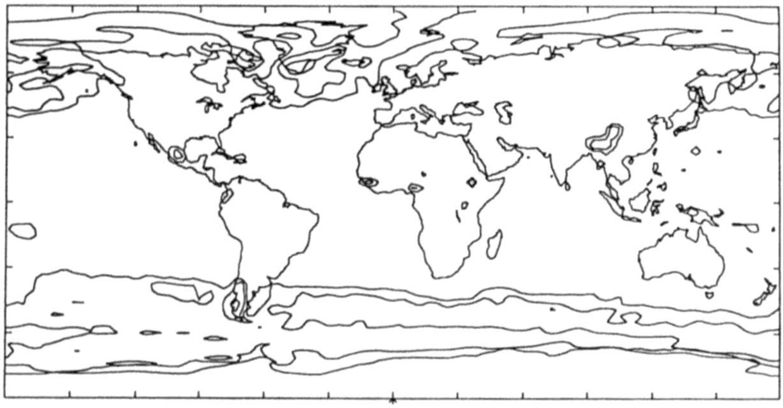

(b)

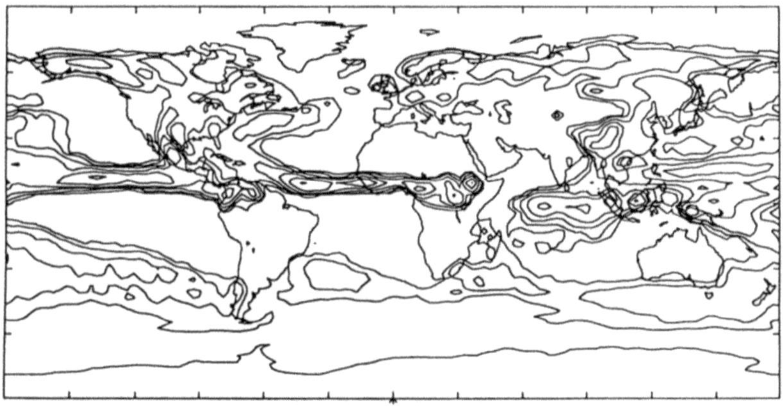

(c)

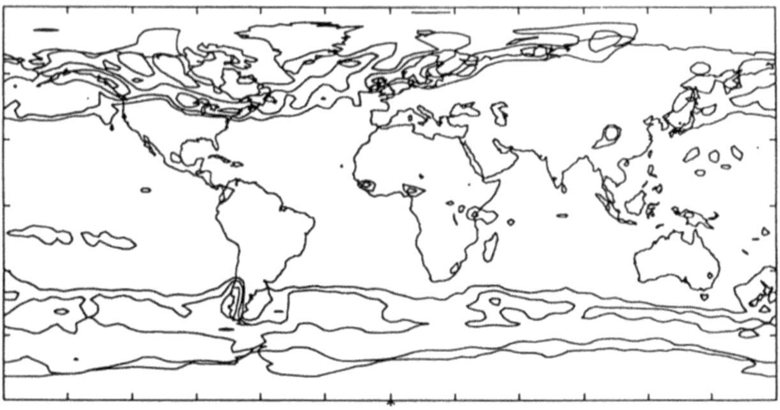

(d)

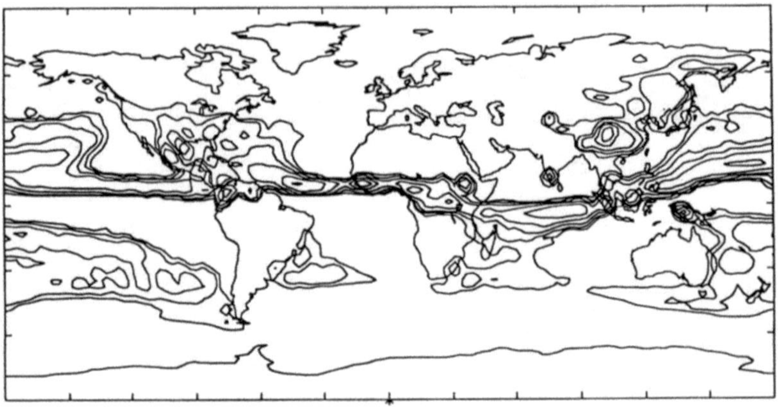

Figure 8: Simulated precipitation for perpetual August boundary conditions, (a) Large scale processes, present day, (b) Convective processes, present day (c) Large scale processes, LGM and (d) Convective processes, LGM. Contour intervals as in figure 4.

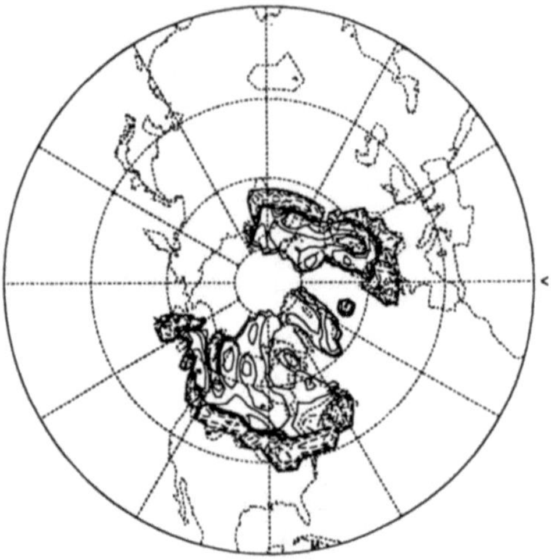

Figure 9: Net accumulation of snow (in metres of water equivalent) in the final 270 days of the integration for prepetual August conditions at the LGM. Contour intervals as in figure 5.

This is occurring in similar regions to that for accumulation in winter. Thus the total annual accumulation in these regions cannot be estimated from these model results.

5 Summary

In this paper we have described some of the first results from our relatively high resolution GCM simulations of the last glacial maximum. The increased resolution improves our model simulation of mid-latitude depressions. It should be noted however, that in general the higher resolution does not necessarily significantly improve the accuracy of our climate simulations for present day boundary conditions.

We have shown that there are substantial changes in the mid-latitude storm tracks which strongly affect the supply of heat and moisture into the continental ice sheets. Of particular interest, the simulations for August suggest that the Scandinavian ice sheet accumulated ice even during summer, and that in winter, the accumulation was considerably less. These results can be compared to those of Gates (1976), Manabe and Broccoli (1985) and Kutzbach(1987). In particular, Manabe and Broccoli also found that the annually averaged net ice accretion rate is positive in the centre of the ice sheets.

Much further work is required before we can have full confidence in these results. Clearly one of the most important aspects is the necessity to include a seasonal cycle, and we are currently in the process of making such a simulation. We have also made no attempt in this paper to compare our simulations with the geological data or previous model studies. The former will be done once the seasonal cycle simulation is complete. The latter will be performed as part of the Palaeoclimate Model Intercomparison Project.

Although we can use the GCM to learn about some aspects of the moisture and heat balance on the ice sheets, a complete analysis will only be possible if the GCM results are included into a more sophisticated dynamic ice sheet model. With this end in mind it will be of interest to investigate the sensitivity of these results to ice sheet reconstructions prior to 21Kyrs BP.

the sensitivity of these results to ice sheet reconstructions prior to 21Kyrs BP.

References

Betts A.K. and M.J. Miller, 1986: A new convective adjustment scheme. Part 1: Observational and theoretical basis. *Quart. J. Roy. Met. Soc.* 112, 677-691.

Broccoli, A.J., and S. Manabe, 1987: The influence of continental ice, atmospheric CO_2 and land albedo on the climate of the last glacial maximum. *Clim. Dyn.* 1, 87-99.

Gallee, H., J.P. Van Ypersele, T. Fichefet, C. Tricot and A. Berger, 1991: Simulation of the last glagial cycle by a coupled sectorially averaged climate -ice sheet model I: The climate model. *J. Geophys. Res.* 96, 13139-13161.

Gates, W.L., 1976: The numerical simulation of ice age climate with a global general circulation model. *J. Atmos. Sci.* 33, 1844-1873.

Hoskins, B.J., H.H. Hsu, I.N. James, M. Masutani, P.D. Sardeshmukh and G.H. White, 1989: *Diagnostics of the global atmospheric circulation.* WRCP report 27.

Hoskins, B.J. and P.J. Valdes, 1990: On the existence of storm tracks. *J. Atmos. Sci.* 47, 1854-1864.

James, I.N. and D.L.T. Anderson, 1983: The seasonal mean flow and distribution of large scale weather systems in the southern hemisphere: The effects of moisture transports. *Quart. J. Roy. Met. Soc.* 110, 943-966.

Joussaume, S., 1991: Paleoclimatic tracers: An investigation using an atmospheric general circulation model under ice age conditions. Part 1: Destert dust. *J. Geophys. Res.* In press.

Kuhle, M., 1987: Subtropical mountain and highland glaciation as ice age triggers and the waning of the glacial periods in the Pleistocene. *GeoJournal* 14, 393-421.

Kutzbach, J.E., 1987: Model simulations of the climatic patterns during the deglaciation of North America. In *North America and the adjacent oceans during the last deglaciation.* W.F. Ruddiman and H.E. Wright, Eds. The Geology of North America, *Geol. Soc. Am.* v. K-3, Boulder, Col., pp 425-446.

Kutzbach, J.E. and P.J. Geutter, 1986: The influence of changing orbital parameters and surface boundary conditions on climate simulations for the past 18000 years. *J. Atmos. Sci.* 43, 1726-1759.

Lautenschlager, M. and K. Herterich, 1990: Atmospheric response to ice age conditions: Climatology near the Earth's surface. *J. Geophys. Res.* 95, 22547-22557.

Manabe,S. and A.J.Broccoli, 1985: The influence of continental ice sheets on the climate of an ice age. *J. Geophys. Res.* 90, 2167-2190.

Morcrette, J-J, 1990: Impact of changes to radiative transfer parameterisations plus cloud optical properties in the ECMWF model. *Mon. Wea. Rev.* 118, 847-873.

Palmer, T.N., G.J. Shutts and R. Swinbank, 1986: Alleviation of a systematic westerly bias in general circulation and numerical weather prediction models through an orographic gravity wave parameterisation. *Quart. J. Roy. Met. Soc.* 112, 1001-1040.

Peltier, W.R. and W. Hyde, 1984: A model of the ice age cycle. In *Milankovitch and Climate, understanding the response to orbital forcing* Berger,

Imbrie, Hays, Kukla and Saltzman Eds. NATO ASI Series C vol. 126, Reidel Publ. Company, Holland, 895pp.

Pollard, D., 1984: Some ice-age aspects of a calving ice sheet model. In *Milankovitch and Climate, understanding the response to orbital forcing* Berger, Imbrie, Hays, Kukla and Saltzman Eds. NATO ASI Series C vol. 126, Reidel Publ. Company, Holland, 895pp.

Rind, D., 1987: Components of the ice age circulation. *J. Geophys. Res.*, 92, 4241-4281.

Rind, D., and D. Peteet, 1985: Terrestrial conditions at the last glacial maximum and CLIMAP sea-surface temperature estimates: Are they consistent ? *Quaternary Res.* 24, 1-22.

Saltzman, B. and M.Y. Verbitsky, 1992: Asthenospheric ice-load effects in a global dynamical- system model of the Pleistocene climate. *Climate Dynamics* 8, 1-11.

Schneider, S.H., and S.L. Thompson, 1979: Ice ages and orbital variations: Some simple theory and modeling. *Quaternary Res.* 12, 188-203.

Schutz, C. and W.L. Gates, 1971: Global climatic data for surface, 800 mb, 400 mb: January. Rand, Santa Monica, R-1029-ARPA, 180 pp.

Shinn, R.A., and E.J. Barron, 1989: Climate Sensitivity to Continental Ice Sheet Size and Configuration. *J. Climate* 2, 1517-1537.

Slingo, J.M., 1987: The development and verification of a cloud prediction scheme for the ECMWF model. *Quart. J. Roy. Met. Soc.* 113, 899-927.

Slingo, J.M. and M. Blackburn, 1992: Implementation of the Betts/Miller convective adjustment scheme in the UGCM. *U.K. Universities Global Atmospheric Modelling Programme technical report 25.*

Suarez, M., and I.M. Held, 1979: The sensitivity of an energy balance climate model to variations in the orbital parameters. *J. Geophys. Res.* 84, 4825-4836.

Thuburn, J., 1993; Use of a flux limited scheme for vertical advection in a GCM. *Quart. J. Roy. Met. Soc.* 119, 469-487.

MAUNDER MINIMUM AND THE LITTLE ICE AGE : IMPACT OF A LONG-TERM VARIATION OF THE SOLAR FLUX ON THE ENERGY AND WATER CYCLE

Robert Sadourny
Laboratoire de Météorologie Dynamique du CNRS
École Normale Supérieure
75231 Paris Cedex 05
France

1. Introduction

Long-term variations of solar luminosity are one of the most natural candidates for explaining observed climate fluctuations, provided that their amplitude and time scale are large enough : the global climate "oscillator" will react significantly to external perturbations only when these are able to warm up or cool down the ocean by an appreciable amount. This is hardly the case for the eleven-year cycle, as it takes decades for the ocean to warm up or cool down. On the other hand, the amplitude of the eleven-year cycle is quite weak : 0.5 Wm^{-2} for the last cycle, which corresponds to a range of variation of only 0.18 Wm^{-2} for the solar energy absorbed by the earth system, a value probably too small to yield appreciable climate changes even if the corresponding positive or negative perturbation were sustained long enough. The eighty-year Gleissberg cycle (the envelope of the 11-year cycle amplitudes) would be a more plausible candidate (Reid, 1987), at least due to its longer time scale ; but its amplitude remains of the same order. The most likely relation so far between solar luminosity and climate is the simultaneous occurrence of the Maunder Minimum, which lasted at least from the mid-XVIIth Century to the early XVIIIth Century and the colder-than-now Little Ice Age (for an account of the possible connection between the two events, see for example Eddy, 1977). In fact, even in this case, the connection is difficult to prove in a quantitative way. On the one hand, the Little Ice Age concept is not always clearly defined, for example in terms of duration ; on the other hand, the reconstruction of the solar output deficit during the Maunder Minimum is a difficult matter. Recently, Nesme-Ribes et al (1993) and Nesme-Ribes (1993), using a combination of surrogate data and simplified solar dynamics theory, have estimated this deficit of the order of 5 to 6 Wm^{-2}. In connection to this, numerical experiments have been designed to study the response of terrestrial climate to a long-term solar luminosity change of that magnitude.

These numerical experiments have been described in Nesme-Ribes et al (1993). I shall quickly summarise them in Section 2. In Section 3, I shall study in some detail a few of the basic mechanisms which control the climate response to a global radiative perturbation such as variations of solar luminosity or greenhouse gas loading in low, middle and high latitudes. Finally, in Section 4, I shall give a brief comparison of model results with Little Ice Age climatic data over Europe, showing that the Maunder Minimum could have been a significant factor of the observed cooling. This does not

NATO ASI Series, Vol. I 22
Long-Term Climatic Variations
Edited by J.-C. Duplessy and M.-T. Spyridakis

mean of course that other factors, like thermohaline circulation changes in the oceans, or atmospheric loading in volcanic aerosols, have not played their part in the process (for an account of the relative rôles of solar luminosity and volcanic aerosols in the dynamics of climate fluctuations, see for example Hansen *et al.* (1990).

2. Numerical experiments

The model used in the experiment related by Nesme-Ribes *et al.* (1993) was the LMD atmospheric general circulation model (Sadourny and Laval, 1984) ; in the version considered there, it contains an interactive cloud-radiation scheme (Le Treut and Li, 1991) and it is coupled with a simplified slab ocean. We define the global climate sensitivity to some external forcing parameter π as

$$\sigma_\pi = \left\| \frac{\mathrm{d}\mathbf{C}}{\mathrm{d}\pi} \right\| , \tag{1}$$

where $\mathbf{C}$ refers to the *N*-dimensional vector variable defining the climate state, while the double bar refers to some appropriate measure of the vector derivative. The simplest, most usual definition of σ_π is the mean thermal surface sensitivity or mean surface warming

$$\sigma_\pi = \frac{\mathrm{d}\{T_s\}}{\mathrm{d}\pi} , \tag{2}$$

where $\{T_S\}$ refers to the globally and seasonally average surface air temperature. The assumptions made when modelling the cloud radiative properties in the version of the LMD model used in the experiment yield a relatively large value of σ_π, due to relatively strong cloud-radiation feedback ; on this point, see the model comparison performed in Cess *et al.* (1990). The mean global warming for the 2 x CO_2 case, for example, is 5.3 K (Li and Le Treut, 1992).

The experiments were designed in the following way. Two simulations were performed in parallel, where the current solar constant L_0 was modified to $L = L_0 \pm \Delta L$, with $\Delta L/L = .0167$). The corresponding perturbation of the solar energy absorbed by the climate system is then estimated as $\Delta S = \Delta L \times 0.25 \times (1-\alpha) = \pm 4\ W\,m^{-2}$, after taking earth's sphericity and planetary albedo ($\alpha = 0.30$) into account. Such perturbations are strong enough to guarantee significant responses in spite of the internal variability of the climate system. Moreover, their magnitude corresponds to the perturbation of the upwelling infrared flux at tropopause level induced by instantaneous CO_2 doubling ($\Delta R = 4\ W\,m^{-2}$), which allows to directly compare the respective sensitivities to infrared and solar forcings : $\Delta\mathbf{C}/\Delta S$ and $\Delta\mathbf{C}/\Delta R$. The derivative $d\mathbf{C}/dS$ is approximated to second order as

$$\frac{d\mathbf{C}}{dS} = \frac{\mathbf{C}(S_0 + \Delta S) - \mathbf{C}(S_0 - \Delta S)}{2\,\Delta S} , \tag{3}$$

which defines both the sensitivity of climate to solar luminosity σ_S and the change of climate ΔC from Maunder Minimum conditions to present, if one assume a relative change of solar luminosity

$$1-\frac{S_{Maunder}}{S_{Present}}=0.004 \qquad (4)$$

consistent with the estimate of Nesme-Ribes *et al.* (1993), based on solar dynamo theory and XVIIth Century observations ; it corresponds indeed to a radiative perturbation of 1 $W\,m^{-2}$ for the absorbed solar energy flux, so that in that case, $\Delta C = dC/dS$.

The time evolution of the globally average air surface temperature for the two perturbation experiments (starting from present conditions) is shown in Figure 1. Equilibrium obtains after approximately eight years ; the equilibrium thermal response is calculated by comparing the two perturbed climate states, averaged from year 10 to year 20. The mean thermal response is slightly asymmetric : $\Delta\{T_S\}=+5.3K$ for $\Delta S = +4Wm^{-2}$, $\Delta\{T_S\}=-6.3K$ for $\Delta S = -4Wm^{-2}$; in addition, the mean seasonal cycle is weaker by nearly one degree in the cold case.

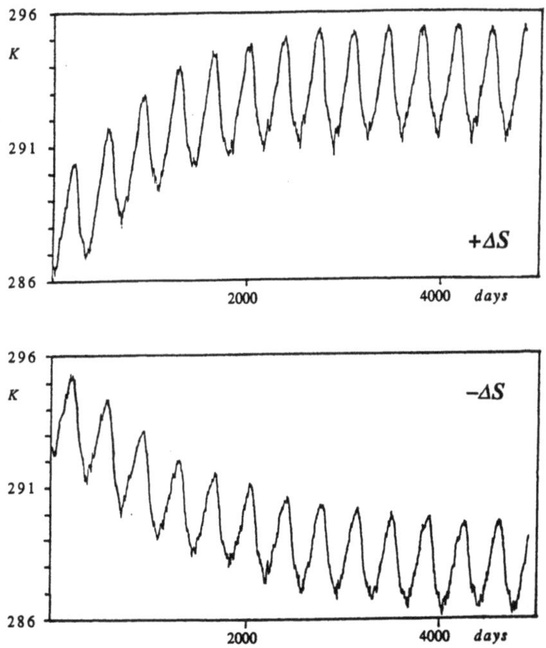

Figure 1. *Time evolution of globally averge surface temperature showing convergence to new equilibria in the $+\Delta S$ (above) and $-\Delta S$ (below) perturbation experiments.*

3. Dynamics of climate warming

3.1. *Temperature, moisture and mean circulation*

The climate response to a variation $+\Delta S$ of the solar flux is hardly distinguishible from the response to a doubling of atmospheric CO_2. This is perhaps to be expected, as both perturbations have identical magnitude and no geographical signature (at least in the horizontal), the structure of the response being imposed by the internal feedbacks. The zonally average warmings $\partial([T](\varphi,p))/\partial S$ (where φ is latitude and p pressure) for the whole year, boreal winter and boreal summer are described in Figure 2. Surface warming in the tropics induces strong moistening there, due to increased sensitivity exhibited by the Clausius-Clapeyron relation at warm temperatures ; this in turn decreases the vertical saturation lapse rate, inducing stronger warming at upper tropospheric levels. On the other hand, the atmospheric moistening induced by warming increases the greenhouse effect of water vapour and reduces the upwelling infrared flux : this explains the simulated stratospheric cooling occurring also in the case of $+\Delta S$, which might look somewhat paradoxical at first sight.

Figures 3 and 4 show the charts of surface air temperature and moisture changes for the whole year and winter and summer seasons. The most conspicuous feature there is the fact that the tropical belt corresponds to minimum warming and maximum moistening. Minimum warming in the tropics is explained by the mechanism of snow albedo feedback which enhances the warming response at high latitudes : as temperature increases, snowmelt expands and the resulting decrease of albedo enhances solar energy absorption, which feeds back positively on the initial warming. Maximum moistening, on the other hand, results from the above mentioned dependence of the Clausius-Clapeyron relation between saturation mixing ratio and temperature ($\partial^2 q_S/\partial T^2 > 0$) ; even though, as was just mentioned, ΔT is less in the tropics, $\Delta q_S = \partial q_S/\partial T \, \Delta T$ results stronger due to the Clausius-Clapeyron effect, reaching 10 % in the tropics.

Figure 5 shows the response of the upper tropospheric flow in terms of the yearly and seasonally average 200 *hPa* streamfunction. Taking into account the zonally average thermal wind relation

$$[u_T](\varphi,p) = \frac{R}{2a\Omega} \frac{\partial}{\partial cos\varphi} \int_p^{p_S} T(\varphi,p')\, dLnp' \quad , \tag{5}$$

(where p_S is surface pressure, R the perfect gas constant, a the earth's radius and Ω the earth's angular velocity), the increase of upper tropospheric jets in both winter and summer hemispheres is consistent with the upper tropospheric increase of the pole-to-equator temperature gradient associated with strong tropical moistening (Figure 4).

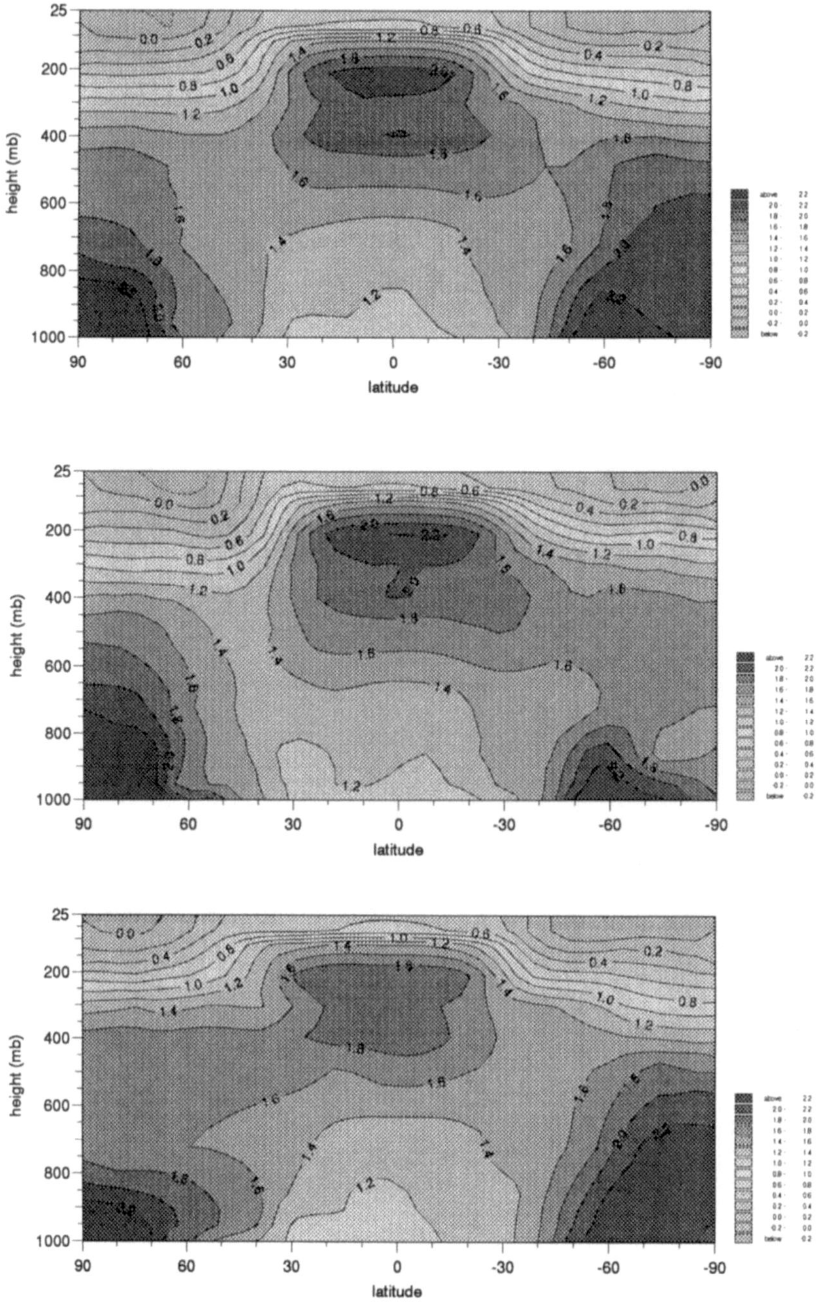

Figure 2. *Zonally average temperature tendencies for the whole year (above), boreal winter (middle) and boreal summer (below). Unit :* $K\ (Wm^{-2})^{-1}$.

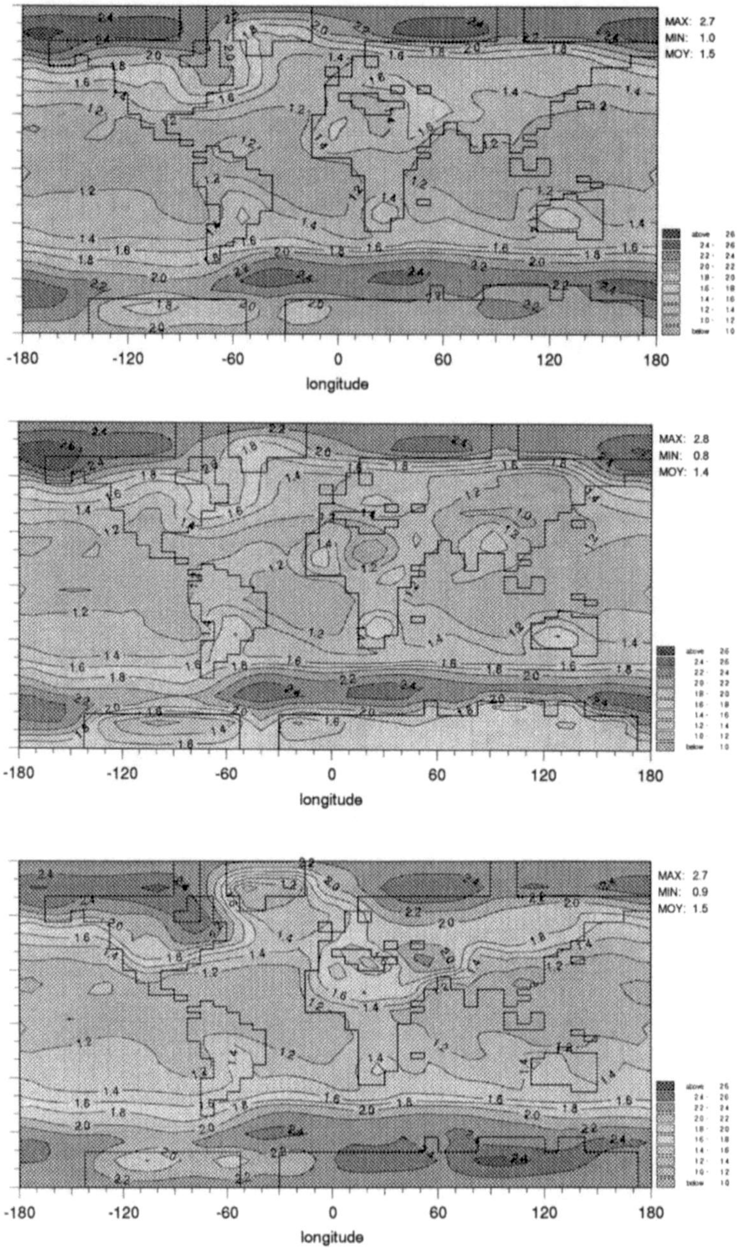

Figure 3. *Charts of surface temperature tendencies for the whole year (above), boreal winter (middle) and boreal summer (below). Unit : K* $(Wm^{-2})^{-1}$.

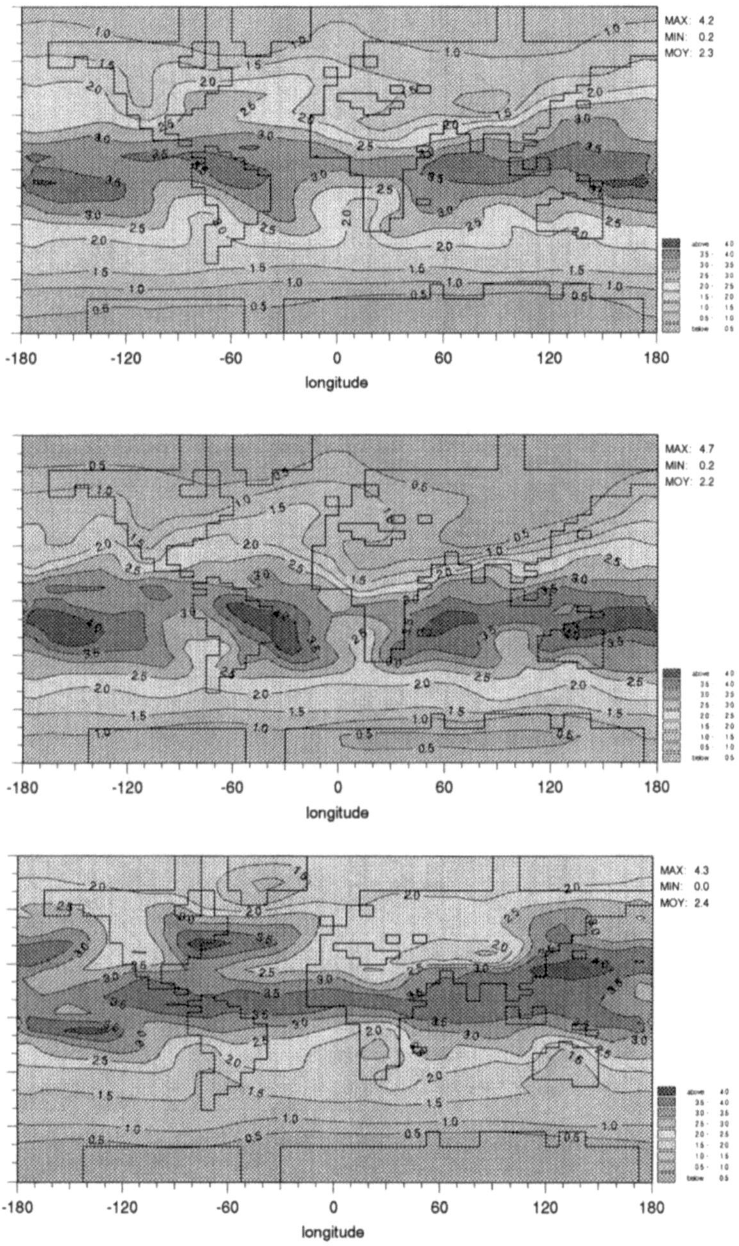

Figure 4. *Charts of precipitable water tendencies for the whole year (above), boreal winter (middle) and boreal summer (below). Unit : mm* $(Wm^{-2})^{-1}$.

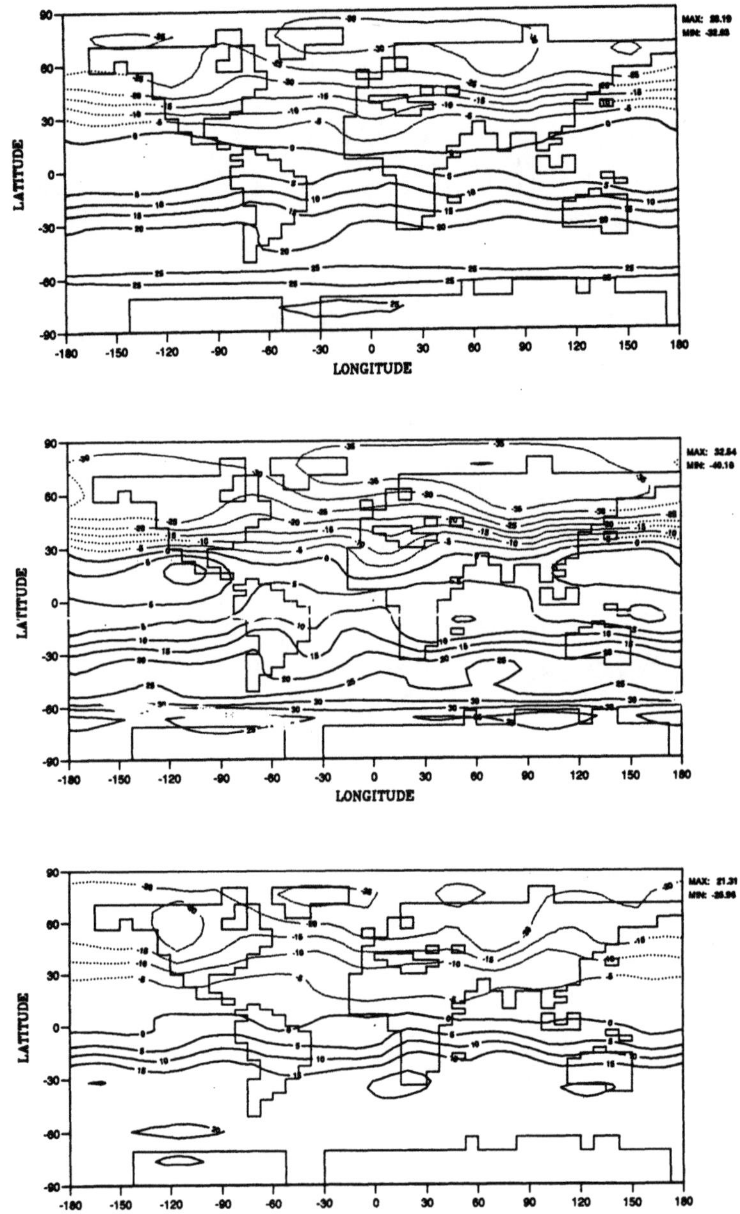

Figure 5. *Charts of tendencies of the 200 hPa streamfunction for the whole year (above), boreal winter (middle) and boreal summer (below). Unit : $10^6 m^2 s^{-1} (Wm^{-2})^{-1}$.*

3.2. *Response of the Hadley circulation*

A matter of special interest is the response of the Hadley cells. Hadley cells are zonally symmetric rolls whose role is to relax the meridional thermal gradient in the tropics. Their dynamics are best analysed by looking at the zonal component of the zonally average vorticity equation ; there we see that the Hadley circulation is maintained against frictional dissipation by a baroclinic source term $s([\omega_\lambda])$, where ω_λ refers to the zonal component of vorticity and the bracket to zonal averaging. This source term can be written in Jacobian form

$$s([\omega_\lambda]) = -\frac{1}{a}\left[\frac{\partial(\alpha, p)}{\partial(\varphi, z)}\right] = -\frac{\mathrm{R}}{a}\left[\frac{\partial(T, Ln\, p)}{\partial(\varphi, z)}\right] , \tag{6}$$

where α refers to specific volume and z to height. In the tropics, we may replace zonally average products by products of averages without making appreciable errors. Expanding the latter expression in (6), we get

$$s([\omega_\lambda]) = \frac{R}{a}\left[\frac{\partial T}{\partial z}\right]\left[\frac{\partial Ln\, p}{\partial \varphi}\right] + \frac{g}{a}\left[\frac{\partial Ln\, T}{\partial \varphi}\right] , \tag{7}$$

where we have used the hydrostatic approximation to simplify the second term on the right-hand side. To evaluate the strength of the Hadley cell, we have to integrate (7) over the Hadley cell domain, i. e. from surface to tropopause level and from the equatorial pressure trough to the subtropical ridge. To fix ideas, we may consider here the boreal Hadley cell, which, having easterly vorticity, must have an easterly vorticity source $s([\omega_\lambda])$. This is indeed the case, as both terms on the right-hand side give a negative contribution : $[\partial Ln\, p/\partial\varphi]$ is positive throughout the easterlies domain, and both $[\partial T/\partial z]$ and $[\partial Ln\, T/\partial\varphi]$ are negative.

How does the Hadley cell strength react to an external perturbation ? First, we may relate the meridional pressure gradient to the strength of the Hadley cell. A more vigorous cell deepens the equatorial trough and strengthens the subtropical high : the meridional pressure gradient increases throughout the easterlies domain. Then we may model (7) in the form

$$\{s\} = \left\{\frac{\partial T}{\partial z}\right\}F(s) + \frac{g}{a}\left\{\frac{\partial Ln\, T}{\partial \varphi}\right\} , \tag{8}$$

where curly brackets refer to integral estimates over the Hadley cell domain, and $F(s)$ is a positive, decreasing function of s. Taking the derivative of (8) with respect to the external parameter π (here the solar flux), we get

$$\frac{\partial\{s\}}{\partial \pi} = \left(1 - F'(\{s\})\left\{\frac{\partial T}{\partial z}\right\}\right)^{-1}\left(\frac{R}{a}\frac{\partial Ln\, p}{\partial \varphi}\frac{\partial}{\partial \pi}\left\{\frac{\partial T}{\partial z}\right\} + \frac{g}{a}\frac{\partial}{\partial \pi}\left\{\frac{\partial Ln\, T}{\partial \varphi}\right\}\right) , \tag{9}$$

or, expanding the meridional derivative of the logarithm of temperature,

$$\frac{\partial\{s\}}{\partial\pi}=\left(1-F'(\{s\})\left\{\frac{\partial T}{\partial z}\right\}\right)^{-1}\left\{F(s)\frac{\partial}{\partial\pi}\left(\frac{\partial T}{\partial z}\right)-\frac{g}{a}\frac{1}{T^2}\frac{\partial T}{\partial\varphi}\frac{\partial T}{\partial\pi}+\frac{g}{a}\frac{1}{T}\frac{\partial}{\partial\pi}\left(\frac{\partial T}{\partial\varphi}\right)\right\} \quad . \tag{10}$$

From this expression we can deduce four mechanisms for the Hadley cell response :

(i) *A pressure effect*, described by the first factor on the right-hand side : $F'(s)$ and $\partial T/\partial s$ being negative, this term only amplifies the response due to the conjugation of the three other effects.

(ii) *A vertical temperature lapse rate effect* (the first term in the last bracket) : $F(s)$ being positive, this term has the sign of the change in vertical lapse rate. For a warming perturbation, this term is positive due to moistening of the lower layers : this means weakening of the Hadley cell.

(iii) *A warming effect* (the second term in the last bracket) : $\partial T/\partial\varphi$ being negative, this term is positive in case of warming : again this means weakening of the Hadley cell.

(iv) *A meridional temperature gradient effect* (the last term in the last bracket) : this last effect is more difficult to analyse. The question is the sign of the change in meridional temperature gradient. In lower layers, the meridional temperature gradient tends to respond to warming by increasing its algebraic value, due to increased moist convection cooling the equatorward side of the Hadley cell : the effect there is thus to weaken the Hadley cell. But the reverse occurs in upper layers, with stronger magnitude : $1/T$ is larger and the change in meridional gradient of temperature stronger in magnitude because of the release of latent heat. Due to these compensations, the net effect should be weak, but perhaps in the sense of strengthening the Hadley cell.

Figure 6 shows that the June-July-August boreal Hadley cell actually weakens in the LMD model as warming occurs : the mean circulation is divided by a factor two when the absorbed solar flux increases by 8 $W\,m^{-2}$. This means that the last effect above is indeed weak, a fact corroborated by the weakness of the response of the meridional temperature gradient illustrated in Figure 2 : the temperature response appears strongly homogenized throughout the tropical belt by the Hadley circulation.

Another characteristic of the Hadley cell response, as also seen in Figure 6, is the meridional expansion towards the winter hemisphere associated with warming. This behaviour can be explained from simple arguments. We have observed in Figure 5 an increase of the zonal wind in the upper troposphere, due to the upper tropospheric tropical warming and the resulting ingrease of pole-to-equator temperature gradient in upper layers. According to the theory of Held and You (1980), the meridional extension of the Hadley cell depends on the latitude φ_0 at which the homogenized angular momentum flow corresponding to vanishing zonal velocity at the equator

$$u_{max}(j)=a\Omega\, sin\varphi\, tg\varphi \quad , \tag{11}$$

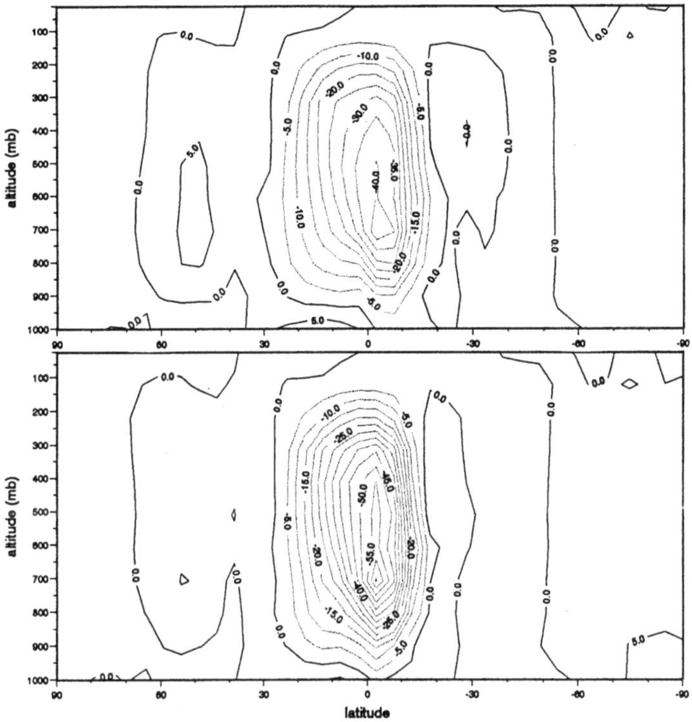

Figure 6. *Zonally average Hadley cells, visualised by their mass streamfunction in the* (φ,p) *plane, for the warming* $(+\Delta S$, *above) and cooling* $(-\Delta S$, *below) experiments.* *Unit*: $10^{14}\ Wm^{-2}\ (Wm^{-2})^{-1}$.

equilibrates the upper tropospheric thermal wind u_T (5). Under warming conditions, u_T increases in the subtropics due to the strong upper tropospheric tropical warming, and φ_0 moves poleward as observed in Figure 6.

3.3. *Global energetics*

The energy equation for the atmosphere reads

$$\frac{\partial}{\partial t}\left(c_v T + Lq + gz\right) = div\left(\left(c_p T + Lq + gz\right)\mathbf{U} + \mathbf{Q}\right) \ , \tag{12}$$

where c_v is the specific heat at constant volume, c_p the specific heat at constant pressure, L the vaporisation heat, T the temperature, q the water vapour mixing ratio, g the gravity acceleration, $\mathbf{U}$ the three-dimensional wind velocity vector and $\mathbf{Q}$ the energy flux due to radiation and exchange of temperature and moisture at the surface. Taking the time average of (12) over one month or more, the accumulation term on the left-hand side can be neglected. If one takes in addition the vertical average, atmospheric energetics reduce to

$$div\left\langle \overline{\left(c_pT + Lq + gz\right)\mathbf{U}_H} \right\rangle = \overline{Q}_S - \overline{Q}_T \ , \tag{13}$$

where the bar refers to time averaging, the brackets to vertical integration, $\mathbf{U}_H$ is horizontal velocity, Q_S the vertical energy flux across the surface, Q_T the net radiation flux at the top of the atmosphere, both counted positively upward. Thus the horizontal transport of energy must equilibrate the net energy input in a fluid column : the atmospheric circulation exporting energy from energy excess regions ($\overline{Q}_S - \overline{Q}_T > 0$) and *vice-versa.*

How do atmospheric energetics respond to a warming due to a variation of the solar constant or an increase of greenhouse gases ? In midlatitudes where $div\mathbf{U}_H \cong 0$ and the wind is approximately geostrophic, (13) reduces to

$$\left\langle \overline{\mathbf{U}_H \cdot \mathbf{grad}_H\left(c_pT + Lq\right)} \right\rangle = \overline{Q}_S - \overline{Q}_T \ , \tag{14}$$

where the left-hand side, corresponding to down-gradient energy mixing by baroclinic transients or stationary planetary waves, is dominated by the contributions from the lower layers where both temperature and moisture gradients are strongest. We have seen above that, as warming occurs, the pole-to-equator lower tropospheric temperature gradient weakens while the pole-to-equator moisture gradient increases (Figure 4). The net effect is a weakening of the poleward enthalpy flux and a strengthening of the poleward moisture flux in midlatitudes, in association with mixing (Figure 7).

In the tropics where temperature, moisture and pressure are well mixed in the horizontal, and transients have little impact on the energetics, (13) can be approximated by

$$\left\langle \overline{c_pT + Lq + gz}\ div\overline{\mathbf{U}}_H \right\rangle = \overline{Q}_S - \overline{Q}_T \ , \tag{15}$$

while conservation of mass imposes

$$\left\langle \ div\overline{\mathbf{U}}_H \right\rangle = 0 \ . \tag{16}$$

Relation (16) means that lower level convergence is equilibrated by upper level divergence, and *vice versa*. On the other hand, we have $\partial T/\partial z < 0,\ \partial q/\partial z < 0$, and

$$\frac{\partial}{\partial z}\overline{c_pT + Lq + gz} > 0 \ , \tag{17}$$

because the atmosphere is, on the average, statically stable. Then one readily sees that convergence of enthalpy and moisture occur in energy excess regions ($\overline{Q}_S - \overline{Q}_T > 0$) and are associated with low-level convergence, upper-level divergence and divergence of potential and total energy. This summarises the energetics of the Hadley-Walker circulation, which transports energy from energy excess regions

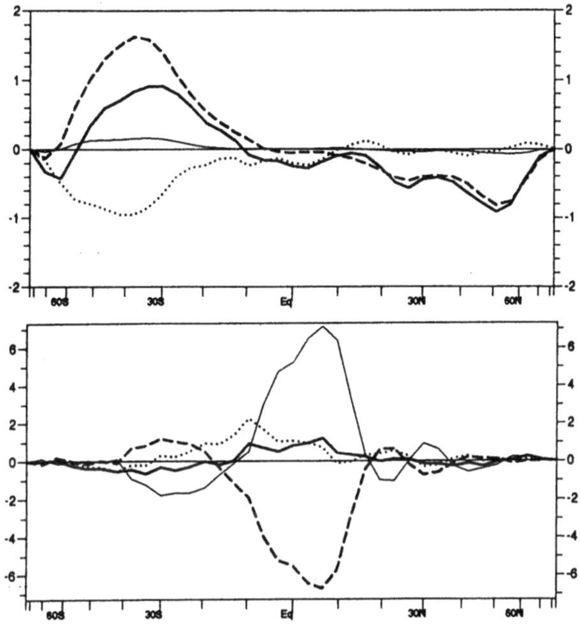

Figure 7. *Seasonally average, vertically and zonally integrated meridional energy flux tendencies for boreal summer, as a function of the sine of latitude. Above, due to transient eddies ; below, due to mean meridional circulation. Total energy in heavy continuous line, latent energy in dotted line, enthalpy in discontinuous heavy line, potential energy in thin continuous line. Unit* : 10^{14} $W\,(Wm^{-2})^{-1}$.

(the "meteorological equator") to energy deficient ones (the subtropical deserts), by transporting enthalpy and latent energy from the latter to the former along its lower branches ; these are in turn transformed by moist convection into potential energy in the ascending branches, potential energy being finally exported along the upper branches. The mean stability of the atmosphere implies that the Hadley cells effectively export total energy in the process. Convergence of moisture towards the meteorological equator in the lower branches of the Hadley-Walker circulation contributes to the modulation of the Monsoon rainfalls.

How does a warming process modify Hadley cell energetics ? From (15), one sees that the decrease of the Hadley mass flux and the decrease of the vertical temperature lapse rate associated with warming in the tropical band will both contribute to a weakening of the equatorward enthalpy flux (Figure 7). The situation is not so clear for the latent heat flux : Figure 7 shows that, at least with the LMD model, the increase of the vertical moisture gradient in the tropics overcomes the decrease of the Hadley mass flux, yielding a stronger equatorward convergence of the latent heat flux. The increase, however, is weak and likely to depend on modelling assumptions (in other

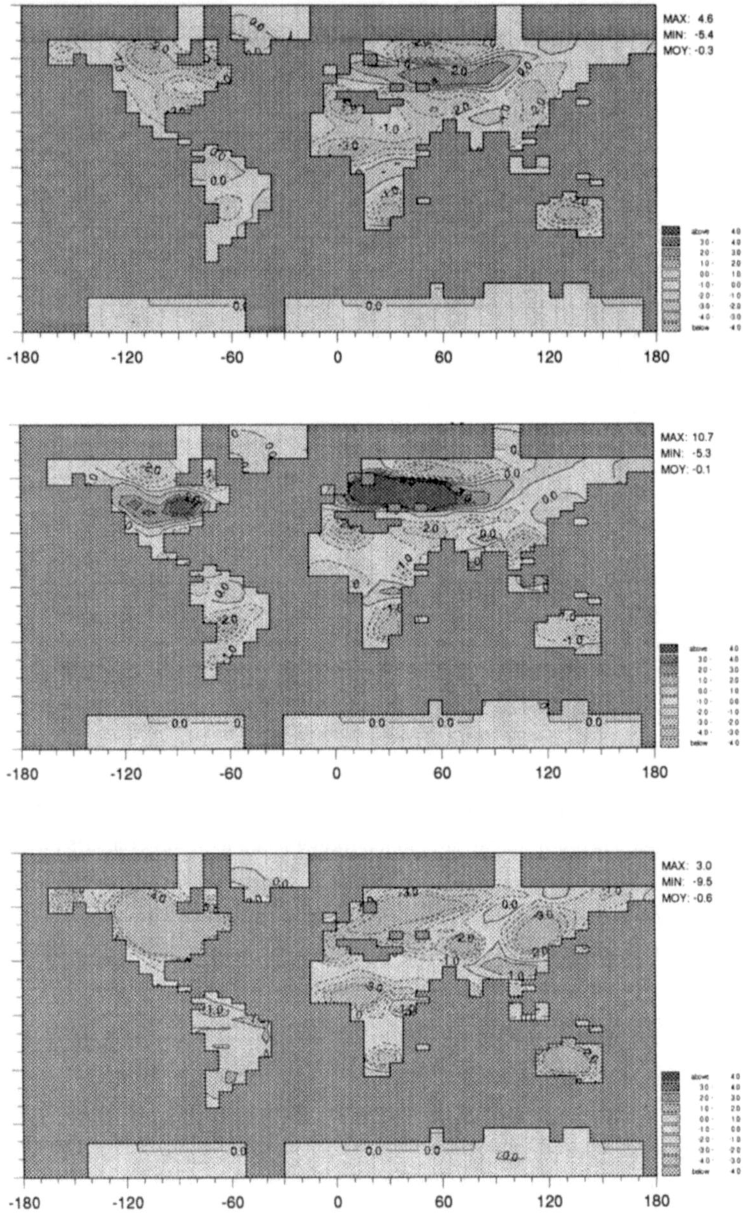

Figure 8. *Soil moisture tendency for yearly averages (above), boreal winter (middle) and boreal summer (below). Unit : mm $(Wm^{-2})^{-1}$.*

words, the Monsoon precipitations may not be strongly sensitive to variations of the solar constant or greenhouse gas loading).

3.4. *Additional remarks on Monsoon rainfall and soil wetness*

This behaviour of the Hadley cell and associated Monsoons is only partly confirmed by Figure 8, showing the soil moisture tendency for the whole year, boreal winter and boreal summer : we observe a slight increase of soil moisture in summer over South-East Asia, indicating a slightly more active Monsoon ; but Sahel, on the other hand, tends to dry out. Monsoon rainfall, of course, is not simply related to the Hadley cell moisture flux, as it strongly involves land-sea contrast effects. It is interesting to compare the present experiment with a simulation of the climate at 9 000 years B.P. using the same model (Maes, 1991). In that case the climate response is driven by seasonal changes in the distribution of insolation due mainly to precession. As total insolation was about the same as present, the ocean temperatures were assumed to also remain the same, and the summer response was mainly an enhancement of the land-sea contrast, with significant warming of continental areas in summer (Figure 9). A striking feature of Figure 9, however, is the cooling of the Monsoon areas over Sahel and India, indicating a coherent, significant increase of Monsoon rainfall, quite different in structure from the Monsoon response of the present experiments. Thus the Monsoon system, because it is mainly driven by land-sea contrasts, responds much more strongly to regional insolation changes related to variations of orbital parametres, than to a variation of solar luminosity.

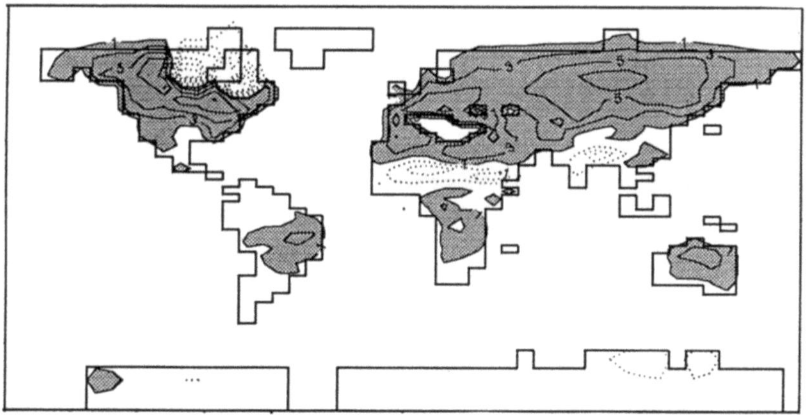

Figure 9. *Surface air temperature change over continents in boreal summer between 9 000 B.P. and present, as simulated by the same version of the LMD model. Unit : K.*

Another conspicuous feature in Figure 8 is the response of soil wetness in middle latitudes of the Northern Hemisphere, showing strong drying in summer and even stronger moistening in winter. The moistening in winter — and the resulting overall moistening obesrved in yearly averages — is consistent with the already mentioned increase of the northward transport of latent heat by the mixing of stronger atmospheric humidity gradients. Summer drying, on the other hand, is basically driven by land surface processes.

4. Simulations vs Little Ice Age conditions in Europe

We may have a short look at the simulated climate change in Europe corresponding to the Maunder Minimum solar perturbation, and compare it with Little Ice Age reconstructions. There is a wealth of quantitative and proxy data in Europe dating back to the XVIIth Century, which has been the source of many temperature and precipitation reconstructions (e.g., Lamb, 1977 ; Legrand et al., 1990, Matthews, 1976 ; Pfister, 1978 ; Sweingruber et al., 1978 ; van den Dool et al., 1978). These cooling reconstructions are summarized and compared to simulated values in Table 1.

Location	*Author(s)*	*Reconstructed*	*Simulated*
British Isles	Lamb (1977)	0.5 to 1	1.4
Netherlands	van den Dool et al. (1978)	1	1.5
Paris	Legrand et al. (1990)	0.8	1.5
South Norway	Matthews (1976)	1.6	1.6
Switzerland	Pfister (1978)	0.2 to 2	1.6

Table 1.
Reconstructed vs simulated cooling (in K)

Due to coarse resolution, model values are much smoother but the order of magnitude is not unrealistic. Keeping all caution necessary in such comparisons, one may conclude that the weakening of solar luminosity may have played a significant rôle in the occurrence of the Little Ice Age.

5. Conclusion

Analysing the general features of climate change induced by a variation of solar luminosity, we have shown that a weakening and a slight meridional extension of the Hadley cell was plausible when the solar constant increases ; simultaneously, the energetic efficiency of the Hadley cell should decrease, but this weakening does not necessarily mean a weakening of monsoon rainfall : in fact, the LMD model even predicts a slight increase of the meridional water vapour flux towards the meteorological equator. In midlatitudes, the poleward total energy flux decreases due to the relative enhancement of solar energy absorption at high latitudes, decreasing the energy deficit there ; the weakening of the meridional temperature gradient yields a weakening of the meridional enthalpy flux by transient eddies, partly compensated by an enhancement of the latent heat flux. The enhancement of the poleward moisture flux might play an important rôle in controlling both the snow-albedo and the thermohaline feedbacks. Also, it drives an increase of winter-time soil moisture in the middle latitudes of the Northern Hemisphere, which, on annual averages, overcomes the summer drying. The response of the Monsoon rainfall, on the other hand, is rather weak. On the one hand, the weakening of the Hadley cell counteracts the atmospheric moistening in the tropices ; on the other hand, the land-sea contrast appears much less enhanced by a solar luminosity increase than it is by insolation changes in the Milankovitch cycle.

To summarise, model results tend to indicate that the Maunder Minimum can indeed have played a significant rôle in the occurrence of the Little Ice Age. One should keep in mind, however, the limitations of this kind of "equilibrium" simulation using a highly simplified ocean, and recall that he notion of climate equilibrium is a convenient but artificial assumption. Realistic modelling of the Little Ice Age climate should take into account, in addition to the weakening of the solar constant, volcanic aerosol loading and explicit prediction of the slow oceanic transients associated with the thermohaline circulation. This requires not only three-dimensional coupled ocean-atmosphere models, but in addition, realistic initialisation of the ocean circulation which remains for the moment out of reach.

References

Cess, R.D., G.L. Potter, J.P. Blanchet, G.J. Boer, A.D. Del Genio, M. Déqué, V. Dymnikov, V. Galin, W.L. Gates, S.J. Ghan, J.T. Kiehl, A.A. Lacis, H. Le Treut, Z.X. Li, X.Z. Liang, B.J. McAvaney, V.P. Meleshko, J.F.B. Mitchell, J.J. Morcrette, D.A. Randall, L. Rikus, E. Roeckner, J.F. Royer, U. Schlese, D.A. Sheinin, A. Slingo, A.P. Sokolov, K.E. Taylor, W.M. Washington, R.T. Wetherald, I. Yagai and M.H. Zhang (1990) : Intercomparison and interpretation of climate feedback processes in nineteen atmospheric general circulation models. *J. Geophys. Res.*, **95**, 16601-16615.

Eddy, J.A. (1977) : Climate and the changing sun. *Climatic Change*, **1**, 173-190.

Hansen, J. & A.A. Lacis (1990) : Sun and dust versus greenhouse gases : an assessment of their relative roles in global climate change. *Nature*, **346**, 713-719.

Held, I.M. & A.Y. Hou (1980) : Nonlinear axially symmetric circulations in a nearly inviscid atmosphere. *J. Atmos. Sci.*, **37**, 515-533.

Lamb, H.H. (1977) : Climate : Present, Past ans Future, Vol. 2 : Climatic History and the Future, *Methuen & Co. Ltd*, London.

Legrand, J.-P., M. le Goff & C. Mazaudier, 1990 : On the climatic changes and the sunspot activity during the XVIIth century. *Annales Geophysicae*, **8**, 637-644.

Le Treut, H. & Z.X. Li (1991) : Sensitivity of an atmospheric general circulation model to prescribed SST changes : feedback effects associated with the simulation of cloud optical properties. *Climate Dynamics*, **5**, 175-187.

Li, Z.X. & H. Le Treut (1992) : Cloud-radiation feedback in a general circulation model and its dependence on cloud modelling assumptions. *Climate Dynamics*, **7**, 133-139.

Maes, Ch. (1991) : Étude de la variabilité du modèle de circulation générale du LMD en réponse à des forçages externes : simulation du climat de l'Holocène. *Internal Report, Laboratoire de Météorologie Dynamique.*

Matthews, J.A. (1976) : 'Little Ice Age' palaeotemperatures from high altitude tree growth in S. Norway. *Nature*, **264**, 243-245.

Pfister, C. (1992) : Monthly temperature and precipitation in central Europe 1525-1979 : quantifying documentary evidence on weather and its effects. In : *Climate since A.D. 1500*. R.S. Bradley & P.D. Jones, eds., *Routledge London*; 118-141.

Reid, G.C. (1987) : Influence of solar variability on global sea surface temperature. *Nature*, **329**, 142-143.

Ribes, E. (1993) : Surrogate indicators for total irradiance : application to a cycle of 11 years or more. Palæoclimate Research, *Palaoclimate Research*, in press.

Ribes, E., E.N. Ferreira, R. Sadourny, H. Le Treut & Z.X. Li (1993) : Solar dynamics and its impact on solar irradiance and the terrestrial climate. *J. Geophys. Res.*, in press.

Sadourny, R. & K. Laval (1984) : January and July performance of the LMD general circulation model. In : *New perspectives in climate modelling*, A. Berger & C. Nicolis, eds., *Elsevier*.

Schweingruber, F.H., H.C. Fritts, O.U. Bräker, L.G. Drew & E. Schar (1978) : The X-ray technique as applied to dendrochronology. Tree Ring Bulletin 38, 61-91.

Van den Dool, H.M., H.J. Krijnen & C.J.E. Schuurmans (1978) : Average winter temperatures at de Bilt (the Netherlands) : 1634-1977. Climatic Change 1, 319-330.

CLIMATE MODEL SENSITIVITY, PALEOCLIMATE AND FUTURE CLIMATE CHANGE

A. J. Broccoli
Geophysical Fluid Dynamics Laboratory/NOAA
Princeton University
P. O. Box 308
Princeton, NJ 08542

ABSTRACT

Radiatively active trace gases such as CO_2, methane, nitrous oxide and chlorofluorocarbons warm the surface-troposphere system by increasing the infrared opacity of the atmosphere. Climate models have been used to estimate the warming associated with future increases in these gases. An important issue is the sensitivity, or the amount of climate response associated with a given amount of radiative forcing. Models currently used for climate change studies differ substantially in their sensitivity, and these differences contribute prominently to the uncertainty in estimating the magnitude and rate of future climate change. The observed climate record can be used to calibrate the sensitivity of climate models. Calibration based on the instrumental record (the last ~100 years) has the advantage of using direct observations of climatic parameters with relatively dense spatial coverage, but suffers from uncertainties about sources of climate forcing other than greenhouse gases and the possibility that natural climate variability may be comparable in magnitude to the greenhouse gas-induced change. In contrast, calibration using paleoclimate data from the late Quaternary involves larger changes in forcing, most of which are relatively well-documented, but is subject to the inherent uncertainties of reconstructing past climate from a variety of sometimes sparsely-distributed proxy data. While better monitoring of ongoing changes in climate forcing and response may be the best way to calibrate climate model sensitivity, paleoclimate information can provide much-needed insights into climate model performance, particularly as improvements occur in the quantity and quality of paleoclimate data.

1. INTRODUCTION

Climate models are among the primary tools used to estimate the possible changes in climate due to anthropogenic greenhouse gases. Because of their importance in this

NATO ASI Series, Vol. I 22
Long-Term Climatic Variations
Edited by J.-C. Duplessy and M.-T. Spyridakis

regard, a great deal of recent attention has been focused on the issue of climate model sensitivity, or the amount of climate response associated with a given amount of radiative forcing. Some of the work of the Intergovernmental Panel on Climate Change (IPCC) Working Group I concerned the differences in sensitivity among the various climate models that have been used to study anthropogenic climate change (Mitchell et al., 1990).

The development of climate models has generally proceeded toward increased spatial resolution, the addition of interactions with other parts of the climate system (e. g., ocean dynamics, land surface processes) and more detailed, physically-based representations of subgrid scale processes. While these developments are predicated on the belief that they will lead to improved model performance, it is often difficult to determine whether they lead to a more realistic sensitivity, even if they can be shown to improve the simulation of the present climate. After all, the perfect simulation of the present climate is not a sufficient condition for the accurate simulation of climate change associated with a given forcing. Even the simulation of interannual variability, though a very important test of climate model performance (Gates, 1992), may not test model sensitivity to changes in forcing comparable in magnitude to the change in greenhouse gas forcing expected during the next century.

One possible technique for evaluating climate model sensitivity is the use of information about past climate. This essay addresses the viability of this method and attempts to identify its strengths and weaknesses. To provide some perspective, it begins with a brief discussion of the physics of the so-called "greenhouse effect." This is followed by a review of results from studies of the climate response to enhanced greenhouse forcing conducted with a hierarchy of climate models. With that background, the evaluation of climate model sensitivity using the instrumental climate record and the paleoclimate record are discussed. The paper concludes with a discussion of the prospects for improved calibration of model sensitivity.

2. PHYSICS OF THE GREENHOUSE EFFECT

A number of trace constituents of the earth's atmosphere are radiatively active in the thermal infrared portion of the electromagnetic spectrum. The most important of these is CO_2; others include methane, nitrous oxide, and the chlorofluorocarbons CFC-11 and CFC-12. The evidence is strong that human activities are increasing the atmospheric concentrations of these gases. (This is obvious in the case of the CFCs, since they have no natural sources.) While the volumetric increase of CO_2 is much larger than those of the other greenhouse gases, the combined radiative effect of the other gases is comparable due to their infrared absorption properties. Wuebbles and Edmonds (1991) provide a review of the sources (both natural and man-made), sinks and radiative effects of the greenhouse gases.

Increases of greenhouse gases produce a warming of the troposphere by the following mechanism. Assume that the earth is in thermal equilibrium, and thus the incoming solar radiation is balanced by outgoing longwave radiation of equal magnitude. The Stefan-Boltzmann law allows the calculation of an effective radiating temperature (T_e) for the earth that depends only on the amount of incoming radiation and the planetary albedo. The computed T_e of ~255 K is lower than the earth's mean surface temperature of ~288 K; since the tropospheric temperature decreases with height, this implies that the effective radiating level lies well above (5-6 km) the surface. The elevated effective radiating level is due to the presence of greenhouse gases in the atmosphere, which act to shield (by virtue of their infrared opacity) the warmer, lower atmosphere from emitting to space. If greenhouse gas concentrations are increased, the atmosphere becomes more opaque to thermal infrared radiation. This increases the degree of shielding, and raises the altitude of the effective radiating level further still. Thus the longwave emission will take place at a lower temperature, resulting in a decrease in the amount of outgoing radiation. This perturbs the thermal equilibrium. A warming of the troposphere results that raises the temperature at all levels, including the effective radiating level, and thus restores thermal equilibrium (Manabe, 1983).

Feedbacks involving other parts of the climate system can amplify this direct warming. For instance, the increase of temperature results in an increase in saturation vapor pressure and, accordingly, the moisture-holding capacity and water vapor content of the troposphere. Since water vapor is also a greenhouse gas, this leads to a further increase in the infrared opacity of the atmosphere and a further warming by the above mechanism. The existence of this water vapor feedback, which provides an important amplification of climate model sensitivity (Manabe and Wetherald, 1967; Hansen et al., 1984), has been supported by an analysis of observational data (Raval and Ramanathan, 1989). Another important feedback is the snow-ice-albedo feedback, in which the warming results in a decrease of the area of snow and ice cover. Since snow and ice have relatively high surface albedos, more solar radiation is absorbed as a result, leading to a further increase in temperature (Budyko, 1969; Sellers, 1969). The snow-ice-albedo feedback is also positive. Other feedbacks may also be important, such as those involving cloud amount and height (Hansen et al., 1984; Wetherald and Manabe, 1985), but there is more uncertainty surrounding both the sign and magnitude of these feedbacks.

3. CLIMATE MODEL ESTIMATES OF SENSITIVITY TO CO_2

Most climate model studies of the effects of increased greenhouse gases on climate have considered only changes in CO_2. This practice arose at a time when the radiative significance of other trace gases was not widely known. While the combined effects of the other greenhouse gases are important and their increases constitute a significant radiative perturbation, the practice of representing their effect in terms of "equivalent CO_2" continues. Although the vertical distribution of the radiative forcing associated with the other trace gases is different from that of CO_2 (Wang et al., 1991), the rough similarity of their climatic effects in the troposphere suggests that this approximation is not too bad when considering changes in surface air temperature. Thus the climate response to a doubling of atmospheric CO_2 is used in this paper as a benchmark for climate model sensitivity to greenhouse gas perturbations.

Climate models have been used to investigate the impact of greenhouse gases on the earth's climate for more than 25 years. The state of the art has progressed dramatically during this period, from relatively simple one-dimensional models of radiative-convective equilibrium to three-dimensional models of the coupled atmosphere-ocean system. The development of a hierarchy of climate models for the study of climate change due to anthropogenic increases in greenhouse gases has led to a variety of research studies. Most of these can be separated into two categories based on their experimental design. Using terminology taken from the IPCC Scientific Assessment (Cubasch and Cess, 1990), these are equilibrium response studies and time-dependent response studies. Such experiments will be the primary subjects of this section.

Equilibrium response studies typically consist of two climate model integrations, each with a different prescribed greenhouse gas content. These integrations are continued until the simulated climate of each is in balance with its radiative forcing, then the two simulated climates are compared to obtain the response to the change in greenhouse gases. The early studies of the impact of anthropogenic increases in greenhouse gases were equilibrium response studies.

The first of these studies was performed with a radiative-convective model, which treats the climate system as a one-dimensional column. Results from the earliest radiative-convective model experiments (Manabe and Wetherald, 1967) illustrated two of the most fundamental aspects of the climate response to increased CO_2: tropospheric warming and stratospheric cooling. They also identified the role of water vapor feedback in amplifying the climate sensitivity. This and similar experiments also yielded sensitivities of roughly 1.5-2.5 K for the change in surface air temperature resulting from a doubling of CO_2 (Manabe, 1971; Wang et al., 1976; Augustsson and Ramanathan, 1977).

More recent equilibrium response studies have been performed with three-dimensional atmospheric general circulation models coupled with simple models of the ocean mixed layer (e. g., Manabe and Stouffer, 1979, 1980; Hansen et al., 1984; Washington and Meehl, 1984; Wilson and Mitchell, 1987; Schlesinger and Zhao, 1989). These tend to give larger values for the sensitivity to doubled CO_2, ranging from 1.9-5.2 K (Mitchell et al., 1990). The larger sensitivities can be attributed to the ability of these models to represent a wider array of feedbacks, many of which are positive, such as the snow-ice-albedo feedback. Detailed intercomparisons involving climate models from different research groups indicate that differences in their depictions of some of these feedbacks are responsible for the wide range of sensitivities (Cess et al., 1989, 1991).

Time-dependent response studies use models of the coupled atmosphere-ocean system to simulate a change in climate in response to gradually increasing greenhouse gas concentrations. Such gradual increases are similar to what is occurring in the real climate system. A number of time-dependent response studies have been performed, primarily using general circulation models of the atmosphere coupled with dynamical ocean models (e.g., Stouffer et al., 1989; Washington and Meehl, 1989, Cubasch et al., 1992). In addition to providing information about the global rate of warming that occurs in response to a given rate of increase in greenhouse gas concentrations, an important result from these studies has been the identification of important regional variations in the warming rate that are related to the mixing of heat into the deep ocean.

Because only a few modeling groups presently are capable of performing these computationally-demanding time-dependent response studies, simple models of the coupled system have been created for use in parameter studies of the time-dependent response (e. g., Hoffert et al., 1980). As part of the IPCC 1990 Scientific Assessment, one such model was tuned to mimic the behavior of three-dimensional coupled atmosphere-ocean models (Bretherton et al., 1990). This work affirmed that the equilibrium climate model sensitivity exerts a major influence on the rate of warming in response to a given rate of increase of greenhouse gases, and thus remains an important parameter in the study of anthropogenic climate change.

4. USING THE CLIMATE RECORD FOR MODEL CALIBRATION

The determination of climate sensitivity remains an important uncertainty (although certainly not the only one) in estimating the magnitude and rate of future climate change. At present, climate models yield a relatively large spread in their estimates of sensitivity.

In fact, the range of variation of climate model results may represent a lower limit to the true uncertainty, since all models could be lacking some important feedback (either positive or negative) that could move the sensitivity outside this range. The construction of more comprehensive models does not guarantee an improvement in their estimates of climate sensitivity, since many physical processes must be parameterized out of necessity, and the validation of such parameterizations presents a formidable problem.

Thus it is useful to develop methods of calibrating the sensitivity of climate models against the observed climate record. Such calibration requires a knowledge of the history of climate forcing as well as adequate climatic data. Two possible avenues involve the use of the instrumental record of the past ~100 years and the use of paleoclimate data for much older periods. The former has the advantage of using direct observations of climatic parameters with relatively dense spatial coverage, while the latter allows consideration of cases in which the climate forcing was drastically different from the present.

4.1 The Instrumental Record

Systematic observations of temperature were undertaken at some locations as early as the eighteenth century. The spatial coverage of observations increased enough to allow estimates of global or hemispheric temperature by the mid- to late-nineteenth century. Several groups have constructed time series of global or hemispheric temperature extending back to that time (Jones, 1988; Hansen and Lebedeff, 1988; Vinnikov et al., 1990). While these data sets are hampered by limited spatial coverage and other uncertainties, particularly in their early decades, they contain similar low-frequency variations. Thus the data are adequate and have been used in a number of comparisons with results from time-dependent model experiments in attempts to estimate the actual climate sensitivity (Wigley and Barnett, 1990).

The second requirement, a knowledge of the history of climate forcing, is more challenging. A variety of forcing functions can be important on the time scales represented in the instrumental record (i.e., years to decades) in addition to greenhouse gas concentrations. These include stratospheric aerosols of volcanic origin, variations in solar radiation, and anthropogenic sulfate aerosols.

Large volcanic eruptions can inject sulfur dioxide gas into the stratosphere, forming sulfuric acid aerosols that can scatter incoming solar radiation, increase planetary albedo, and thus have a significant impact on the earth's radiation balance. The combination of a reduction of total solar radiation with the warming due to the thermal infrared effects of the aerosols results in a small net cooling effect at the earth's surface, which can persist for a few years following an eruption (Mass and Portman, 1989). The inclusion of stratospheric aerosols in climate model simulations has yielded similar results (Hansen et al., 1992). Determining the history of volcanic aerosol forcing for the instrumental period is quite difficult, however. Measurements of volcanic aerosols in the stratosphere and their optical properties have only been available in recent years. Historical records provide clues to earlier eruptions, although it can be quite difficult to reconstruct the sulfur content of the erupted material and whether or not it reached the stratosphere, which is essential for it to have persisted long enough to have a climatic effect. Increased acidity in glacial ice cores has also been linked to sulfate aerosol-producing eruptions (Porter, 1986), but again there are difficulties in interpreting these proxy data.

Little is known about changes in insolation on time scales comparable to the period of instrumental records (10 to 10^3 years), since precise monitoring of solar irradiance by satellites has existed for little more than a decade. Satellite evidence indicates that irradiance varies in conjunction with the 11-year cycle of solar magnetic activity (i. e., the sunspot cycle), with a total variation of approximately 0.1 percent (Willson and Hudson, 1988). Proxies of solar magnetic activity, such as sunspot numbers, carbon-14 and beryllium-10, indicate substantial variations on time scales of decades to centuries (Stuiver and Quay, 1980; Beer et al., 1988). There has been some speculation that irradiance changes may be associated with these variations (Reid, 1991; Friis-Christensen and Lassen, 1991), but observational evidence and physical mechanisms have been lacking. However, some evidence has been offered for irradiance variations of up to 0.5 percent on these time scales based on solar modeling and the behavior of other stars similar to the sun (Lean, 1991). But at present the primary difficulty in documenting these relationships, or past changes in solar irradiance in general, is the short period of precise observations.

Anthropogenic sulfur emissions may also have potential climatic effects through two mechanisms. A direct effect is the scattering and absorption of incoming solar radiation by sulfate aerosols. An indirect effect is the potential of aerosol particles to increase cloud albedo by acting as cloud condensation nuclei. While both of these effects are extremely difficult to assess quantitatively, recent work suggests a net cooling effect at the surface with a magnitude that may be comparable to the change in greenhouse gas forcing over the period of instrumental records (Charlson et al., 1990, 1991, 1992). While estimates of the history of sulfur emissions may be available for use in chemistry/transport models to estimate the historical variations in aerosol concentrations, the large uncertainties in the magnitude of both their direct and indirect effects makes a quantitative estimate of the time variation of anthropogenic sulfate aerosol forcing quite difficult.

This suggests that efforts to calibrate climate model sensitivity using the instrumental climate record are somewhat handicapped by an inability to quantify the variations in forcing over the past century. Only if one were to assume that increases in greenhouse gas concentrations were the only changes in forcing would it be possible to do so with reasonable accuracy. In addition, the internal variability inherent to the climate system may produce variations in climate comparable in magnitude to the forced variations, further compounding the problem. Substantial variability appears to occur in the climate system even in the absence of external forcing, spanning a wide range of time scales. Much of this variability involves interactions between the atmospheric and oceanic components of the climate system. The well-known El Nino-Southern Oscillation phenomenon is an example, with climatic effects in widespread areas around the globe. More recently, models of the coupled ocean-atmosphere system have displayed variability of oceanic overturning (with accompanying atmospheric variations) on interdecadal time scales (Delworth et al., 1993). On shorter time scales, climate model experiments have demonstrated that substantial variability can occur even without changes in the ocean (Manabe and Hahn, 1981; Lau, 1981). Interactions with the land surface and internal atmospheric dynamics are likely sources of this variability (Delworth and Manabe, 1989).

It would clearly be too pessimistic to suggest that the instrumental climate record provides little useful information for calibrating the sensitivity of climate models. With proper caution, it should be possible to narrow some of the uncertainties about past climate forcing, and it may be possible to make some estimates of the magnitude of the unknown forcing. But given the contribution of internal climate variability acting as a "wild card," a case can be made for the pursuit of other, parallel approaches for calibrating climate sensitivity.

4.2 The Paleoclimate Record

Considerable amounts of paleoclimate data are available for climate modeling studies, particularly for the period of the last glacial maximum (LGM). The pioneering effort of the CLIMAP Project (1981) produced reconstructions of continental ice extent and thickness, sea ice extent, sea surface temperature (SST) and land surface albedo. Chemical analysis of ice core data has provided information about the atmospheric composition during the LGM, including the concentrations of CO_2 and other trace gases and the presence of airborne dust. In addition, past variations of the spatial and seasonal distribution of solar radiation have been calculated from the periodic variations in the earth's orbital parameters (e. g., Berger, 1978).

These paleoclimate data have been used in a number of climate model studies of past climates. Data from earlier time periods have also been used, but their relative abundance during and since the LGM have made the late Quaternary perhaps the most common subject of climate model studies. The goal of some of these studies has been to reconstruct a more complete picture of past climates, using climate models to generate information about an array of climatic variables that is physically consistent with the available paleoclimate data (e. g., Gates, 1976; Manabe and Hahn, 1977; Kutzbach and Wright, 1985). In such experiments, land ice extent and elevation, SSTs, sea ice extent, atmospheric composition and incoming solar radiation are used to force the climate models. Similar studies have attempted to objectively evaluate model performance through the use of additional paleoclimate data that were not used in forcing the models. For example, Webb et al. (1987) used temperatures and precipitation from a climate model to simulate the spatial distribution of vegetation at various times during the late Quaternary. The simulated vegetation distributions were then compared to those reconstructed from fossil pollen data in a test of model performance.

The abundance of paleoclimate data also makes the LGM a good subject for studies that evaluate climate model sensitivity. An important issue in the design of these experiments concerns how the oceans are treated in the models. This, in turn, is related to the role of the CLIMAP reconstructions of ocean surface conditions (SST and sea ice). In the experiments described above, SST and sea ice were prescribed; such experiments are relatively inexpensive to run, but require the reconstructed ocean surface data as input. This need is eliminated if SST is predicted by coupling an atmospheric GCM with some sort of ocean model. The simplest of these consider only the oceanic mixed layer and its role as a reservoir of heat on seasonal and interannual time scales. More comprehensive models are constructed by coupling an atmospheric GCM with an oceanic GCM, thereby accounting for the role of ocean circulation in influencing climate. Thus reconstructions of past SSTs can serve either as surface boundary conditions for integrations of atmosphere-only climate models, or as an independent data set against which simulated SSTs can be compared. Both of these approaches have been used in efforts to calibrate climate model sensitivity.

Hansen et al. (1984) conducted climate model experiments which utilized all available reconstructed variables for the LGM (SSTs, sea ice, land ice, land albedo, atmospheric CO_2). Climate model sensitivity was calibrated by examining the radiation balance at the top of the atmosphere. They found a negative radiation balance when the model was forced by the CLIMAP distributions of SST, sea ice, land ice and land albedo (with no change in CO_2), implying that the model would cool further if SSTs were allowed to change. A second experiment in which the boundary conditions were modified in accordance with then-recent research results (CLIMAP sea ice scaled back, reduced CO_2) yielded an even larger radiative imbalance, again indicating that the climate model would have cooled further. They suggested that this apparent overestimation of climate model sensitivity could be attributed to a number of possible causes, including an overestimate of cloud feedback in the model or a warm bias to the CLIMAP reconstructed SSTs.

A somewhat different approach was taken by Manabe and Broccoli (1985). They performed two experiments using different versions of an atmosphere-mixed layer ocean model. Each experiment consisted of two climate model integrations: one using the LGM land ice, land albedo and atmospheric CO_2, and the other a control. The two versions differed in their treatment of cloud cover. One prescribed clouds as a function of latitude and height, based on observations, but invariant over time. The other predicted cloud cover based on the simulated relative humidity at each grid point.

Manabe and Broccoli (1985) concluded that both versions of the model yielded sensitivities roughly in agreement with paleoclimatic estimates, and that it was difficult to resolve the difference between the two versions based on existing paleoclimate data. Breaking this down by latitude, there was good general agreement with paleotemperatures in the Northern Hemisphere extratropics for both versions of the model. In the Southern Hemisphere extratropics, there was reasonable agreement with the CLIMAP SST data, especially for the model with prescribed clouds. Both versions seemed to underestimate the terrestrial cooling there, although the continental data used were very limited. In addition, there was reasonable agreement between the Antarctic surface air temperature change and that reconstructed from isotopic data in the Vostok ice core (Jouzel et al., 1987), although this agreement may have been fortuitous since more recent ice sheet reconstructions suggest that the LGM Antarctic ice elevation used in the Manabe and Broccoli experiments was overestimated. In the tropics, both versions of the model indicated more cooling than the CLIMAP SSTs, while at the same time underestimating the terrestrial cooling reconstructed from pollen and mountain snowlines. The disparity between the oceanic and continental estimates of low-latitude cooling during the LGM was greater than the difference between the sensitivities of the two models.

These studies are examples of the way in which paleoclimate data can be utilized in calibrating climate model sensitivity. Obviously there are some weaknesses in this approach, the largest of which is probably the substantial uncertainty that surrounds reconstructions of past climates and forcings. For example, considerable uncertainty exists about the role of windblown dust as a source of climate forcing at the LGM. Large increases in dust are recorded in polar ice cores (e. g., De Angelis et al., 1987); if these increases are representative of a widespread increase in atmospheric dust loading, the climatic effects could have been considerable (Harvey, 1988). A recent tracer study using a climate model does not simulate a global increase in dust loading, and suggests that local or regional sources were responsible for the dust increases in polar ice cores (Joussaume, 1993). Further work will be required to resolve this issue.

But as new techniques for interpreting proxy data are developed and perfected it may be possible to improve our confidence in quantitatively reconstructing past climates and forcings. Such improvements will also be useful in future efforts to estimate climate sensitivity from paleoclimate data without the direct use of climate models (Hoffert and Covey, 1992). As the number of locations at which climate has been reconstructed increases, it may also be possible to distinguish regional variations in past climate change (for example, the presence or absence of polar amplification of warming or cooling) and compare these with variations simulated by climate models.

5. PROSPECTS FOR IMPROVED CALIBRATION OF MODEL SENSITIVITY

Despite the current limitations of paleoclimatic reconstructions, a review of previous work suggests that a quantitative assessment of a climate model's ability to simulate the LGM climate can be an important tool in evaluating model sensitivity. While it may not be possible to resolve modest differences in sensitivity, it should be possible to determine whether or not the sensitivity is in approximate agreement with that of the real climate system. In addition, previous assessments of this kind have identified important questions for further study, such as the difference between terrestrial and oceanic estimates of the tropical temperature change since the LGM. Questions such as this have implications for both climate modeling and understanding past climates.

One of the requirements for improved calibration of climate model sensitivity is the continued collection and analysis of the proxy data required for reconstructing past climates. The construction of more comprehensive data sets (combining both terrestrial and oceanic data) that describe spatial patterns of climate change for a particular time would be of particular use to climate modelers, since most of the original sources of paleoclimate data describe variations over time at a single location or a relatively small area. The efforts of the CLIMAP Project (1976, 1981) serve as a prototype for this strategy. Peterson et al. (1979) constructed one such data set, and also described different degrees of analysis for such data (borrowing terminology from the initialization of numerical weather prediction models). Another paleoclimate data set was assembled by the Cooperative Holocene Mapping Project (COHMAP Members, 1988). The creation of expanded data sets of this kind will require the integration of data of different types and from disparate sources and the expertise of a variety of paleoclimate specialists.

As further improvements in climate models occur (e. g., increased horizontal and vertical resolution, ocean dynamics, interactive vegetation), it will be useful to re-assess the calibration of their sensitivity. The simulation of a particular past time period such as the LGM can serve as a valuable benchmark, particularly since information on past climate forcing and response would be expected to become more reliable over time.

Perhaps the best way to calibrate climate model sensitivity would be to simulate ongoing changes in climate. With the current relatively rapid increase in greenhouse gas concentrations, a concomitant rapid change in climate forcing is also underway. By carefully monitoring both the changes in forcing and resulting response, the necessary information for hindcasts/forecasts by comprehensive climate models would be available. But since the history of forcing and response is replete with uncertainties at present, it is difficult to use the recent climate record for this purpose, since those uncertainties (along with the internal variability of the climate system) are of comparable magnitude to the expected climate response. Improved systems for climate monitoring (Hansen et al., 1990, Karl et al., 1993) would greatly improve our ability to use the recent climate record to calibrate climate models. But even if such systems were in place today, several more decades might be required to gather enough data. In the meantime, paleoclimate information can provide some much-needed insights into climate model performance.

ACKNOWLEDGEMENTS

I thank S. Manabe for renewing my interest in the study of past climates upon my arrival at GFDL, for the many discussions we have had on the topics addressed herein, and for his comments on this paper. I also thank J. Mahlman and N.-C. Lau for their valuable suggestions.

REFERENCES

Augustsson T, Ramanathan V (1977) A radiative-convective model study of the CO_2 climate problem. J Atmos Sci 34:448-451

Beer J, Siegenthaler U, Bonami G, Finkel RC, Oeschger H, Suter M, Wölfli W (1988)

Information on past solar activity and geomagnetism from ^{10}Be in the Camp Century ice core. Nature 331:675-680

Berger AL (1978) Long-term variations of caloric insolation resulting from the earth's orbital elements. Quat Res 9:139-167

Bretherton FP, Bryan K, Woods JD (1990) Time-dependent greenhouse gas-induced climate change. In Houghton JT, Jenkins GJ, Ephraums JJ (eds) Climate Change: The IPCC Scientific Assessment. Cambridge University Press Cambridge

Budyko, MI (1968) The effect of solar radiation variations on the climate of the earth. Tellus 21:611-619

Cess RD, Potter GL, Blanchet J-P, Boer GJ, Ghan SJ, Kiehl JT, Le Treut H, Li ZX, Liang X-Z, Mitchell JFB, Morcrette J-J, Randall DA, Richies MR, Roeckner E, Schlese U, Slingo A, Taylor KE, Washington WM, Wetherald RT, Yagai I (1989) Interpretation of cloud-climate feedback as produced by 14 general circulation models. Science, 245:513-516

Cess RD, Potter GL, Zhang M-H, Blanchet J-P, Chalita S, Colman R, Dazlich DA, DelGenio AD, Dymnikov V, Galin V, Jerrett D, Keup E, Lacis AA, Le Treut H, Liang X-Z, Mahfouf J-F, McAvaney BJ, Meleshko VP, Mitchell JFB, Morcrette J-J, Norris PM, Randall DA, Rikus L, Roeckner E, Royer J-F, Schlese U, Sheinin DA, Slingo JM, Sokolov AP, Taylor KE, Washington WM, Wetherald RT, Yagai I (1991) Interpretation of snow-climate feedback as produced by 17 general circulation models. Science 253:888-892

Charlson RJ, Langner J, Rodhe H (1990) Sulfate aerosols and climate. Nature 348:22

Charlson RJ, Langner J, Rodhe H, Leovy CB, Warren SG (1991) Perturbation of the Northern Hemisphere radiative balance by backscattering from anthropogenic sulfate aerosols. Tellus 43AB:152-163

Charlson RJ, Schwartz SE, Hales JM, Cess RD, Coakley, Jr. JA, Hansen JE, Hofmann DJ (1992) Climate forcing by anthropogenic aerosols. Science 255:423-430

CLIMAP Project (1976) The surface of the ice age earth. Science 191:1131-1137

CLIMAP Project (1981) Seasonal reconstructions of the earth's surface at the last glacial maximum. Geol Sci Amer Map Chart Series MC-36

COHMAP Members (1988) Climatic changes of the last 18,000 years: observations and model simulations. Science 241:1043-1052

Cubasch U, Cess RD (1990) Processes and modeling. In Houghton JT, Jenkins GJ, Ephraums JJ (eds) Climate Change: The IPCC Scientific Assessment. Cambridge

University Press Cambridge

Cubasch U, Hasselmann K, Maier-Reimer E (1992) Time-dependent greenhouse warming computations with a coupled ocean-atmosphere model. Clim Dyn 8:55-69

De Angelis M, Barkov NI, Petrov VN (1987) Aerosol concentrations over the last climatic cycle (160 kyr) from an Antarctic ice core. Nature 325:318-321

Delworth T, Manabe S (1989) The influence of soil wetness on near-surface atmospheric variability. J Clim 2:1447-1462

Delworth T, Manabe S, Stouffer RJ (1993) Interdecadal variations of the thermohaline circulation in a coupled ocean-atmosphere model. J Clim in press

Friis-Christensen E, Lassen K (1991) Length of the solar cycle: an indicator of solar activity closely associated with climate. Science 254:698-700

Gates WL (1976) The numerical simulation of ice-age climate with a global general circulation model. J Atmos Sci 33:1844-1873

Gates WL (1992) AMIP: The Atmospheric Model Intercomparison Project. Bull Amer Meteor Soc 73:1962-1970

Hansen J, Lacis A, Ruedy R, Sato M (1992) Potential climatic impact of Mount Pinatubo eruption. Geophys Res Lett 19:215-218

Hansen J, Lacis A, Rind D, Russell G, Stone P, Fung I, Ruedy R, Lerner J (1984) Climate sensitivity: analysis of feedback mechanisms. In Hansen J, Takahashi T (eds) Climate processes and climate sensitivity. American Geophysical Union Washington.

Hansen J, Lebedeff S (1988) Global surface air temperatures: update through 1987. Geophys Res Lett 15:323-326

Hansen J, Rossow W, Fung I (1990) The missing data on global climate change. Issues Sci Technol 7:62-69

Harvey LDD (1988) Climatic impact of ice-age aerosols. Nature 334:333-335

Hoffert MI, Callegari AJ, Hsieh C-T (1980) The role of deep sea heat storage in the secular response to climate forcing. J Geophys Res 85:6667-6679

Hoffert MI, Covey C (1992) Deriving global climate sensitivity from paleoclimate reconstructions. Nature 360:573-576

Jones PD (1988) Hemispheric surface air temperature variations: recent trends and an update to 1987. J Clim 3:1193-1203

Joussaume S (1993) Paleoclimatic tracers: an investigation using an atmospheric general circulation model under ice age conditions. 1. Desert dust. J. Geophys Res 98:2767-2805

Jouzel J, Lorius C, Petit JR, Genthon C, Barkov NI, Kotlyakov VM, Petrov VM (1987) Vostok ice core: a continuous isotope temperature record over the last climatic cycle (160,000 years). Nature 329:403-408

Karl TR, Quayle RG, Groisman PYa (1993) Detecting climate variations and change: new challenges for observing and data management systems. J Clim 6:1481-1494

Kutzbach JE, Wright HE (1985) Simulation of the climate of 18,000 yr bp: results for the North American/North Atlantic/European sector and comparison with the geologic record. Quat Sci Rev 4:147-187

Lau N-C (1981) A diagnostic study of recurrent meteorological anomalies appearing in a 15-year simulation with a GFDL general circulation model. Mon Wea Rev 109:2287-2311

Lean J (1991) Variations in the Sun's radiative output. Rev Geophys 29:505-535

Manabe S (1971) Estimates of future changes of climate due to increase of carbon dioxide concentration in the air. In Matthews WH, Kellogg WW, Robinson GDV (eds) Man's impact on the climate. MIT Press Cambridge, Massachusetts

Manabe S (1983) Carbon dioxide and climatic change. In Saltzman B (ed) Theory of Climate. Academic Press New York

Manabe S, Broccoli AJ (1985) A comparison of climate model sensitivity with data from the last glacial maximum. J Atmos Sci 42:2643-2651

Manabe S, Hahn DG (1977) Simulation of the tropical climate of an ice age. J Geophys Res 82:3889-3911

Manabe S, Hahn DG (1981) Simulation of atmospheric variability. Mon Wea Rev 109:2260-2286

Manabe S, Stouffer, RJ (1979) A CO_2-climate sensitivity study with a mathematical model of the global climate. Nature 282:491-493

Manabe S Wetherald RT (1967) Thermal equilibrium of the atmosphere with a given distribution of relative humidity. J Atmos Sci 24:241-259

Mass CF, Portman DA (1989) Major volcanic eruptions and climate: a critical evaluation. J Clim 2:566-593

Mitchell JFB, Manabe S, Meleshko V, Tokioka T (1990) Equilibrium climate change and its implications for the future. In Houghton JT, Jenkins GJ, Ephraums JJ (eds) Climate Change: The IPCC Scientific Assessment. Cambridge University Press Cambridge

Peterson GM, Webb III T, Kutzbach JE, van der Hammen T, Wijmstra TA, Street FA

(1979) The continental record of environmental conditions at 18,000 yr B.P.: an initial evaluation. Quat Res 12:47-82

Porter SC (1986) Pattern and forcing of northern hemisphere glacier variations during the last millenium. Quat Res 26:27-48

Raval A, Ramanathan V (1989) Observational determination of the greenhouse effect. Nature 342:758-761

Reid GC (1991) Solar total irradiance variations and the global sea surface temperature record. J Geophys Res 96:2835-2844

Schlesinger ME, Zhao ZC (1989) Seasonal climatic change introduced by doubled CO_2 as simulated by the OSU atmospheric GCM/mixed-layer ocean model. J Clim 2:429-495

Sellers, WD (1969) A climate model based on the energy balance of the earth-atmosphere system. J Appl Meteor 8:392-400

Stouffer RJ, Manabe S, Bryan K (1989) On the climate change induced by a gradual increase of atmospheric carbon dioxide. Nature 342:660-662

Vinnikov KYa, Groisman PYa, Lugina KM (1990) Empirical data on contemporary global climate changes (temperature and precipitation). J Clim 3:662-677

Wang W-C, Dudek MP, Liang X-Z, Kiehl JT (1991) Inadequacy of effective CO_2 as a proxy in simulating the greenhouse effect of other radiatively active gases. Nature 350:573-577

Wang W-C, Yung YL, Lacis AA, No T, Hansen JE (1976) Greenhouse effect due to man-made perturbation of trace gases. Science 194:685-690

Washington WM, Meehl GA (1984) Seasonal cycle experiments on the climate sensitivity due to a doubling of CO2 with an atmospheric GCM coupled to a simple mixed layer ocean model. J Geophys Res 89:9475-9503

Washington WM, Meehl GA (1989) Climate sensitivity due to increased CO_2: experiments with a coupled atmosphere and ocean general circulation model. Clim Dyn 4:1-38

Webb III T, Bartlein PJ, Kutzbach JE (1987) Climatic change in eastern North America during the past 18,000 years: comparisons of model data with pollen results. In Ruddiman WF, Wright Jr HE (eds) North America and adjacent oceans during the last glaciation. Geological Society of America Boulder

Wetherald RT, Manabe S (1988) Cloud feedback processes in a general circulation model. J Atmos Sci 45:1397-1415

Wigley TML, Barnett TP (1990) Detection of the greenhouse effect in the observations. In Houghton JT, Jenkins GJ, Ephraums JJ (eds) Climate Change: The IPCC Scientific Assessment. Cambridge University Press Cambridge

Willson RC, Hudson HS (1988) Solar luminosity variations in solar cycle 21. Nature 332:810-812

Wilson CA, Mitchell JFB (1987) Simulated climate and CO_2 induced climate change over western Europe. Clim Change 10:11-42

Wuebbles DJ, Edmonds J (1991) Primer on greenhouse gases. Lewis Publishers Chelsea, Michigan

The ASI Series Books Published as a Result of Activities of the Special Programme on Global Environmental Change

This book contains the proceedings of a NATO Advanced Study Institute held within the activities of the NATO Special Programme on Global Environmental Change, which started in 1991 under the auspices of the NATO Science Committee.

The volumes published as a result of the activities of the Special Programme are:

Vol. 1: **Global Environmental Change.** Edited by R. W. Corell and P. A. Anderson. 1991.
Vol. 2: **The Last Deglaciation: Absolute and Radiocarbon Chronologies.** Edited by E. Bard and W. S. Broecker. 1992.
Vol. 3: **Start of a Glacial.** Edited by G. J. Kukla and E. Went. 1992.
Vol. 4: **Interactions of C, N, P and S Biogeochemical Cycles and Global Change.** Edited by R. Wollast, F. T. Mackenzie and L. Chou. 1993.
Vol. 5: **Energy and Water Cycles in the Climate System.** Edited by E. Raschke and D. Jacob. 1993.
Vol. 6: **Prediction of Interannual Climate Variations.** Edited by J. Shukla. 1993.
Vol. 7: **The Tropospheric Chemistry of Ozone in the Polar Regions.** Edited by H. Niki and K. H. Becker. 1993.
Vol. 8: **The Role of the Stratosphere in Global Change.** Edited by M.-L. Chanin. 1993.
Vol. 9: **High Spectral Resolution Infrared Remote Sensing for Earth's Weather and Climate Studies.** Edited by A. Chedin, M.T. Chahine and N.A. Scott. 1993.
Vol. 10: **Towards a Model of Ocean Biogeochemical Processes.** Edited by G. T. Evans and M.J.R. Fasham. 1993.
Vol. 11: **Modelling Oceanic Climate Interactions.** Edited by J. Willebrand and D.L.T. Anderson. 1993.
Vol. 12: **Ice in the Climate System.** Edited by W. Richard Peltier. 1993.
Vol. 13: **Atmospheric Methane: Sources, Sinks, and Role in Global Change.** Edited by M. A. K. Khalil. 1993.
Vol. 14: **The Role of Regional Organizations in the Context of Climate Change.** Edited by M. H. Glantz. 1993.
Vol. 15: **The Global Carbon Cycle.** Edited by M. Heimann. 1993.
Vol. 16: **Interacting Stresses on Plants in a Changing Climate.** Edited by M. B. Jackson and C. R. Black. 1993.
Vol. 17: **Carbon Cycling in the Glacial Ocean: Constraints on the Ocean's Role in Global Change.** Edited by R. Zahn, T. F. Pedersen, M. A. Kaminski and L. Labeyrie. 1994.

Vol. 18: **Stratospheric Ozone Depletion/UV-B Radiation in the Biosphere.** Edited by R. H. Biggs and M. E. B. Joyner. 1994.

Vol. 19: **Data Assimilation: Tools for Modelling the Ocean in a Global Change Perspective.** Edited by P. O. Brasseur and J. Nihoul. 1994.

Vol. 20: **Biodiversity, Temperate Ecosystems, and Global Change.** Edited by T. J. B. Boyle and C. E. B. Boyle. 1994.

Vol. 21: **Low-Temperature Chemistry of the Atmosphere.** Edited by G. K. Moortgat, A. J. Barnes, G. Le Bras and J. R. Sodeau. 1994.

Vol. 22: **Long-Term Climatic Variations – Data and Modelling** Edited by J.-C. Duplessy and M.-T. Spyridakis. 1994.

Springer-Verlag and the Environment

We at Springer-Verlag firmly believe that an international science publisher has a special obligation to the environment, and our corporate policies consistently reflect this conviction.

We also expect our business partners – paper mills, printers, packaging manufacturers, etc. – to commit themselves to using environmentally friendly materials and production processes.

The paper in this book is made from low- or no-chlorine pulp and is acid free, in conformance with international standards for paper permanency.